Advancements in Micronutrient Research

A. Hemantaranjan

Department of Plant Physiology
Institute of Agricultural Sciences,
Banaras Hindu University,
Varanasi - 221 005, INDIA

SCIENTIFIC PUBLISHERS

Published by
SCIENTIFIC PUBLISHERS (INDIA)

Jodhpur –

5 A, New Pali Road
P.O. Box 91
Jodhpur - 342 001 INDIA

Delhi –

- 4806/24, Ansari Road
Daryaganj
New Delhi - 110 002 INDIA

Print: 2018

ISBN: 978-81-7233-575-5

Visit the Scientific Publishers (India) website at
http://www.scientificpub.com
Printed in India

Advancements in Micronutrient Research

PREFACE

The treatise "Advancements In Micronutrient Research" has been designed with a view to a thorough apprehension of multidisciplinary research, in the most fascinating area of agriculture, the micronutrients—their physiological and biochemical significance, carried out in the last few decades. Recent achievements in global agricultural productivity have had no historical parallel. We are in the midst of the greatest food production revolution the world has known. The revelation of the existence, activity, and benefits of mineral nutrients with special reference to trace elements or micronutrients is indeed an important milestone towards modern agriculture, and has created a subject of great interest to plant researchers. Contributions in this direction have unfolded many facts about the physiology of nutrition. The growing knowledge of micronutrient applications and their effects in plants have made us realize that enormous changes are possible through the judicious supply of certain inorganic substances. Impressive progress has been achieved during the last two decades in our understanding of the mechanisms of nutrient uptake and their function in plant metabolism, simultaneously with corresponding advances made in increasing crop yields by the supply of various micronutrients inspite of a wide range of variability within cultivars of a given species.

The essentiality of micronutrients was proved long before their role in crop production was recognized. The microelements are referred to as micronutrients and are required in quantities often less than 1 ppm in solution and not in plants. These elements are also known as oligo elements or spurene elements or trace elements. At present, about twentyfour elements are considered as essential for all plants or for a limited number of plant species. These elements are the following : carbon, hydrogen, oxygen, nitrogen, calcium, phosphorus, potassium, magnesium, sulphur, iron, manganese, zinc, copper, boron, chlorine, molybdenum, sodium, cobalt, nickel, vanadium, silicon, iodine, selenium and aluminium. This list of essential elements can not be considered to be complete as with continuing improvement in technology of purification, cleaning of materials used in culture work, exclusion of dust and other contents of atmosphere, analysis of nanogram quantities of elements and detection of the minute physiological and biochemical reactions which can be attributed to the presence of an element, many new elements may be added in future to the list of essential elements. Of course, the requirement for new element will be lower than that for the known micronutrients and beneficial elements.

This treatise is a summary of important new findings on trace elements nutrition and interactions, and of basic and applied needs in microelement nutrition. It is nearly impossible to compile all new data and ideas generated in micronutrient researches during the last few decades, but, this treatise, may provide a useful perspective of important progress and fruitful research opportunities. For new workers in this field, a specialised book on micronutrient research consisting of manifold information and references has been realized to be extremely important to quickly learn about the total area of trace element nutrition including the hardly known trace elements, and obtain help in a precise manner (including genetic materials needed for research).

The treatise entitled "Advancements In Micronutrient Research" is a treasure for interdisciplinary exchange of information which is a valuable contribution to the understanding of trace elements nutrition in plants. As can be seen from the chapters, the approach to the topic ranges from theoretical to applied, molecular to organismic, and single to multivariable systems. The first chapter deals with the general overview on the development of trace element research elucidating that over sixty five trace elements have an important role in physiological processes of biosphere coupled with the concentrations of different elements in earth weight-ratio, sea water, plants, animals and human beings. In addition, an interrelationship among different elements have been also elaborated along with the concept of the interdisciplinary trace-element research. Besides these in the second chapter, the same author has nicely discussed the physiological importance of a number of hardly known trace elements resulted from several plant and animal experiments. Subsequently, some insights in the precise determination of selenium in biological and environmental samples by graphite furnace atomic absorption spectrophotometry have been detailed in chapter three followed by an overview of the determination of beryllium in environmental and biological samples by atomic absorption and atomic emmission spectrometries in chapter four, doubtless, enriching this treatise in part. Further, an attempt to systematically compile significant pieces of information on the different well-recognised nutrient media (with particular reference to the micronutrients)/related methods proposed/modified by various teams for raising successful laboratory cultures of the major representatives of the free-living and symbiotic cyanobacteria has been made in chapter five, which is of much interest in terms of prospects of using the method of mix culturing, based on proper manipulation of macro- and micronutrients for raising plant-cyanobacterial co- cultures, in biotechnology. Chapter six includes a comprehensible detailed review on legume/rhizobial interactions within the legume root nodule dealing with the role of plastid in metabolic reactions involving iron nutrition where the U.S. authors have examined the intricacies on iron partitioning in the root nodule/*Rhizobium* symbiosis focussing two specific aspects related to : (a) roles which iron acquisition and storage may play in this interaction; (b) biosynthesis of the tetrapyrrole component of leghaemoglobin.

In view of the physiological significance of zinc and its poor presence in the soils all over the world, in general, and Indo- Gangetic soils, in particular, it appears that there is enough scope of improvement of crops through zinc fertilization, hence, chapter seven reviews various physiological aspects exclusively related to zinc nutrition in the light of recent advances. Nevertheless, chapter eight provides a deep insight into mechanistics of micronutrients regulating the metabolic processes, with the inclusions of a brief survey on bioinorganic aspects of micronutrients and recent advancements in biochemical physiology of boron, copper, iron, manganese, molybdenum and zinc. Finally, in this sequence, the physiology and biochemistry of all possible micronutrients (including the beneficial elements) have been very systematically illustrated by the most renowned and pioneer Indian Scientists in the field of plant nutrition, Professor S.C. Agarwala and his associate, Lucknow University, in the ninth substantial chapter of this treatise.

Chapter ten exclusively throws some light on the relationship of vesicular-arbuscular (VA) mycorrhizae and micronutrient nutrition which is, indeed, a novel piece of information compiled beautifully. This review highlights the need for more research in order to develop a better understanding of the interactions between micronutrients and VA mycorrhizae and VA mycorrhizal fungi. After this follows a chapter in which area of micronutrient stress in relation to genetic variability in plants coupled with the response of plants to such stress in relation to altered gene expression in terms of synthesizing novel polypeptides and protein profiles have been elucidated scientifically. The last chapter contains a detailed account on various important techniques used in micronutrient researches. Some of these methods have facilitated the discovery of new micronutrient elements and have helped in the elucidation of their role in different facets of plant metabolism.

Although this treatise has been edited by one person, it is nevertheless the product of cooperation at various levels. I am ecstatic to mention that my interest in trace-element nutrition and my scientific career in this field are due to the consistent encouragement and inspirations of esteemed Dr. S.C. Agarwala, Lucknow University; Dr. O.K. Garg, Banaras Hindu University; Dr. M.M. R.K. Afridi, Aligarh Muslim University; Dr. Y.C. Panchal, University of Agricultural Sciences (Dharwad); Dr. D.N. Tyagi, Banaras Hindu University, Dr. S. Kannan, B.A.R.C. (Bombay), Dr. A. Narayanan, A.P.A.U. (Bapatla), Dr. B.B. Singh, N.D.U.A.T. (Faizabad) and above all Dr. Arthur Wallace, University of California, U.S.A., Dr. Istvan PAIS, University of Horticulture, Budapest (Hungary) and Dr. R.B. Clark, University of Nebraska, U.S.A. I am gratefull thankful to them. My sincere thanks are also due to the Director, Institute of Agricultural Sciences, and the Dean, Faculty of Agriculture, Banaras Hindu University for their consistent moral support and kind cooperation as and when required. I have my special thanks to my all learned colleagues, Dr. D.N. Tyagi, Dr. (Mrs.) Chandra Prabha Singh, Dr. (Mrs.) Bandana Bose and Dr. J.P. Srivastava; staff members and students of the Department of Plant Physiology for their good wishes and willing help at every step of this academic work.

Hopefully, this treatise has to be tremendously helpful for researchers in the various fields of agricultural, biological and environmental sciences, who already have a profound knowledge of plant physiology, plant biochemistry, soil science and genetics. In the literature, preference has been given to more recent publications in addition to representative examples of classical contributions cited in various sections apart from a number of tables, graphs and figures.

Last but not the least, I am thankful from the core of my heart to all eminent scientists and co-authors who have been actively engaged in micronutrient researches in India, U.S.A., Hungary, Spain and West Indies for taking enormous pains in the fulfilment of the noble objective by contributing relevant chapters, in addition to my own, of this treatise. Their true cooperation coupled with patience, indeed, has enabled me to accomplish this gigantic task of international importance. In the end, I feel immense pleasure to be grateful to my affectionate parents and all other family members for their thorough encouragement and best wishes *ab initio* which I can never forget.

October 2nd, 1995
Varanasi, 221 005

Dr. A. Hemantaranjan

CONTENTS

3. SOME INSIGHTS IN THE DETERMINATION OF SELENIUM IN BIOLOGICAL AND ENVIRONMENTAL SAMPLES BY GRAPHITE FURNACE ATOMIC ABSORPTION SPECTROMETRY

A. Javier Aller

4. OVERVIEW OF THE DETERMINATION OF BERYLLIUM IN ENVIRONMENTAL AND BIOLOGICAL SAMPLES BY ATOMIC ABSORPTION AND ATOMIC EMISSION SPECTROMETRIES

A.J. Aller

6. LEGUME/RHIZOBIAL INTERACTIONS WITHIN THE LEGUME ROOT NODULE : ROLE OF THE PLASTID IN METABOLIC REACTIONS INVOLVING IRON NUTRITION

James A. Guikema, William R. Odum, and Peter P. Wong

7. PHYSIOLOGICAL AND BIOCHEMICAL SIGNIFICANCE OF ZINC NUTRITION IN PLANTS

A. Hemantaranjan

8. MICRONUTRIENTS–BIOCHEMICAL AND PHYSIOLOGICAL ACTIVITIES

Indravadan Chhelshankar Dave

9. PHYSIOLOGY AND BIOCHEMISTRY OF MICRONUTRIENT ELEMENTS

S.C. Agarwala and Chitralekha Chatterjee

10. VESICULAR–ARBUSCULAR MYCORRHIZAE AND MICRONUTRIENT NUTRITION

Susan B. Persad–Chinnery and Louis E. Chinnery

11. MICRONUTRIENT STRESS AND GENETIC VARIABILITY

Indravadan Chhelshankar Dave

12. TECHNIQUES IN MICRONUTRIENT RESEARCH

S.C. Agarwala and Chitralekha Chatterjee

Contents

Advancements In Micronutrient Research
Editor: A. Hemantaranjan
1996, Scientific Publishers, Jodhpur, pp. 1–18.

1

GENERAL OVERVIEW ON THE DEVELOPMENT OF TRACE ELEMENT RESEARCH

ISTVAN PAIS
KERTESZETI ES ELELMISZERIPARI EGYETEM, KEMIAI ES BIOKEMIAI TANSZEKE
Department of Chemistry and Biochemistry of the
University For Horticulture and Food Industry,
H-1502 Budapest, Villanyi ut 29- 31,
HUNGARY

CONTENTS

1. INTRODUCTION

The term" trace element" came from the fact that these elements are present in the living systems only in traces, i.e., their concentration is 1 μg/g or lower. In spite of the above mentioned fact that their concentration ratio is so low, their importance in physiological processes is equal to or in some correlations greater than those elements which represent 10% or higher concentration ratio.

The Earth and according to our recent knowledge, the solar system contain 88 permanent chemical elements.The noble gases and the strong radio-active elements may be excluded from the chemical elements which can have positive physiological role (Table 1). Now there remained 76 elements. It is well known that seven non- metals and four metals are called macroelements, because their concentration in most living beings is higher than 0.1 per cent. These elements are collected in Table 2.

TABLE 1.: Element excluded from positive role in biosphere

Noble gases	Radioactive elements
Argon	Actinium
Helium	Polonium
Krypton	Protactinium
Neon	Radium
Radon	Thorium
Xenon	Uranium

TABLE 2.: Essential macroelements

Symbol	Name
H	Hydrogen
O	Oxygen
C	Carbon
N	Nitrogen
P	Phosphorus
S	Sulphur
Cl	Chlorine
K	Potassium
Na	Sodium
Ca	Calcium
Mg	Magnesium *

* = Some authors classify this element in family of trace elements or it is called" meso-element"

If we count their number, we get 65 elements which are called trace elements. This family of chemical elements is demonstrated in Figure 1.

To avoid any misunderstanding, we accentuate that it does not prove that all 65 trace elements have an important role in the physiological processes of biosphere (PAIS, 1985), but it may mean that it is not so easy to exclude any of them to have this type of the role in life processes. We recall a sentence of Schwarz(1970) as the basis of all future research work in the field of trace elements science: to demonstrate that an element is essential, is a very hard task, but to exclude categorically from being essential or important, has not any real basis.

2. CRITERIA OF ESSENTIALITY AND BENEFICIALITY

The right rules of essentiality given by Arnon and Stout (1939) were published more than fifty years ago. These rules are as follows :

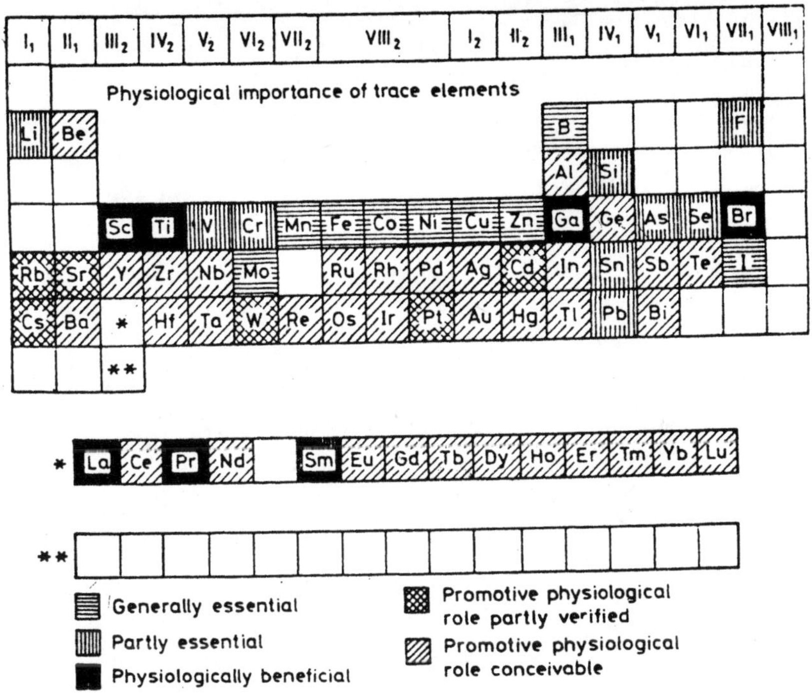

Fig. 1. The trace elements in the periodical system

(a) The organism can neither grown or complete its life-cycle in the absence of the element.

(b) The element cannot be replaced completely by any other element.

(c) The given element has direct influence on the organism and is involved in its metabolism.

We would like to underline that these rules were accepted worldwide and in main relations are valid now in spite of the fact that in the last 2-3 decades some changes and reevaluation occurred in the whole problem. For example Bowen proposed also some changes in his excellent book, (Bowen, 1966). The author of this chapter never accepted the widespread classification saying:there are two groups of elements=essential and non-essential.

In the late sixties and in seventies some new trace elements became accepted as essential for some froms of life: selenium, arsenic, nickel, chromium etc. If we look over the recent literature (Horowitz, 1988; Schmidt, 1989; Gibson, 1990;Nielsen,1990), now a days we can calculate for the living systems with 9 generally and 9 partly essential trace elements. These elements are listed in Table 3. It should be emphasized that all classification of elements has an uncertainty and it depends on personal opinion.

TABLE 3.: The essential trace elements

Generally	Partly
Boron	Arsenic
Cobalt	Chromium
Copper	Fluorine
Iodine	Lithium
Iron[x]	Lead
Manganese	Selenium
Molybdenum	Silicon
Nickel	Tin
Zinc	Vanadium

x= This element used to be called"Meso-element"

The essentiality of elements included in Table 3, in some cases was demonstrated under specific experimental conditions: synthetic foods, specially cleaned" plastic houses" for animals etc. We are convinced that this classification will be changed time to time and the list of these elements may grow in the next future.

Concerning the new scientific results in the last 2-3 decades, we have seen the necessity to create the term of beneficiality and in 1990 we published (PAIS, 1990) the criteria of beneficiality. One element may be regarded as beneficial, if the following criteria exist:

(a) By application of the given element we can register promotive role in plant growth, in body weight gain of domesticated animals or in the health conditions of humans.

(b) After the application of the given element the activity of one or more enzymes with positive physiological effects, is increased.

(c) By application of the given element the plants, animals and human beings show better health conditions or better immunological responses against diseases.

(d) Applying the given element, the toxical effects of other elements are decreased or can diminished("antidotum" effect).

(e) By application of the given element the uptake and utilization of other essential elements can be increased.

(f) Applying the given element some physiological parameters show an advantageous change, e.g., the cholesterol-level in blood will be lower, etc.

According to the above mentioned criteria, which element can be regarded as member of the family of beneficial elements? As we collected our knowledge on three hardly known trace elements in an other chapter of this book, titanium, gallium and zirconium should be regarded as beneficial elements.

It was also demonstrated (Liu, 1988) that different rare earth elements can play beneficial role in different plant experiments, therefore lanthanum, praseodymium and samarium can be included in this family of elements. Scandium, platinum and bromine moreover rubidium were demonstrated to have different, physiologically positive effects, therefore- at this moment-we can count with 10 beneficial elements.

In the case of the remained 37 trace elements we should express our conviction that sooner or later, but in the near future we can classify them in this or that group of elements.

3. Criteria of Toxicity

The problem of toxicity is also a very manysided question to which to give a correct answer is not an easy task. Some hundred years ago Paracelsus (Theophrastus Bombastus von Hohenheim, 1493- 1541) gave the main basis which materials should be considered as toxic. On the basis of this old opinion we should like to say something which can be "unusual and surprising": there are not any materials(elements and compounds) which are "eo ipso" toxic and on the other hand all materials(elements and compounds) are toxic. The main point of view is the concentration or the quantity of the substance.

We would like to illuminate this situation with the help of the Figure 2., which is wellknown in the trace element literature. As it is clearly demonstrated, all substances can cause a deficiency, which appear in form of deficiency diseases, in severe cases occur lathality and on the other hand in a higher concentration cause a toxicity and lethality.

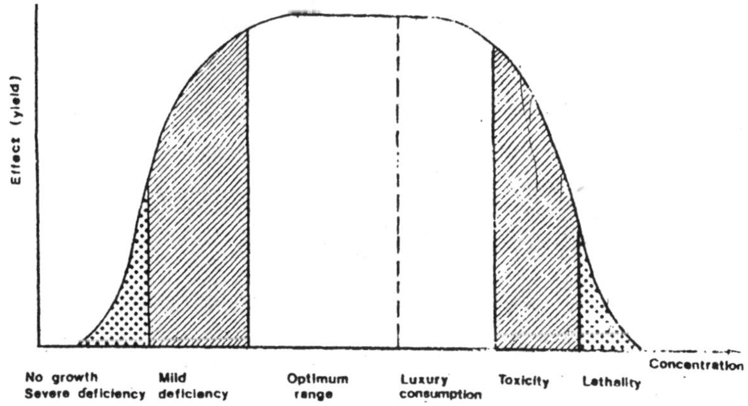

Fig. 2. Correlations between concentration and effect of elements

We try now to classify the most important parameters which are meaningful to decide what is toxical:

(a) The first parameter is the compound form. The living organisms can take chemical elements in elemental status very rarely: mostly we intake them in compound-form. The compound which is water-soluble, or soluble in week acid medium, or in digestion system which can form uptakeable chelates, these are some criteria that make a form really toxic, while the others are usually not harmful. We should give some easily acceptable examples.
The mercury chloride and much more the methyl-mercury is - between given concentration ratios-highly toxic, but the HgS form, because it is unsolvable, does not mean any danger for the living organisms. The soluble barium compounds are usually very toxic, but the $BaSO_4$ has no real toxicity: this compounds is used in X-ray diagnoses as contrast material in stomach.

(b) The second parameter is the oxidation state. The international literature gave a lot of arguments that the trivalent chromium has a very low toxicity, but the hexavalent form, i.e., chromates are usually highly toxic under common circumstances. Usually by uptake the reduction form Cr(VI) form to Cr(III) form is rapid enough, therefore the toxicity is not so serious. The same situation is vaild to As(III) and As(V): the latter has generally higher toxical effects than the former one. Here we have a similarly important parameter; the organic compound-form as the source of more uptakeable matter.

(c) The third factor is the presence or absence of other (synergistic or antagonistic) elements. A very demonstrative example is the toxicity of cadmium by smoking. According to the literature one cigarette contains on an average 1.5 µg, easily absorbable cadmium. It means that a common smoker has a cadmium uptake of 30-50 µg/day which is the limit of a strong danger. If the organism of the above mentioned smoker has a quite good zinc-and calcium-supply, in this case the mentioned quantity of cadmium is not so harmful. If the zinc- or the calcium-level is lower than what is needed, the toxicity increases exponentially.
The Wilson-disease is known from literature as an enrichment of braincells with copper. If we apply a higher zinc- uptake in daily nutrition, the copper-level will decrease, although this way of cure is not the best one: more appropriate the chelate- therapy with a simple tripeptide.

(d) The fourth factor is the mode and way of intake. Injection is mostly more dangerous than digestion via the intestine-system: we all know from the literature that some elements show an excellent, near 100 per cent absorption from the foods while some others show only some per cent. These parameters change person to person, food to food etc., therefore it is totally impossible to give generally valid laws of element-uptake These uptake processes are in deep correlation with the pH, with the exchange-possibilities of ligands, generally to say: with the coordination-chemical situation in the food-chain.

(e) The fifth one is the time of action. Usually a single dose of the toxical compounds is not so dangerous, but if the uptake is going on regularly during a long time-period, this latter one is much more harmful. It is quite understandable that the time and the doses of the compound are together so meaningful.

(f) The last factor is the defending-mechanisms of living systems. We have now no possibility to deal in details here with the metallothioneins in human and animal organisms, or with the role of newly found (Liu, 1988) phytochelatins in the plants, but these mentioned factors can also mitigate or diminish the toxical effects. We will also not deal here with the activity of immune system of different human beings: under influence of the same microorganism one part a human beings incline, others have response to diseases.

The author does not think that this given picture of toxicity is totally complete, but it was always clear that to qualify this or that compound as harmful, we should also add a lot of different parameters under which the term of toxicity is valid or not.

Finishing the consideration of toxicity, we should pay some attention to a general overview what and when is too little or too much.

In Table 4 we collected the chemical composition of Earth, the sea-water and the same of land-plants, land-animals and for human beings. These data are from different kinds of literature, therefore these numbers are not absolutely accurate. We compiled the most frequent and physiologically most important elements. As we can see from these data, in most cases the chemical composition of the inorganic nature differs very much from the organic one. We would focus now our attention on the main differences:

(a) The quantity of hydrogen is 10 to 50 times higher in the organic than in the inorganic world. – The ratio for carbon is some hundred times higher in the living organisms than in the Earth's crust.- The greatest difference can be seen in the quantity of nitrogen: the enrichment factor is about one thousand.

(b) Most metals which have a concentration near or over 1 per cent in the Earth's crust, their ratio decreased with a factor five to thousand in the organic world. On the other hand the greater part of the essential trace metals, show an enrichment in the organic world. This correlation is valid not only for some metals like copper, nickel and molybdenum, but for boron, iodine and selenium as non-metals, too.

(c) The selection in food-chain produced even more surprising samples. The sea-water, as it is demonstrated in Table 4., is relatively very poor in different chemical elements, excepted sodium and chlorine. Some mussels and oysters can accumulate a special group of elements, e.g. chromium, iron, cadmium and zinc with an over ten thousand or more than one hundred thousand times. It is valid also for other sea-animals, which can be good sources of important elements but in the same minute dangerous sources for toxic heavy metals, too.

It is well known from international literature that some Ascidian species can have in their blood 0.1-0.2 per cent vanadium and the highest concentration can be 1.5 per cent : i.e., the accumulation rate in this latter example is near to ten million. In some relation this is similar to the thyroid gland which contains over 80 per cent of the iodine-content of the human organism.

The mentioned enrichments were natural processes during the long time of the biological evolution and now a days are also the consequences of the developed genetical determination of the element enrichment. These changes mostly have no any correlation with toxicity.

The "Homo sapiens" as a part of the nature, started to initiate more and more changes in the original equilibria of nature. These changes were not so great before the industrialization, therefore the start of anthropogenic changing-era can date to the French revolution. Chemical industry started at that time, the use of natural energy-resource developed more and more rapidly, the motorization etc.

In our century and most of all in the last two-three decades some hundred thousand tons "new" compounds came to the environment as earlier unusual and because living beings have not had enough time to attempt accomodation to them, these pollutants are the sources of very dangerous contamination. These toxical contaminations has not always a direct effect, but usually indirect effects. The equilibria in the food-chain are namely very vulnerable as we tried to give some demonstrative examples.

Generally to say : these contaminations from anthropogenic sources should be as low as possible, otherwise the original or earlier- for living systems tolerable- equilibria will change for the worse in a tragical direction. The direct or indirect toxical effects to trace elements therefore have so great importance in the food- chain and finally in the future of our globe.

TABLE 4 : The concentration of different elements

Name	Earth weight - ratio	Sea-water mg/l	Plants mg/kg	Animals mg/kg	Human beings mg/kg
Oxygen	466 000	857 000	410 000	400 000	624 000
Silicon	277 200	3.0	200-5000	100-6000	0.3-6.0
Aluminium	81 300	0.01	500	4-100	< 0.8
Iron	50 000	0.01	140	160	100
Calcium	36 300	400	18 000	200 - 85000	19000
Sodium	28 300	10 500	1200	4000	800
Potassium	25 900	380	14000	7400	2000
Magnesium	20 900	1350	3200	1000	300
Titanium	4400	0.001	1.0	0.2	< 0.02
Hydrogen	1400	108000	55 000	70 000	99 000
Phosphorus	1180	0.07	2300	17000-40000	9000
Manganese	1000	0.002	120	0.2	0.3
Sulphur	520	885	3400	5000	4000
Carbon	320	28	454 000	465000	221000
Chlorine	314	19000	2000	2800	800
Fluorine	300	1.3	1-40	350-800	600
Chromium	200	< 0.001	0.2	0.07	0.07
Vanadium	150	0.002	1.6	0.1	0.2
Zinc	132	0.01	100	0.3	30
Nickel	80	0.005	3.0	0.8	0.15
Copper	70	0.003	14.0	2.4	1.6
Lithium	65	0.18	0.1	0.02	0.02
Nitrogen	40	0.5	30000	100000	31000
Cobalt	23	< 0.001	0.5	0.03	0.02
Molybdenum	15	0.01	0.9	0.2	0.2
Boron	3.0	4.6	50.0	0.5	< 1.0
Iodine	0.3	0.06	0.42	0.4	0.2
Selenium	0.09	< 0.01	0.2	1.7	0.3

4. THE IMPORTANCE OF TRACE ELEMENTS IN ENVIRONMENT

The importance of the trace elements has a similarity to the chemical bonds. The first order of chemical bonds (ionic, metallic and covalent) have much higher valence-energy (200- 800kJ pro mol), while the second order bonding varieties (hydrogen bond, Van der Waals bonds) have a much lower (0.1 to 40 kJ pro mol) energy-level. These weaker bonding possibilities has much more importance in living systems,than the first order one, because the macromolecules(carbohydrates, proteins, lipids and nucleic acids) receive their fine-structure from these week forces and in the biochemical processes the fine-structure, in other words the structure-geometry is the basis of the mean processes.

The appropriate level of trace elements, the good supply of the living with them, is a cardinal question of life. As it is well-known, the enzymes and enzymes-systems direct the biochemical processes of the living and these enzymes contain trace elements as inbuilt parts of their structure or they need them as activator agents for their biochemical role. In other words: the "optimal" concentration of these trace elements is the basis for the health of the living, including human beings, too.

The different geographical areas of the world-before the modern, industrialized society- had a certain concentrations of these elements, therefore the living had the possibility to accomodate themselves to the original circumstances. But by the dramatical changes of these equilibria this accomodation is not so easy, or needs a much longer time, therefore the living beings including man - have infeasible conditions. These great changes in the original equilibria are the main sources of tragical problems of environmental contaminations.

As we mentioned, the balance of the trace elements supply of the living is not the best now a days. In some regions of the world some trace elements have a minimum-level and the same elements in other districts have an excess-level. We should like to give some examples.

The selenium has been found in some regions(south-west to north- east) of the big country, China in deficiency level (it is the basis of the wellknown Heshan-disease) and in the same country we can find districts with superflous selenium-levels. The same situation is valid for the United States. We can find there seleniferous districts, where the cultivated wheat contains 150- 200 μg Se/g and in other districts of USA was found a deficient selenium-supply of agricultural plants. Selenium-deficiency was demonstrated in Scandinavia, also in New Zealand and the Se-supply in some other European countries is not quite near to the optimum level.

In the country of the author, Hungary, geographical circumstances are relatively very good for agricultural production. In some districts we have deficiency, in others superfluous supply in different nutritive elements. We have in some countries (west) which are poor in iodine, while some other countries we have enough and some mineral waters are relatively very rich in iodine.

Generally to say: for all elements we can find deficient and toxic districts, geographical region on our planet. It means in other words: some parts of the world need supply from this or that nutritive element, or on the other hand, in some districts we must take care of the toxical symptoms and we should do our best to avoid the toxical symptoms which are the basis of very dangerous diseases.

The war against contamination of the environment is worldwide accepted "scientific religio", but most of the peoples do not know exactly what is our duty to be successful in this war. The pollution of the nature is really a manysided problem, we would like to deal here only with the cycle of the nutritive elements, first of all, with trace elements.

Pollution has sometimes geological (geographical) origin: e.g., the high arsenic content of drinking water in some regions of Taiwan, or in south-east part of Hungary. If we have these type of pollution, we can organise some chemical methods(application of ion-exchange resins, etc.) to avoid or diminish the pollution and insure there for the inhabitants a good, or as a minimum: tolerable quality of drinking water. The problem is exactly the same in final solution of other cases: of the pollution has an andropogenic origin(in the most cases that is the reality!), we should switch out the sources of pollution: we should find new technological ways in the industrial processes or we should find new materials to aviod the application of toxical parts of the food-chain and exclude them as chemical compounds.

If we can find a deficiency situation in this or that district of the world, we should try to give a support from the given element in the nutritional chain of the living beings: it means to apply trace-element containing additives in animal nutrition and also trace-element supply (pills) or roborative drinks in human nutrition.

This last way of solution is gravid with different problems: first of all because some trace elements have a very narrow limits between the daily required and the toxical levels, therefore all supplementation must be well investigated and submit to a regular and very accurate medical control. In some cases this supplementation is very easy: e.g., in iodine-deficient regions of world we use a salt enriched with different level of sodium-iodine and in this way we can exclude the iodine-deficiency from the daily life of some hundred million peoples.

Generally to say: these cases of deficiency are worldwide known problems to be solved. The author is fully convinced that in the next future this will be one of the most quizzing problem of human nutrition. In this correlation we should cite some sentences of paper of Beisel(1990), who wrote the following: "The different functional impairments of the immune system caused by malnutrition can be classified as acquired immunodeficiency syndromes. MAIDS (Malnutrition Associated Immune Deficiency Syndrome) distinguish the problems used by malnutrition from the recent, and more newsworthy epidemic of AIDS caused by a human immunodeficiency virus".

We can calculate also with NAIDS (Nutritionally Acquired Immunodeficiency Syndrome). "Compared with the relatively small number of patients with AIDS, the global incidence of NAIDS is orders of a magnitude greater, and contribute to far more deaths".- "However, deficiencies of essential micronutrients may contribute to, or to be sole cause of NAIDS". - "Further, any advances in basic knowledge about the role of micronutrients on the function of the immune system or about micronutrients malnutrients must be translated into practical and effective prophylactic measures and then applied in public health education and practice".

The MAIDS, accepting the arguments of Beisel is now or will be in the future the most common nutritional problem of the whole humanity. The better developed (therefore richer) part of the world, has quite different problems from the poor, underdeveloped countries : namely the problem in the former part of world is too high a consumption of carbohydrates and fats, the disadvantages of industrialisation, the lack of physical movement and training etc.

5. Interrelationships Among Different Elements

About 200 years ago it was well-known that the nutritive elements can increase or may decrease the physiological effect of another one. The first effect is known by term of synergism, the other one by the term of antagonism. The most well-known interrelationship is between the alkali and earthalkali metals: the sodium and potassium are in antagonistic correlation, and the same is valid for calcium and magnesium.

In the field of trace elements the first and most known antagonism was demonstrated between copper and molybdenum (and also some sulphur compounds) in Australia (Underwood, 1971). It was also demonstrated very early that molybdenum and tungsten are antagonists to each other (Notton and Hewitt,1971).

If we simplify these interrelationships, we can look at the binding-sites of different enzymes and enzyme-systems. Namely the different metal-ions are in competition for these binding-places of enzymes. The main parameter of this competition can be understood by the laws of coordination chemistry, i.e., which complex or chelate is more stable under the given circumstances, like pH, the ionic strength, etc. The exchange of ligands of the given chelate; this is the theoretical (and the practical) basis to from new compounds. The uptake of trace elements, their crossing through the cell-membranes usually has also the same basis. Generally to say : the uptake of the given element from the environment by the different living organisms, the transport from cell to cell, finally the physiological role of it, is in deep correlation with the complex-stability, with different carrier-molecules, etc.

Coming back to the basic problem, the antagonism and the synergism are based on the binding forces to the active places of enzymes or enzyme-system. The results of this competition can be understood calculating the stability constant of the complex between the enzymes and different trace elements. For example, the so called "toxic heavy metals", like cadmium, can have a more stable complex with some enzymes than the essential trace element; zinc. This stability depends in some cases on the actual concentration. If the essential trace element has had enough high concentration to occupy the active places of enzymes, in this case the toxic heavy metal has not enough chance to bind itself into the structure of the biologically important molecules and in this way destroy or hinder their basic role in the living organism.

Some essential elements are able to show "antidotum" effect against the most toxical heavy metals, as it is demonstrated by a figure of Chowdhury and Chandra (1987), which is illustrated in Figure 3.

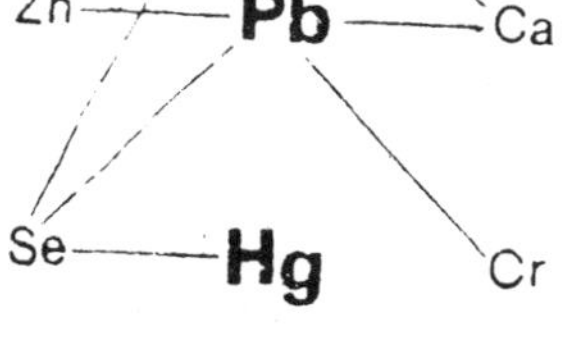

Fig. 3. Interactions of toxic metals and essential elements

The above mentioned processes are in deep correlation with the order of facts in time. In spite of the fact that cadmium forms a stronger chelate with some binding sites of some enzymes, if these places are occupied. by zinc-ions, the change of zinc-ions for cadium-ion is much slower, therefore the cadmium toxicity is not so dangerous. This effect can have a correlation with the induction of the metallothionein biosynthesis, which may also be against cadmium-toxicity. To express it in a simpler way: if our organism is well supplied with zinc-ions, the toxical level of cadmium is much lower, or by application of zinc-this fact was demonstrated by different scientists-we can fight against the cadmium toxicity.

Our environment contains only rarely too highly toxical pollutants, like cadmium, mercury, lead, thallium, beryllium etc. These elements are all enzyme-inhibitors. If the concentration of the most important essential trace elements, like zinc, iron, selenium, etc. is in minority, in this case we can calculate with severe deficiency symptoms, i.e., the malnutrition with these important factors of the living can be the basis of different diseases. As it was demonstrated by different authors (Chandra, 1985; Chandra and Chandra, 1986; Bridges *et al* ., 1980; Bess *et al.*,1992), the human immunological system is also in deep correlation with the supply of different essential elements in our food-chain.

The environment-as a consequence of industrialisation,increased chemical activity in agriculture, the very rapid growth of motorization, the acidic rains and some other reasons-contains more and more chemical elements which were not members of the food-chain earlier, therefore the genetically determined immunological system has not the possibility to accomodate to these compounds and other materials. The so called "civilisation diseases" are sometimes also the consequences of these toxic elements and compounds.

According to the literature (Björnberg,1987; Chiba *et al.*, 1987; McKenna *et al.*,1992), applying food-stuffs or medicaments, which contain the daily required amount of zinc, iron, selenium, etc., the dangerous effects of cadmium, lead and mercury can be lower or calculating optimal circumstances, we can counteract them totally.

It is also demonstratted that the great number of "modern" diseases, like Wilson's disease, Azheimer's disease, atherosclerosis, different malignancies, etc. are also in deep correlation with the changed equilibria between different nutritive elements. Therefore, we can state that human health-generally has two main basis : the genetic line what is independent from us and the daily nutrition, which is-in most cases-depends on our eating habits. We can and we should find the main laws of healthy human nutrition and we should work out the general rules for the daily element uptake.

The Recommended Daily Allowances(RDA) values published lastly (Havel, 1989) contain the quantities only for 8 trace elements and most essential and beneficial elements are not included in them. This is the fact we mentioned earlier, that our up-to-date knowledge about 35-45 elements is very poor in spite of the fact that these elements are now and in the next future will be more and more members of the daily food-chain of the whole humanity.

What way should be proposed to solve these basic problems of the future of the human kind? In the next point we should like to deal with the advantages of the interdisciplinary trace elements research.

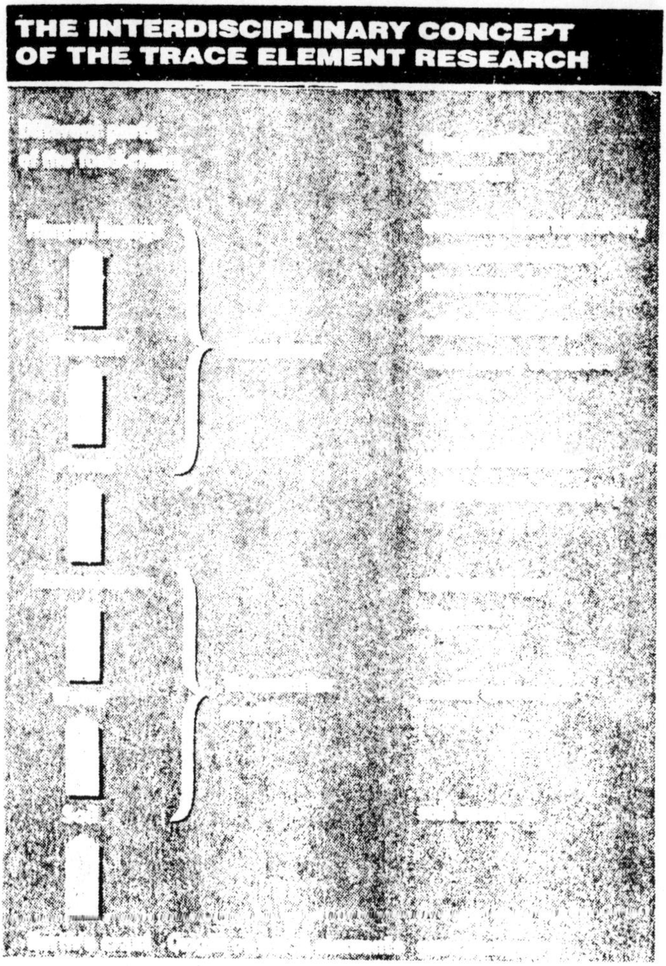

Fig. 4. The interdisciplinary concept of the trace element research

6. The Concept of The Interdisciplinary Trace — Element Research

The extremely rapid differentiation in different parts of science give now less and less possibility for the scientists to have enough information on the neighbouring fields of their specific scientific territory. As a consequence of this situation, the team-work started about a half of century ago and its importance is now days higher than once.

The author was under the first proposing the interdisciplinary trace -

element research work (PAIS, 1971) and according to this concept organised in his country, Hungary, the Trace Element Committee of the Hungarian Academy of Sciences. In this Committee 9-10 scientific fields are repre-sentated and in the Proceedings of five International Symposia the advantage of the manysided trace-element science (PAIS, 1984; PAIS, 1987; PAIS, 1988; PAIS, 1990; PAIS, 1992) can be realized.

The same concept seems to be accepted by more and more scientists, as recently was published by Iyengar (1991).

The trace element science is really a specific field having an interdisciplinary character. If we look only at the food-chain, which is started from geochemical sources and has three transmitter media: soil, water and atmosphere. This series of scientific fields needs the knowledge as minimum of 4 scientific fields: geochemistry, soil science, water science and some parts of physics.

When we turn from the geosphere to the biosphere, we see first the knigdom of plants which needs the knowledge of plant growers, including the modern aspects of plant physiology and plant biochemistry, too. Thereafter comes the kingdom of animals, which needs the knowledge of animal breeders and veterinary doctors, too. The last station of the biosphere there are the human beings and our health needs the very broad knowledge of the medical sciences.

The geosphere and biosphere equivocally need the knowledge of analytical chemistry, because the correct and sensitive analysis of chemical composition of different parts of geo- and biosphere is the original basis for all knowledge and consequences. If we know the backgrounds of living processes, we should use the knowledge of biochemistry and since trace elements are inbuilt in enzymes and enzymes-systems, we should use the knowledge of coordination chemistry: as we mentioned it earlier, the nutrition of the living is based namely to changes in chelate binding and constitution.

The simple overview of these correlations is given in Figure 4 (PAIS,1992). As it is quite clear from this scheme, we need altogether 10 different scientific fields to fully understand the issue of the whole trace - element science. We are convinced that the team-work and the interdisciplinary cooperation is the only basis to solve the manysided problems of environment and to ensure the future of human race on our globe.

REFERENCES

Arnon, D.I. and Stout, P.R. 1939. The essentiality of some elements in minute quantity for plants and special reference to copper. *Plant Physiol.*, **14** : 371-375.

Beisel, W.R. 1990. Future role of micronutrient on immune functions. *Annals NY. Acad. Sci.*, **58** : 7 267-274

Bess, J.W., Powell, P.J., Issaq, H.I., Schumack, L.J., Grimes, M.K., Henderson, L.E. and Arthur, L.O. 1992. Tightly bound zinc in human immunodeficiency virus type I., human T-cell leukemia virus type I., and other retroviruses. *J. Virolog.*, **66:** 840- 847.

Björnberg, A.A. 1987. Removal of mercury in fish (*in vivo*) in so-called "black-listed lakes" in Sweden by addition of selenium. *Proc. Internat. Conf., New Orleans* Vol. **2**. 67-69.

Bowen, H.J.M. 1966. The biochemistry of trace elements. Academic Press, New York, 1-241 pp.

Bridges, R.B., Chow, Ch.K. and Rehm , S.R. 1990. Micronutrient status and immune function in smokers. *Ann NY. Acad. Sci.,* **58** *:* 7218-231.

Chandra, R.K. 1985. Effect of macro-and micronutrient deficiencies and excesses on immune response. *Food Technol.,* **39** (2) : 91-93.

Chandra, S. and Chandra, R.K. 1986. Nutrition, immune response and outcome. *Progr. Food Nutr. Sci.,* **10** : 1-65.

Chiba, M., Fujimoto, N. and Kikuchi, M. 1987. Experimental study on interaction between selenium and lead. *Proc. VI . UOEH Internat. Symp., Kitakyush* 113-114.

Chowdhury, B.A. and Chandra, R.K. 1987. Biological and health application of toxic heavy metals and essential element interactions. *Progr. Food Nuts. Sci.*, **11** : 55-113.

Gibson, R.S. 1990. Essential trace elements and their nutritional importance in 1990s. *J.Can. Diet. Assoc.,* **51** : 292-296.

Havel, R.J. (Ed.) 1989. Recommended daily allowances. *Nat. Acad. Press,* 10th edition, 1-285pp.

Horowitz, C.T. 1988. Is the major part of the periodic system really in-essential for life? *J. Tr. El. Electr. Hlth. Dis.,* **2** : 135-144.

Iyengar, G.V. 1991. Milestones in biological trace element research. *Sci. Total Envir.,* **100** : 1-15.

Liu, Z. 1988. The effects of rare earth elements in growth of crops. *Proc III. Internat. Tr. El. Symp.,* Budapest (Ed.:I.Pais) 23-35.

McKenna, I.M., Chaney, R.L., Tao, S.H., Leach, R.M. and Williams, F.M. 1992. Interactions of plants zinc and plant species on bioavailability of plant cadmium to Japanese quail fed lettuce and spinach. *Environ. Res.,* **57** : 73-87.

Nielsen, F.H. 1990. New essential trace elements for the life science. *Biol. Tr.El. Res.,* **26** : 599-611.

Notton, B.A. and Hewitt, E.J. 1971. The role of tungsten in the inhibition of nitrate reductase activity in spinach (*Spinacea oleracea* L.) leaves. *Biochem. Biophys. Res. Commun.,* **44** : 702- 710.

PAIS, I. 1971. Chemistry, agriculture and man (in Hungarian). *Tud. Mezögazd.,* **9** : 9-13.

PAIS, I. (Ed.) 1985. New results in the research of hardly known trace elements. *Proc. I. Internat. Tr. El. Symp.,* Budapest 1-299 pp.

PAIS, I. 1985. Some aspects of the development in the research of the hardly known trace elements and their importance in the interdisciplinary trace element research work. *Proc. I. Internat. Tr. El. Symp.,* Budapest (Ed. : I. PAIS) 3-14.

PAIS, I. (Ed.) 1987. New results in the research of hardly known trace elements and the analytical problems of trace element research. *Proc. II. Internat. Tr. El. Symp.,* Budapest 1-309 pp.

PAIS, I. (Ed). 1988. New results in the research of hardly known trace elements and their role in the food-chain. *Proc III. Internat. Tr. El. Symp.*, Budapest. 1-219 pp.

PAIS, I. (Ed.) 1990. New results in the research of hardly known trace elements and their importance in the International Geosphere-Biosphere Programme. *Proc IV. Internat. Tr. El. Symp.,* Budapest 1-350 pp.

PAIS, I. 1990. The problematic of essentiality and beneficiality of trace elements. *Proc. IV. Internat. Tr. El. Symp.,* Budapest. (Ed: I. PAIS) 191-204.

PAIS, I.(Ed.) 1992. New perspectives in the research of hardly known trace elements. *Proc. V. Internat. Tr. El. Symp.,* Budapest 1-240 pp.

PAIS, I. 1992. The role of trace elements in food production. *Hung. Agric. Res.,* **1** : 26-29

Rauser, W.E. 1990. Phytochelatins. *Annu. Rev. Biochem.,* **59** : 61-86.

Schmidt, L.H. 1989. Zur Essentialität der Spurenelemente. *Pharmaize,* **44** : 377-380.

Schwarz, K. 1970. Control of the environmental conditions of trace element research: and experimental approach to unrecognized trace element requirements. In: MILLS, C.F. (Ed.). *Proc. Internat. Symp.,* Aberdeen. 25-52.

Underwood, E.J. 1971. Trace elements in human and animal nutrition, 3rd edition, Academic Press, 1-543 pp.

Advancements In Micronutrient Research
Editor : A. Hemantaranjan
1996, Scientific Publishers, Jodhpur, pp. 19-44.

2

THE IMPORTANCE OF HARDLY KNOWN TRACE ELEMENTS

ISTVAN PAIS

Kerteszeti ES Elemliszeripari Egyetem, Kemiai Esbiokemiai Tanszeki
Department of Chemistry and Biochemistry of the University for Horticulture and Food Industry, H - 1502 Budapest, Villanyi ut 29-31, HUNGARY.

CONTENTS

1. INTRODUCTION

As we discussed in the general introduction of the trace element research, in the family of 65 trace elements there are about 45 elements which are considered and may be called as hardly known trace elements.

Since our knowledge on their physiological role is very incomplete and due to the pollution of nature these elements can be the member of the food-chain, we should make scientific efforts to improve our knowledge on their physiological role if we want to avoid the undesirable consequences of their effects. We have always been convinced that it is our important task to imporve our knowledge on their physiological role.

Having these considerations in mind, we started with the research of some hardly known trace elements at the beginning of the seventies : on a much larger scale on the role of titanium, later on gallium and recently on zirconium. In the next some subchapters we would like to recapitulate the most important parts of our research.

2. THE PHYSIOLOGICAL IMPORTANCE OF TITANIUM

It is not quite generally known that titanium is not a rare but a relatively very common element occurring in Earth : it is exactly the tenth in the line of 88 permanent elements which constitute the matter of our planet.

It was mostly documentated before 1970 that titanium seems to be a physiologically inactive, and for the living systems not an important element. This opinion may generally be accepted as true, because titanium compounds occurring in nature (Ti-dioxide, Ti-silicates and some titanates) are usually - calculating the common pH-range in the living systems : pH 6-8- insoluble in water, therefore not uptakeable for the plants and does not have the possibility to cross the membranes of the living cells.

In spite of the fact that in the literature we can find some details (Konishi and Tsuge, 1936; Schroeder *et al.,* 1963; Dobroljubszkij *et al.,* 1981; Rutskaya, 1976; Mrowca, 1973) in connection with the physiological role of the titanium, before the systematic research work of the Chemical Department, University of Horticulture, Budapest, Hungary (recent name is : Department of Chemistry and Biochemistry, University of Horticulture and Food Industry, Budapest, Hungary) nobody dealt really deeply and as comprehensively with the role of titanium in living organisms.

The main success of our research work was to find a pH-stable and not unnatural titanium-chelate : the chelating agent is ascorbic acid, which gives the required type chelate compound with titanium. This compound was demonstrated to be very successful in plant growing as we reported in our great review publication (PAIS, 1983), in which we summarized our results in connection with 14-16 different agricultural plant species. This publication summarized our results closing with the year 1981. As a comprehensive information we give now Figure 1., which shows the average data of six years' experiments showing that the yield-increase was usually 10-15 per cent and the quality of products was also better.

2.1. Plant Experiments

Before giving the report of our new results, we should mention that two Belgian and one Spanish research group also did experiments with titanium compounds, mostly with our patented Ti- ascorbates (trade name : TITAVIT).

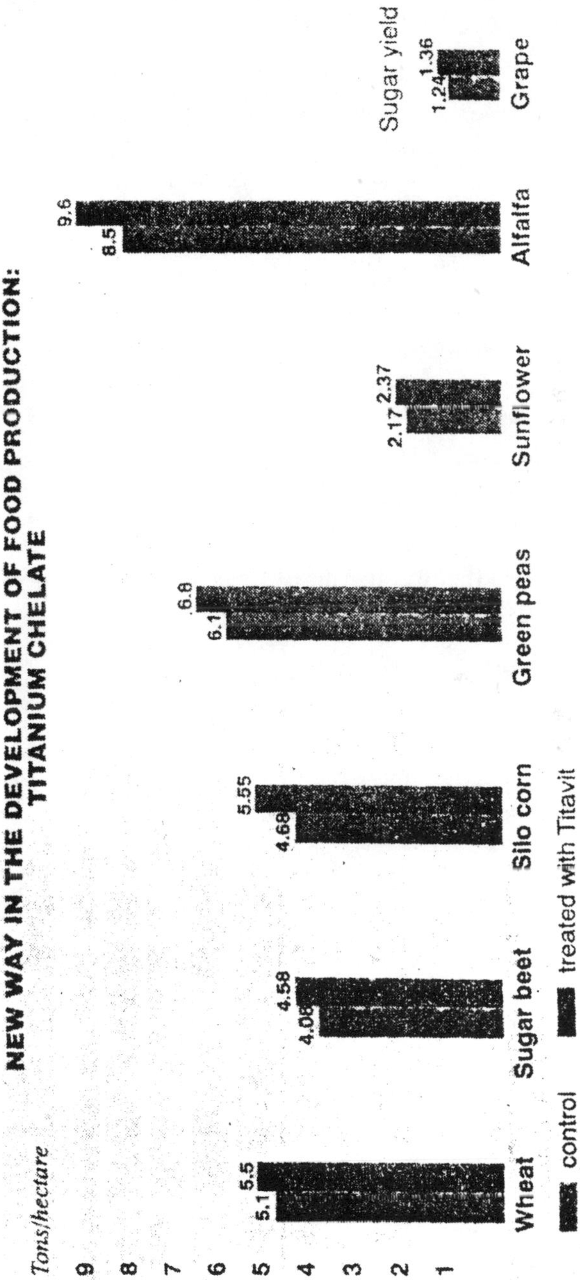

Fig. 1. Average data of field-experiments with different plants (1975-1981)

Ram *et al.* (1983) published a paper on response of bean to foliar spray of titanium. They applied inorganic Ti-tetrachloride and organic "titanocene" (dicyclo-penta-dienyl-titanium dichloride). The bean plants were sprayed at 4 and 6 leaf stage with the solutions. They registered the following data : (a) The total chlorophyll-content of the leaves showed an increase of 10- 40 per cent if the organic form was applied and 6-20 per cent if the inorganic form was applied respectively. (b) Titanium- application caused a significant increase in the nutrient element content of the leaves.

The same team wrote in their second publication (Ram *et al.,* 1988) that the foliar application of titanocene resulted in a 20 per cent higher dry-matter production in tomato plants.

Kiekens and Camerlynck applied Ti-ascorbate solution on maize and cowpea (1986) grown in nutrient solution. Both plants showed a dry-weight increase as 10-30 per cent versus the control plants.

Martinez-Sanchez *et al.* published three papers in the year 1990 in connection with paprika (*Capsicum annuum* L.) Plants. They applied our TITAVIT-solution and they found up to 20 % yield-increase and the analysis of leaves showed higher uptake of N, P, K, Fe, Mn and Zn, as we registered the same fact earlier. They found in dip irrigation fields a yield-increase of 25 % and in flood irrigation system over 40 %, respectively.

2.1.1. Light-chamber experiments

In the light-chamber of our Department we carried out tomato- experiments under hydroponic circumstances. We applied a nutritonal solution which we worked out 18-20 years ago containing all essential macro- and microelements. The indifferent plastic dishes with "plastic soil" from fumed polystyrol – without the possibility of ion-exchange and contamination – were divided into six groups. The first was the control, the second was sprayed with a Ti-ascorbate solution containing 5 mg Ti/ml, the third and fourth contained 0.2 and 0.5 ppm titanium as inorganic $TiCl_4$ in root zone, while the fifth and sixth contained the same concentrations of the titanium but in Ti-ascorbate form. Starting with this experiment we planted in all dishes 4 tomato seedlings germinated in ion-free water. The foliar application was made at 4 leaves stage.

One week after the treatment we finsihed the experiment and measured the dry matter production and the element content of whole plants by ICP-method, only the N-content was measured by Kjeldahl-method. As the data collected in Tabel 1 show, in all cases the dry matter production was better in relation to control. The concentration-values in Table 2 demonstrate that the application of titanium increased the uptake of essential nutritive elements. In dry matter production the inorganic form, in the element content the foliar application gave better results.

As a part of the earlier detailed tomato experiment, from the titanium-treated (5 ppm Ti-ascorbate solution) and control leaves we isolated 3 cell-fractions and after wet-ashing we determined the trace element content of different fractions with ICP- technique Feher *et al.* (1987).

In Table 3, we collected the analytical data of isolated chloroplasts from the treated and control plants. As the data clearly show, the titanium had higher concentration in chloroplasts and two, in the photosynthetic pathways very important trace elements : the iron and manganese had also higher levels than in the control samples. These data may be accepted as indirect documentation of the increased photosynthetic activity.

TABLE 1. Dry matter production of tomato plants in solution culture

Treatment in root zone	Whole plant g/dish	Difference %
$TiCl_4$ 0.2 ppm	1.007	+ 54
$TiCl_4$ 0.5 ppm	0.821	+ 26
Ti-ascorbate 0.2 ppm	0.807	+ 19
Ti-ascorbate 0.5 ppm	0.915	+ 40
Control	0.652	

TABLE 2. Nutritive element content of tomato plants in solution culture

Treatment	N	Pmg/disk	K	Ca	Cu	Fe µg/disk	Zn
Control	79.4	23.7	104.1	33.8	76.8	253.9	188.4
Sprayed	82.9	26.7	125.1	42.2	88.8	714.2	266.7
Root zone	81.2	24.7	112.9	37.2	76.2	780.3	199.5

TABLE 3. Analytical data in isolated chloroplasts

Treatment	Ti	Fe /µg/g	Mn
Control	0.4	102.6	37.3
Ti-sprayed	1.1	154.4	46.5

We treated tomato-plants with 5 mg Ti/ litre Ti-ascorbate solution and as it was mentioned earlier, we isolated the chloroplast- fraction and in suspension of it, we searched the Hill-reaction, the reduction of $NADP^+$ and the change of pH. The control plants were of two sorts : sprayed with ion-free water or the same concentration of ascorbic acid as in TITAVIT solution.

From the literature we know that from photolysis of water originated hydrogen-ions are oxidized by $NADP^+$. We can follow this process by application of NADP-reductase enzyme and the changes of this we can measure by photometric way. The absorbance values of the two controls were very similar, while the treated samples showed a significantly higher absorbance values : the difference was over 20 per cent.

We measured the oxidation of dichlorophenol-indophenol (DPIP) in photometric way and calculated the oxygen-production of enlighted chloroplasts to 1 mg chlorophyll. As the data in Table 4 show, the developed oxygen in the treated samples was significantly higher than in water-sprayed and in ascorbate-sprayed samples. It clearly demonstrates the higher photosynthetic activity as a consequence of the effect of titanium-ascorbate.

We would like to mention that the enlighted chloroplasts showed a decreasing tendency in pH-values as a consequence of titanium- treatment. We received very similar results in increased oxygen- evaluation in photosynthetic algae cultures.

In Figure 2. we would like to show the changes in NADP⁺ reductase activity in different treatments.

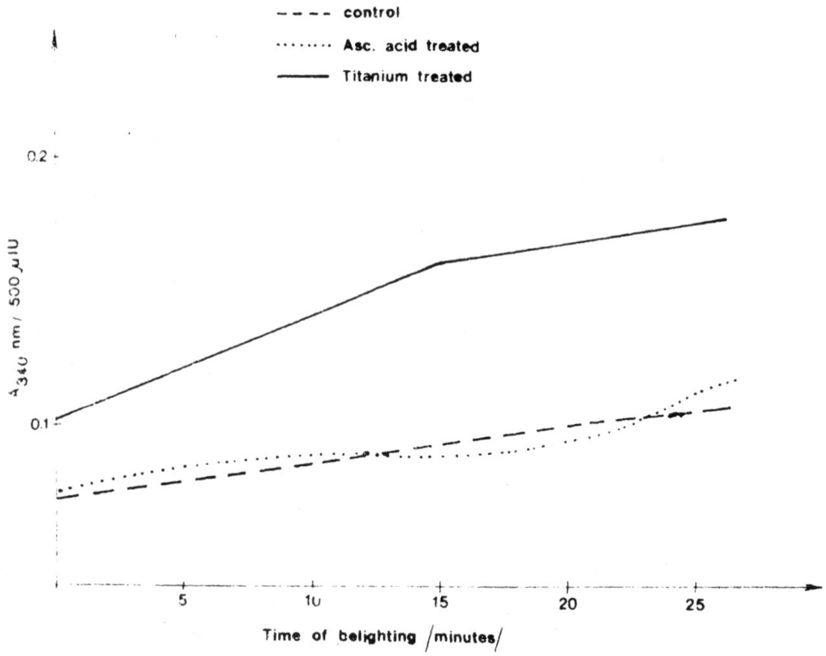

Fig. 2. The activity of photosynthetic NADP⁺ –reductase in chloroplast-suspension of the titanium-treated and control tomato leaves

TABLE 4. Oxygen production from enlighted chloroplast fractions

Treatment	Calculated O_2/1 mg chlorophyll
Control (ion free water)	0.805
Control (ascorbic acid)	0.810
Titanium–ascorbate	1.015

2.1.2. Different field-experiments, 1982-1986

As we published earlier (PAIS 1985), we organised field- experiments with winter-apples, green-pea sorts and winter- wheat. In all cases the titanium-treatment showed a great increase of yiled and also the improvement of the quality of the products.

In these experiments we applied a new composition, called "Super-Titavit", which is combined solution containing as the main factor Ti-ascorbate and some trace elements (iron, zinc, manganese, molybdenum) usually in a concentration of 0.5-1.0 ppm. We applied these new compositions because the different soils may be poor in different essential trace elements and in these cases the beneficial effect of Ti-ascorbate to mobilize these elements for plants is not always so successful : it seems easier if we give some elements together with the titanium-ascorbate.

We organised these experiments partly in the Experimental Farm of our University (Halasztelek) and in the Experimental Farm of the University of Agriculture (Gôdôllõ). We collected our results in Tables 5– 11. We should also consider that in green pea and later in winter-wheat experiment we used different sorts, therefore the value of our results is very much higher for practical application.

TABLE 5. Experiments with Jonathan apple in Halásztelek, 1982-1984

Year and treatments		Yield kg/parcel	Difference %	Calculated Yield t/ha
	Control	202		14.5
1982	Titavit	248 **	+ 22.8	17.8
	Super-Titavit	253 **	+ 25.2	18.2
	Control	270		19.4
1983	Titavit	290	+7.4	20.9
	Super-Titavit	305 ***	+ 13.0	22.9
	Control	218		16.3
1984	Titavit	251	+ 15.1	18.8
	Super-Titavit	269*	+ 23.4	20.1

* at p= 5%, ** at =1% abd *** at p=0.1% level significance

TABLE 6. The internal content and storability of apples, 1982- 1984

Year and treatments		Acid % 0	Reducing sugar %	Storability defects%
	Control	9.3	8.2	26.9
1982	Titavit	9.7	9.2	19.1
	Super-Titavit	9.8	8.5	18.7
	Control	11.0	8.6	8.4
1983	Titavit	11.7	8.4	2.9
	Super-Titavit	13.1	8.8	4.5
	Control	10.4	5.0	–
1984	Titavit	9.3	9.6	–
	Super-Titavit	8.7	10.1	–

– =In this year we have not data

TABLE 7. Farm experiments with green pea varieties in Gôdôllõ, 1982-1984

Year and treatments		Number of varieties	Yield in average t/ha	Difference %
	Control		2.16	
1982	Titavit	2	2.38	+ 10.2
	Super-Titavit		2.37	+ 9.7
	Control		5.25	
1983	Titavit	22	5.50	+ 4.8
	Super-Titavit		5.58	+ 6.3
	Control		7.39	
1984	Titavit	28	7.83	+ 6.0
	Super-Titavit		7.84	+ 6.1

TABLE 8. The changes in tenderness of green pea by titanium–treatment

Year and treatments		Tenderness T°	Difference %
	Control	148.2	
1982	Titavit	139.8	- 5.7
	Super-Titavit	142.5	- 3.8
	Control	173.2	
1983	Titavit	165.0	- 4.7
	Super-Titavit	156.6	- 9.6
	Control	146.5	
1984	Titavit	133.3	- 9.0
	Super-Titavit	137.3	- 6.3

2.1.3. Different experiments with pigments and enzymes

In the last ten years we organised new experiments to collect experiences in correlation with the effect of titanium-ascorbate on pigment biosynthesis and also on activity of different enzymes.

TABLE 9. Farm experiments with winter wheat in Gôdôllõ, 1982-1986

Year and treatments		Weight of 100 Seeds g	Yield t/ha	Difference %
	Control	40.5	4.08	
1982	Titavit	43.5	4.30	+ 5.5
	Super-Tîtavit	42.5	4.36	+ 6.9
	Control	35.9	5.99	
1983	Titavit	37.8	6.44	+ 7.5
	Super-Titavit	38.4	6.53	+ 9.0
	Control	41.0	6.00	
1984	Titavit	43.2	6.41**	+ 6.8
	Super-Titavit	43.3	6.45**	+ 7.5

** at p=1 % level significance

TABLE 10. Farm-experiments with given pea varieties in Godollo, 1985-1986

Year and treatments		Number of varieties	Yield in average t/ha	Difference %	Diameters in mm and in % 8.5	8.5-10.0	10.0
	Control		5.66		19.0	51.8	29.2
1985	Titavit	34	6.13	+ 8.3	4.7	68.2	27.1
	Control		4.17		3.3	69.2	27.5
1986	Titavit	6	4.70	+ 12.7	0.8	79.8	19.4

TABLE 11. Farm-experiment with winter wheat in Gödöllõ, 1985

Treatments	Number of varieties	Weight of 1000 seeds g	Yield t/ha	Difference %
Control		37.9	6.13	
Titavit	5	40.0	6.61	+ 7.8

In Hungary the fruits of *Lycopersicon esculentum* "Ventura" variety are used in tomato processing. Treating plants with titanium- ascorbate leads to many desired changes in the quality and quantity characteristics of tomato fruit. We investigated the changes in pigment composition of these frutis (Daood *et al.,* 1987; Biacs *et al.,* 1988).

Thin layer chromatography (TLC) and high performance liquid chromatography (HPLC) were applied to investigate the changes taking place in the pigment composition of tomato fruits harvested from Ti-ascorbate treated plants during the last stages of ripening.

It was found that Ti-ascorbate treatment increased the stability of chlorophyll and lutein against photo-oxidative damage, while the development of the major colour substances like lycopene, neurosporin and violet xanthophyll was disturbed to some extent at the final stage of ripening. This disturbance was found to be related to high activation of lipoxygenase enzyme at the final stage of ripening under the influence of titanium-treatment.

Stability of different pigments of Titavit-treated and control fruit towards lipoxygenase-catalyzed linoleic acid oxidation and storage of different conditions was investigated by means a newly elaborated HPLC system. The pigments of Ti-ascorbate treated fruits were more stable in comparison with those of control samples.

Higher organic acid content of treated fruits was found to protect several pigment from auto-oxidative damage during freeze storage.

Lipoxygenase was extracted (Daood *et al.*, 1988) from untreated tomato fruits and those treated with titanium- ascorbate at the last stage of ripening. It was found that titanium has an activating effect on lipoxygenase so that the development of red pigments was inhibited to some extent. The addition of Ti-ascorbate to the reaction mixture resulted in a lag period of about 1 minute to lipoxygenase-catalyzed linoleic acid oxidation, whereas, incubation of this titanium-chelate with enzyme extract for 10 minutes significantly increased enzyme activity. Activation of lipoxygenase by Ti-ascorbate was confirmed when addition of titanium in inorganic form ($TiCl_4$) also increased the enzyme activity.

This effect of titanium on lipoxygenase activity during ripening may be due to : (a) Decreased biosynthesis of the enzyme or the availability of some essential components such as iron at the initial stage of ripening. (b) Increased antioxidant content of treated fruits at the beginning of ripening which affects the utilization of molecular oxygen by the enzyme in the course of linoleic acid oxidation. (c) High availability of some essential elements and cofactors, like Fe and Mn, at the final stage of ripening which activate the enzyme when they exist at low concentration, but may be more effective when at higher concentration, such as that found in Ti-ascorbate treated fruit. (d) By direct interaction between the enzyme and titanium, similar to what occurs with iron, as titanium absorbed by the plant is found absorbed on the surface of some macromolecules (like proteins) since there is no big difference in the Ti-content of Titavit treated and untreated fruits.

We have confirmed that titanium has an activating effect on lipoxygenase both *in vivo* and *in vitro*. The activated enzyme may induce changes in various metabolic pathways, such as ethylene production, lipid oxidation, chlorophyll and caroteniod synthesis and degradation etc. That is a reason why we are convinced that titanium-application has a great physiological importance.

In 1988 and 1989 open field experiments were done in the Agrobotanical Garden of Teacher's Training College in Nyiregyhaza, Hungary. We selected three tomato cultivars used in canning industry. Two times 2 mg Ti/L solution of titanium-ascorbate was sprayed on the leaves of tomato plants before the flowering and two weeks after the first treatment. The foliar application was given with four replications in randomized block design (Simon *et al.*, 1990).

We measured the dry matter content of fruits, the sugar and acid content by HPLC technique and the activity of phosphofructokinase (PFK) enzyme. In the first year we found a yield - increase of 8 % as average, while in the second year it was found near 30 % as average of the three cultivars. The dry matter content was determined as refractive index : we found 10-12 % improvement and these results showed significance at P = 5 and P = 0.1 % level.

As Figure 3 shows, the sugar-content was higher during the ripening and after freeze storage (-24°C). Phosphofructokinase enzyme, which plays key role in the regulation of glycolysis, had lower activity during matration. Since titanium content of the fruits is low and there is no difference between control and treated fruits, this inhibitory effect of Ti-treatment seems to be rather indirect.

Treatment of tomato plants can act through higher content of citrate and malate in fruits (negative effectors of PFK). A new pyrophosphate-dependent

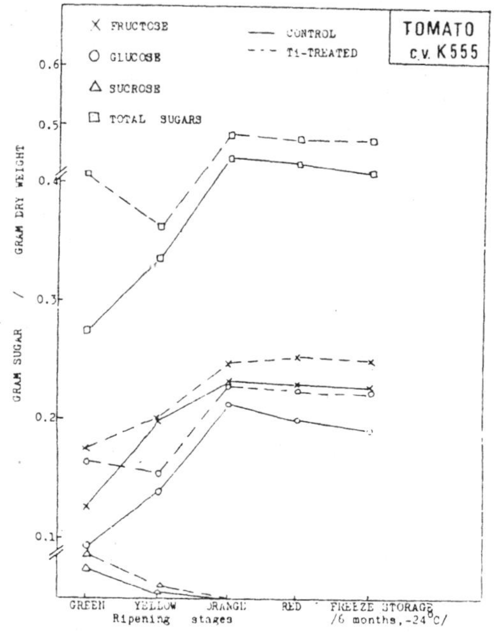

Fig 3. Changes in the carbohydrate content of control and titanium-treated tomato fruits (cultivar "K 555") during ripening and after freeze storage

form of phosphofructokinase was discovered several years ago in plants. This enzyme is under the control of a new regulatory metabolite : fructose 2,6-bisphosphate. We try to search the effect of titanium-treatment on the regulation of different forms of PFK present in tomato.

2.2. Animal Experiments

Based on our successful plant-experiments, in 1979 we applied titanium-ascorbate, in solid form, in animal feeding. Because our first experiment with broiler-chickens showed promising results, we organised new feeding experiments with cattle, fatted lambs, rabbits and pigs. The results of these farm-experiments are collected in Table 12.

TABLE 12. Some data of animal feeding experiments in large large scale conditions (1979-1983).

Breed	Number of animals	Duration of experiment day	Body weight gain %
Swine 1.	288	76	+ 14.3
Swine 2.	504	47	+ 7.9
Rabbit	300	68	+ 5.6
Lamb	100	61	+ 10.9
Heifer	24	123	+ 5.4
Bulls	24	123	+ 10.5

The experimental animals received in their daily feed 30-45 mg Ti/kg. As the data show that feeding with Ti-ascorbate favourably influenced the body weight gain. During these experiments no unfavourable clinical symptoms were observed among the experimental animals.

Having this basis, animal feeding experiment was organised to examine the possible toxic effect of titanium-ascorbate, its accumulation within the organism and its excretion. This experiment was begun with 300 broiler-chickens.

After the selection from the flock, 1 control and 4 experimental groups were formed, each containing 56 cockelers. In the fattening period we formed a new group with extremely high Ti- content in feed. We applied by the common control group a second control, in which the birds received the same quantity of ascorbic acid as in the Ti-ascorbate additive.

The usual examinations were executed during the experiment and at the end of the first, fourth and seventh week 4 animals were bled to death immediately after hatching and the trace element content was determined in the liver and kidneys by ICP-technique. Finally faeces was collected from the different groups and analysed for titanium. Our results can be seen in Table 13. and 14.

TABLE 13. Results of body weight gain

Experimental group	Average body weight gain			
	14 day		49 day	
	g	%	g	%
Control	260	100.0	1930	100.0
Ascorbic acid	251	96.4	1940	100.5
18 mg Ti/kg	257	98.8	1920	99.5
45 mg Ti/kg	268	103.0	2060	106.7
450 mg Ti/kg	239	92.0	1790	9.27

Neither the animals of control nor those of the experimental groups presented pathological clinical symptoms (Bokori *et al.,* 1985). There was no death even in special group fed with fodder containing 2250 mg Ti/kg. The body weight gain was the best applying 45 mg Ti/kg, lower and higher Ti-ascorbate did not show so favourable results.

Evaluating the analytical data we can state that the titanium- content was very low in liver and kidney : there is no danger of higher accumulation. As a consequence of titanium-application the iron-content of liver increased showing that titanium has a positive (synergistic) effect on iron-metabolism, which we have found in other experiments, too.

Concerning the excretion, the applied titanium was found mostly in faeces. We should like to emphasize that the excreted iron was lower in faeces of Ti-treated chickens : it means not only the uptake, but also the retaining of iron was higher than in control birds. This observation has a great importance in animal nutrition, because the iron-anemia is a very common disease in animal feeding.

In the year 1984 we organised a pig-feeding experiment in a big State Farm in Szekszard (Nagy *et al.,* 1986). On 25 January we started with the first part of the experiment with 254 control and 254 Ti-treated piglets which were 38-40 days old. The percentage of female and the barrow pigs was the same. In the second part of the experiment, two weeks later, we started with 262-262 piglets with the same sex-proportion.

The members of the control groups received the same, in Hungary mostly used pig-food, while the feed of pigs in the experimental groups was mixed with solid Ti-ascorbate in 30 mg Ti/kg feed concentration. The first experiment was finished after 148 days, the second one after 155 days.

As the data collected in Table 15 show, the average body weight gain of titanium-treated pigs was in the first experiment by 3.8 kilograms, in the second experiment by 1.6 kilograms greater than that of the control pigs. The difference is not extremely large, but deserves interest.

TABLE 14. Microelement contents of samples referred to dry matter, in mg/kg

Sample	Symbol	Control (Weeks)			45 mg Ti/Kg (weeks)			450 mg Ti/kg (weeks)		
		1.	4.	7.	1.	4.	7	1	4.	7.
	Fe	200	250	350	210	200	500	95	240	680
Liver	Ti	0	0	0	0	0	0.2	0.9	1.2	2.5
	Zn	100	100	110	270	90	450	80	90	140
	Fe	75	260	150	148	220	300	80	130	190
Kidney	Ti	0	0	0	0.6	0.2	0.5	1.2	3.1	2.9
	Zn	58	120	80	110	75	100	80	90	100
	Fe	270	340	300	190	135	169	31	41	45
Faeces	Ti	0.9	2.6	7.3	10	13.2	13.1	57	35	27
	Zn	520	190	440	390	270	300	570	310	405

If we calculate the sold mass of pigs, which is to be seen in Table 16, our results can be evaluated as excellent. The value of these experiments is not only in selling + 13.2 per cent more pork, because it has not only the body weight gain as source. The number of sold pigs was in titanium-treated groups about 10 per cent greater than in control groups. We can conclude that titanium ascorbate should have a promotive role in the health- status of the pigs and this fact should have a great physiological importance.

Having these experiences, in 1984 and 1985 we organised new pig experiments in the same Farm to search in which way the Ti- ascorbate is able to promote the vigour of young pigs. We selected 48-48 sows in three groups : the sows in the first group received the common pig-food, in the second we added to this common food 500 g. ascorbic acid pro 1 ton and in the third we mixed 1 kilogram solid Ti-ascorbate which contained 3 per cent active titanium.

We measured in both cases the body weight of the young pigs on the 21st day, and the iron-content in the blood-serum of the young pigs in the first and in the fifth day of life. As the results in Table 17 shows, the ascorbic acid alone has some positive effect, but the titanium-ascorbate effect is much more characteristic.

TABLE 15. The data of the pig experiments in Szekszard, 1984

Treatment	The average body weight of pigs, kg			
	1. experiment	Difference	2. Experiment	Difference
Control	88.36		95.56	
Ti-treated	92.16	3.80	97.14	1.58

TABLE 16. The sold pork in the experiments in Szekszard,1984

Treatment	The number of saled pigs	Saled pork kg	Difference%
Control	436	40040	
Ti-treated	479	45310	+ 13.2

TABLE 17. The data of the pig experiments in Szekszard,1985-1986

Treatment	Average body weight, kg		Iron content	μM/litre
	First bearing	Second bearing	1day old	5days old
Control	5.07	4.86	9.1	8.1
Asc. acid	5.46	5.11	12.6	8.6
Ti-ascorbate	5.81	6.16	15.6	21.5

In 1988 we organized an experiment in three replications with five groups of laying hens to examine the effect of Ti-ascorbate on egg-production and on quality of eggs. We mixed different quantity of solid titanium-ascorbate in the daily feed of hens and we formed a second control group beside the general control group with common feed where the daily feed was mixed with the same quantity of ascorbic acid as present in Ti-ascorbate (Halmagyi-Valter *et al.*, 1988).

We collected our results comprehensively in Table 18. As it is clearly demonstrated that the ascorbic acid has no meaningful effects but the Ti-ascorbate treatment resulted in positive changes in quantity and quality in all parameters, as well.

In 1988 we organised a reproduction-expèriment in Szekszard with pigs. We formed three groups containing 32 sows in farrow each and during the three weeks before the farrow the sows of two groups received a feed mixed with solid Ti-ascorbate and solid Ti- ascorbate + additional chelate forming compound which is cheaper than the pure ascorbic acid. This second composition was called : 'New Titavit". The active titanium-concentration in both groups was 45 mg Ti/kg in diet (PAIS *et al.*, 1989).

In Summer the first generation was born and then we gave to the Sows and to the weaned pigs the feed mixed with mentioned two types of Ti-ascorbate. After the separation the sows were again artificially inseminated and we counted the success of insemination and some other parameters. The members of the first generation were partly fattened and after the slaughtering the quality of meat was also measured. These results were collected in Table 19 and Table 20.

Summarizing our results, Ti-ascorbete has a positive effect in reproduction and also in quality of the meat-production.

TABLE18. The results of the laying hen experiment

Treatment	Average egg production%	Average difference in weight of eggs%	Average difference In eggshells hardness%
Control	67.1		
Ascorbic acid	67.9	-0.10	+3.7
30mg Ti/kg	71.6	+1.42	+6.9
45 mg Ti/kg	69.7	+2.76	+11.4
60 mg Ti/kg	69.1	+2.48	+8.0

TABLE19. Data of pig-reproduction experiment

Parameters	Control	Titavit	New Titavit
Success of insemination after separation, %			
1-10days	31.25	53.57	53.13
11-30days	15.62	10.72	18.75
after 30 days	6.25	0.00	0.00
non pregnant sows	31.25	32.14	18.85
Number of liverborn piglets			
I. Bearing	152	150	150
II. Bearing	147	164	176
Average litterweight, kg			
I. Bearing	13.36	15.08	14.42
II. Bearing	12.23	12.39	13.42

TABLE 20. Data of slaughtered pigs at the end of fattening

Parameter	Control	Titavit	NewTitavit
Body weght gain, kg	67.76	69.18	71.51
Average% in white meat	42.0	44.3	45.2

In 1990 we organised reproduction experiment in the same State Farm with cows with a system similar to the pig-experiment. We found a milk-production 6 % higher in the Ti-treated groups. The insemination ratio in young cows as a consequence of Ti-ascorbate was more favourable than in animals of the control group.These data were not published (PAIS *et al.*, 1990).

2.3. Effects of Titanium on Algae and Yeast

Our first experiment with blue-green algae was made in 1983 (Toth *et al.*, 1985). Earlier we had success with different pesticides (Farkas *et al.*, 1981) where titanium-ascorbate treatment showed a favourable curative effect against pesticide toxicity.

We used *Anacystis nidulans* cells which were grown under sterile conditions in glass vessels of 1000 cm^3 containing Kratz-Myers C nutrient solution in 10000 lux fluorescent white light at 37°C. Air conditioning five per cent carbondioxide was bubbled through the solution. At the beginning of the experiment the initial number of cells was standarized at $5x10^7$ cells/cm^3. The applied treatments can be seen in Figure 4.

We registered the increase of the cell-number from the beginning of treatments on the basis of the optical density measurement at 650 nm. The

Treatments

1. TITAVIT" (1ppm)
2. Atrazine (0.1 ppm) (Hungazin PK-50)
3. Atrazine (1 ppm) (Hungazin PK-50)
4. TITAVIT" (1ppm) and Atrazine (0.1 ppm)
5. Control

Fig. 4. The different treatment in algae experiment, 1983

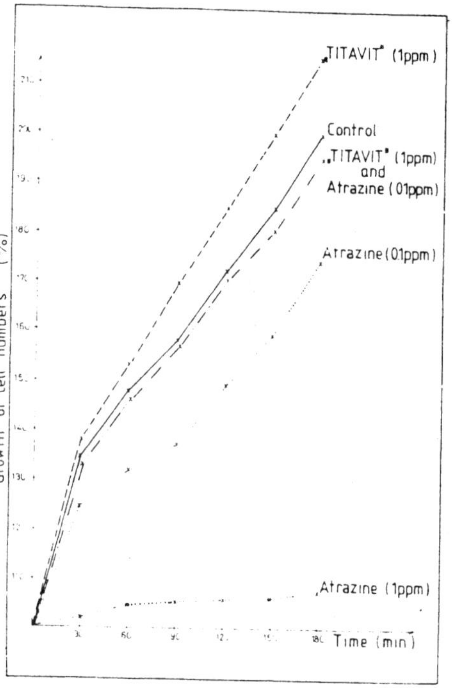

Fig. 5. The growth of cell number by different treatments

oxygen evolution was also measured for three hours with the apparatus measuring the dissolved oxygen. The dry weight of cells was determined after twenty-four hours. We determined also the chlorophyll-*a* content of cells after methanol extraction and measuring at 665 nm.

If we see the rate of cell-number increase in Figure 5, it becomes gradually lower in the Hungazin-treated cultures than in control culture, depending on the Hungazin concentration. Under the influence of titanium-ascorbate treatment the increase of the number of cells gathered speed compared with the control. The application of 1 mg Ti/liter after 0.1 ppm atrazine showed similar results than in control (curative effect). The situation by the oxygen evolution was in most cases quite similar. We can state that the titanium-ascorbate application can give an inhibiting effect against the Hungazin toxicity.

Some years later (Simon *et al.*, 1988) we renewed our experiments with *Chlorella pyrenoidosa*. We found that the Ti- ascorbate in 1 μm concentration can increase the dry-mass production and also the protein-content of algae. Lower titanium- concentrations are not effective enough, while in higher concentrations we found a strong inhibition effect. The favourable concentration of Ti-ascorbate enhanced the photosynthetic pigment concentration of the algae cells.

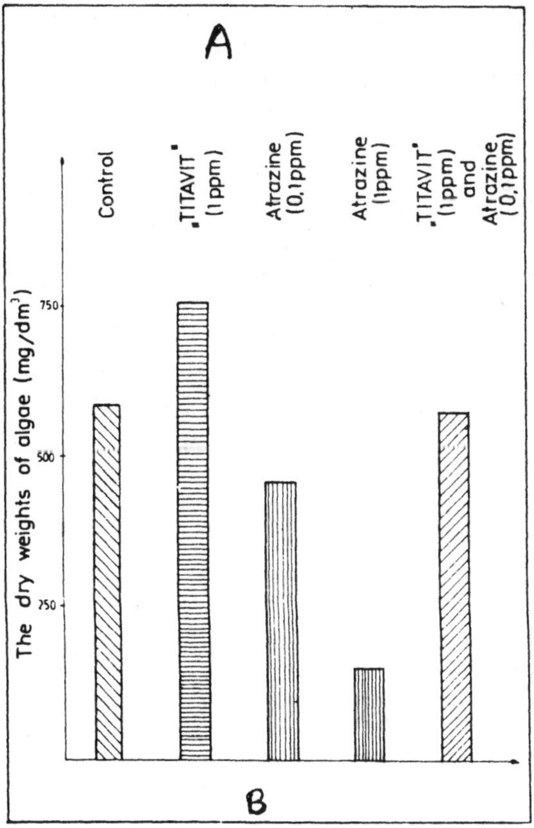

Fig. 6. Demonstration of effect of titanium on *Anacystis nidulans* by dry weight

In 1985 we published a paper (Kiss *et al.*, 1985) in conncetion with common effect of titanium and gallium on photosynthetic rate of *Anacystis nidulans*. The oxygen productivity was highly enhanced by gallium, but the titanium-treatment was also quite effective. The dry-weight increase was much higher by titanium application as it is demonstrated in Figure 6. In the same paper it was demonstrated that titanium has a favourable effect on the activity of fructose 1,6-bisphosphatase.

In 1991 we started with experiments to search for the effect of titanium ascorbate on baker-yeast. The earlier experiments (Janzso *et al.*, 1990) demonstrated fully that under given conditions yeast are able to accumulate large amounts of trace elements in their cells and incorporate them into organic compounds.

In our investigation titanium has been introduced into *Saccharomyces cerevisiae* cells. The nutrient media we have supplemented with titanium-ascorbate solution : at 20-50 μg Ti/cm^3 concentration the titanium uptake resulted 1500-6000 μg Ti/g dried yeast. Addition of titanium-ascorbate is more suitable in the exponential phase of yeast growth : in this case the titanium is bound to the cell-components instead of adsorption on the surface (PAIS *et al.*, 1992).

According to our results, the protein biosynthesis is increased by 20-25 % and in correlation with the control samples the amino acid composition of protein is changed to a great extent, as it is demonstrated by Table 21.

This titanium-enriched yeast was given for toxicological control-experiment and it was found that the LD-50 value is over 20 g titanium/kg : i.e., the titanium has no toxical effects. This last result is in good agreement with our earlier mentioned chicken-experiment, when after application more than 2000 mg Ti/kg feed, we did not find any toxical clinical symptoms in the experimental animals.

3. Beneficial Effects of Gallium

Although gallium is widespread in nature, its minerals are extremely rare and gallium is generally associated with aluminium in common minerals. The average content of gallium in the earth's crust is estimated 10 mg/kg. The mean gallium content of soils is some mg/kg with a range 0.5-200 mg/kg. Gallium is present in most soils where it occurs mainly in aluminium-silicates. Gallium is found in terrestrial plants at a concentration range of 0.01-0.20 mg/kg, in marine algae at 0.5 mg/kg and in marine fishes at 0.15 mg/kg.

Gallium had no known function in plant or animal metabolism and has not been identified as being essential for growth. Presumably gallium is not completely indifferent in biological systems as it can probably be a substitute for iron since their ionic radii are very similar (Fe^{3+} = 0.064 nm and Ga^{3+} = 0.062 nm), and both elements bind ligands with octahedric geometry.

TABLE 21. Differnces in amino-acid content

Amino acid	Control	Ti-treated	Difference %
Asp	42.8	58.9	+ 37.6
Ser	18.5	29.0	+ 56.7
Cys	3.0	4.1	+ 36.6
Lys	38.4	56.4	+ 40.8
Phe	18.2	15.3	– 15.9
Glu	63.5	56.8	– 10.6

Our research group demonstrated first that photosynthetic oxygen evolution and fructose-1,6-bishosphatase enzyme activity in synchronized cultures of blue-green algae (*Anacystis nidulans*) can be enhanced by treatment with 0.5 μM gallium (Kiss *et al.*, 1985). In another experiment, concentration of 0.1 μM to 10 μM gallium was found to enhance dry weight of *Anacystis nidulans,* while higher gallium concentrations activated glucose-6-phosphate dehydrogenase and peroxidase enzymes (Kiss *et al.*, 1987).

Concentration of 1 μM gallium produced higher dry weights and chlorophyll content for *Scenedesmus obliquus* and *Chlorella vulgaris* green algae (Balogh *et al.*, 1987).

In our new experiments (Simon *et al.*, 1989) we treated *Chlorella pyrenoidosa* in *iron deficient* and normal iron supplied nutrient solution with 1 μM gallium-nitrate solution. This application enhanced the growth rate and chlorophyll-content of algae cultures. Significant changes were recorded in individual photosynthetic pigment composition of treated cultures (Shorrocks, 1990).

In both iron-deficient and normal iron supplied cultures, gallium treatment enhances peroxidase enzyme activity. Partially purified peroxidase showed different kinetic properties in the sample treated with gallium. An iron-gallium interaction is proposed that affects chlorophyll metabolism and peroxidase activity in green algae. To demonstrate the mentioned fact, we publish now the peroxidase enzyme activity in Figure 7.

According to these experiences, we organised out-door experiments with tomato (K 555 cultivar) in Nyireguhaza. We applied a gallium- nitrate solution in 0.07 g Ga/ha concentration in three times repeated foliar application (Simon *et al.*, 1991). As a consequence of this treatment, the yield increased, but this change showed no significance. The dry matter content of fruits and their glucose and fructose content showed a significant enhancement : over 30 per cent.

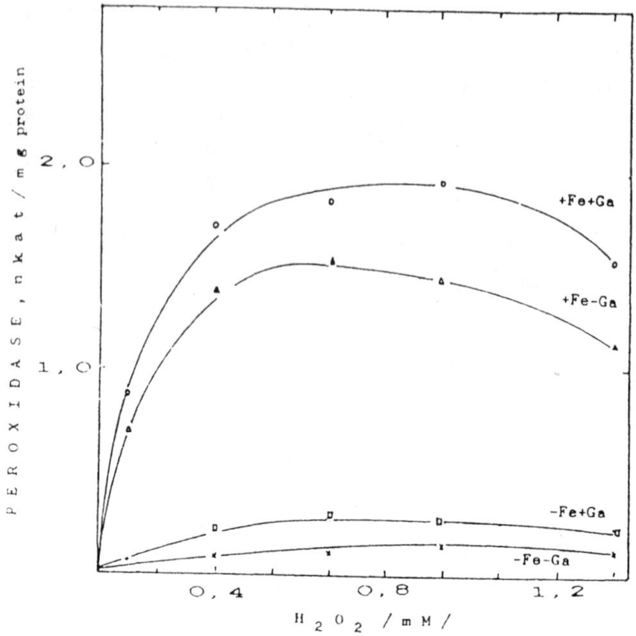

Fig. 7. Effect of iron and gallium treatments on peroxidase enzyme activity of Chlorella

We measured also the fructose-1,6-bisphoshpatase activity of tomato and we found a great increase in the enzyme activity. The increase in carbohydrate synthesis can be attributed to this increased activity.

In the leaves of tomato we analysed the concentration of different nutritive elements by ICP technique. Some element concentrations showed a small decrease but in iron-level we found a meaningful increase.

4. Beneficial Effects of Zirconium

Zirconium is the twentieth most common element of the earth's crust, but our knowledge about its physiological effects is rather poor. The reason of this fact is very similar to the case of titanium – that the solubility of its compounds in nature is mostly very low, therefore it is usually not available for living organisms.

This was the starting point of our idea : what physiological effect can be recognised if we apply it (similarly to Titavit) in organically bound form ?

Our experiments started with lower forms of the living : bacteria, algae any yeasts. Now we give a short survey of algae (Simon *et al.*, 1992) and yeast-strains (Novak-Fodor *et al.*, 1992). 0.1-1.0 μM Zr-ascorbate slightly stimulated, while 5-50 μM of Zr slightly inhibited the growth rate and dry matter accumulation of Chlorella green algae. According to our HPLC-measurements the zirconium-treatment changed the ratio of different photosynthetic pigments.

We analysed the dry material by ICP-technique and recognised a high accumulation of zirconium by algal cells and also some changes in elemental composition of algae : zirconium enhances the uptake of some essential nutritive elements. This effect is similar to that which we compiled in connection with titanium.

In yeast-experiment we used two organic chelate : Zr-ascorbate and Zr-citrate as treating forms and two yeast-strains : *Saccharomyces cerevisiae* and *Candida utilis*. We can state that neither Zr-ascorbate, nor Zr-citrate inhibit the cell-reproduction of *Saccharomyces cerevisiae* and of *Candida utilis* and both strains can accumulate considerable amount of zirconium (Figure 8.).

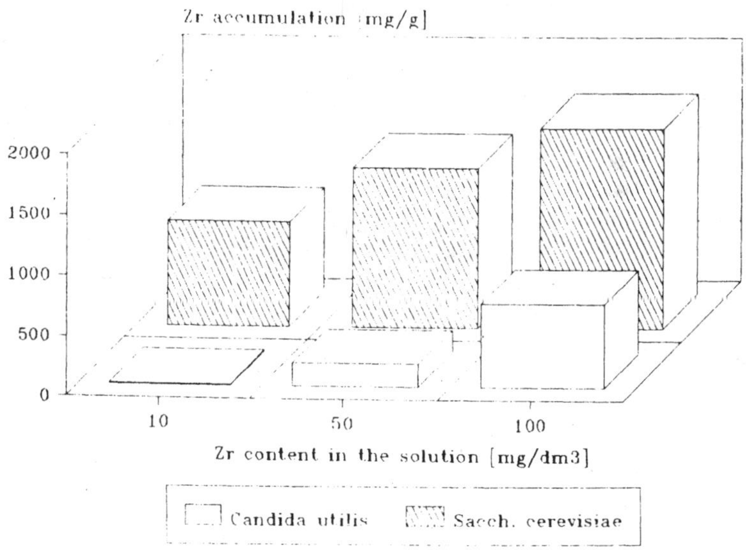

Fig. 8. Accumulation of ziroconium by yeast-strains

The zirconium-treatment enhanced the protein-synthesis and changed the amino-acid composition of proteins : this effect is also very similar to the earlier mentioned effect of titanium- ascorbate.

5. Summary

The author is convinced that three hardly known trace elements : titanium, gallium and zirconium, according to the described results of his research-group, may be regarded as beneficial trace elements, responding to the criteria which were given in the general part of this book/(chapter1).

References

Antòn, E., Jara, J.R., Martinez-Sànchez, F., Giménez, J.L. and Alcaraz, C.F. 1990. Titanium in plant nutrition. I. A. rapid method for titanium determination in plant tissues and soils. *Proc. III. Nat. Symp. Min. Nutr. Plants, Palma,* 109-113.

Balogh, A., Kiss F. and Szabolcsi, L. 1987. Influence of gallium on growth of green and blue-green algae. *Proc. II. Internat. Tr. El. Symp., Budapest* (Ed. : I. PAIS), 127-132.

Biacs, P., Daood, H.G., PAIS, I., Feher, M. and Hajdu, F. 1988. Chromatographic investigation of tomato fruit metabolism during ripening. *Chromotography' 87* (Eds. : H. Kalàsz & L.S. Ettre), 39-59.

Bokori, J., Nagy, B. and PAIS, I. 1985. Application of titanium- ascorbate in animal feeding. *Proc. I. Internat. Tr. El. Symp.,* Budapest (Ed. : I. PAIS), 212-221.

Daood, H.G., Biacs, P., Féher, M., Hoschke, À., Hajdù, F. and PAIS, I. 1987. Effect of titanium treatment on the development and stability of pigments in tomato fruit. *Proc. II. Internat. Tr. El. Symp.,* Budapest (Ed. : I. PAIS), 69-100.

Daood, H.G., Biacs, P., Fehér, M., Hajdù, F. and PAIS, I. 1988. Effect of titanium on the activity of the lipoxygenase. *J. Plant Nutr.,* **11** : 505-516.

Dobroljubszkij, O.K., Tanurkov, G.R. and Sztrakhov, V.G. 1981. Effectiveness of zinc, manganese and titanium use in vineyards (in Russian). *Sadov. Vinograd, Vinedel. Mold.,* **36** (6) : 33-35.

Farkas, E., Tóth, A. and PAIS, I. 1981. The effect of titanium on plants damaged by herbicides. *Acta Phytopath. Hung.,* **16** : 259-261.

Fehér, M., Papp, Kl. and PAIS, I. 1985. New results in the application of titanium-ascorbate as a plant-conditioning product. *Proc. I. Internat. Tr. El. Symp.,* Budapest (Ed. : I. PAIS), 197-211.

Fehér, M., Papp, Kl.. Fodor, P. and PAIS, I. 1987. The effect of titanium on the uptake of other nutritive elements. *Proc. II. Internat. Tr. El. Symp., Budapest (Ed. : I. PAIS),* 113-126.

Giménez, J.L., Martinez. Sànchez, F., Moreno, A., Fuentes, J.L. and Alcaraz, C.F. 1990. Titanium in plant nutrition. III. Effect of Ti (IV) on yield of *Capsicum annuum* L. *Proc. III. Nat. Symp. Min. Nutr. Plants, Palma,* **123-128.**

Halmàgyi-Valter, T., Gippert, T. and PAIS, I. 1988. The effect of titanium ascorbate on reproduction. *Proc. III. Internat. Tr. El. Symp.,* Budapest (Ed. : I. PAIS), 145-152.

Janzsò, B., Suhajda, À., Hollò, J., Szabò, G., Extende, S., Marossy, K. and Hadady, K. 1990. Accumulation of different elements in yeasts. *Proc. V. Eur. Congr. Biotechnol.*, Copenhagen.

Kiekens, L. and Camerlynck, R. 1987. Influence of TITAVIT on growth of maize and cowpea grown in nutrient solution. *Proc. II. Internat. Tr. El. Symp.*, Budapest (Ed. : I. PAIS), 101- 112.

Kiss, F., Deàk, Gy., Fehér, M., Balogh, A., Szabolcsi, L. and PAIS, I. 1985. The effect of titanium and gallium on photosynthetic rate of algae. *J. Plant Nutr.*, **8** : 825-831.

Kiss, F., Vincze, Gy., Kalucza, Gy., Szabolcsi, L. and Balogh, A. 1987. Study of the effect of gallium on *Synechococcus* AN 6301 blue-green alga. (in Hungarian). *Acta Acad. Pedag. Nyiregyhàza,* **11** : 63-69.

Konishi, K. and Tsuge, T. 1956. On the mineral matters of certain leguminous crops. II. Nodule formation and titanium supply. *Mem. Coll. Agr., Kyoto Imp. Univ.*, **37** : 24-35.

Martinez-Sànchez, F., Giménez, J.L., Frutos, M.J., Carvajal, M. and Alcaraz, C.F., 1990. Titanium in plant nutrition. II Effect of Ti (IV) on germination and initial plant development of *Capsicum annum* L. *Proc. III. Nat. Symp. Min. Nutr. Plants, Palma,* 115-121.

Mrowca, J.J. 1973. Sulfinato cyclopentadienyl-titanium and- zirconium complexes. *US Patent,* **3**.728.365.

Nagy, B., PAIS, I. and Bokori, J. 1986. The application of titanium-chelate in pig-feeding. *Proc. V. Internat. Tr. El. Symp.*, Jena (Eds. : M. ANKE *et al.,*), 1254-1259.

Novàk-Fodor, M., Janzsò, B., Suhajda, À. and PAIS, I. 1992. Preparation of zirconium-containing yeasts. *Proc. V. Internat. Tr. El. Symp.*, Budapest (Ed. : I. PAIS), 53-60.

PAIS, I. 1983. The biological importance of titanium. *J. Plant Nutr.*, **6** : 3-131.

PAIS, I. 1985. The importance of hardly known trace elements. *Proc. I. ISAMA Conf.*, Ghent (Ed. : E.E. Van de Gehuchte), 52-62.

PAIS, I. 1988. Importance of hardly known trace elements and convincing data on the beneficial character of titanium. *Proc. I. Internat. Congr. Plant Physiol.*, New Delhi (Eds. : S.K. Sinha *et al.,*), 1172-1177.

PAIS, I., Bokori, J., Nagy, B., Szilàgyi, M., Fekete, S. and Kolonics, F. 1989. The role of titanium-ascorbate in animal nutrition and reproduction. *Proc. VI. Internat. Tr. El. Symp.*, Jena (Eds. : M. Anke *et al.*), 1356-1360.

PAIS, I., Nagy, B., Bokori, J., Fekete, S. and Szilagyi, M. 1990. The improvement of animal breeding by application of titanium- ascorbate. (Unpublished data).

PAIS, I., Janzsò, B. and Suhakda, A. 1992. Production of yeast enriched with titanium (in Hungarian). *Proc. Sci. Session "Lippay Jànos"*, Budapest, 259-262.

Ram, N., Verloo, M. and Cottenie, A. 1983. Response of bean to foliar spray of titanium *Plant Soil,* **73** : 285-290.

Ram, N., Verloo, M. and Cottente, A., 1988. Influence of foliar and soil applied titanium on tomato. *Acta Agron. Hung.*, **37** : 43-46.

Rutskaya, S.I. 1976. Effect of titanic salt on sugar-beet of the varieties jaltuskovszkaja monospermous and polyspermous, B 031 (in Russian). *Fiz. Biochim. Kult. Rast.*, **8** (1) : 72-76.

Schroeder, H.A., Balassa, J.J. and Tipton, I.H. 1963. Abnormal trace metals in man : titanium. *J. Chron. Dis.*, **16** : 55- 69.

Shorroks, V. 1990. Gallium goes green ? After titanium comes gallium. *Micronutr. News*, **10** (2) : 18.

Simon, L., Kiss, F., Balogh, A. and PAIS, I. 1988. Effect of titanium on growth and photosynthetic pigment composition of Chlorella pyrenoidosa I. Effect of titanium-ascorbate on growth and protein content of Chlorella. *Proc. III. Internat. Tr. El. Symp.*, Budapest (Ed. : I. PAIS), 76-86.

Simon, L., Kiss, F., Bakò A., Hajdù, F., Hörcsik, T., Zs. Balogh, A. and PAIS, I, 1989. Effect of gallium on photosynthetic pigments and peroxidase activity of Chlorella pyrenoidosa. *J. Plant Nutr.*, **12** : 1123-1140

Simon, L., Balogh, À. Hajdù, F. and PAIS, I., 1990. Effect of titanium on the carbohydrate content and phosphofructokinase enzyme activity of tomato. *Proc. IV. Internal. Tr. El. Symp.*, Budapest (Ed. : I. PAIS), 49-84.

Simon, L., Kiss, F., Kàdàr-Papp, Kl., Vass, A., Balogh, A. and PAIS, I. 1991; Effect of gallium on growth and carbohydrate content of tomato (in Hungarian). *Agrokém. Talajtan*, **40** : 218-224.

Simon, L., Novàk-Fodor, M., Papp, T., Hajdù, F., Balogh, À. and PAIS, I. 1992. The effect of zirconium on the *Chlorella pyrenoidosa* green algae. *Proc. V. Internat Tr. El. Symp.*, Budapest (Ed. : I. PAIS), 61-74.

Tòth, À., Farkas, E., Deàk, Gy., Kiss, F., and PAIS, I. 1985. The interaction of titanium-ascorbate with herbicides. *Proc. I. Internat. Tr. El. Symp.* Budapest (Ed. : I. PAIS), 222-233.

Advancements In Micronutrient Research
Editor : A. Hemantaranjan
1996, Scientific Publishers, Jodhpur, pp. 45-81.

3

SOME INSIGHTS IN THE DETERMINATION OF SELENIUM IN BIOLOGICAL AND ENVIRONMENTAL SAMPLES BY GRAPHITE FURNACE ATOMIC ABSORPTION SPECTROMETRY

A. JAVIER ALLER
Department of Biochemistry and Molecular Biology,
University of Leon, E-24071 Leon, SPAIN.

1. INTRODUCTION

1.1. Biological and environmental significance of selenium

Before 1957, selenium was mainly known as an element whose compounds have toxic properties. However, since it was discovered as an essential trace element for humans and animals (Schwarz and Foltz, 1957) increasing attention has been paid to this chemical element. Selenium was recognized as a component of the enzyme glutathione peroxidase (EC 1.11.1.9, GSH/Px) (Rotruck *et al.*, 1973, Oh *et al.*, 1976) and consequently as a cellular antioxidant (Xia *et al.*, 1989). The need of Se for higher plants is still questioned (Aller *et al.*, 1990), although some Se-accumulator plants growing on seleniferous soils may concentrate this element up to 1000 ppm, thus making them potentially toxic to animals (Fishbein, 1977). Signs of poisoning (seleniosis) caused by intake of various selenium compounds have become known, especially in animals (Selenium, 1976).

Selenium deficiency together with additional pathogenic factors (such as vitamin E deficiency, excessive food intake, large amounts of polyunsaturated fatty acids, as well as cholesterol, heavy metals {mercury, cadmium, etc.} intoxication, xenobiotics and carcinogens) commonly cause potentially fatal conditions in humans and animals. Thus, increased incidences of various diseases (Keshan Disease Research Group, 1979a, b and c) (Keshan or white muscle disease, possibly Kaschin-beck's syndrome, cataracts and a higher prevalence of mastitis in cattle and sheep, as well as disturbed fertility and illthrift in various animal species) were found to be correlated with selenium deficiency in humans and animals. Selenium deficiency is also of vital importance in livestock farming being associated with instances of meat, wool and milk underproduction.

Correlations between the occurrence of disease and the presence of selenium deficiency or excess have not yet been clarified. This is mainly due to the following reasons: (i) symptoms of selenium deficiency can often only be detected in the second generation and (ii) the deficiency and toxic levels of Se in both humans and animals are in a very short concentration range, which can easily altered by fluctuations of the Se concentration in biological and environmental samples.

Current knowledge of Se in aquatic systems is sparse and heterogeneous, probably due to the difficulties in obtaining rapid and accurate measurements of the element at low concentrations (usually <10 ng/L and rarely exceeding 200 ng/L), which makes the element difficult to quantify. However, in some cases Se may occur in toxic levels, causing long term ecological unbalances, or even death of organisms. This is of vital importance when considering quality and use of drinking water for human and animals, because the maximum allowable concentration of selenium in drinking water is about 10 μg/L, requiring for its determination a detection limit of about 1 5 μ/L. Analysing of Se in biological and environmental samples is further complicated by the fact that different bioavailable forms of the element must be quantified in relation to its total content for understanding of various metabolic pathways (Lindstrom, 1980). In limnology and oceanography data have been presented concerning concentration levels of Se, as well as transport, cycling, and bioaccumulation (Lindstrom, 1983a and b), but its ecological relevance is still unclear or unexplored (Aller *et al.,* 1988; Aller and Deban, 1989).

Hence, the growing interest in the knowledge of concentration levels, bioavailability, and cycling characteristics of this element in various ecosystems are the most important goals. As a result, the increasing implications of Se in nutrition and medicine support the need for adequate means of assessing the low selenium concentrations in biological and environmental samples.

1.2. Monitoring of Selenium status in living systems

According to some researches, the use of dietary data results in considerable inaccuracy in the assessment of the Se status of individuals (Levander, 1986). To this purpose, several methods have been developed, based either on the determination of Se concentration in tissues and biological fluids or on the measurement of biochemical and functional selenium-dependent indices, mainly the activity of the only known seleno-enzyme in humans, glutathione peroxidase (Pleban *et al.,* 1982; Nève, 1991). Knowledge about these indices has increased tremendously in the last few years but, simultaneously, diagnosis of Se status and its biochemical significance has become more and more complex, because it is present at extremely low concentrations in the material to be investigated. Major difficulties are due to the necessity for detecction of minor disturbances in Se status (the so-called "marginal" deficiency) that appear to have deleterious effects on human health.

Determination of Se in biological materials, mainly serum or plasma, which have a similar Se content (Oster *et al.,* 1988), are the most commonly used indices of Se status in humans. However, plasma or serum levels are short-term indicators having limitations with regard to the evaluation of increasing Se status, partly because of differences in Se homeostasis in relation to the chemical form of the element.

Blood Se levels have been widely used as an index of Se status in both animals and humans, and in general there is a good association between levels and the dietary intake of this element over a wide range (Yang *et al.,* 1989). Because of the relatively long life span of erythrocytes, the red blood cell Se is employed to monitor the long-term Se status. Therefore, whole blood values may be more instructive in interpreting Se status as an indicator of long-term exposure. Nevertheless, misuse and overuse of certain indices such as plasma

Se concentration or erythrocyte glutathione peroxidase activity are also frequent and lead to diagnostic errors, while selenium determination in whole blood by some techniques suffer from considerable analytical problems.

1.3. Methods for the determination of Selenium

The concentrations of selenium in biological and environmental materials were able to be measured over many years only with elaborate and time-consuming analysis, requiring besides large sample quantities. However, considerable efforts have been undertaken during the last decades to improve the performance of analytical methods which, in the past, suffered from poor sensitivity and inaccuracy. Thus, a lot of different instrumental methods have been developed for the determination of selenium (Reamer and Veilon, 1981a, 1981b; Goulden *et al.*, 1981; Bem, 1981; Verlinden *et al.*, 1981; Nygaard and Lowry, 1982; Roden and Tallman, 1982; Robberecht and Vangrieken, 1982; Siu and Berman, 1983; Raptis *et al.*, 1983; Robberecht and Deelstra, 1984; Lewis, 1988; Robberecht *et al.*, 1990; Aoyama *et al.*, 1991; Ni *et al.*, 1991). Among them, techniques such as spectrophotometry (Liu *et al.*, 1990; Safavi *et al.*, 1990), cathodic stripping voltametry (Peerzada and Pessina, 1990; Vandenberg and Khan, 1990), neutron activation analysis (Chatt *et al.*, 1990; Lavi and Alfassi, 1990; Langenauer and Krahenbuhl, 1991), X-ray emission and fluorescence spectrometries (Robberecht and Vangrieken, 1983; Mino and Ota, 1990), differential pulse polarography (Batley, 1986; Ferri *et al.*, 1989; Hasdemir and Somer, 1990), proton induced X-ray emission (PIXE) (Liu *et al.*, 1990), molecular-emission cavity analysis (Belcher *et al.*, 1980; Konimtzis *et al.*, 1981), chromatography (Al-Attar and Nickless, 1988; Handelmen *et al.*, 1989; Nakagawa *et al.*, 1989; Tang *et al.*, 1989; Sarzanini *et al.*, 1990), mass spectrometry (Heumann and Wachsmann, 1989; Heumann and Grosser, 1989), inductively coupled plasma-atomic emission spectrometry (Deoliveira *et al.*, 1983), and inductively coupled plasma-mass spectrometry (Ting *et al.*, 1989) have been largely used.

Although fluorimetry produces very satisfactory results (Petersson *et al.*, 1988; Handelman *et al.*, 1989), this oldest method is rather tedious for metals determination. However, recent developments in the instrumentation of atomic absorption spectrometry (AAS) provide improved methodologies for the determination of selenium. Thus, the introduction of the double beam instruments constituted an important advance where the interfering background absorption can be compensated for in part or completely. As a result of this development, the Se sensitivity is almost always improved. Flame atomic absorption spectrometry (FAAS) shows poor sensitivity for Se forcing to extract or to concentrate it, or to look for more sensitive methods. Extraction opens up the possibility of contamination, while the use of new instrumentation developments, such as hydride generation(HG) (Sanz *et al.*, 1988; Hansson, *et al.*, 1989a, b; Mclaughlin *et al.*, 1990; Schramel and Xu, 1991; Arikawa and Iwasaki, 1991) and graphite furnace (GF) AAS methods, have shown to provide more rapid and reliable Se determinations than extraction-FAAS. Although HG is the most largely used method for the determination of Selenium, GFAAS is normally the method of choice when only micro-samples are disposable to be analysed. In GFAAS, the residence time of the atoms to be determined in the absorption cell is very much longer and the atoms are present in a very much smaller volume compared to either the FAAS or HGFAAS. For this reason, the atomic absorption signal for a given amount of selenium in the sample is correspondingly stronger, which has the parallel effect of increasing the sensitivity.

Lamps used in Se GFAAS are usually either hollow cathode lamps (HCL) or electrodeless discharge lamps (EDL). Sensitivity for selenium is better for EDL to HCL (Fuavo *et al.*, 1989), although some differences in sensitivity can be obtained for different lamp geometries (Rayson and Bass, 1991). The breakthrough of GFAAS methods with powerful background correction systems such as the Zeeman or Smith-Hieftje effects resulted from efforts to overcome interferences and to improve the sensitivity of atomic absorption spectrometry for selenium. Another auxiliary procedure is the use of platforms within the graphite tube, by which significantly less gas phase interference are observed. As the absolute detection limits (at the picograms level) are lowest with the graphite furnace, this technique has been extensively applied in both biological and environmental analyses.

2. Determination Of Selenium By GFAAS

Steps in the determination of selenium by GFAAS need particular care in decomposition and preparation of samples, preservation of the element from interferences (mainly due to Se losses in the ashing step), and finally the detection stage that has to be sensitive enough for the low selenium levels usually found in this kind of samples (Nève *et al.*, 1982).

2.1. Sample introduction into the atomizer

Although many efforts have been performed to develop new reproducible systems for the sample direct introduction as solid (Headridge and Nicholson, 1982; Lindberg *et al.*, 1988; Durnberger *et al.*, 1987; Esser and Durnberger, 1987), slurry (Ebdon and Parry, 1988; Bradshaw and Slavin, 1989), or gas phase (Heinrichs and Keltsch, 1982; Muangnoicharoen *et al.*, 1986) into the graphite tube, the traditional and most reproducible way of sample introduction is as solution which continues being the most popular choice for the determination of selenium.

2.2. Sample preparation

Any sample preparation method must take in mind that the behaviour of selenium during the atomization step from both samples and calibration standards might be similar. In environmental samples, Se is usually present as Se(IV) and Se(VI). Thus, Se(VI) is normally found in surface waters, while in ground-waters predominates Se(IV). However, in biological materials Se is typically found as Se^{2-}, although other oxidation status can also be present (Welz *et al.,* 1984). Thus, biological species such as selenomethionine, selenourea, selenocystine and selenocysteine may or may not be oxidized to selenite by the nitric acid digestion methods, but they must, with effective chemical modification, give the same analytical response as the selenite. By using oxidizing acids to digest the organic material Se can be oxidized to Se(IV) and/or Se(VI) with the subsequent recovery losses (Yang *et al.,* 1990). Inasmuch, in some organic matrices, all or part of the Se may be in a volatile form. The temperature at which Se compounds are volatile is dependent on the oxidation state of Se. As a result, most procedures for the determination of Se in organic matrices use a digestion-oxidation procedure to convert Se^{2-} to Se(IV) or Se(VI). Plasma, and especially blood, require harsh conditions for the liberation of their selenium content (Verlinden, 1982). Thus, graded severity of destruction with different reagents must be used. Although reduction of Se(VI) to Se(IV) is difficult, it is recommended heating HCl in autoclaves or boiling HCl (4-6 M) with final sample solutions. Reduction of Se(VI) depends on the acid concentration, reaction time, sample volume, and nature of the sample. Losses of Se in reductions depend on the type of material of the reduction vessels. Oxidation of Se and native selenides by HNO_3 or aqua regia proceeds only to the Se(IV) state (Kuldvere, 1989). Reducing of Se(VI) by aqua regia or reversed aqua regia proceeds to Se(IV). However, chlorine generated in mixtures using HCl is a strong oxidising agent acting as follows:

$$H_2SeO_3 + H_2O + Cl_2 \xleftrightarrow{K = 10^{6.8}} SeO_4^{2-} + 4H^+ + 2Cl^-$$

Low concentrations (10 μM) of iodide also interfere because Se(IV) is reduced to elemental Se (Ericzon *et al.,* 1990) in accordance with the following reaction:

$$H_2SeO_3 + 4H^+ + 4I^- \longleftrightarrow Se_{(s)} + 2\,I_{2(aq)} + 3H_2O$$

This reaction, however, has been used to develop an indirect method for the determination of selenium (Vijayakumar *et al.,* 1982) by AAS.

2.3. Hightlights in the determination of Se by GFAAS

2.3.1. Calibration procedures

The calibration method of standard additions is frequently applied to unknown samples in an effort to matrix match and thereby overcome chemical or physical interferences. However, there are certain problems with standard additions calibration related to the inherent error in the extrapolated concentration of the unknown sample based on the number and concentrations of the addition standards chosen (Miller, 1992). In addition, standard additions method does not overcome spectral interferences. As a result, it would be more convenient to use or develop a method where calibration should be performed against aqueous standard solutions. In this case, it is necessary to assume that if selenium in both the sample and standards does not give the same analytical response in terms of appearance time and peak shape, then there still exists the possibility of an analyte interference which could compromise accuracy.

2.3.2. Factors affecting Selenium atomization: Interferences

Factors affecting Se atomization can provide from the type and number of analyses of the atomization tube (atomization tube material and atomization type), the analytical conditions (ashing programme), as well as from the composition and states of chemical species in the matrix (matrix composition).

2.3.2.1. Atomization tube material

Material generally used to make the atomization tube is graphite, although tungsten tube atomizers have also been used (Xiao-quan *et al.*, 1992) for the determination of selenium. Many of the conflicts in the literature concerning the determination of Se deal with the properties of the graphite tubes. Thus, Alexander *et al.* (1980) report that the graphite tubes change the Se sensitivity as they aged. In addition, purity of graphite tube lowers sensitivity and causes also a source of interferences. As a result, pyrolytic graphite coatings are now being used to overcome these troubles; however, other materials, such as tantalum carbide (Hocquellet and Candillier, 1991), have also been occasionally used as coatings. In some cases, such as from samples with phosphorous in the presence of nickel (Maage *et al.*, 1991), uncoated graphite tubes show less interferences than the coated ones. Sensitivity obtained for pyrolytic graphite tubes is higher than that for uncoated tubes (Fuavo *et al.* 1989). Pyrolytic graphite shows two important peculiarities when comparing with electrographite. Thus, the former is a much more inert material and chemical reductions on its surface proceeds significantly slower. Consequently, any catalytic effect in the pyrolytic graphite tube should be more pronounced. The pyrolytic graphite shows a higher degree of crystalline order and a low concentration of active sites on its surface, but these peculiarities are lost with firing. Se atomic absorption peak profiles depend on the age of the graphite tube. Thus, the integrated area lowers with the life-time of the graphite tube, while the appearance time is delayed (Figure 1). Assuming that interferences and sensitivity strongly depend on the quality of the pyrolytic graphite coating and this varies during the lifetime of the tube (Figs. 1 and 2), selenium sensitivity depends on matrix composition because of interferences present.

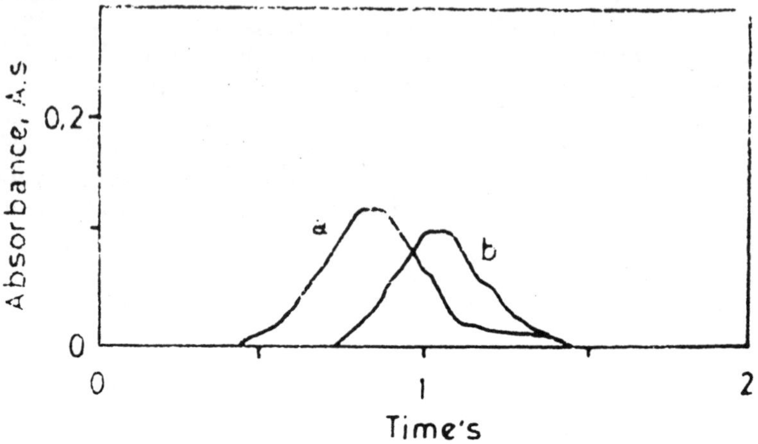

Fig. 1. Peak-profiles of the Se (0.2 ng) atomic absorption signals (wall atomization) in the presence of 0.1% $Mg(NO_3)_2$ for different life-times of the graphite tube. (a) 10 firings and (b) 200 firings. Ashing temperature = 900 °C.

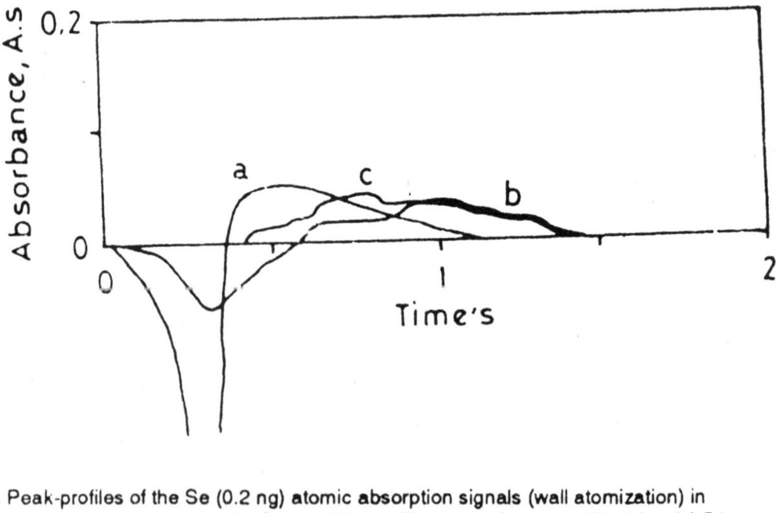

Fig. 2. Peak-profiles of the Se (0.2 ng) atomic absorption signals (wall atomization) in the presence of 0.1% $MgCl_2$ for different life-times of the graphite tube. (a) 54 firings, (b) 100 firings and (c) 200 firings. Ashing temperature = 700 °C. This figure shows that the interference of $MgCl_2$ decreases using old tubes.

2.3.2.2. Atomization type

Atomization can be performed on the wall of the graphite tube, on a platform, and on a ribbon. Thus, the following atomization modes have been named: wall atomization, platform atomization and rod or ribbon atomization. The selenium sensitivity obtained for both the platform and ribbon atomization is generally better than the corresponding one for the wall atomization (Fig. 3). However, wall atomization appears to show best results for the determination of Se in chloride matrices.

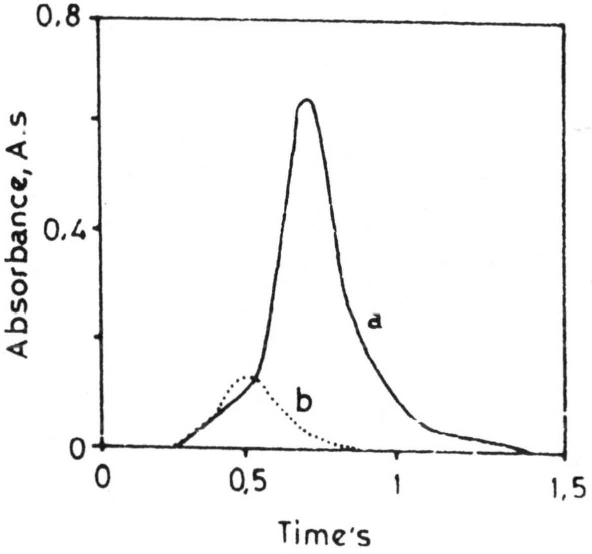

Fig. 3. Peak-profiles of the Se (2.5 ng) atomic absorption signals with (a) and without (b) the cloth ribbon. Ashing temperature = 900 °C.

2.3.2.3. Ashing programme

In opposition to the FAAS, in GFAAS the analyte atomization takes place in a discontinuous way and a variable temperature programme can be used depending on both the analyte and matrix composition. The temperature programme is usually constituted by the following stages: dry, ashing, atomization and cleaning. Certainly, an ashing programme is much more effective when the analyte signal appeared at higher temperatures. Thus, a direct relationship between the ashing temperature and the effectiveness of the analyte atomization is always assumed to be true in certain temperature range. By the ashing step the matrix components are previously evaporated and thus separated from the analyte.

In some ashing programmes, a cool step, named pre-atomization cool-down step, is introduced before atomization (Table 1). By using a pre-atomization cool-down step the tube is heated more uniformly and the effect of the cool tube ends can be avoided almost completely. The background peaks obtained in some cases and appeared earlier than the Se signal (due probably to the Rayleigh scattering) disappear completely using the pre-atomization cool-down (Bozsai *et al.*, 1990). The selected pyrolysis temperature at which Se loss sets depends on matrix components.

TABLE 1. A typical ashing programme with a pre-atomization cool-down step ($T_{Ash1} > T_{Ash2}$) for the determination of Se.

Parameter	Dry	Ash 1	Ash 2	Atomozation	Clean
Temp, °C	150	1100	300	2350	2350
Ramp, s	2	10	10	0	-
Hold,s	3	5	0	4	1
Purge	Low	Medium	Medium	Off	Medium

2.3.2.4. Matrix composition

Physical and/or chemical factors affecting atomization of Se provide usually from the main matrix constituents (acids and salts). However, the influence of acidity on the Se atomization is much lower than that the counter-ion present (Table 2). It is interesting to note that generally a higher sensitivity is obtained for the nitric acid solution, as it has already been cited (Chung *et al.*, 1984). This enhancement in the sensitivity could be due to the breakage of the pyrolytic coating leading to a much more porous surface. This porosity increases crystal imperfections of graphite favouring the formation of stable graphite-selenium lamellar compounds. In addition, nitric acid could also favourize the formation of selenium oxides preventing losses of this volatile element.

TABLE 2. Integrated absorbances (PA), peak heights (PH) and PH/PA ratios obtained for selenium (0.2 ng) in different inorganic acids matrices.

pH		Matrix		
		HCl	HNO_3	H_2SO_4
2	PA (A.s)	0.081	0.120	0.100
	PH (A)	0.016	0.057	0.018
	PH/PA	0.190	0.470	0.180
4	PA (A.s)	0.070	0.109	0.102
	PH (A)	0.029	0.032	0.019
	PH/PA	0.410	0.300	0.190
6	PA (A.s)	0.051	0.109	0.112
	PH (A)	0.022	0.033	0.028
	PH/PA	0.420	0.310	0.250

Even at low concentrations, perchloric acid reduces the formation of analyte atoms during the atomization process and it also reduces substantially the life time of the graphite tube (Maage *et al.*, 1991). Other matrix components show also important effects on the determination of Se. Thus, many salts such as the nitrates of calcium, lanthanum and palladium (>100 µg Pd) deteriorate the tube graphite by the intercalation or carbide formation (Docekalova *et al.*, 1991). However, salts can have other particular effects on the atomization of selenium. Thus, some of the matrix components can be implied in the formation of some compounds with Se showing an increased Se molecular content of the gas phase and restricting the element detection limits. These analytical problems refer to matrix interferences and can be divided in three major classes: thermal pre-atomization losses, chemical interferences and spectral interferences.

2.3.2.4.1. Loss of analyte

Selenium, as many other volatile elements, is usually lost during the first stage of the atomization process. Selenium can be lost as Se atoms or some other volatile species, but principal losses of Se at low temperatures occur in the form of oxides, dimers, and hydroxides. Expulsion of selenium (as SeO_2) together with the violently volatilized matrix (SO_3) early in the atomization stage has been suggested (Welz *et al.*, 1992) to be behind the interference of sulfate in the determination of selenium.

2.3.2.4.2. Chemical interferences

The appearance of atoms in a graphite furnace is the result of various chemical reactions both on the surface and in the gas phase of the graphite furnace. Depending on the physical and chemical properties of the analyte and on the analysis conditions, either atoms or molecules of the element to be determined evaporate into the gas phase. The formation of a Se compound results in a chemical interference.

Interferences from metals have been recently shown to be also strongly dependent on the counter ion present (Garcia-Olalla, 1991). It has been indicated that chlorine interferes in the determination of Se as a result of the formation of a Se chloride. However, chlorine only interferes at very high concentrations (Frech *et al.*, 1985). The interfering effect of sulphur (Figure 4) has also been explained by a reaction in the gas phase (Esser and Durnberger, 1987) through the formation of the SeS_2 compound (Cedergren *et al.*, 1986). Thus, doubts about the interfering mechanism behind the behaviour of sulfur remains still.

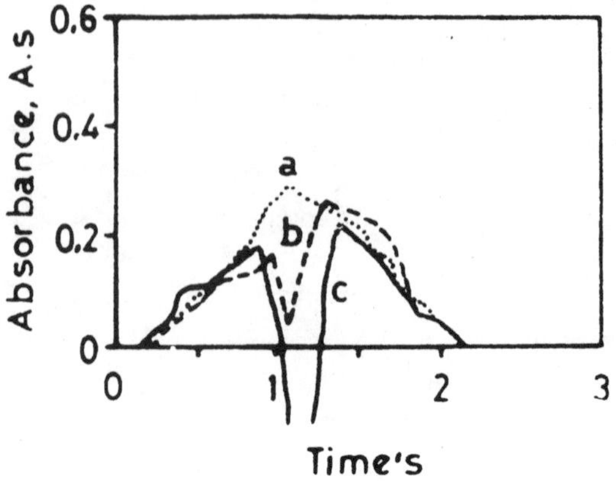

Fig. 4. Peak-profiles of the Se (200 ng) atomic absorption signals alone (a) and in the presence of 2000 mg/L (b) and 5000 mg/L (c) of sulfur (as sulphate). Ashing temperature = 500 °C.

2.3.2.4.3. Spectral interferences

The excess of free atoms, ions and molecules of accompanying components (matrix) may be also present along with selenium atoms in the gas phase during atomization stage. These species can absorb radiation at or very near to the wavelength of Se leading to spectral interferences. Some of the problems related to the Se determination at the 196 nm wavelength result from the severe background effects, especially in the presence of high concentrations of alkali (NaCl, KCl) or alkaline earth ($MgCl_2$, $CaCl_2$) metals (Aller and Garcia-Olalla, 1992). These metals (probably as a metal chloride) provide a large background absorbance behind the Se determination. This is a type of spectral interference, but there are some special spectral problems that pertain to Se. Usually, spectral interferences in the determination of Se arise from the iron and phosphate present in biological materials. Iron, Ca orthophosphate and Ca hydrogen orthophosphate lines absorbs continuum radiation within the spectral window of the monochromator, set at 196 nm. This caused overcorrection and produced a negative signal in the absence of Se. The spectral interferences of phosphate, which can occur even when Zeeman-effect background correction (Radziuk and Thomassen, 1992), are caused by absorption of structured background radiation by molecules of PO or P_2 (Radziuk and Thomassen, 1992; Saeed and Thomassen, 1992) appeared in the phosphorous atomization process. Other spectral interferences provided from the $MnCl_2$ (Fig. 5), $AlCl_3$ and Al_2O_3 (Aller and Garcia-Olalla, 1992). Cobalt shows an spectral interference at the Se 196.0 nm wavelength (while Ni and Cr at 204.0 nm Se line) using continuum source background correction (Fernandez and Beaty, 1984). However, this interference is not observed using either Zeeman-effect or Smith-Hieftje background correction systems.

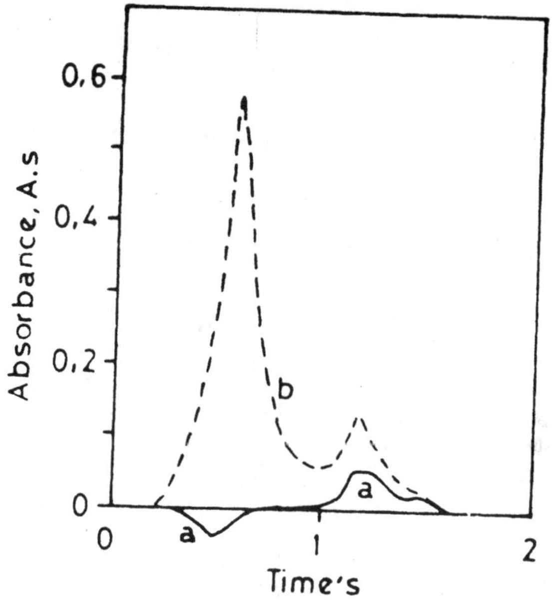

Fig. 5. Peak-profile of the Se (0.2 ng) atomic absorption signal in the presence of 0.1% $MnCl_2$ (a). The line (b) represent the Se atomic absorption signal plus the background. Ashing temperature = 500 °C.

2.3.3. Elimination of interferences

Interferences have been intending to overcome by using the traditional separation procedures {solvent-extraction (Hocquellet and Candillier, 1991; Chambers and McClellan, 1976; Kamada *et al.*, 1978; Nève *et al.*, 1980; Kamada and Yamamoto, 1980; Chung *et al.*, 1984; Demidova *et al.*, 1989; Parsley, 1991), gas chromatography (Al-Attar and Nickless, 1990; Gui-bin *et al.*, 1992), co-precipitation with arsenic (Kujirai *et al.*, 1983), HPLC (Blais *et al.*, 1991), adsorption on thiol cotton fibre (Xiao-Quan and Kai-Jing, 1985), charcoal (Muangnoicharoen *et al.*, 1986) or silica gel (Sturgeon *et al.*, 1985), ion exchange (Chiou and Manuel, 1984) } or the combined effect of a variable ashing programme (ashing and atomization temperatures, and number of steps) and the use of chemical modifiers. Thus, it is generally accepted that the use of integrated area measurements, chemical modification, platform atomization and stopped gas flow reduce interferences to a great extent. However, the ashing of some kind of samples (plasma and blood) is more effective with wall atomization and carbon build-up is minimised (Voth-Beach and Shrader, 1987).

Selenium lost during the dry cycle in the presence of hydrogen can be retained if an activated graphite surface is used together with a Ni modifier

(Dedina *et al.*, 1987). In addition, Pd also inhibits formation of the dimer and hydroxide and reduces oxide losses at low temperatures. Chemical interferences are generally minimised by the use of platform atomization (microboat), an aerosol deposition system (Dulude and Sotera, 1984), and higher ashing temperatures via improved thermal stability.

Spectral interferences due to both broadband and structured absorption (like presents by the high concentrations of iron) can be eliminated using either the Zeeman-effect or the Smith-Hieftje background correction systems (Fig. 6). We can see from this figure that the iron interference was overcome and iron increases selenium sensitivity when overcompensation problems are eliminated, as it is also shown in Table 3. Nevertheless, carbon monoxide has been used to remove oxygen from the furnace in order to prevent the formation of iron oxide which has been suggested to be implicated in causing Fe spectral

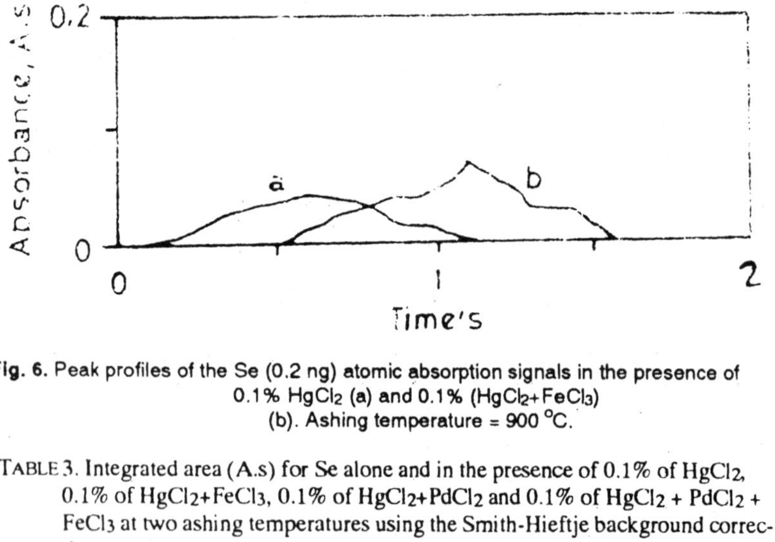

Fig. 6. Peak profiles of the Se (0.2 ng) atomic absorption signals in the presence of 0.1% $HgCl_2$ (a) and 0.1% ($HgCl_2$+$FeCl_3$) (b). Ashing temperature = 900 °C.

TABLE 3. Integrated area (A.s) for Se alone and in the presence of 0.1% of $HgCl_2$, 0.1% of $HgCl_2$+$FeCl_3$, 0.1% of $HgCl_2$+$PdCl_2$ and 0.1% of $HgCl_2$ + $PdCl_2$ + $FeCl_3$ at two ashing temperatures using the Smith-Hieftje background correction system.

Sample	Ashing temperature, °C	
	900	800
Se	0.007	0.005
Se + Hg	0.025	0.027
Se + Hg + Fe	0.045	0.055
Se + Hg + Pd	0.034	0.032
Se + Hg + Pd + Fe	0.061	0.073

interference even using deuterium-arc background correction system (Peile *et al.*, 1989). The modifier Pd/$Mg(NO_3)_2$ can reduce the spectral interference originated by phosphate (Schnipper and Jons, 1990). This is supported by the fact that Se sensitivity is always the same (Se plus modifier and Se plus phosphate plus modifier). In conclusion, this modifier has greater affinity to phosphate than the Se. In these cases, it is important to know the concentration level of phosphate (interferent) in samples when determining Se by GFAAS. Additionally, this matrix modifier will do not modify the charring temperature of Se because only if the modifier reacts with Se charring could be stated at much higher temperatures. Other spectral interferences from metal chlorides are overcome by using higher ashing temperatures, where the metal-chloride species are dissociated. However, chlorides and oxides of some accompanying components are very difficult to eliminate and, in some cases, even the use of modifiers and platforms fails to produce desirable results (Aller and Garcia-Olalla, 1992).

2.3.3.1. Chemical modification

In general, chemical modifiers are added to the sample in order to prevent losses of selenium during the ashing process. Chemical modifiers are usually transition metals which form stable compounds (selenide?) with selenium up to about 1200 °C or even higher. Thermal stabilization of Se in graphite furnace atomizers has long been the subject of scientific researchs. Thus, many chemical modifiers have been used in the Se stabilization (Table 4), although, the most largely used are Ni and Pd. The role of modifier is to maintain the Se atoms in a stable environment through the formation of some physical or chemical bonds. The sensitivity of Se atomization will depend on both the energy of these bonds and the amount of modifier (Se/modifier ratio) (Table 5). As a result, different role is assumed by chemical modifiers as depending on matrix composition (Chung *et al.*, 1984), but the Se sensitivity is firstly increased and then reduced with the mass of modifier (Fig. 7).

TABLE 4. Main modifiers used in the determination of Se.

Ag	Ni	Cu	Mo
Zn	Pd	Pt	Cr
Mg	Fe	Mn	Tl
W	Zr	Ce	Co
La	Li	KI	Ir-Mg
Cu-Fe	Pd-Cd	Pd-Hg	Ni-Mg $(NO_3)_2$
Cu-Mg $(NO_3)_2$	Pd-mg $(NO_3)_2$	Pd-Ascorbic acid	Pd-NH_2 OH. HCl
Pd-Triton	Ni-Triton	Ni-Mg-Triton	Pd-Cu-Triton

The nickel nitrate was immediatly accepted as a chemical modifier for the determination of Se, becoming one of the most popular (Neve *et al.*, 1980; Kirkbright *et al.*, 1980; Dedina *et al.*, 1987; Camrick *et al.*, 1983). Cedergren

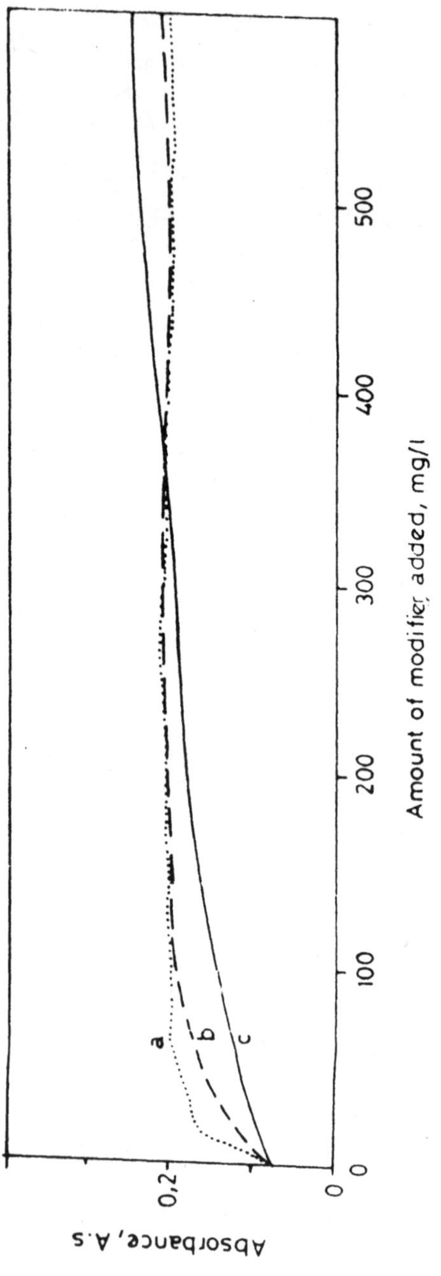

Fig. 7. Effect of the amount of chemical modifier [(a) $Mg(NO_3)_2$; (b) $PdCl_2$, and (c) $CdCl_2$] on the integrated area of the Se (0.2 ng) atomic absorption signal.

and coworkers (Cedergren *et al.*, 1986) note that Ni is significant in stabilizing Se(IV) and Se(VI) in the presence of several inorganic species but the stabilizing effect is lost in the presence of an organic matrix. However, Ni in methanol was also used as an efficient modifier for Se. The use of Ni as chemical modifier have the additional advantage that Ni converts the interfering PO^+ or P_2 molecules into free phosphorus atoms (Radziuk and Thomassen, 1992). The thermal stabilization effect of Ni on Se is not unusual, and the presence of Cu results also in the formation of a copper selenide which is less volatile than Se. In the formation of inter-metallic compounds (Se-Cu, Se-Ni), the order of volatility is Se > Se-Cu > Se-Ni (Khalighie *et al.*, 1982). Ni works as chemical modifier only in its elemental form and with an excess between thousand and million times higher than that selenium (Table 5).

TABLE 5. Selenium/modifier ratio for different chemical modifiers in various matrices and using either wall or platform atomization.

Se/ modfier ratio	Modifier	Matrix	Atomizer	Ref
2.10^{-4}	Ni	Aqueous	Platform	Carnick *et al.* (1983)
3.3×10^{-5}	Ni	Urine	Platform	Carnick *et al.* (1983)
10^{-3}	Ni	Arsenic	Wall	Kujrai *et al.* (1983)
1.7×10^{-3}	Ni	Piazsele	Wall	Neve *et al.* (1980)
10^{-4}	Pd + Cd	Aqueous, Coal Fly Ash	Wall	Garcia-Olalla (1991)
10^{-4}	Pd + Mg	Aqueous	Platform	Parsley (1991)
2×10^{-3}	Pd	Aqueous	Wall	Xiao-Quan & Kai-Jing (1985)
5×10^{-4}	Zn	Arsenic	Wall	Kujrai *et al* (1983)
10^{-3}	Cu	Arsenic	Wall	Kujrai *et al* (1983)
10^{-4}	Pd+Mg	Mineral Water	Platform	Bozsai *et al.* (1990)
4.2×10^{-5}	Pd + Cu	Bovine Liver	Platform	Martin & Williams (1989)
2×10^{-4}	Pd+Asc.	Blood	Wall	Knowles & Brodie (1988)
8×10^{-6}	Ni	Marine	Platform	Maage *et al.* (1991)

Benefical properties of palladium as chemical modifier were also recently shown. The Pd appears to be ideal for stabilizing Se allowing thermal pretreatments very higher than in the absence of it. The interest in Pd as a general modifier continues to expand, and examples of its use include the determination of Se in different biological and environmental samples. Nevertheless, the use of palladium alone as modifier is strongly dependent upon the matrix composition shown by the differences in the Se atomic absorption signals (Tables 6 and 7). It is known that Pd is most effective as a modifier when it is reduced to the zero oxidation state at low temperatures. Voth-Beach and Shrader (1987) discuss about the advantages of a "reduced" Pd modifier, where the reduction of Pd on the graphite surface is critical for its effective utilization in delaying atomization (Teague-Nishimura *et al.*, 1987).

TABLE 6.- Peak time (tp) of the Se atomic absorption signals obtained for different chemical modifiers and matrices.

Reference	tp, s	Modifier	Matrix
Martin & Williams (1989)	2.84	Pd	Aqueous
Martin & Williams (1989)	1.69	Pd	Bovine Liver
Martin & Williams (1989)	2.22	Pd + Mg $(NO_3)_2$	Aqueous, Bovine Liver
Martin & Williams (1989)	2.89	Pd+Mg+Trit	Aqueous
Martin & Williams (1989)	2.35	Pd+Mg+Trit	Bovine Liver
Martin & Williams (1989)	2.49	Pd+Cu+Trit	Aqueous
Martin & Williams (1989)	2.17	Pd+Cu+Trit	Bovine Liver
Knowles & Brodie (1988)	1.82	Pd	Aqueous
Knowles & Brodie (1988)	1.76	Ni	Aqueous
Knowles & Brodie (1988)	1.85	Pd+Ascorbic Acid	Aqueous
Maage *et al.* (1991)	1.21	Ni	Plaice
Garcia-Olalla *et al.* (1991)	1.58	Cd + $PdCl_2$	Aqueous, Coal Fly Ash
Parsley (1991)	1.81	Pd + Mg $(NO_3)_2$	Phosphate
Bozsai *et al.* (1990a)	1.75	Pd + Mg	Mineral Water
Bozsai *et al.* (1990b)	1.15	Pd + Mg	Mineral Water
Carnick *et al.* (1983)	1.75	Ni	Aqueous
Carnick *et al.* (1983)	1.67	Ni	Urine
Carnick *et al.* (1983)	2.1	Cu	Aqueous
Carnick *et al.* (1983)	2.8	Ni	Aqueous
Carnick *et al.* (1983)	2.5	Cu	Aqueous
Carnick *et al.* (1983)	2.7	Ni	Urine
Carnick *et al.* (1983)	2.3	Cu	Urine+H_2 SO_4 + NaCl
Carnick *et al.* (1983)	2.7	Ni	Urine + H_2 SO_4 + NaCl
Carnick *et al.* (1983)	2.05	Mg	Urine
Carnick *et al.* (1983)	2.2	Mg	Aqueous
Bradshaw & Slavin (1989c)	2.55	Pd + Mg	Fly Ash
Esser & Durnberger (1987d)	5	None	Aqueous, Coal Ash
Esser & Durnberger (1987d)	7	Ni + C	Aqueous, Coal Fly Ash

a) With pre-atomization cool-down step b) Without pre-atomization cool-down step

c) Slurry introduction d) Solid sampling

TABLE 7. Some characteristic times (appearance time, t_{app}; atomization time, t1; and residence time, t_2) of the Se atomic absorption signals for different chemical modifiers and matrices.

Ref	t_{app}, s	t_1, s	t_2, s	Modifier	Sample
Martin & Williams (1989)	1.36	1.06	1.83	Pd + Cu + Triton	Aqueous
	1.36	0.68	1.19	Pd + Cu + Triton	Bovine Liver
	1.36	0.77	1.34	Pd + Cu	Bovine Liver
	1.89	0.88	1.41[a]	Pd	Aqueous
	1.89	0.88	1.71[b]	Pd	Aqueous
	1.23	0.51[c]	1.14	Pd	Bovine Liver
	1.23	0.42[d]	1.36	Pd	Bovine Liver
	1.92	0.97	1.77[e]	Pd + Mg+ Triton	Aqueous
	1.92	0.97	1.88[f]	Pd + Mg + Triton	Aqueous
	1.35	0.27,0.97g	1.50	Pd + Mg + Triton	Bovine Liver
	1.33	0.31,1.02g	1.55	Pd + Mg	Bovine Liver
	1.54	0.66	1.18	Cu + Mg	Aqueous
	1.54	0.59	1.05	Cu + Mg	Bovine Liver
Carnick *et al.* (1983)	0.92	0.76	2.54	Ni	Aqueous
	0.49	1.15	2.97	Ni	Urine
	1.47	0.66	2.2	Cu + Mg	Aqueous
	1.76	1.03	5.7	Ni + Mg	Aqueous
	1.43	1.14	2.6	Cu + Mg	Urine
	1.43	1.28	3.28	Ni + Mg	Urine
	1.45	0.86	2.03	Cu + Mg	Urine + H_2SO_4 + NaCl
	1.45	1.16	3.19	Ni + Mg	Urine + H_2SO_4 + NaCl
Knowles & Brodie (1988)	1.51	0.26	0.55	Ni	Aqueous
	1.51	0.32	0.55	Pd	Aqueous
	1.52	0.26	0.55	Pd + Asc	Aqueous

[a]) 150 pg Se; [b]) 250 pg Se; [c]) 80 pg Se; [d]) 230 pg Se; [e]) 150 pg Se; [f]) 300 pg Se; [g]) Two peaks.

Although the graphite of the tube can reduce Pd(II) ions to the elemental state, addition of an auxiliary reducing agent has also been suggested for many applications. Thus, a chemically (H_2, ascorbic acid, hydroxylamine hydrochloride, albumin) or thermally (treating the modifier at ca. 1270 K after injection into the furnace) reduction has been recommended. Addition of

5-10% H_2 to the Ar purge gas has been successfully used for the determination of Se in biological samples. However, the H_2 gas need to be purged prior to the Se atomization because on the contrary the Se signal is reduced by the formation of SeH_2 (Creed *et al.*, 1992). In addition, ascorbic acid and 1% hydroxylamine hydrochloride have also been used as other auxiliary reducing agents for Pd in the determination of Se in blood and aqueous solutions (Knowles and Brodie, 1988). The Pd modifier and ascorbic acid solutions need to be separately introduced into the graphite tube because a mixing of reducing agent and the Pd solution causes a precipitation of the Pd ions. In addition, in acidic aqueous solutions ascorbic acid can reacts with Se forming a colloidal suspension which give not atomic absorption signal for Se. Hence, ascorbic acid is of limited use as a reducing agent of matrix modifiers for Se in aqueous solutions. Albumin is other additive which appears to be converted into graphite and thus Pd could produce better sensitivities for Se when present in an organic matrix (Teague-Nishimura *et al.*, 1987).

The use of mixing or complex chemical modifiers, where their efficiency could be very better than that reported by Pd alone, is a recently goal in order to improve the accuracy of the Se GFAAS methods. The efficiency of Pd as chemical modifier can be improved by combining its use with other compounds, particularly high amounts of $Mg(NO_3)_2$ (Docekalova *et al.*, 1991). Welz *et al.* (Schlemmer and Welz, 1986; Welz *et al.*, 1988a, 1988b) found improved results with Pd using Mg nitrate as an additional chemical additive and did not attempt to reduce Pd suggesting that Pd combined with Mg nitrates provides a broad-based modification system useful for a number of elements. Thus, the use of complex chemical modifiers does not need a step for the reduction of Pd. However, assuming that the Pd work very better as its elemental step, we must conclude that some of the mixing components could act as a reducer for Pd. This mixed modifier prompts a more uniform drying trapping the analyte in a stabilizing environment. In addition, these authors reported that with this mixed modifier [($Pd(NO_3)_2$ and $Mg(NO_3)_2$] Se can be sucessfully determined in airborne particulates (Welz *et al.*, 1988a) and waters (Welz *et al.*, 1988b) using matrix-free standard solutions for calibration. The Pd/$Mg(NO_3)_2$ chemical modifier was also optimized in the determination of Se in the presence of phosphate (in concentrations normally found in human body fluids) (Schnipper and Jøns, 1990).

Chemical modifiers are generally used as nitrate: $Mg(NO_3)_2$, $Ni(NO_3)_2$, $Pd(NO_3)_2$. In this case, the micro-environment of Se is probably very oxidant and Se will be surely as SeO_2. Contrarily, if modifiers are used as chlorides, the micro-environment of Se would be less oxidant and Se would probably be as selenide, whatever the chloride concentration should not be enough to form volatile Se chlorides. With this in mind, we have recently used some metal (Hg and Cd) chlorides, alones and together with Pd chloride, for the determination of Se (Garcia-Olalla *et al.*, 1991; Garcia-Olalla and Aller, 1992a, 1992b). Using these chemical modifiers, Pd found a less oxidant environment being more easily reduced to the elemental state. As a result, the use of adapted chemical modifiers now offer many possibilities for an accurate and rapid determination of selenium.

Several other additives have been found to be useful in certain analyses. Active gases such as methane or oxygen promote better combustion or oxidation of sample constituents or of the metal modifier itself. Surfactants may be used to decrease surface tension and promote better dispensing from the autosampler capillary, while anti-foam reagents prevent foaming and splattering of certain biological samples in the furnace.

2.3.4. Peak profile characteristics and Se atomization mechanisms

For the atomization process to be controlled, it is necessary that the mechanism be understood by which free atoms of selenium are formed. Selenium peak profile characteristics can show some suggestions about its atomization mechanism. Thus, broader half-widths would suggest strong Se-graphite interactions, while if peak-profiles are characterized by a rapid atoms generation gas-phase reactions are accounted for. In addition, a comparison between the peak profile characteristics for the Se alone and Se with the chemical modifier could bring some lights about the action of chemical modifier upon the Se atomization. Tables 6 and 7 show the peak time (t_p), appearance time (t_{app}), atomization time (t_1) and residence time (t_2), from different papers which are a few of the parameters used to deduce or to compare atomization mechanisms.

Atomization profiles for the selenium atomic absorption signals, and consequently atomization mechanisms of selenium, are dependent upon the atomization type (wall, platform, or cloth ribbon), matrix composition and chemical modifier used. A delaying is observed in the Se atomic absorption signal using either the ribbon or platform atomization. This delaying has been attributed to the time lag occurred between the heating of tube and of ribbon or platform. Both ribbon and platform are heated by radiation from the tube wall creating a more stable or steady-state condition for volatilization of the sample. The role of both the graphite cloth ribbon and platform is to cause a large contact area with selenium greatly enhancing the sensitivity. This increased sensitivity is usually attributed to a modification of the graphite surface leading to a more efficient atomization. Chemical modifiers lead also in general to a delay in the selenium atomization process in either solution, slurry or solid sampling (Tables 6 and 7).

Selenium at the tube temperatures below 2000 °C in the gas phase exists mainly in a molecular form (Esser and Durnberger, 1987). Thus, Se dimers,

H_2Se and SeS are some of the stable species theoretically calculated (Cedergren *et al.*, 1986). However, volatile Se dimers, oxides (SeO, SeO_2) and hydroxide appears to be produced as precursors to the atom formation, being Se firstly formed by gas-phase dissociation of these compounds. Thus, the atomization mechanism for Se alone can be resumed as follows (Styris, 1986; Droessler and Holcombe, 1987; Styris *et al.*, 1991). Selenium oxides desorb from an untreated graphite surface and thermally dissociates in the gas phase at elevated temperature (approximately 1175 K) forming atomic Se.

In spite of the great number of chemical modifiers used, little evidence has been presented in order to explain their actuation mechanism. Certainly, the true behaviour of chemical modifiers will be dependent upon several factors: analyte, chemical modification method, matrix composition, and some operational parameters controlling atomization process.

Using Mo as chemical modifier the formation of a high molecular weight heteropoly molybdate anion has been suggested at isolating the selenium. On the other hand, palladium and nickel are known to form intermetallic compounds and solid solutions with analyte in the graphite furnaces, although the formation mechanism of these compounds is still rather vague. Nevertheless, it has also been speculated that Ni and Pd modifiers catalyse the reduction of analyte oxide with H_2, CO or some hydrocarbons.

The enhancement due to the Ni chemical modifier could be explained by either the formation of a high boiling selenide or the reduction of SeO_2. The Styris (1986) and Droessler and Holcombe (1987) works show conflicting conclusions about the Ni/Se system. Droessler and Holcombe (1987) conclude in their paper that there is no indication of the formation of a nickel selenide. This unability of Ni and Ni+Mg to stabilize Se(IV) had already been anticipated by Welz *et al.* (1984) and more recently suggested by others (Subramanian, 1988). In opposition, other authors suggest that a nickel selenide is formed (Styris, 1986; Terui *et al.*, 1991) after the reduction of SeO_2, according to the following reaction,

$$SeO\ (s) + 2\,Ni\ (s) \text{ ---------- } NiSe\ (s) + NiO_2\ (s)$$

In aqueous solutions sampling the reaction between Ni and Se is a homogeneous one, probably occurring at elevated temperatures during the drying phase. However, in the solid sampling technique a heterogeneous reaction must be supposed between the evaporating volatile Se compounds and the Ni surface, where the surface area plays a dominant role during reaction (Durnberger *et al.*, 1987). In the presence of Ni, the SeO_2 molecules remain on the graphite surface untill higher temperatures, and then the remaining SeO_2 species are attached to NiO on the graphite surface until they are reduced. Although large quantities of Ni inhibit carbide formation by blocking active sites of graphite (Styris, 1986), the formation of a Se elemental state also occurs, probably by Ni reducing the SeO_2. The formation of gaseous Se dimers is favoured by the $NiCl_2$, and inhibited by the $Ni(NO_3)_2$.

Mixed oxide formation is also a possible mechanism for the stabilization of highly volatile elements. Such a mechanism for stabilization of selenium can also be possible using Ni chloride as the modifier, because $NiCl_2$ is stable in an argon atmosphere in the temperature range 230-630 °C (Dedina *et al.*, 1987). Above this temperature $NiCl_2$ is decomposed and clorine evolves in the form of HCl.

The high efficiency and universality of Pd modifiers can be explained not only by the ease of formation of Pd metal from its compounds but also by the unique catalytic properties of Pd. In particular the catalytic properties of Pd are only slightly dependent on the particle size, because Pd electron configuration depends weakly on cluster size.

Selenium stabilization results from the formation of a compound or solid solution with palladium (Oishi *et al.*, 1991), while the Se atomization occurs by a dissociative adsorption process. Selenium is stabilized by palladium at ambient temperatures through the formation on the graphite surface of a stoichiometric ternary compound with oxygen. This compound thermally decomposes at temperatures between 1070 and 1770 K (Majidi and Robertson, 1991). Released elemental Se is trapped or adsorbed at high energy retention sites created by the presence of Pd, and release from these sites occurs later at temperatures greater than 1500 K, depending on the quantity of Pd employed (Styris *et al.*, 1991). As a result, surface reactions are induced by the modifier, while heterogeneous gas-solid reactions are of major concern in controlling the atomization of selenium.

If palladium chloride is used as chemical modifier, the following considerations can be underlyed. When a reductant (e.g., activated carbon such as the graphite of the atomizer) is present, $PdCl_2$ is partially reduced to the metal even at room temperature (Morikawa *et al.*, 1969) and consequently palladium can stabilize selenium at the relatively low temperatures used during the ashing stage. Additionally, $PdCl_2$ sharply decreases the reduction temperature of some metal oxides (PbO and Ga_2O_3) with the graphite (Volynsky *et al.*, 1991).

Metal chlorides, others than the $PdCl_2$, have also Aller, been used as chemical modifiers (Garcia-Olalla *et al.*, 1991; Garcia-Olalla, 1992a, 1992b). It can be generally accepted that if chloride is eliminated during the pyrolysis stage, Se can probably be atomized as selenide. However, we must take in mind that some metal chlorides can be present at the gas phase even at high temperatures. Thus, metal halides, such as $HgCl_2$ and $CdCl_2$, show a very high saturation pressure at low temperatures. In addition, the reaction conditioned pressure of gaseous chlorides for the reaction

$$Me_yO_z + 2zHCl \text{ ------- } yMeCl_{\frac{2z}{y}} + yH_2O \quad \text{(Where M = Hg, Cd)}$$

is also very high even at low temperatures. As a result, these metals (Hg and Cd) will be firstly present in the graphite tube as gaseous chlorides. Nevertheless, the equilibrium constant for the atomization reaction of these chlorides

$$MeCl_2(g) + H_2(g) ------- + Me(g) + 2HX(g)$$

is very large (106 for Hg between 1000 and 2500 K, and 10^3-10^6 for Cd between 1800-2500 K) and consequently they will be dissociated for the increased temperatures. The formation of these metal chlorides protects from the losses of these volatile elements at the ashing temperatures, favouring also the reduction of Pd. However, the dissociation of these metal chlorides take place at higher temperatures forming a selenide, dissolving also the elemental Pd and so favouring the stabilization of selenium.

TABLE 8. Analytical characteristics of a few methods used for the determination of Se in different matrices.

Ref	Detection Limit	Sensitivity	Characteristic mass	Modifiers	Matrix
Kujrai et al. (1983)	0.3 mg/L			Zn	Ni, Co alloys
Chambers & McClellan (1976)	2 μg/l	(2-14) μg/L			
Gui-bin *et al.* (1992)	0.005 ng		0.018 ng (Me_2Se)		
Gui-bin *et al.* (1992)	0.005 ng		0.012 ng (Et_2Se)		
Gui-bin *et al.* (1992)	0.020 ng		0.030 ng (Me_2Se_2)		
Muangnoicharoen *et al.* (1986)	0.1 ng m^{-3}	0.92-3.05 ng m^{-3}			Air
Sturgeon *et al.* (1985)	7 ng/L			Ni	Seawater
Forrer *et al.* (1991)	0.04 μmol/L	0.04-4 μmol/l		Pd	Biological
Bozsai *et al.* (1990)	1.5 μ/L			Pd-Mg	Miner-water
Teague-Nishimura *et al.* (1987)	17 μg/L				
Carnrick *et al.* (1983)	10 μg/L		25 pg	Ni	Urine
Garcia-Olalla *et al.* (1991)	7 μg/L	(1.37-100) μg/L	13.7 pg	$CdCl_2$-$PdCl_2$	Coal Fly Ash
Manning & Slavin (1983)	0.8 μg/L	(10-4000) μg/L	30 pg	Ni nitrate	Standard
Garcia-Olalla & Allar (1992)	7.45 μg/L	4.10; (7.45-120) μg/L		$HgCl_2$ $PdCl_2$	Coal Fly Ash

2.4. Analytical characteristics and applications

Analytical characteristics (limits of detection, sensitivity, characteristic mass) of some of the developed method for the determination of Se in different matrices are shown in Table 8. We can see from this table that the results depend on both the matrix composition and the modifier used.

Many methods for the determination of selenium in different matrices have been developed untill now, but the most frequent ones have been applied to biological (Martin and Williams, 1989; Lewis *et al.*, 1986; Van Cauwenbergh *et al.*, 1990; Hoenig, 1991; Ni *et al.*, 1991; Jin *et al.*, 1990; Knowles and Brodie, 1989; Macpherson *et al.*, 1988; Jacobson and Lockitch, 1988; Suistomaa *et al.*, 1987; Sampson, 1987; Shand and Ure, 1987; Saeed, 1987; Aadland *et al.*, 1987; Pohl *et al.*, 1987; Krynitsky, 1987; Nhve and Molle, 1986; Hoenig and vanhoeyweghen, 1986; Welz and Schlemmer, 1986; Cutter, 1985; Mailer and Pratley, 1983; Forrer *et al.*, 1991; Van Cauwenbergh *et al.*, 1990; Johannessen *et al.*, 1993) and environmental (Welz *et al.*, 1986;Manning and Slavin, 1983; Bettinelli *et al.*, 1989) samples.

3. CONCLUSIONS

This paper reviews those parameters affecting on the determination of selenium by GFAAS. Main remarks are focused on sample preparation, interferences in the selenium atomization and its elimination, as well as on the atomization mechanisms.

From the above discussion we must conclude that very low concentrations of selenium can now be accurately determined in biological and environmental complex samples. At present, sample preparation procedures for the determination of selenium are well known and they can avoid differences in the behaviour of the various Se oxidation states usually found in these kind of samples.

Interferences in the Se GFAAS are in general well stablished and their elimination by the use of adapted chemical modifiers is in an advanced stage. However, new researches are needed in order to obtain better understanding of the behaviour in the ghraphite tube of both selenium and modifiers together with different complex matrices. These studies could permit adapting modifiers, or modifiers group, to every kind of sample.

REFERENCES

Aadland, E., Aaseth, J., Radziuk, B., Saeed, K. and Thomassen, Y. 1987. Direct electrothermal atomic absorption spectrometric analysis of biological samples and its application to the determination of selenium in human-liver biopsy specimens. *Fresenius' Z. Anal. Chemie.*, **328**: 362-366.

Al-Attar, A.F. and Nickless, G. 1988. 1,2-diaminobenzene derivatives for the determination of selenium by gas chromatography. *J. Chromatogr.*, **440**: 333-352.

Al-Attar, A.F. and Nickless, G. 1990. Determination of selenium by gas chromatography and comparison with graphite furnace atomic absorption spectrometry. *Analyst.*, **115**: 1441-1446.

Alexander, J., Saeed, K. and Thomassen, Y. 1980. Thermal stabilization of inorganic and organic selenium compounds for direct electrothermal atomic absorption spectrometry. *Anal. Chim. Acta.,* **120**: 377-382.

Aller, A.J., Bernal, J.L. and del Nozal, M.J. 1988. Geochemistry of trace elements. Commun. INIA (Instituto Nacional de Investigaciones Agrarias), Serie Tecnologia Agraria, 34 pp.

Aller, A.J., Bernal, J.L., del Nozal and M.J., Deban, L. 1990. Effects of selected trace elements on plant growth. *J. Sci. Food Agric.,* **51**: 447-479.

Aller, A.J. and Deban, L. 1989. Total and extractable contents of trace metals in agricultural soils of the Valderas area, Spain. *The Science of the Total Environment.,* **79**: 253-270.

Aller, A.J. and Garcia-Olalla, C. 1992. Spectral interferences on the determination of selenium by electrothermal atomic absorption spectrometry. *J. Anal. At. Spectrom.,* **7**: 753-760.

Aoyama, E., Nakagawa. T., Suzuki, Y. and Tanaka, H. 1991. Selenium in human serum-determination by rapid wet-ashing and high-performance liquid chromatographic-fluorimetric method. *ACS Symposium Series.,* **445**: 240-245.

Arikawa, Y. and Iwasaki, M. 1991. Determination of selenium in biological samples by hydride generation AAS after combustion in high-pressure oxygen. *Nippon Kagaku Kaishi. Iss,* **2**: 120-124.

Batley, G.E. 1986. Differential-pulse polarographic determination of selenium species in contaminated waters. *Anal. Chim. Acta.,* **187**: 109-116.

Belcher, R., Bogdanski, S.L., Henden, E. and Townshend, A. 1980. Molecular emission cavity analysis.14.-Determination of selenium utilizing hydrogen selenide generation. *Anal. Chim. Acta.,* **113**: 13-20.

Bem, E.M. 1981. Determination of selenium in the environmental and in biological material. *Environmental Health Perspectives.,* **37**: 183-200.

Bettinelli, M., Baroni, U. and Pastorelli, N. 1989. Determination of arsenic, cadmium, lead, antimony, selenium and thallium in coal fly ash using the stabilised temperature platform furnace and Zeeman-effect background correction. *J. Anal. At. Spectrom.,* **3**: 1005-1011.

Blais, J.-S., Huyghues-Despointes, A., Momplaisir, G.M. and Marshall, W.D. 1991. High-performance liquid chromatography - atomic absorption spectrometry interface for the determination of selenoniocholine and trimethylselenonium cations: application to human urine. *J. Anal. At. Spectrom.,* **6**: 225-232.

Bozsai, G., Schlemmer, G. and Grobenski, Z. 1990. Determination of arsenic, cadmium, lead, and selenium in highly mineralized waters by graphite furnace atomic absorption spectrometry. *Talanta.,* **37**: 545-553.

Bradshaw, D. and Slavin, W. 1989. Rapid slurry analysis of solid coal and fly ash samples. *Spectrochim. Acta.,* **44B**: 1245-1256.

Carnrick, G.R., Manning, D.C. and Slavin, W. 1983. Determination of selenium in biological materials with platform furnace atomic-absorption spectroscopy and Zeeman background correction. *Analyst.,* **108**: 1297-1312.

Cedergren, A., Lindberg, I., Lundberg, E., Baxter, D.C. and Frech, W. 1986. Investigation of reactions involved in graphite furnace atomic absorption procedures.12.-A study of some factors influencing the determination of selenium. *Anal. Chim. Acta.*, **180**: 373-383.

Chambers, J.C. and McClellan, B.E. 1976. Enhancement of atomic absorption sensitivity for copper, cadmium, antimony, arsenic and selenium by means of solvent extraction. *Anal. Chem.*, **48**: 2061-2066.

Chatt, A., Rao, R.R., Jayawickreme and C.K., Mcdowell, L.S. 1990. Determination of sampling constants for selenium in biological reference materials by neutron-activation. *Fresenius' J. of Anal. Chem.*, **338**: 399-407.

Chiou, K.Y. and Manuel, O.K. 1984. Determination of tellurium and selenium in atmospheric aerosol samples by graphite furnace atomic absorption spectrometry. *Anal. Chem.*, **56**: 2721-2723.

Chung, C.H., Iwamoto, E., Yamamoto, M. and Yamamoto, Y. 1984a. Selective determination of arsenic(III, V), antimony(III, V), selenium(IV, VI) and tellurium(IV, VI) by extraction and graphite furnace atomic absorption spectrometry. *Spectrochim. Acta.*, **39B**: 459-466.

Chung, C.H., Iwamoto, E., Yamamoto, M., Yamamoto, Y. and Ikeda, M. 1984b. Graphite furnace atomic absorption spectrometry with graphite cloth ribbon for selenium determination. *Anal. Chem.*, **56**: 829-831.

Creed, J., Martin, T., Lobring, L. and O'Dell, J. 1992. Minimizing chloride interferences produced by combination acid digestion using palladium and hydrogen as a matrix modifier in graphite furnace atomic absorption spectrometry. *Environ. Sci. Technol.*, **26**: 102-106.

Cutter, G.A. 1985. Determination of selenium speciation in biogenic particles and sediments. *Anal. Chem.*, **57**: 2951-2955.

Dedina, J., Frech, W., Cedergren, A., Lindberg, I. and Lundberg, E. 1987a. Determination of selenium by graphite furnace atomic absorption spectrometry. 2.- Role of nickel for analyte stability. *J. Anal. At. Spectrom.*, **2**: 435-439.

Dedina, J., Frech, W., Lindberg, I., Lundberg, E. and Cedergren, A. 1987b. Determination of selenium by graphite furnace atomic absorption spectrometry.1.-Interaction between selenium and carbon. *J. Anal. At. Spectrom.*, **2**: 287-291.

Demidova, M.G., Torgov, V.G. and Kosolapov, A.D. 1989. Determination of selenium in the form of an organoselenium compound with 1-hexene by extraction and atomic-absorption in a graphite-furnace. *J. of Analytical Chemistry of the USSR.*, **44**: 1013-1017.

Deoliveira, E., Berman, S.S. and Mclaren, J.W. 1983. Simultaneous determination of arsenic, antimony, and selenium in marine samples by inductively coupled plasma atomic emission spectrometry. *Anal. Chem.*, **55**: 2047-2050.

Docekalova, B., Docekal, B., Komarek, J. and Novotny, I. 1991. Determination of selenium by electrothermal atomic absorption spectrometry. Part 1.- Chemical modifiers. *J. Anal. At. Spectrom.*, **6**: 661-668.

Droessler, M.S. and Holcombe, J.A. 1987. Mass spectral and atomic absorption studies of selenium vaporization from a graphite surface. *Spectrochim. Acta.*, **42B**: 981-994.

Dulude, G.R. and Sotera, J.J. 1984. Dealing with interferences in furnace atomizer AAS-Se in nickel alloys. *Spectrochim. Acta.*, **39B**: 511-518.

Durnberger, R., Esser, P. and Jan ßen, A. 1987. Determination of selenium with solid sampling and direct Zeeman-AAS. Part 1. *Fresenius' Z. Anal. Chem.*, **327**: 343-346.

Ebdon, L. and Parry, H.G.M. 1988. Direct atomic spectrometric analysis by slurry atomization.4.-Determination of selenium in coal by electrothermal atomization atomic absorption spectormetry. *J. Anal. At Spectrom*, **3** : 131-134.

Ericzon, C., Pettersson, J. and Olin, A. 1990. Interference from iodide in the determination of trace level selenium. *Talanta*, **37**: 725-730.

Esser, P. and Durnberger, R. 1987. Determination of selenium with solid sampling and direct Zeeman-AAS. Part 2. *Fresenius' Z. Anal. Chem.*, **328**: 359-361.

Fernandez, F.J. and Beaty, M.M. 1984. Elimination of spectral interferences in the determination of Se in nickel-base alloys using Zeeman background correction. *Spectrochim. Acta.*, **39B**: 519-523.

Ferri, T., Morabito, R., Petronio, B.M. and Pitti, E. 1989. Differential pulse polarographic-determination of arsenic, selenium and tellurium at Mu-G levels. *Talanta.*, **36**: 1259-1263.

Fishbein, L. 1977. In "Toxicology of Trace Elements, Advances in Modern Toxicology", Goyer, R.A., Mahlman, M.A. (eds.), J. Wiley, London, pp. 191-240.

Forrer, R. and Gautschi, K., Lutz, H. 1991. Comparative determination of selenium in the serum of various animal species and humans by means of electrothermal atomic absorption spectrometry. *J. Trace Elem. Electrolytes Health Dis.*, **5**: 101-113.

Frech, W., Lundberg, E. and Cedergren, A. 1985. Investigations of some methods used to reduce interference effects in graphite furnace atomic absorption spectrometry. *Prog. Anal. At. Spectrom.*, **8**: 257-370.

Fuavo, V.A., Porter, B.L. and Sneddon, J. 1989. Direct determination of selenium in horse blood by graphite furnace atomic absorption spectroscopy. *Anal. Letters.*, **22**: 71-82.

Garcia-Olalla, C. and Aller, A.J. 1992a. Peak profile characteristics and atomization mechanisms for selenium in the presence of mercury for graphite furnace atomic absorption spectrometry. *Fresenius' J. Anal. Chem.*, **342**: 70-75.

Garcia-Olalla, C. and Aller, A.J. 1992b. Alternative mercury-palladium chemical modifier for the determination of selenium in coal fly ash by graphite furnace atomic absorption spectrometry. *Anal. Chim. Acta.*, **259**: 295-303.

Garcia-Olalla, C., Robles, L.C. and Aller, A.J. 1991a. The effect of metal chlorides, sulfates and nitrates on atomization of selenium in graphite furnace atomic absorption spectrometry. *Anal. Sci.*, **7**: 611-615.

Garcia-Olalla, C., Robles, L.C. and Alemany, M.T., Aller, A.J. 1991b. Determination of Selenium in coal fly ashes by graphite furnace atomic absorption spectrometry using a cadmium/palladium chemical modifier. *Anal. Chim. Acta.*, **247**: 19-25.

Goulden, P.D., Anthony, D.H.J. and Austen, K.D. 1981. Determination of arsenic and selenium in water, fish, and sediments by inductively coupled argon plasma emission spectrometry. *Anal. Chem.*, **53**: 2027-2029.

Gui-bin, J., Zhe-ming, N., Li, Z., Ang, L., Heng-bin, H. and Xiao-quan, S. 1992. Determination of alkylselenides by gas chromatography-electrothermal atomic absorption spectrometry. *J. Anal. At. Spectrom.*, **7**: 447-450.

Handelman, G.J., Kosted, P., Short, S. and Dratz, E.A. 1989. Determination of selenium in Human blood by High-performance Liquid Chromatography with fluorescence detection. *Anal. Chem.*, **61**: 2244-2249.

Hansson, L., Pettersson, J., Eriksson, L. and Olin, A. 1989a. Atomic absorption spectrometric determination of selenium in human blood components. *Clin. Chem.*, **35**: 537-540.

Hansson, L., Pettersson, J. and Olin, A. 1989b. Determination of selenium in fish flesh by hydride generation atomic absorption spectrometry. *Analyst.*, **114**: 527-528.

Hasdemir, E. and Somer, G. 1990. Polarographic-determination of selenium in the presence of copper ion. *Analyst.*, **115**: 297-299.

Headridge, I.B. and Nicholson, R.A. 1982. Determination of arsenic, antimony, selenium and tellurium in nickel-base alloys by atomic absorption spectrometry with introduction of solid samples into furnace. *Analyst.*, **107**: 1200-1211.

Heinrichs, H. and Keltsch, H. 1982. Determination of arsenic, bismuth, cadmium, selenium and thallium by atomic absorption spectrometry with a volatilization technique. *Anal. Chem.*, **54**: 1211-1214.

Heumann, K.G. and Grosser, R. 1989. Negative thermal ionization mass spectrometry of selenium. 2.-Selenite and selenate species determination in ground waters with the isotope-dilution technique. *Fresenius' Z. fur Analyt. Chemie.*, **332**: 880-883.

Heumann, K.G. and Wachsmann, M. 1989. Negative thermal ionization mass spectrometry of selenium.4.-Selenium trace determination in sediments and related samples. *Fresenius' Z. fur Analyt. Chemie.*, **335**: 755-758.

Hocquellet, P. and Candillier, M.-P. 1991. Evaluation of microwave digestion and solvent extraction for the determination of trace amounts of selenium in feeds and plant and animal tissues by electrothermal atomic absorption spectrometry. *Analyst.*, **116**: 505-509.

Hoenig, M. 1991. Determination du selenium dans le sang par spectromietrieed' absorption atomique electrothermique avec effect Zeeman: discussion des parametres critiques. *Analusis.*, **19**: 41-46.

Hoenig, M. and Vanhoeyweghen, P. 1986. Determination of selenium and arsenic in animal tissues with platform furnace atomic absorption spectrometry and deuterium background correction. *Int. J. Environ. Anal. Chem.*, **24**: 193-202.

Jacobson, B.E. and Lockitch, G. 1988. Direct determination of selenium in serum by graphite furnace atomic absorption spectrometry with deuterium background correction and a reduced palladium modifier - Age specific reference ranges. *Clin. Chem.*, **34**: 709-714.

Jin, Z., Shougui, J., Shikun, C., Desen, J. and Chakraborti, D. 1990. Direct determination of lead in human blood and selenium, cadmium, copper, zinc in serum by electrothermal atomic absorption spectrometry using Zeeman effect background correction. *Fresenius' J. Anal. Chim.*, **337**: 877-881.

Johannessen, J.K., Gammelgaard, B., Jons, O. and Hansen, S.H. 1993. Comparison of chemical modifiers for simultaneous determination of different selenium compounds in serum and urine by Zeeman-effect electrothermal atomic absorption spectrometry. *J. Anal. At. Spectrom.*, **8**: 999-1004.

Kamada, T., Shiraishi, T. and Yamamoto, Y. 1978. Differential determination of selenium (IV) and selenium (VI) with sodium diethyldithiocarbamate, ammonium pyrrolidinedithiocarbamate and dithizone by atomic-absorption spectrophotometry with a carbon-tube atomizer. *Talanta.*, **25**: 15-19.

Kamada, T. and Yamamoto, Y. 1980. Use of transition elements to enhance sensitivity for selenium determination by GFAAS combined with solvent-extraction with the APDC-MIBK system. *Talanta.*, **27**: 473-476.

Kaschin-Beck Disease Research Team. 1979a. Curing Kaschin-Beck disease with selenium and vitamin E in 224 patients with X-ray observation and causation research. *Chin. Med.*, J. **59**: 169-171.

Keshan Disease Research Group, 1979b. Research on the relation between selenium and Keshan Disease. *Acta Acad. Med. Sin.*, **1**: 75-84.

Keshan Disease Research Group, 1979c. Observations on effect of sodium selenite in prevention of Keshan disease. *Chinese Med.*, **J. 92**: 471-476.

Khalighie, J., Ure, A.M. and West, T.S. 1982. Atom-trapping absorption spectrometry with water-cooled metal collector tubes. *Anal. Chim. Acta.*, **134**: 271-281.

Knowles, M.B. and Brodie, K.G.. 1988. Determination of selenium in blood by Zeeman graphite furnace atomic absorption spectrometry using a palladium-ascorbic acid chemical modifier. *J. Anal. At. Spectrom.*, **3**: 511-516.

Knowles, M.B. and Brodie, K.G. 1989. Determination of selenium in serum by Zeeman graphite furnace atomic absorption spectrometry using a palladium-ascorbic acid chemical modifier. *J. Anal. At. Spectrom.*, **4**: 305-305.

Kirkbright, G.F., Hsiao-Chuan, S. and Snook, R.D. 1980. An evaluation of some matrix modification procedures for use in the determination of mercury and selenium by atomic absorption spectroscopy with a graphite tube electrothermal atomizer. *At. Spectrosc.*, **1**: 85-89.

Konimtzis, T.A., Papadoyannis, I.N. and Sofoniou, M.C. 1981. Determination of selenium in water samples by molecular emission cavity analysis after coprecipitation. *Anal. Chim. Acta.*, **123**: 315-317.

Krynitsky, A.J. 1987. Preparation of biological tissue for determination of arsenic and selenium by graphite furnace atomic absorption spectrometry. *Anal. Chem.*, **59**: 1884-1886.

Kujirai, O., Kobayashi, T., Ide, K. and Sudo, E. 1983. Determination of traces of selenium in heat-resisting alloys by graphite-furnace atomic-absorption spectrometry after co-precipitation with arsenic. *Talanta*, **30**: 9-14.

Kuldvere, A. 1989. Extraction of geological materials with minerals acids for the determination of arsenic, antimony, bismuth and selenium by hydride generation atomic absorption spectrometry. *Analyst.*, **114**: 125-131.

Langenauer, M. and Krahenbuhl, U. 1991. Determination of selenium in food and minerals by neutron-activation analysis. *Chimia.*, **45**: 8-10.

Lavi, N. and Alfassi, Z.B. 1990. Determination of trace amounts of cadmium, cobalt, chromium, iron, molybdenum, nickel, selenium, titanium, vanadium and zinc in blood and milk by neutron-activation analysis. *Analyst.*, **115**: 817-822.

Lewis, S.A. 1988. Determination of selenium in biological matrices. *Methods in Enzymology.*, **158**: 391-402.

Lindberg, I., Lundberg, E., Arkhammar, P. and Berggren, P.O. 1988. Direct determination of selenium in solid biological materials by graphite furnace atomic absorption spectrometry. *J. Anal. At. Spectrom.*, **3**: 497-501.

LindstrOm, K. 1980. Peridinium cinctum bioassays of Se in Lake Erken. *Arch. Hydrobiol.*, **89**: 110-117.

LindstrOm, K. 1983a. Selenium as a growth factor for plankton algae in laboratory experiments and in some Swedish lakes. *Hydrobiologia.*, **101**: 35-48.

LindstrOm, K. 1983b. Changes in growth and size of Keratella cochlearis (Gossc) in relation to some environmental factors in cultures. *Hydrobiologia.*, **104**: 325-328.

Liu, Q.O., Shao, H.R. and Moschini, G. 1990a. Determination of the total selenium in different materials by PIXE. *J. of Radioanalytical and Nuclear Chemistry - Letters.*, **144**: 47-55.

Liu, S.P., Zhou, G.M. and Huang, Z.G. 1990b. A highly sensitive color-reaction for Se(IV) with the iodide Rhodamine B-PVA system-spectrophotometric determination of trace amounts of selenium in water. *Talanta.*, **37**: 749-752.

Levander, O.A. 1986. The need for measures of selenium status. *J. Am. Coll. Toxicol.*, **5**: 37-44.

Lewis, S.A., Hardison, N.W. and Veillon, C. 1986. Comparison of isotope dilution mass spectrometry and graphite furnace atomic absorption spectrometry with Zeeman background correction for determination of plasma selenium. *Anal. Chem.*, **58**: 1272-1273.

Maage, A., Julshamn, K. and Andersen, K.-J. 1991. Determination of selenium in acid digested marine samples by electrothermal atomic absorption spectrometry with continuum source background correction and nickel as a chemical modifier. *J. Anal. At. Spectrom.*, **6**: 277-281.

Macpherson, A.K. Sampson, B., and Diplock, A.T. 1988. Comparison of methods for the determination of selenium in biological fluids. *Analyst.*, **113**: 281-283.

Mailer, R.J. and Pratley, J.E. 1983. Evaluation of selenium determination in biological material by atomic absorption spectroscopy. *Analyst.*, **108**: 1060-1066.

Majidi, V. and Robertson, J.D. 1991. Investigation of high temperature reaction on solid substrates with Rutherford backscattering spectrometry: interaction of palladium with selenium on heated graphite surfaces. *Spectrochim. Acta.*, **46B**: 1723-1733.

Manning, D.C. and Slavin, W. 1983. The determination of trace elements in natural waters using the stabilized temperature platform furnace. *Appl. Spectrosc.*, **37**: 1-11.

Martin, C.K. and Williams, J.C. 1989. Determination of selenium in bovine liver by Zeeman-effect atomic absorption spectrometry using a palladium-copper chemical modifier. *J. Anal. At. Spectrom.*, 4: 691-695.

Miller, J.N. 1992. The method of standard additions. Spectroscopy Europe., **4/6**: 26-27.

Mclaughlin, K., Dadgar, D., Smyth, M.R. and Mcmaster, D. 1990. Determination of selenium in blood-plasma and serum by flow-injection hydride genaration atomic-absorption spectrometry. *Analyst.*, **115**: 275-278.

Mino, Y. and Ota, N. 1990. Inorganic chemical approaches to pharmacognosy.8.-Determination of selenium in crude drugs by energy-dispersive X-ray fluorescence spectrometry. *Chemical & Pharmaceutical Bulletin.*, **38**: 2208-2211.

Morikawa, K., Shirasaki, T. and Okada, M. 1969. In "Advances in Catalysis and Related Subjects", ed., Eley, D.D., Academic Press, N.Y., Vol. **20**, p. 98.

Muangnoicharoen, S., Chiou, K.Y. and Manuel, O.K. 1986. Determination of selenium and tellurium in the gas phase using specific columns and atomic absorption spectrometry. *Anal. Chem.*, **58**: 2811-2813.

Nakagawa, T., Tanaka, H., Aoyama, E., Hasegawa, N. and Kobayashi, N. 1989. High-performance liquid-chromatography fluorometric-determination of selenium based on selenotrisulfide formation reaction. *Anal. Chem.*, **61**: 233-236.

Nakamura, M., Takamura, N., Akagi, Y., Tanaka, S. and Hashimoto, Y. 1990. Determination of selenium in the atmosphere by fluorescence HPLC with selenium-2,3-diaminonaphtalene complex. *Bunseki Kagaku.*, **39**: T49-T53.

Nève, J. 1991. Methods in determination of selenium states. *J. Trace Elem. Electrolytes Health Dis.*, **5**: 1-17.

Neve, J., Hanoeq, M. and Molle, L. 1980. Etude critique de la determination par spectrometrie d'absorption atomique sans flame du selenium dans les milieux complexes apres extraction a l'aide d'o-diamines aromatiques et ajout de nickel (II). *Anal. Chim. Acta.*, **115**: 133-141.

Nève, J., Hanocq, M., Molle, L. and Lefebvre, G. 1982. Study of some systematic errors during the determination of the total selenium and some of its ionic species in biological materials. *Analyst.*, **107**: 934-941.

Neve, J. and Molle, L. 1986. Direct determination of selenium in human-serum by graphite furnace atomic absorption spectroscopy - Improvements due to oxygen ashing in graphite tube and Zeeman effect background correction. *Acta Pharmacologica et Toxicologica.*, **59**:606-609.

Ni, Z.M., Shan, X.Q., Jin, L.Z., Luan, S., Zhang, L. and Subramanian, K.S. 1991. Arsenic, bismuth, copper, lead, nickel and selenium in some biological samples - Determination by graphite furnace atomization atomic absorption spectrometry. *ACS Symposium Series.*, **445**: 206-214.

Nygaard, D.D. and Lowry, J.H. 1982. Sample digestion procedures for simultaneous determination of arsenic, antimony and selenium by inductively coupled argon plasma emission spectrometry with hydride generation. *Anal. Chem.*, **54**: 803-807.

Oh, S.H., Pope, A.L. and Hoekstra, W.G. 1976. Dietary selenium requirement of sheep fed a practical-type diet as assessed by tissue glutathione peroxidase and other criteria. *J. Anim. Sci.*, **42**: 984-992.

Oishi, K., Yasuda, K. and Hirokawa, K. 1991. Beginning of atomization in presence of matrix modifier in graphite furnace atomic absorption spectrometry. *Anal. Sci.* **7**: 883-887.

Oster, O., Schmiedel, G. and Prellwitz, W. 1988. The organ distribution of selenium in German adults. *Biol. Trace Element Res.,* **15**: 23-45.

Parsley, D.H.. 1991. Determination of copper, cobalt, selenium and molybdenum in liver by flame and electrothermal atomic absorption spectrometry. *J. Anal. At. Spectrom.,* **6**: 289-293.

Peerzada, N. and Pessina, L. 1990. Determination of selenium(IV) in biological sample by cathodic stripping voltammetry. *Anal. Letters.,* **23**: 2027-2035.

Peile, R., Grey, R. and Starek, R. 1989. Interference-free determination of selenium in the presence of iron using a graphite furnace with carbon monoxide-platinum modification. *J. Anal. At. Spectrom.,* **4**: 407-410.

Petersson, J., Hansson, L. and Ornemark, U, Olin, A. 1988. Fluorimetry of selenium in body fluids after digestion with nitric acid, magnesium nitrate hexahydrate, and hydrochloric acid. Clin. Chemistry., **34**: 1908-1910.

Pleban P.A., Beachum J. and Munyani A. 1982. Determination of selenium concentration and glutathione-peroxidase activity in plasma and erythrocytes. *Clin. Chem.* **28**: 311-316.

Pohl, B., Grund, A. and Knowles, M. 1987. Selenium determination in blood with atomic absorption spectrometry (AAS) using Zeeman background correction and palladium ascorbic acid chemical modification. *Fresenius' Z. Anal. Chemie.,* **327**: 20-21.

Radziuk, B. and Thomassen, Y. 1992. Chemical modification and spectral interferences in selenium determination using Zeeman-effect electrothermal atomic absorption spectrometry. *J. Anal. At. Spectrom.,* **7**: 397-403.

Raptis, S.E., Kaiser, G. and Tolg, G. 1983. A survey of selenium in the environment and a critical review of its determination at trace levels. *Fresenius' Z. fur Analyt. Chemie.,* **316**: 105-123.

Rayson, G.D. and Bass, D.A. 1991. A simultaneous comparison of four Se hollow cathode lamps used in graphite furnace atomic absorption spectrometry. *Anal. Chem.* **45**: 1049-1050.

Reamer, D.C. and Veilon, C. 1981. Determination of selenium in biological materials by stable isotope dilution gas chromatography mass spectrometry. *Anal. Chem.,* **53**: 2166-2169.

Reamer, D.C. and Veillon, C. 1981. Preparation of biological materials for determination of selenium by hydride generation - atomic absorption spectrometry. *Anal. Chem.,* **53**: 1192-1195.

Robberecht, H.J. and Deelstra, H.A. 1984. Selenium in human urine. Determination, speciation and concentration levels. *Talanta.,* **31**: 497-508.

Robberecht, H., Deelstra, H. and Vangrieken, R. 1990. Determination of selenium in blood components by X-ray emission spectrometry - Procedures, concentration levels and health implications. *Biol. Trace Element Research.,* **25**: 149-185.

Robberecht, H. and Vangrieken, R. 1983. Determination of selenium in biological materials by pre-concentration and X-ray emission spectrometry. *Anal. Chim. Acta.,* **147**: 113-121.

Robberecht, H. and Vangrieken, R. 1982. Selenium in environmental waters–Determination, speciation and concentration levels. *Talanta.,* **29**: 823-844.

Roden, D.R. and Tallman, D.E.. 1982. Determination of inorganic selenium species in groundwaters containing organic interferences by ion chromatography and hydride generation atomic-absorption spectrometry. *Anal. Chem.,* **54**: 307-309.

Rotruck, J.T., Pope, A.L., Ganther, H.E., Swanson, A.B., Hafeman, D.G. and Hockstra, W.G. 1973. Biochemical role as a component of glutathione peroxidase. *Science.,* **179**: 588-590.

Saeed, K. 1987. Direct electrothermal atomization atomic absorption spectrometric determination of selenium in whole-blood and serum with continuum source background correction. *J. Anal. At. Spectrom.,* **2**: 151-155.

Saeed, K. and Thomassen, Y. 1981. Spectral interferences from phosphate matrices in the determination of arsenic, antimony, selenium and tellurium by electrothermal atomic absorption spectrometry. *Anal. Chim. Acta.,* **130**: 281-287.

Safavi, A. Afkhami, A., and Massoumi, A. 1990. Spectrophotometric catalytic determination of ultra-trace amounts of selenium based on the reduction of resazurin by sulfide. *Anal. Chim. Acta.,* **232**: 351-356.

Sampson, B. 1987. Determination of selenium in serum by electrothermal atomization atomic absorption spectrometry with deuterium-arc background correction. J. *Anal. At. Spectrom.,* **2**: 447-450.

Sanz, J., Gallarta, F., Galban, J. and Castillo, J.R. 1988. Determination of selenium by hydride generation ultraviolet-visible molecular absorption spectrometry with diode-array detection. *Analyst.,* **113**: 1387-1391.

Sarzanini, C., Abollino, O. Mentasti, E., and Porta, V. 1990. Simultaneous ion chromatographic determination of selenium and metal-ions in waters at trace level. *Chromatographia.,* **30**: 293-297.

Schlemmer, G. and Welz, B. 1986. Palladium and magnesium nitrates, a more universal modifier for graphite-furnace atomic absorption spectrometry. *Spectrochim. Acta.,* **41B**: 1157-1165.

Schnipper, A. and Jons, O. 1990. Influence of modifier in selenium determination by GFAAS in phosphate rich matrices. Presented at Euroanalysis VII, Vienna, August 26-31.

Schramel, P. and Xu, L.Q. 1991. Determination of arsenic, antimony, bismuth, selenium and tin in biological and environmental-samples by continuous-flow hydride generation ICP-AES without gas-liquid separator. *Fresenius' J. Anal. Chem.,* **340**: 41-47.

Schwarz, K. and Foltz,C.M. 1957. Selenium as an integral part of factor 3 against dietary liver degeneration. *J. Am. Chem. Soc.,* **79**: 3292-3293.

Selenium. 1976. Committee on Medical and Biological effects of environmental pollutants, National Academy of Sciences, Washington DC.

Shand, C.A. and Ure, A.M. 1987. Graphite furnace atomic absorption spectrometric determination of selenium in plant materials following combustion in a stream of oxygen. *J. Anal. At. Spectrom.*, **2**: 143-149.

Siu, K.W.M. and Berman, S.S. 1983. Determination of selenium in marine sediments by gas chromatography with electron capture detection. *Anal. Chem.* 55: 1603-1605.

Sturgeon, R.E. Willie, S.N., and Berman, S.S. 1985. Preconcentration of selenium and antimony from seawter for determination by graphite furnace atomic absorption spectrometry. *Anal. Chem.*, **57**: 6-9.

Styris, D.L. 1986. Elucidating atomization mechanisms by simultaneous mass spectrometry and atomic absorption spectrometry. *Fresenius' Z. Anal. Chem.*, **323**: 710-715.

Styris, D.L., Prell, L.J., Redfield, D.A., Holcombe, J.A., Bass, D.A. and Majidi, V. 1991. Mechanisms of selenium vaporization with palladium modifiers using electrothermal atomization and mass spectrometric detection. *Anal. Chem.*, **63**: 508-517.

Subramanian, K.S. 1988. Determination of trace metals in blood by graphite furnace atomic absorption spectrophotometry: Recent studies. *At. Spectrosc.*, **9**: 169-178.

Suistomaa, U., Saaranen, M. and Vanhaperttula, T. 1987. Determination of selenium in human-spermatozoa and prostasomes using base digestion and electrothermal atomic absorption spectrophotometry. *Clin. Chim. Acta.*, **168**: 323-328.

Tang, Y.B., Hu, Z.D. and Liu, M.C. 1989. Rapid reversed-phase chromatographic-separation and determination of selenium and tellurium. *Chromatographia.*, **27**: 292-296.

Teague-Nishimura, J.E., Tominaga, T., Katsura, T. and Matsumoto, K. 1987. Direct experimental evidence for *in situ* graphite and palladium selenium formations with improvement on the sensitivity of selenium in graphite furnace atomic absorption spectrometry. *Anal. Chem.*, **59**: 1647-1651.

Terui, Y., Yasuda, K. and Hirokawa, K. 1991. Metallographical consideration on the mechanism of matrix modifier in graphite furnace atomic absorption spectrometry. *Anal. Sci.*, **7**: 397-402.

Ting, B.T.G., Nooers, C.S. and Janghorbani, M. 1989. Isotopic determination of selenium in biological materials with inductively coupled plasma mass spectrometry. *Analyst,*. **114**: 667-674.

Van Cauwenbergh, R., Robberecht, H. and Beelstra, H. 1990. Selenium concentration levels in whole blood of Belgian blood bank donors, as determined by direct graphite furnace atomic absorption spectrometry. *J. Trace Elem. Electrolytes Health Dis.*, **4**: 215-224.

Van Cauwenbergh, R., Robberecht, H., Van Dael, P. and Deelstra, H. 1990. Direct selenium determination in human whole blood by graphite furnace atomic absorption spectrometry with deuterium correction using a L'vov platform, a Pd/Mg matrix modification and appropriate dilution. *J. Tracc Elem. Electrolytes Health Dis.* **4**: 127-131.

Vandenberg, C.M.G. and Khan, S.H. 1990. Determination of selenium in sea-water by adsorptive cathodic stripping voltammetry. Anal. Chim. Acta. **231**: 221-229.

Verlinden, M., Deelstra, H. and Adriaenssens, E. 1981. The determination of selenium by atomic-absorption spectrometry: A Review. *Talanta.*, **28**: 637-646.

Verlinden, M. 1982. On the acid decomposition of human blood and plasma for the determination of selenium. *Talanta.*, **29**: 875-882.

Vijayakumar, M., Ramakrishna, T.V. and Aravamudan, G. 1982. Determination of trace quantities of selenium by indirect atomic-absorption spectrophotometry. *Talanta.*, **29**: 61-63.

Volynsky, A., Tikhomirov, S. and Elagin, A. 1991. Proposed mechanism for the action of palladium and nickel modifiers in electrothermal atomic-absorption spectrometry. *Analyst.*, **116**:145-148.

Voth-Beach, L.M. and Shrader, D.E. 1987. Investigations of a reduced palladium chemical modifier for graphite-furnace atomic absorption spectrometry. *J. Anal. At. Spectrom.*, **2**: 45-50.

Welz, B., Bozzai, G., Sperling, M. and Radziuk, B. 1992. Palladium nitrate - Magnesium nitrate modifier for electrothermal atomic absorption spectrometry. *J. Anal. At. Spectrom.*, **7**: 505-509.

Welz, B., Mudakavi, J.R. and Schlemmer, G. 1988. Palladium nitrate - Magnesium nitrate modifier for graphite furnace atomic absorption spectrometry. 1.- Determination of arsenic, antimony, selenium and thallium in airborne particulate matter. *J. Anal. At. Spectrom.*, **3**: 93-97.

Welz, B., Mudakavi, J.R. and Schlemmer, G. 1988. Palladium nitrate - Magnesium nitrate modifier for graphite furnace atomic absorption spectrometry. 2.- Determination of arsenic, cadmium, copper, manganese, lead, antimony, selenium and thallium in water. *J. Anal. At. Spectrom.*, **3**: 695-701.

Welz, B. and Schlemmer, G. 1986. Determination of arsenic, selenium and cadmium in marine biological tissue samples using a stabilized temperature platform furnace and comparing deuterium are with Zeeman-effect background correction atomic absorption spectrometry. *J. Anal. At. Spectrom.*, **1**: 119-124.

Welz, B., Schlemmer, G. and Mudakavi, J.R. 1992. Palladium nitrate - Magnesium nitrate modifier for electrothermal atomic absorption spectrometry. Part 5. Performance for the determination of 21 elements. *J. Anal. At. Spectrom.*, **7**: 1257-1271.

Welz, B., Schlemmer, G. and Voellkopf, U. 1984. Influence of the valency state on the determination of selenium in graphite furnace atomic absorption spectrometry. *Spectrochim. Acta*, **39B**: 501-510.

Xia, Y., Hill, K.E. and Burk, R.F. 1989. Biochemical studies of a selenium deficient population in China: measurement of selenium, glutathione peroxidase and other oxidant defense indices in blood. *J. Nutr.*, **119**: 1318-1326.

Xiao-Quan, S. and Kai-Jing, H. 1985. Matrix modification for determination of selenium in geological samples by graphite-furnace atomic-absorption spectrometry after preseparation with thiol cotton fibre. *Talanta.*, **32**: 23-26.

Xiao-quan, S. Radziuk, B., Welz, B. and Sychra, V. 1992. Application of palladium as a chemical modifier in electrothermal atomic absorption spectrometry with a tungsten tube atomizer. *J. Anal. At. Spectrom.*, **7**: 389-396.

Yang, J.Y., Yang, M.H. and Lin, S.M. 1990. Effect of wet decomposition methods on the determination of cobalt, copper, selenium and zinc in biological samples using electrophoresis. *Anal. Chem.*, **62**: 146-150.

Yang, G., Zhou, R., Yin, S., Gu, L., Yan, B., Liu, Y. and Li, X. 1989. Studies on safe maximal dietary selenium intake in a seleniferous area in China. Selenium intake and tissue selenium levels of the inhabitants. *J. Trace Elem. Electrolytes Health Dis.*, **3**: 77-87.

Advancements In Micronutrient Research
Editor : A. Hemantaranjan
1996, Scientific Publishers, Jodhpur, pp. 83-102

4

OVERVIEW OF THE DETERMINATION OF BERYLLIUM IN ENVIRONMENTAL AND BIOLOGICAL SAMPLES BY ATOMIC ABSORPTION AND ATOMIC EMISSION SPECTROMETRIES

A.J. ALLER

Department of Biochemistry and Molecular Biology,
University of Leon, E-24071 Leon, SPAIN

1. INTRODUCTION

Although, beryllium can stimulate the growing of the fungus "*Aspergillus niger*" (Steinberg, 1946) and of the green alga "*Chlorella pyrenoidsa*" (in the presence of low concentrations of Mg at pH levels above 11.2) (Hoagland. 1952), certain concentrations of this element may show severe toxic effects on plants life. Thus, contents of Be untill 16 ppm in culture solutions reduce the yields (up to the 50 %) of peas (*Pisum*), soybean (*Glycine max*), alfalfa (*Medicago*), and lettuce (*Latuca*) (Romney and Childress, 1965). Beryllium in land cultures can show a different behaviour because soils fix Be in such a way that it could be unavailable to plants. In spite of these benefical effects, beryllium is considered as one element of the greatest concern to the environmental quality and health, being its increased release during coal utilization and waste disposal (Merian, 1984) and its use in the modern nuclear and aerospace industries (Aller, 1985, 1986) the most important ways of entry in the environment.

Beryllium toxicity depends on the way of entry into organims. Ingestion of Be does not shows harmly effects to animals and man. Thus, poultry can tolerate concentrations of Be in feed of 1000 mg/Kg over periods of several months (Hapke, 1991). However, contradictory results were found for the sensitivity of fishes to Be in surrounding water (Griffitts and Skilleter, 1991; Hapke, 1987). These results could be explained by synergistic interactions between Be and other cations.

The acute toxicity of Be compounds on the skin (dermatitis, ulcerations, and granulomas) (Epstein, 1967) is mainly originated by contact with water-soluble Be compounds ($BeSO_4$, $BeCl_2$, BeF_2, etc.) and less by contact with Be metal, alloys or ceramic materials. Otherwise, beryllium particles could be introduced in subcutaneous tissues from cuts or puncture wounds.

The largest toxicity of beryllium is shown when taken by inhalation (fumes, mists or dust particles with a size below about 10 μm), causing acute lung and pulmonary diseases (bronchitis, pneumonitis, and alveolar interstitial oedema) (Frieman and Hardy, 1970). Berylliosis (chronic beryllium disease) is characterized by a granulomatous interstitial pneumonitis (inflamation and granuloma formation within the supportive lung tissues) (McConnochie *et al.*, 1988). Total Be concentration are not the main factor causing berylliosis because higher concentrations than that reported for this disease have also been occassionally found in perfectly normal lungs. This suggests a different and individual susceptibility, which is related with the adjuvant role of Be compounds (Hall, 1988). The adjuvant action has been given in cases of chronic Be disease, increasing serum concentrations of immunoglobulins (IgG) (Spencer and Hall, 1984).

Beryllium compounds show also carcinogenic, tauteratogenic, and mutagenic properties to animals, being also regarded as a suspect carcinogen in man (Aller, 1990) because exposures to their ambients pose an excess risk of contracting lung cancer. Beryllium decreases also the fidelity of the DNA synthesis, interfering with gene transcription, and inhibiting *in vivo* and in cultured cells DNA synthesis and cell division (Witschi, 1970). The G_1 phase of the cell cycle is the most sensitive to beryllium, unlike other toxic metals that cause an S phase block. Beryllium can act as a macromolecular complexing agent between nucleic acids and proteins, being selectively bound with high affinity to the regulatory non-histone nuclear proteins. Beryllium can also inhibit certain protein or enzyme phosphorylation reactions (Bassler, 1965).

The high toxicity of beryllium is the main reason for its determination in environmental and biological samples. The threshold value for the concentration of beryllium in these samples is below μg/l level and, consequently, very sensitive methods appear to be worthwhile to control Be contamination. There are many methods (Fishbein, 1984), including molecular absorption and emission spectrometries (Aller, 1990; Gladilovich and Stolyarov, 1981; Urner *et al.*, 1981), activation analysis (Friedli *et al.*, 1987), and others (Aller, 1992; Dinsdade and Bourdillon, 1982; Friedli *et al.*, 1987; Kozlenkov and Belov, 1982; Ottolini *et al.*, 1993; Williams and Wallach, 1989), occasionally used for the determination of Be in many samples. However, these methods do not satisfy generally the detection limits required in these tasks, being atomic spectrometries, mainly the graphite furnace atomic absorption spectrometry (GFAAS), the preferred methods.

2. Handling Of Samples

Final results of an analysis are as reliable as the procedures outlined by the sample manipulation method. Sampling representativity is a critical aspect, but homogeneity of sample as well as contamination and/or losses during sample manipulation and storage are also of the greatest concern.

2.1. Sample preparation

Solid samples must be accurately treated before their introduction in the atomizer. Acid mixtures have been usually employed for the dissolution of solid samples (estuarine sediments, coal fly ashes, etc) (Aller and Bonilla, 1990; Miller, 1991; Sturgeon, *et al.*, 1982) prior to the determination of beryllium, but occassionally they may fail to achieve complete disolution with underestimation of the content of beryllium. Consequently, other dissolution methods such as the fusion procedures (Grognard, 1987; Owens and Gladney 1976) have been extensively used. These digestion methods employ many disintegrants, usually a mixture of sodium peroxide/sodium hydroxide, lithium methaborate, lithium tetraborate and a mixture of lithium methaborate/lithium tetraborate. However, all these methods are usually time consuming, while some volatile compounds could be lost at the high fusion temperatures. As a result, other dissolution procedures including pressure-bomb acid digestions or microwave digestions have also been developed.

Liquid samples usually require minimum treatments unless Be is strongly bound to organic matter. In this case, stable organic Be complexes can be completely decomposed by a photochemical oxidation of the sample (Burba, *et al.*, 1993; Rigin, 1984) using an UV-irradiation. Total decomposition of organic and inorganic samples without losses of Be can be achieved treating the sample with halogen or inert gas fluoride (xenon tetrafluoride) by autoclaving. Sample storage must be made in teflon containers in order to diminish contamination from walls and losses by fixation in the bottle walls.

2.2. Solid sampling

Chemical processing of sample for quantitative atomic spectroscopy is a time-consuming procedure generally involving a potential contamination of dissolution. So, procedures with a minium of preparation of sample before flame, plasma or graphite atomization have been intended for developing. Thus, direct determination of Be in some samples (usually coal and coal fly ashes) by a solid or slurry sampling technique has been reported by a few methods (Bendicho and de Loss-Vollebregt, 1991; Haraldsen and Bruno-Pougnet, 1989; Nakamura *et al.*, 1992). The advantage of this technique lies in its rapidity and simplicity since very little sample preparation is required apart from grinding samples into a fine powder. However, the calibration procedure is a difficult step.

An important aspect in solid sampling is the particle size of sample powder. This is due to the heterogeneous distribution of elements which either form solid solutions with rock-forming minerals or are located at the boundary of the minerals as oxides, sulfides, etc. The best grinding time to obtain a good analytical repeatability for the determination of Be is longer than 20 min (Nakamura *et al.*, 1992), while the powder characteristics must be as follows: particle size range, 0.3-25 μm (median 2.4 μm; mode 2.0 μm), the size of the 12% of the particles larger than 10 μm. Mixing graphite powder with solid powders provides sharp spectral lines and enhances their intensity owing to the improved atomization.

3. Calibration Methods

All atomic spectroscopic methods, like others physico-chemical methods of analysis, need the comparison of the analyte signal from unknown samples with that obtained for the standard solutions. However, different alternatives can be possible. Standard additions method is largely applied in atomic spectrometry, but this is not a general calibration method because during the signal detection the Be added to sample need to show a behaviour similar to that observed for the Be present in unknown sample, and this condition is not always fulfilled. The best calibration method for any atomic spectrometry is comparing with standard Be single solutions (Nakamura *et al.*, 1992; Schramel and Li-Qiang, 1982), but many times matrix-matched standards are needed (Thulasidas *et al.*, 1988).

Oxide mixtures blended in graphite are satisfactory for the preparation of calibration standards for trace analysis by spark optical emission spectrometry (OES) (Mills, 1983). In spite of source variations, incorporation of an internal standard element during sample preparation is an almost universal practice in OES to obtain improved precision.

4. APPRAISAL OF THE ATOMIC METHODS FOR THE DETERMINATION OF BERYLLIUM

Although Be is some times determined by OES, so called atomic emission spectrometry (AES), the most largely used technique for the determination of this metal at trace levels is the atomic absorption spectrometry (AAS), mainly GFAAS. In spite of the good sensitivity shown by the atomic fluorescence spectrometry, this technique has been scarcely used for the determination of Be (Rigin, 1984).

In the AAS methods, it is of the greatest concern that Be concentration in the atomizer (furnace, flame, plasma, etc.) should be the greatest possible. This condition is accomplished by decomposition of sample through many dissociation reactions of Be compounds. Mechanisms of the beryllium atomization is not quite known, but from the researchs do in the graphite furnace it is possible to conclude that the BeO adsorbate seems to be the first precursor of the free Be atoms (Styris and Redfield, 1987), according to the following scheme,

$$BeO_{(ad)} \xrightarrow{2000\ K} Be_{(g)} + O_{(g)} \qquad \{1\}$$

However, the reduction of the adsorbed oxide could also be an alternative source of the Be atoms (Styris and Redfield, 1987). In opposition, the atomic emission techniques need a high concentration of the ionic species of Be in the atomizer, which are formed from the Be atoms through the following reaction,

$$Be \longleftrightarrow Be^{+} + 2e^{-}\ (\text{or } Be^{2+} + 2e^{-}) \qquad \{2\}$$

In order to obtain the best sensitivity for each technique, the reactions {1} and {2} must be largely extended to the right. However, chemicals, instrumental and operating conditions can affect on the extension of these reactions, so appearing interferent effects.

4.1. Atomic absorption spectrometries

The most developed techniques in AAS are those using a flame and a graphite tube for free atom generation, being respectively known as flame atomic absorption spectrometry (FAAS) and GFAAS. Every one shows some peculiarities, which are later outlined.

4.1.1. Instrumental and operating conditions

In GFAAS, the type of material used to made atomizer shows important effects on the determination of Be. Pyrolytic graphite coated graphite tubes, but not the uncoated tubes, enhance the Be atomic absorption signal, but coated tubes are unsuitable to be used for the determination of this metal because of both the extremely poor reproducibility of the Be atomization signals and the memory effects (Burba *et al.,* 1983). In order to overcome this unsuitability, graphite tubes have been coated with La or Ti carbide (Brajter and Klejny, 1985) and the analytical characteristics (sensitivity, reproducibility and detection limit) for the determination of Be are improved using these carbide-coated tubes (better with the Ti carbide) (Brajter and Klejny, 1985). However, a memory effect appears when the La carbide coating is used. This could be explained by an incomplete volatilization of Be during the atomization step. Other coatings also used included zirconium (Schmidt and Dietl, 1988) and thorium nitrate (Vanhoe *et al.,* 1988). The zirconium coating seems to be better because graphite tubes increase both their durability and resistance against chemicals, showing also reduced interferences.

Platform atomization shows generally better sensitivity than the wall atomization mode. Atomization under pressure (Fazakas, 1983) shows some advantages over the atomization at atmospheric pressure because sensitivity, detection limit and a reduction of interferences in the determination of Be are improved. The determining factors in the effect of pressure on GFAAS signals have been outlined as due to the diffusion rate of the analyte and the degree of Lorentz broadening and shift.

In FAAS, the most largely used atomization system for the determination of Be is the dinitrogen oxide-acetylene flame (Castillo *et al.,* 1982; Robles *et al.,* 1991) because of its higher temperature to atomize the refractory oxide of Be. However, the air-acetylene flame was also recently used for its determination in combination with the addition of Mg (500-700 μg ml^{-1}) to the Be aqueous test solution in a H_2SO_4 medium (pH 1 to 7) (Zhao and Yao, 1991). The composition of these flames affect on the atomization degree of beryllium, being necessary in every case to adjust the combustion conditions of fuels in order to obtain the best sensitivity.

4.1.2. Interferences

In GFAAS, positive interferences have been found for some elements (Al, Ca, Ce, Cr, Fe, La, Mg, Mn, Mo, Si, Sr, and H_2SO_4) (Ellis *et al.,* 1988; Hurlbut, 1978), although other authors also found minor effects for some of these elements (Ca, Fe, Si, W and Zn) (Goyal *et al.,* 1987). A negative effect was observed for Th (Thulasidas *et al.*, 1988), while some contradictory results have been derived from W because both negative (Ellis *et al.*, 1988) and positive (Hurlbut, 1978) interferences have been found. Using integrated measurements, only a little enhance from Mo and a slow decrease from W have been noted in the absorbance of Be (Ellis *et al.*, 1988). The HF, $HClO_4$ and HCl acids also show some interferent effect (Hurlbut, 1978; Luo *et al.*, 1990) and they will have to be removed prior to the determination of beryllium.

The mechanism being behind any interference seems to be very complex and strongly dependent on the atomizer type. Thus, using pyrolytic graphite coated graphite tubes Al enhances the Be signal, but using uncoated tubes Al decreases the Be signal (Schmidt and Dietl, 1987). Although many interferences are eliminated using metal or metallic salts as coating for the graphite tubes, Ca still causes depression on the atomic absorption of Be using Zr-coated graphite tubes, but not using pyrolytically coated tubes. Platinum also interferes in the determination of Be with either untreated or La-treated tubes (Brajter and Klejny, 1985), due mainly to the presence of a spectral interference. Other interferences derived from losses in the ashing stage prior the atomization of beryllium. This is due to the relatively large volumes of water in the atomizer, which can lead to losses of Be by the formation of the gaseous dihydroxide {$Be(OH)_2$} at temperatures as low as 1200 K (Frech *et al.*, 1985).

In FAAS, Al(III) and Ti(IV) diminish severely the atomic absorption signal of beryllium (Castillo *et al.*, 1982), while Pd, Si and F–sligthly supress it only at high concentrations (5000 mg/l) (Castillo *et al.*, 1982; Ellis *et al.*, 1988). Contrarily, other elements (Mg; Na and Ca) cause a slight enhancement on the Be signal (Ellis *et al.*, 1988).

4.1.3. Suppression of interferences

Although, many of the general methods developed for the elimination of interferences in AAS (Aller, 1987) can be used in the case of Be, some peculiarities can be outlined. The determination of beryllium by *GFAAS* is subject to spectral, physical and chemical interferences, which have been routinely overcome by the use of chemical modifiers and a careful choice of the atomization conditions (fast heating, atomizer type, background correction system and platform atomization) (Table 1). Thus, furnace programmes with rapid heating remove the interference from Ca(II) on the Be signal (Burba *et al.*, 1983), while the interference from Al is overcome using Zr-coated tubes (Schmidt and Dietl, 1987). Although lanthanum enhances the absorption signal of Be, it eliminates also interferences from various cations (Hurlbut, 1978).

TABLE 1. Typical instrumental settings for the determination of Be by FAAS and GFAAS.

Wavelength, nm			234.9		
Be line source			Hollow cathode lamp		
FAAS : Flame			Dinitrogen oxide-acetylene		
GFAAS : Temperature programme :					
	Dry	Ashing 1	Ashing 2	Atomization	Cleaning
Temperture, °C	150	900	1000	2400	2500
Ramp time, s	2	20	10	0	–
Hold time, s	0	0	10	5	0
Purge, position	1	2	2	0	3
Integration	–	–	–	5	–

The interference of acids, such as HF in large amounts, can be eliminated using the $Al(NO_3)_3$ as chemical modifier (Luo *et al.,* 1990). The mechanism of action of this chemical modifier is probably based on the formation of an Al fluoride due to the high affinity of Al for the fluoride. The presence of chemical modifiers allows also to work with elevated ashing temperatures without losses in sensitivity. Many chemical modifiers have been used for the determination of Be (Luo *et al.*, 1990; Paschal and Bailey, 1986; Shan *et al.*, 1989; Vanhoe *et al.,* 1988; Yao and Jiang, 1991) (Table 2), but the most largely used are the Mg and Al nitrates.

TABLE 2. Main chemical modifiers used in the determination of Be by GFAAS.

Chemical modifier	Reference
$La(NO_3)_2$	(Luo, Li, Wang, 1990; Shan, Yian, Ni, 1989)
$Mg(NO_3)_2$	(Luo, Li, Wang, 1990; Paschal, Bailey, 1986; Shan, Yian, Ni, 1989; Vanhoe, Vandecasteele, Desmet, Dams, 1988)
$Ca(NO_3)_2$	(Luo, Li, Wang, 1990; Shan, Yian, Ni, 1989; Yao, Jiang, 1991)
$(NH_4)6MO_7O_{24}.4H_2O$	(Shan, Yian, Ni, 1989)
$NH_4H_2PO_4$	(Shan, Yian, Ni, 1989)
$Al(NO_3)_3$	(Luo, Li, Wang, 1990; Shan, Yian, Ni, 1989)
$(NH_4)_3PMo_{12}O_{40}$	(Shan, Yian, Ni, 1989)
$Sr(NO_3)_2$	(Luo, Li, Wang, 1990)
Ascorbic acid	(Shan, Yian, Ni, 1989)

In FAAS, interferences have been routinely overcome adding a counter interferent agent. The interference of Al anf Ti has been removed adding La

(Castillo *et al.*, 1982), while contents of 0.1% HF and 2.5% 8-hydroxyquinoline were also effective to overcome Al interference up to concentrations of 1000 mg/L Al. The use of HF to overcome the Ca interference is prone, because of the formation of CaF_2, with a possible coprecipitation of Be. Other reagents used to compensate the Al interference in the determination of Be have been the KF and ethylenediaminetetraacetic acid, but the most effective has shown to be the quinolin-8-ol (Luecke, 1992).

TABLE 3. Typical extraction/preconcentration procedures used in combination with atomic methods for the determination of Be.

Separation method	System (Reactant)	Technique
Adsorption (activated carbon)	Acetyl-acetonate	GFAAS[a]
Solvent extraction	HBPA (MIBK)	FAAS[b]
Solvent-extraction	Trioctylamine(NH_4SCN) (MIBK)	FAAS[c]
Solvent extraction	Trifluoracetylacetone	AFS,ICP[d]
Solvent extraction	Acetylacetone (benzene, MIBK)	GFAAS[e]
Solvent extraction	DAPM(Chloroform, Dichloroethane)	Emission[f]
Flow-injection or suction-flow-solvent extraction	acetylacetone	CP-AES[g]
Liquid chelating exchanger	β-diketo (cyclohexane)	GFAAS[h]
Ion-exchange resin	HBPA	AAS[i]
Ion-exchange (Dowex 50 W x)	8,20-50 mesh, sodium form	GFAAS[j]
Ion-exchange (cellulose-pyrogallol)		FAAS, GFAAS[k]
Coprecepitation	*Hafnium hydroxide*	GFAAS[l]
Centrifugation		GFAAS[m]
Ultrafiltration		GFAAS[m]
Electrophoresis		GFAAS[m]
Isotacophoresis		GFAAS[m]
Algae	*Chlorella vulgaris*	FAAS, DCP-AES[n]
Bacteria	*Escherichia coli, Pseudomonas putida*	GFAAS[o]

HBPA: N-benzoyl-N-phenylhydroxylamine; MIBK: Methylisobuthyl ketone; DAPM: Diantipyrylmethane.

a) Okutani, Tsuruta, Sakuragawa, 1993; b)Robles, Garcia-Olalla, Alemany, Aller, 1991; c)Castillo, Belarra, Dominguez, Aznarez, 1982; d)Rigin, 1984; e)Schmidt, Dietl, 1987; f)Makhnev, Degtev, Chapko, Kozhenvnikova, Khor'kova, 1986; g)Yamamoto, Obata, Nitta, Nakata, Kumamaru, 1988; h)Bhattacharyya, Das, 1992; i)Das, Pobi, 1991; j)Brajter, Klejny, 1985; k)Burba, Willmer, Betz, Fuchs, 1983; l)Ueda, Kitadani, 1988; m)Stiefel, Schulze, Tolg, Zorn, 1980; n)Gatewood, Sneddon, 1990; o)Robles, Aller, 1994.

In order to improve the analytical characteristics of any procedure for the determination of beryllium, many methods have been used for the separation/preconcentration of this element prior to its determination. However, although separation/preconcentration methods coupled with atomic spectrometries include mainly solvent extraction procedures (Bhattacharyya and Das, 1992; Castillo *et al.*, 1982; Makhnev *et al.*, 1986; Rigin, 1984; Robles and Garcia-Olalla *et al.*, 1991; Schmidt and Dietl, 1987; Yamamoto, *et al.*, 1988), other separation methods have also been used in this task (Burba, *et al.*, 1983; Brajter and Klejny, 1985; Das and Pobi, 1991; Gatewood and Sneddon, 1990; Okutani *et al.*, 1993; Robles and Aller, 1994; Stiefel *et al.*, 1980; Ueda and Kitadani, 1988) (Table 3). These separation methods show strong capabilities for the isolation of Be from many interferents, being Al one of the most important.

TABLE 4. Observed Be lines in AES

Species	λ (nm)	Relative intensity[a]	I_N/I_B(1 mg/l Be)
Be (I)	332.134	1	
Be (I)	234.861	19	45[b]
			96[c]
Be (I)	249.473	3	2[b]
			8[c]
Be (I)	265.045	4	6[c]
Be (II)	313.042	160[d]	46[b]
			110[c]
Be (II)	313.107	160[d]	25[b]
			41[c]

a) Radziemski, Cremers, Loree, 1983; b) Pougnet, Haraldsen, Orren, 1985; c) Schramel, Li-Qlang, 1982; d) Relative intensity refered to the doublet (313.042 - 313.107) nm.

After solvent extraction, a subsequent reextraction procedure using nitric acid is sometimes necessary when a graphite atomizer is used, because the injection of organic solvents in the graphite tube can produce creeping effects around the hole of the sample inlet during the injection procedure. Moreover, other error source can be the result of evaporation of a part of the organic solvent from the beakers of the atomic sample changer in a relatively short time. However, if we introduce the organic solution directly into the atomizer we can obtain an improved analytical sensitivity (Robles *et al.*, 1991).

4.2. Atomic emission spectrometries

4.2.1. General considerations

In AES, sample decomposition can be performed using different systems (sparks, arcs, flames, plasmas, lasers, microwaves, or furnaces) so generating different atomic emission techniques. Although, spark OES (Mills, 1983) has been largely used in the past for metals determination in solid samples (and flame AES for liquid samples), the most important atomic emission technique for the determination of Be include those using plasmas as excitation sources {direct current plasma-atomic emission spectrometry (DCP-AES) (Chang *et al.,* 1982; Kivimaki and Lajunen, 1985) and inductively coupled plasma-atomic emission spectrometry (ICP-AES) (Pougnet *et al.*, 1985; Schramel and Li-Qlang, 1982; Yamamoto *et al.,* 1988)}. However, recently developed techniques, such as laser induced-breakdown spectroscoy (LIBS) (Cremers and Radziemski, 1985; Millard *et al.,* 1986; Radziemski *et al.,* 1983) and furnace atomization plasma emission spectrometry (FAPES) (Sturgeon *et al.,* 1990), provide also a fast and relatively precise analysis of Be with promising analytical characteristics.

In the LIBS method a plasma is also used for decomposition of sample. The plasma is formed by focusing a powerful ns-laser pulse at a sufficiently high irradiance in air at atmospheric pressure causing dielectric breakdown of a carrier gas within the duration of the laser pulse.

In FAPES, the graphite furnace, which has been largely used as an atomizer for atomic absorption, is promisingly employed as an emission source for thermal excitation of atomic vapor prior to the introduction of analyte vapor into hotter excitation sources, such as inductively coupled plasmas.

4.2.2. Interferences

Table 4 shows the Be lines potentially used in AES, but two of them show higher sensitivity. The high temperature of the plasmas (10000 K - 25000 K) significantly reduces chemical and spectral interferences, because molecules are dissociated and consequently molecular emission is not usually observed. However, interferences are still a problem for ICP-AES, mainly spectral interferences from lines of easily ionizable elements. In the LIBS spak technique, electron density is much higher than in conventional plasma or flame spectrometry and ionization effects are expected to be small. However, high electronic concentrations would reduce the ionic species concentration and consequently the corresponding spectral line intensities. The well known potential interferences in the determination of beryllium by AES at the two most sensitive wavelengths are shown in Fig. 1. Iron emits a secondary peak

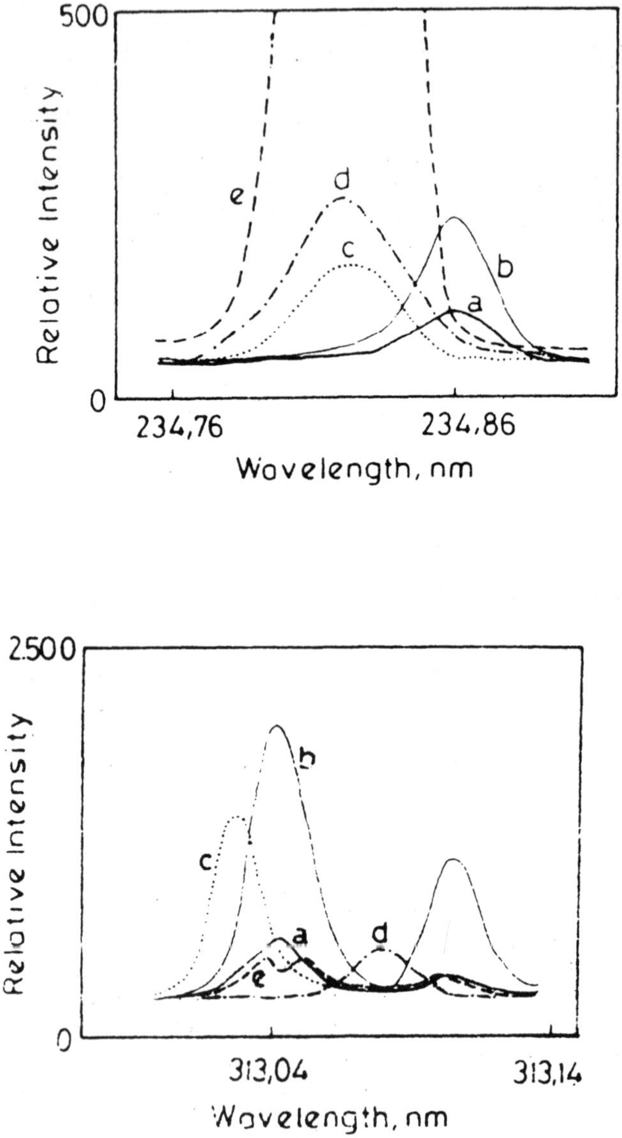

Fig. 1. Interferences by Fe at Be 2345.86nm (A) and by V, Ti and OH^- at Be 313.04nm (B).(A) :: a, 10 µg/l Be; b, 100 µg/l Be; c, 5 mg/l Fe; d, 10 mg/l Fe; and e, 100 mg/l Fe. (B) :: a, 10 µg/l Be; b 100 µg/l Be; c, 10m/l V; d, 10 mg/l Ti; and e water lines (OH emission).

at the 234.86 nm wavelength causing a potential interference with the 234.9 nm line used for the determination of Be, but usually for the Fe/Be ratios greater than 10^3. Vanadium can interfere near the 313.04 nm Be line, while Ti line lies between the two Be lines.

TABLE 5. Analytical characteristics of the atomic absorption spectrometries.

Parameter	FAAS {Matrix}	GFAAS {Matrix}
Detection limits	4 µg/L {Organic}[a]	0.05 µg/L {Urine}[f]
	2 µg/L {Natural waters}[b]	0.6 ng/L {Oganic}[g]
	0.1 mg/L {Waters}[c]	1.5 µg/L {Organic}[h]
	50 ng/L {Natural Waters}[d]	16 ng/L {Alloys}[i]
	5.2 µg/L {Biological}[e]	27 ng/L {Sediments, Fly Ash}[j]
		6.25 ng/L {Organic}[k]
		1 ng/L {Natural waters}[d]
		1.5 µg/L {Soils}[l]
		0.08 µg/L {Biological}[e]
		0.1 ng {Water}[m]
		1 pg {Urine}[n]
		10 pg {Uranium}[o]
		1.6 pg {Sediments}[p]
Characteristic mass/0.0044 A.s		0.73 pg {Sediments}[p]
		1 pg {Urine}[f]
		0.33 pg {Alloys}[i]
Linear Range	< 1 mg/L {Natural waters}[b]	< 10 µg/L {Sediments}[p]
	< 2.6 mg/L {Biological}[e]	< 16 µg/L {Biological}[e]
	0.1 4 mg/L {Waters}[c]	0.5 - 10 mg/L {Water}[m]
	0.02 - 3 mg/L {Organic}[a]	0.1-2 mg {Uranium}[o]
		< 16 µg/L {Urine}[f]

a) Castillo, Belarra, Dominguez, Aznarez, 1982; b) Robles, Garcia-Olalla, Alemany, Aller, 1991; c) Gatewood, Sneddon, 1990; d) Burba, Willmer, Betz, Fuchs, 1983; e) Ellis, Hodge, Darby, Jones, Hinners, 1988; f) Paschal, Bailey, 1986; g) Okutani, Tsuruta, Sakuragawa, 1993; h) Schmidt, Dietl, 1988; i) Brajter, Klejny, 1985; j) Vanhoe, Vandecasteele, Desmet, Dams, 1988; k) Ueda, Kitadani, 1988; l) Schmidt, Dietl, 1987; m) Thulasidas, Kulkarni, Porwal, Page, Sastry, 1988; n) Shan, Yian, Ni, 1989; o) Goyal, Purohit, Dhobale, Patel, Page, Sastry, 1987; p) Luo, Li, Wang, 1990.

Those techniques, such as LIBS, without electrodes to generate the plasma display reduced interferences (many perturbations of the plasma and interferences lines are eliminated), but the LIBS spark technique is sensitive to particle size (Cremers and Radziemski, 1985) when it is used for monitoring airborne Be particles. No interferences occur with Be 313.1 nm from neutral or ionic nitrogen or oxygen lines, although a few only weak lines of O(II) could appear within 0.2 nm of the beryllium wavelengths (Radziemski *et al.*, 1983). Potential interferences could derive from the Hg, Tc and Mo lines, close to the Be(II) 313.1 nm doublet (Radziemski *et al.*, 1983). An essential part in the LIBS technique is time resolution, allowing discrimination against continuum light from the plasma at early times. This is due to the fact that the signal, background and noise levels depend on the delay time between plasma initiation and observation (t_d), and interferences from molecular species, including OH, can be eliminated at $t_d = 4$ μs (Radziemski *et al.*, 1983). In the FAPES technique, there is a highly structured background from NO bands in the vicinity of the analysis line of Be (234.8 nm) (Sturgeon *et al.*, 1990). However, the intensity of the NO bands decrease strongly as the temperature of the furnace increases.

TABLE 6.- Analytical characteristics of the atomic emission spectrometries.

Parameters Technique	FAPES	DCP-AES	LIBS	ICP-AES
Detection limits	5-16 pg [a], α	0.05 mg/L [b], α	0.5 ng/g [e], β	0.15 μg/L [g], β
		1.5 μg/L [c], α	2 mg/L [f], α	0.22 μg/L [g], α
		1.0 μg/L [c], β		0.31 μg/L [g], α
		3.6 ng/sample [d], α		4.8 μg/L [h], χ
				3.8 μg/L [g], χ
				4.7 μg/L [g], δ
				0.14 μg/L [h], ε
				0.27 μg/L [g], ε
				0.40 μg/L [h], β
				0.73 μg/L [g], ε

α) 234.861 nm; β) 313.11 nm; χ) 249.473 nm; δ) 265.045 nm; ε) 313.042 nm.

a) Sturgeon, Willie, Luong, Berman, 1990; b) Gatewood, Sneddon, 1990; c) Kivimaki, Lajunen, 1985; d) Chang, Morse, Harley, Lippmann, Cohen, 1982; e) Radziemski, Cremers, Loree, 1983; f) Millard, Dalling, Radziemski, 1986; g) Burba, Willmer, Betz, Fuchs, 1983; h) Pougnet, Haraldsen, Orren, 1985.

5. ANALYTICAL FIGURES OF MERIT

Analytical characteristics of the atomic spectrometric methods for the determination of Be in biological and environmental samples are shown in Tables 5 and 6. The detection limit is the figure-of-merit usually employed for the comparison of different techniques. However, detection limits appeared in the scientific literature show a wide divergence in the way how they were calculated. As a result, any comparison between detection limits of different developed methods is biased. Nonetheless, in spite of these troubles some conclusions can be done. The best sensitivity is obtained for GFAAS, mainly using a preconcentration procedure. The limits of detection for GFAAS vary between 0.6 ng/L and 1.5 μg/L (relative) and 1 pg and 100 pg (absolute), while for FAAS they are in the range of 0.05-100 μg/L. In general, ICP-AES shows superior detection limits compared with FAAS, but the atomic emission methods developed for the determination of Be show poor sensitivity, ranging between 0.14 μg/L and 4.8 μg/L for ICP-AES, depending on the wavelength used.

6. CONCLUSIONS

In general, AES offers advantages over FAAS, being its multielement capability and faster analysis at both major and trace levels some of the most importants. An additional advantage of AES is that the dynamic range is several orders of magnitude, which is a consequence of the fact that radiation intensity is directly proportional to the atom concentration. The general advantages of the atomic absorption spectrometries compared with emission techniques are the best knowledge of interference mechanisms and the ways of elimination. The GFAAS technique is the most sensitive for the determination of beryllium in biological and environmental samples, although the larger time required for every measurement is a nuisance. Nonetheless, if the condition of sensitivity is not an essential necessity, either FAAS or AES can also provide reliable results. Among the emission techniques, ICP-AES is the most accepted, but LIBS is a good alternative technique, mainly for the determination of Be in airborne particulates. FAPES is still a developing technique.

REFERENCES

Aller, A.J. 1985. Ceramic materials: Compounds types and powder preparation. *Journal of Powder & Bulk Solids Technology,* 9(4): 23-28.

Aller, A.J. 1986. New materials in the aerospace industry. *Proyecto.,* 18: 71-82.

Aller, A.J. 1987. "Analytical atomic absorption spectroscopy". Leon University Press, pp. 178-190.

Aller, A.J. 1990. The Clinical significance of beryllium. *J. Trace Elem. Electrolytes Health Dis.,* 4: 1-6.

Aller, A.J. 1990. Determination of beryllium by infrared spectrometry. *Appl. Spectrosc.,* 44(7): 1159-1162.

Aller, A.J. 1992. The challenge of spectroscopy in the microanalysis of biological surfaces. *Appl. Spectrosc. Reviews.,* 27(3): 223-244.

Aller, A.J., and Bonilla, D. 1990. A digestion procedure for determination of major and minor elements in fly ash by atomic absorption spectrometry. *Anal. Sci.*, **6**: 309-311.

Bassler, R. 1965. Effects of Be on synthesis of nuclear proteins and DNA in fibroblasts cultivated *In vitro. C. R. Seances Soc. Biol. Fil.*, **159**: 1620-1623.

Bendicho, C. and de Loos-Vollebregt, M.T.C. 1991. Solid sampling in electrothermal atomic absorption spectrometry using commercial atomizers. *J. Anl. At. Spectrom.*, **6**: 353-374.

Bhattacharyya, S.S. and Das, A.K. 1992. Determination of beryllium in coal fly ash by electrothermal atomic absorption spectrometry. *J. Anal. At. Spectrom.*, **7**: 417-420.

Brajter, K. and Klejny, K. 1985. Determination of beryllium in Pt-Be alloy by atomic-absorption spectrometry after ion-exchange separation. *Talanta*, **32**(7): 521-524.

Burba, P., Willmer, P.G., Betz, M. and Fuchs, F. 1983. Atomic absorption determination of beryllium traces in natural waters. *Int. J. Environ. Anal. Chem.*, **13**: 177-191.

Castillo, J.R., Belarra, M.A., Dominguez, M. and Aznarez, J. 1982. Flame atomic-absorption determination of beryllium after extraction from NH4SCN medium with trioctylamine in MIBK. *Talanta*, **29**: 485-488.

Chang, A.E., Morse, R., Harley, N.H., Lippmann, M. and Cohen, B.S. 1982. Atomic emission spectrometry of trace levels of beryllium in industrial aerosols. *Am. I id. Hyg. Assoc. J.*, **43**(2): 117-119.

Cremers, D.A. and Radziemski, L.J. 1985. Direct detection of beryllium on filters using the Laser Spark. *Appl. Spectrosc.*, **39**(1): 57-63.

Das, J., and Pobi, M. 1991. Separation of beryllium and aluminium from other elements using an ion-exchange resin with N-benzoylphenylhydroxylamine as a functional group. *Anal. Chim. Acta.*, **242**: 107-111.

Dinsdade, D. and Bourdillon, A.J. 1982. The ultrastructural localization of beryllium in biological samples by electron energy-loss spectrometry. *Experimental and Molecular Pathology*, **36**: 396-402.

Ellis, W.G., Hodge, V.F., Darby, D.A., Jones, C.L. and Hinners, T.A. 1988. Determination of beryllium in liquid and solid waste samples by flame and furnace AAS. *Atom. Spectrosc.*, **9**(6): 181-188.

Epstein, W.L. 1967. Granulomatous hypersensitivity. *Prog. Allergy*, **11**: 36-88.

Fazakas, J. 1983. Sensitivity enhancement of beryllium determination in graphite furnace atomic absorption spectrometry by use of atomization under pressure. *Mikrochim. Acta*, **2**(3-4): 217-224.

Fishbein, L. 1984. Overview of analysis of carcinogenic and or mutagenic metals in biological and environmental samples.-1. Arsenic, beryllium, cadmium, chromium and selenium. *Int. J. Environ. Anal. Chem.*, **17**(2): 113-170.

Frech, W., Lundberg, E. and Cedergren, A. 1985. Investigations of some methods used to reduce interference effects in graphite furnace atomic absorption spectrometry. *Prog. Anal. At. Spectrom.*, **8**(3/4): 257-370.

Friedli, C., Diaco, Th. and Lerch, P. 1987. Lithium and beryllium trace determination using heavy-ion activation analysis. *J. of Radioanal. and Nuclear Chem.*, **112**(2): 403-413.

Frieman, D.G. and Hardy, H.L. 1970. Beryllium disease. The relation of pulmonary pathology to clinical course and prognosis based on a study of 130 cases from the U.S. beryllium case registry. *Human Pathol.*, **1**: 25-44.

Gatewood, P.L. and Sneddon, J. 1990. Removal and recovery of beryllium in waters by *Chlorella vulgaris*. *J. Environ. Sci. Health*, **A25**(1): 21-28.

Gladilovich, D.B. and Stolyarov, K.P. 1981. State of the art of spectrophotometric and luminescent methods of determining beryllium (Review). *Industrial Laboratory-USSR.*, **47**(5): 425-429.

Goyal, N., Purohit, P.J., Dhobale, A.R., Patel, B.M., Page, A.G. and Sastry, M.D. 1987. Electrothermal atomization atomic absorption spectrometric determination of silver, beryllium, calcium, iron, lead and tin in uranium without preliminary separation. *J. Anal. At. Spectrom.*, **2**(5): 459-461.

Griffitts, W.R. and Skilleter, D.N. 1991. "Beryllium". In: Metals and their compounds in the environment. Merian, E. (ed.), VCH, Weinheim, pp. 775-787.

Grognard, M. 1987. Estimation of beryllium concentration in the standard reference material BCR 176 by STPF-Zeeman AAS after lithium metaborate fusion. *Atom. Spectrosc.*, **8**(5): 153-154.

Hall, J.G. 1988. Studies on the adjuvant action of beryllium. IV. The preparation of beryllium containing macromolecules that induce immunoblast responses *In vivo*. *Immunology*, **64**: 345-351.

Hapke, H.-J. 1987. Toxikologie fur Veterinarmediziner, 2nd Ed., Ferdinand Enke-Verlag, Stuttgart.

Hapke, H.-J. 1991. "Effects of metals on domestic animals". In: Metals and compounds in the environment. Merian, E. (ed.), VCH, Weinheim, pp. 531-546.

Haraldsen, L. and Bruno-Pougnet, M.A. 1989. Direct determination of beryllium in coal slurries using graphite furnace atomic absorption spectrometry with automatic injection. *The Analyst*, **114**: 1331-1333.

Hoagland, M.B. 1952. Beryllium and growth. II. The effect of beryllium on plant growth. *Arch. Biochem. Biophys.*, **35**: 249-258.

Hurlbut, J.A. 1978. Determination of beryllium in biological tissues and fluids by flameless atomic absorption spectroscopy. *Atom. Absorpt. Newsl.*, **17**(6): 121-124.

Kivimaki, P.R. and Lajunen, L.H.J. 1985. Determination of trace amounts of beryllium in aqueous solutions, and some biological and geological samples by DCP-AES. *Finnish Chemical Letters.*, (**6**): 246-250.

Kozlenkov, A.I. and Belov, Y.I. 1982. Electron micro-probe determination of beryllium with the use of a diffraction grating monochromator. *Mikrochim. Acta.*, 2(3-4): 159-167.

Luecke, W. 1992. Systematic investigation of aluminium interferences on the alkaline earth elements in flame atomic absorption spectrometry. Part I. Behaviour of beryllium. *J. Anal. At. Spectrom.*, **7**: 765-768.

Luo, F.R., Li, X.Y. and Wang, M.H. 1990. The study of modifiers for the determination of beryllium by graphite-furnace atomic absorption spectrometry. *Atom. Spectrosc.*, **11**(2): 78-82.

Makhnev, Y.A., Degtev, M.I., Chapko, T.A., Kozhenvnikova, L.A. and Khor'kova, M.A. 1986. Atomic emission determination of beryllium in natural and waste-waters after chemical preconcentration. *J. Anal. Chem.-USSR.*, **41**(7): 905-910.

McConnochie, K., Williams, W.R., Kilpatrick, G.S. and Williams, W.J. 1988. Chronic beryllium disease in identical twins. *Br. J. Dis. Chest.*, **82**: 431-435.

Merian, E. 1984. Introduction on environmental chemistry and global cycles of chromium, nickel, cobalt, beryllium, arsenic, cadmium and selenium, and their derivatives. *Toxicological and Environmental Chemistry*, **8**(1): 9-38.

Millard, J.A., Dalling, R.H. and Radziemski, L.J. 1986. Time-resolved Laser-Induced Breakdown Spectrometry for the rapid-determination of beryllium in beryllium-copper alloys. *Appl. Spectrosc.*, **40**(4): 491-494.

Miller, T.J. 1991. Development of a rapid, economical and safe method for the dissolution of particulate beryllium for atomic-absorption spectrophotometry. *Anal. Lett.*, **24**(11): 2075-2081.

Mills, J.C. 1983. Determination of boron, beryllium and lithium in coal ash and geological materials by Spark Optical-Emission Spectrometry. *Anal. Chim Acta.*, **154**(Nov): 227-234.

Nakamura, T., Oka, H., Morikawa, M. and Sato, J. 1992. Determination of lithium, beryllium, cobalt, nickel, copper, rubidium, caesium, lead and bismuth in silicate rocks by direct-atomization atomic-absorption spectrometry. *The Analyst*, **117**(2): 131-135.

Okutani, T., Tsuruta, Y. and Sakuragawa, A. 1993. Determination of a trace amount of beryllium in water samples by graphite furnace atomic absorption spectrometry after preconcentration and separation as a beryllium-acetylacetonate complex on activated carbon. *Anal. Chem.*, **65**: 1273-1276.

Ottolini, L., Bottazzi, P. and Vannucci, R. 1993. Quantification of lithium, beryllium and boron in silicates by secondary ion mass spectrometry using conventional energy filtering. *Anal. Chem.*, **65**: 1960-1968.

Owens, J.W. and Gladney, E.S. 1976. Lithium metaborate fusion and the determination of trace metals in fly ash by flameless atomic absorption. *At. Absorpt. Newsl.*, **15**(4): 95-97.

Paschal, D.C. and Bailey, G.G. 1986. Determination of beryllium in urine with electrothermal atomic absorption using the L'vov platform and matrix modification. *Atom. Spectrosc.*, **7**(1): 1-3.

Pougnet, M.A.B., Haraldsen, L. and Orren, M.J. 1985. Determination of beryllium and Lithium in coal ash by Inductively Coupled Plasma Atomic Emission Spectroscopy. *Int. J. Environ. Anal. Chem.*, **21**(3): 213-228.

Radziemski, L.J., Cremers, D.A. and Loree, T.R. 1983. Detection of beryllium by laser-Induced-breakdown spectroscopy. *Spectrochim. Acta, Part B.*, **38**(1-2): 349-355.

Rigin, V.I. 1984. Extraction-Atomic fluorescence determination of beryllium in coals, ashes, air and water. *J. Anal. Chem. - USSR.*, **39**(5): 642-648.

Robles, L.C. and Aller, A.J. 1994. Preconcentration of beryllium on the outer membrane of *Escherichia coli* and *Pseudomonas putida* prior to determination by graphite furnace atomic absorption spectrometry. *J. Anal. At. Spectrom.* Submitted for publication.

Robles, L.C., Garcia-Olalla, C., Alemany, M.T. and Aller, A.J. 1991. Flame atomic-absorption spectrometric method for the determination of beryllium in natural waters after separation with N-benzoyl-N-phenylhydroxylamine. *The Analyst,* **116**(7): 735-737.

Romney, E.M. and Childress, J.D. 1965. Effects of beryllium in plants and soil. *Soil Sci.*, **100**: 210-217.

Schmidt, W.F. and Dietl, F. 1987. Determination of beryllium in soils by means of flameless atomic absorption with zirconium-coated graphite tubes. *Fresenius' Z. Anal. Chem.*, **326**: 40-42.

Schmidt, W.F. and Dietl, F. 1988. Determination of beryllium in plants by means of flameless atomic absorption with zirconium-coated and non-coated graphite tubes. *Fresenius' Z. Anal. Chem.*, **329**(8): 853-855.

Schramel, P. and Li-Qlang, X. 1982. Determination of beryllium in the parts-per-billion range in three standard reference materials by inductively coupled plasma atomic emission spectrometry. *Anal. Chem.*, **54**: 1333-1336.

Shan, X.-Q., Yian, Z. and Ni, Z.-M. 1989. Determination of beryllium in urine by graphite-furnace atomic absorption spectrometry. *Anal. Chim. Acta,* **217**: 271-280.

Spencer, J.O. and Hall, J.G. 1984. The flow and composition of lymph from the caudal mediastinal lymph node of sheep. *Immunology,* **52**: 7-15.

Steinberg, R.A. 1946. Specificity of potassium and magnesium for growth of *Aspergillus niger*. *Am. J. Bot.*, **33**: 210-214.

Stiefel, T., Schulze, K., Tolg, G. and Zorn, H. 1980. Analysis of trace elements distributed in blood.- Determination of beryllium concentrations greater than or equal to 0.01 ng/g in human and animal blood components by preparative electrophoresis and flameless atomic absorption spectrometry. *Fresenius' Z. Anal. Chem.*, **300**(3): 189-196.

Sturgeon, R.E., Desaulniers, J.A.H., Berman, S.S. and Russell, D.S. 1982. Determination of trace metals in estuarine sediments by graphite-furnace atomic absorption spectrometry. *Anal. Chim. Acta.*, **134**: 283-291.

Sturgeon, R.E., Willie, S.N., Luong, V. and Berman, S.S. 1990. Figures of merit for furnace atomization plasma emission spectrometry. *Anal. Chem.*, **62**: 2370-2376.

Styris, D.L. and Redfield, D.A. 1987. Mechanisms controlling graphite furnace atomization and stabilization of beryllium. *Anal. Chem.*, **59**: 2897 2903.

Thulasidas, S.K., Kulkarni, M.J., Porwal, N.K., Page, A.G. and Sastry, M.D. 1988. Direct determination of beryllium, cadmium, lithium, lead and silver in thorium nitrate solution by electrothermal atomization atomic absorption spectrometry. *Anal. Letters*, **21**(2): 265-278.

Ueda, J. and Kitadani, T. 1988. Separation and concentration of beryllium by coprecipitation with hafnium hydroxide prior to determination by graphite furnace atomic absorption spectrometry. *The Analyst,* **113**: 581-583.

Urner, Z., Kohoutek, Z. and Sucha, L. 1981. Determination of beryllium in airborne pollutants from thermal power stations. *Anal. Chem., (Poland).*, **H16**:149-168.

Vanhoe, H., Vandecasteele, C., Desmet, B. and Dams, R. 1988. Determination of beryllium in environmental samples by electrothermal atomic absorption spectrometry. *J. Anal. At. Spectrom.,* **3**: 703-707.

Williams, W.J. and Wallach, E.R. 1989. Laser microprobe mass spectrometry (LAMMS) analysis of beryllium, sarcoidosis and other granulomatous diseases. *Sarcoidosis,* **6**: 111-117.

Witschi, H.P. 1970. Effects of beryllium on deoxyribonucleic acid synthesizing enzymes in regenerating rat liver. *Biochem. J.,* **120**: 623-634.

Yamamoto, M., Obata, Y., Nitta, Y., Nakata, F. and Kumamaru, T. 1988. Microporous polytetrafluoroethytlene tube separator for the determination of beryllium by flow injection - or suction-flow-solvent extraction followed by inductively coupled plasma atomic emission spectrometry. *J. Anal. At. Spectrom.,* **3**: 441-445.

Yao, J. and Jiang, Y. 1991. Determination of some elements by graphite-furnace AAS using calcium nitrate as a modifier. *Guangpuxue Yu Guangpu Fenxi.,* **11**(3): 45-50.

Zhao, W. and Yao, T. 1991. Determination of beryllium by air-acetylene AAS. *Guangpuxue Yu Guangpu Fenxi.,* **11**(3): 56-58.

Advancements In Micronutrient Research
Editor : A. Hemantaranjan
1996, Scientific Publishers, Jodhpur, pp. 103–126.

5

Mineral Requirements Of The Free-living And Symbiotic Cyanobacteria, With Special Reference To Micronutrition, For Raising Laboratory Cultures

A. Vaishampayan

Laboratory of Microbial Genetics & Biological Nitrogen Fixation,
Department of Genetics & Plant Breeding,
Institute of Agriculture Sciences,
Banaras Hindu University,
Varanasi-221 005, INDIA

Contents

1. Introduction

Cyanobacteria are photosynthetic prokaryotic microorganisms, some of which are capable of nitrogen fixation. Their main photosynthetic pigments are chlorophyll-a, carotenes and xanthophylls together with phycobiliproteins, c-phycocyanin (blue) and c-phycoerythrin (red). Due to the presence of these latter pigments and mucilage, the colour of cyanobacteria in nature may range from dirty yellow, through various shades of blue-green, to brown or black. Their range in vegetative form extends from simple unicells to multiseriate, true branching thalli. Nearly 150 cyanobacterial strains are now known to fix atmospheric dinitrogen.

However, in spite of the variety of cyanobacteria now on record which are ubiquitous in view of their trophic independence from carbon and nitrogen, it is still difficult to maintain the morphological integrity and optimum nitrogen-fixing property of most of these photobiological dinitrogen fixers in a synthetic nutrient culture medium. A number of media have been prescribed from time to time and modified by one and/or the other group(s) of workers taking into consideration either the optimum nitrogen-fixing potential or some other morpho-physiological aspects of this unique group of prokaryotic, but typically oxygenic photosynthetic, microorganisms.

The present article is an attempt to systematically compile significant pieces of information on the different well-recognised nutrient media (with particular reference to the micronutrients)/related methods proposed/modified by various teams for raising successful laboratory cultures of the major representatives of the free-living and symbiotic cyanobacteria.

2. THE NUTRIENT MEDIA

2.1. Free-living Cyanobacteria

Various nutrient media used for culturing cyanobacteria are listed in Table 1. From these, the most suitable for any particular cyanobacterial strain is generally chosen. Temperature and light are the other important factors whose optimum level for a given cyanobacterial strain has to be determined for raising the culture, but this subject is out of the purview of this article and the readers are referred to consult for this aspect elsewhere (Fogg *et al.*, 1973). Rippka *et al.* (1979) recommended BG-11 and its modified (BG-11o nitrate) as good media for most of the cyanobacteria. For marine strains they recommended ASN-III or MN media. In addition, Chu-10 medium, as modified by Gerloff *et al.* (1950), has been found to be quite suitable for the different species of *Nostoc* which are ideal materials for most of the physiological (Fogg *et al.*, 1973; Singh and Singh, 1978; Vaishampayan *et al.*, 1978, 1992a, b; Vaishampayan, 1984a, b; Mishra *et al.*, 1991a) biochemical (Vaishampayan, 1982a, 1984c, 1985d; Rai, 1990; Verma *et al.*, 1990; Prasad *et al.*, 1991) and genetic (Singh and Vaishampayan, 1978; Singh *et al.*, 1979; Vaishampayan and Prasad, 1981, 1982, 1984; Vaishampayan and Singh, 1981a, b; Vaishampayan, 1983a, 1984d-h, 1985a-c, 1988, 1994a, b; Vaishampayan and Mishra,

TABLE 1. Composition of the eight synthetic nutrient media used for the isolation and culture of the various nitrogen-fixing and non-nitorgen fixing Cyanobacterial specis.

Compounds	Conscentration of the salt (mg L^{-1}, except otherwise noted)							
	1 Allen and Arnon	2 ASM	3ASN - III	4ASP - II	3BG-11o	5 Chu- 10	3 MN	6 Kratz and Myers
Biotin	–	–	–	0.001	–	–	–	–
$CaCl_2$	74.00	36.00	500.00	370.00	36.00	55.00	18.00	–
* $Ca(NO_3)_2$ $4H_2O$	–	–	–	–	–	40.00	–	25.00
Capantothenate	–	–	–	0.10	–	–	–	–
Citric acid	–	–	3.00	–	6.00	3.00	3.00	–
$CoCl_2$ - $6H_2O$	–	0.02	–	0.01	–	–	–	–
*$Co(NO_3)_2$. $6H_2O$	0.05	–	0.05	–	0.05	–	0.05	–
Cr $(SO_4)3$ K_2SO_4. $2H_2O$	0.19	–	–	–	–	–	–	–
$CuCl_2$. $2H_2O$	–	0.0002	–	0.0003	–	–	–	–
**$CuSO_4$. $5H_2O$	0.079	–	0.079	–	0.079	0.079	0.079	0.079
Deionized Water	1,000ml	1,000 ml	1,000 ml	1,000 ml	1,000 ml	1,000 ml	250 ml	1,000 ml
Fe Cl_2	–	0.60	–	2.30	–	–	–	–
Fe EDTA	26	–	–	–	–	–	–	–
Ferric Citrate	–	–	–	–	–	3.00	–	–
Fe $(SO_4)_3$. $6H_2O$	–	–	–	–	–	–	–	4.00
Folic Acid	–	–	–	0.002	–	–	–	–
H_3BO_3	2.86	2.50	2.86	34.30	2.86	2.86	2.86	2.86
Inositol	–	–	–	5.00	–	–	–	–
Iron (III) ammoniun citrate	–	–	3.00	–	6.00	–	3.00	–
KCl	–	–	500.00	600.00	–	–	–	–
K_2HPO_4 . $3H_2O$	400.00	20.00	20.00	6.00	40.00	13.00	20.00	1300.00
* KNO_3	2000.00	–	–	–	–	–	–	1000.00
$MgCl_2$. $6H_2O$	–	40.00	2000.00	–	–	–	–	–
Mg SO_4 . $7H_2O$	–	50.00	3500.00	5000.00	75.00	25.00	–	250.00

$MnCl_2 . 4H_2O$	250.00	1.40	1.81	4.30	1.81	1.81	1.81	1.81
$MnSO_4 . 4H_2O$	2.00	–	–	–	–	–	–	–
MoO_3	0.15	–	–	–	–	–	–	0.02
Na citrate. $2H_2O$	–	–	–	–	–	–	–	165.00
NaCl	230.00	–	25000.00	18000.00	–	–	–	–
Na_2CO_3	–	–	20.00	–	20.00	20.00	20.00	–
Na_2 EDTA	–	10.00	0.50	30.00	1.00	–	0.50	–
Na_2HPO_4	–	14.00	–	–	–	–	–	–
$Na_2SiO_3 . 4H_2O$	–	–	–	150.00	–	44.00	–	–
Na_2WO_4	0.02	–	–	–	–	–	–	–
$Na_2MoO_4 . 2H_2O$	–	–	0.39	–	0.39	0.39	0.39	–
*Na NO_3	–	17.00	750.00	50.00	1500**	–	750.00	–
NH_4VO_3	0.02	–	–	–	–	–	–	–
Nicotinic acid	–	–	–	0.01	–	–	–	–
$NiSO_4 . 6H_2O$	0.04	–	–	–	–	–	–	–
p-Aminobenzoic acid	–	–	–	0.01	–	–	–	–
post-Autoclaving pH adjustment	–	–	7.5	7.6-8.6	7.1	–	8.3	7-9.00
Sea water	–	–	–	–	–	–	750 ml	–
Thiamin-HCl	–	–	–	0.50	–	–	–	–
Thymine	–	–	–	30.00	–	–	–	–
TiO (Cr_2O_4)x $7H_2O$	0.01	–	–	–	–	–	–	–
Tris	–	–	–	1000.00	–	–	–	–
Vitamin B_{12}	–	–	–	0.002	–	–	–	–
$ZnCl_2$	–	0.04	–	0.31	–	–	–	–
$ZnSO_4 . 7H_2O$	0.222	–	0.222	–	0.222	0.222	0.222	0.222

* Omitted in the case of N_2 - fixing cynabacteria.

** NO_3 supplemented BG-11o medium, used for non N_2-fixing cyanobacteria, is known as BG-11o medium.

1. Allen and Arnon (1955) ; 2. Gorham *et al.* (1964); 3 Rippka *et al.* (1979); 4. Provasoli *et al.* (1957); 5. Chu (1942); 6. Kratz and Myers (1955)

1990; Mishra *et al.*, 1991b) studies conducted during the last two decades in cyanobacteria.

Of the above, the results deserving special mention are those related to interaction studies on (i) molybdenum (Mo), tungsten (W), chromium (Cr) and vanadium (V), (ii) iron (Fe) and copper (Cu), and (iii) exogenously supplied amino acids in cyanobacteria.

2.1.1. *Mo-W-Cr-V Interactions*

Mo, W and Cr belong to group VI B of the periodic table, forming a series of chemical analogues. Mo is a definite requirement for the growth of the nitrogen-fixing cyanobacteria on molecular nitrogen or nitrate, since the nitrogen-fixing enzyme nitrogenase and nitrate-metabolizing enzyme nitrate reductase, both are molybdo-proteins (Brill, 1980). W is found to be a competitive inhibitor of Mo function *in vivo* in bacteria (Brill *et al.*, 1974; Nagatani and Brill, 1974; Cardenas and Mortenson, 1975) in green alga *Chlorella* (Cardenas *et al.*, 1972) and the higher plant spinach (Notton and Hewitt, 1971). The replacement of Mo by W in molybdo-enzymes is shown to render them biologically functionless (Hewitt, 1974). Nason *et al.* (1971) reported that an Mo-containing sub-unit of nitrate reductase of *Neurospora crassa* can be replaced functionally by a Mo-containing sub-unit of nitrogenase of soybean nodule bacterioids. This and the subsequent studies led to the concept that there is an ammonium-repressible, Mo-containing low molecular weight co-factor of polypeptide nature which is common to different mlybdo-enzymes including nitrogenase and nitrate reductase (Lee *et al.*, 1974; Shah and Brill, 1977). Whether this ammonium-repressible common co-factor which complexes with Mo, is the molecule determining the functional specificity of group VI B elements in the biological activity of nitrogenase and nitrate reductase, is not known. Mutants of the nitrogen-fixing cyanobacterium *Nostoc muscorum* have been isolated (Singh *et al.*, 1978a, b) which require W or Cr but not Mo for growth in molecular nitrogen or nitrate medium. The results have clearly shown the involvement of a single genetic determinant in the control of biological specificity of group VI B elements in molecular nitrogen and nitrate metabolism. This is because of the fact that both W and Cr have proved toxic to the parent organism specifically in molecular nitrogen or nitrate medium, but not in nitrite or ammonium medium, suggesting that the wild cyanobacterial strains are auxotrophic for Mo for growth in molecular nitrogen or nitrate medium. However, W- or Cr-toxicity has been always found to be less in nitrate medium than in molecular nitrogen medium (Table 2). Addition of Mo to Mo-free, W-containing or Cr-containing medium always leads to the quick recovery of the growth of the wild type organism. On the other hand, addition of W or Cr to Mo- containing culture causes an abrupt decline in cyanobacterial growth, suggesting that W or Cr has been acting as a competitive inhibitor of Mo function for growth with molecular nitrogen or nitrate as a nitrogen source.

TABLE 2. Growth of parent *Nostoc muscorum* in N_2, 5mM NO_3^- 5mM NO_2^- or 1 mM NH_4^+ medium with various concentrations of molybdate Mo, tungstate (W) or chromate (Cr)

Mo, W or Cr	Concentration (μg/ml)	Medium			
		N2	NO_3^-	NO_2^-	NH_4^+
Mo	0	0.005	0.005	0.450	0.455
	10	0.370	0.435	0.455	0.455
	50	0.225	0.350	0.410	0.450
	100	0.190	0.355	0.445	0.455
W	0	0.005	0.005	0.445	0.450
	10	0.000	0.015	0.450	0.455
	50	0.000	0.010	0.455	0.445
	100	0.000	0.000	0.455	0.445
Cr	0	0.005	0.005	0.450	0.455
	10	0.105	0.105	0.455	0.445
	50	0.005	0.105	0.455	0.450
	100	0.000	0.000	0.450	0.455

Growth was measured in terms of the increase in optical density after 8 days of inoculation of the cyanobacterium in media containing graded concentrations of molybdate, tungstate or chromate.

Contrary, to the natural behaviour of the cyanobacteria with the trace mineral element, Mo, a W-requiring mutant of *Nostoc muscorum* could not grow without W in molecular nitrogen or nitrate medium but did grow in nitrite or ammonium medium (Singh *et al.,* 1978a). Mo or Cr were unable to substitute for W function in the W-requiring strain, and rather appeared to inhibit W function in this mutant strain as W or Cr does in the Mo-requiring parental strain. Similarly, the Cr-requiring mutant of this organism behaved like the Mo-requiring parent or the W-requiring mutant in all respects for growth in molecular nitrogen or nitrate except that Cr has taken over the function of Mo in this mutant.

The other important point to mention is the fact that the Mo-requiring parent strain does not form heterocysts in nitrate medium containing Mo, but does form heterocysts in nitrate medium lacking Mo, suggesting a requirement of Mo in the inhibition of heterocyst formation by nitrate.

Since active nitrate reductase is a requirement for nitrate inhibition of heterocyst formation in cyanobacteria (Singh *et al.,* 1977a, b), it has been resolved that Mo-requirement for nitrate inhibition of heterocyst differentiation has been operative at the level of Mo-activation of nitrate reductase. The Mo-requiring mutants formed heterocysts in nitrate medium containing W or

Cr, suggesting the W or Cr was unable to substitute physiologically for the role of Mo in the activation of nitrate reductase. Mutation from Mo-requirement for growth to W- or Cr-requirement for growth in molecular nitrogen or nitrate medium also resulted in W- or Cr-requirement for nitrate inhibition of heterocyst differentiation. Apparently, therefore, the three strains were alike with respect to transport system for the three elements.

Further investigations were conducted to test whether vanadium (V), belonging to group V of the periodic table can replace Mo (in the Mo-requiring parent) or W (in the W-requiring mutant) or Cr (in the Cr-requiring mutant) of *N. muscorum* for nitrogen fixation and nitrate metabolization. Any concentration of V could not induce any nitrogenase or nitrate reductase activity/N-free or nitrate-mediated growth and nitrate-inhibition of heterocyst differentiation in the Mo-, W- or Cr-lacking culture of these three strains, respectively. However, with the normal requisite concentration of Mo, W or Cr the respective strain exhibited an increase in the above activities with the increasing V concentrations up to a threshold level of 0.0125 ppm in parent *N. muscorum* (Vaishampayan, 1983b, c) and W- or Cr-requiring mutant of this organism (Vaishampayan and Hemantaranjan, 1984). However, at concentrations above 0.0125 ppm V proved inhibitory to the N-free and nitrate-mediated culture of either of the three strains and induced the formation of heterocysts in nitrate medium, suggesting the exclusive operation of the higher concentrations of V at the level of nitrogenase and nitrate reductase. This unique phenomenon was explained to be due to V replacing Mo, W or Cr because of their close structural similarity (Kula, 1966; 1967; Callis and Wentworth, 1977), and therefore, inactivating the Fe-Mo and Mo-containing co-factors of the nitrogenase and nitrate redutase enzyme systems, in contrast to W or Cr, which is known to functionally replace Mo for these two enzyme activities in this organism. But the growth-promoting efficiency of the lower concentrations of V cannot be due to its functioning in place of Mo. Had it been functioning in place of Mo, it would have supported growth of the organism even in the absence of Mo, W or Cr. The synergistic effect of lower concentrations of V has been explained to be due to its involvement in stabilizing the Mo-protein binding in nitrogenase and nitrate reductase. Earlier biochemical studies have shown that V influences the rate of nitrogen fixation in *Azotobacter* species by stabilizing Mo-protein binding in nitrogenase (Benemann *et al.*, 1972). Existence of a V-nitrogenase (Burns *et al.*, 1971) active in nitrogen fixation in *A. vinelandii* and *A. chrococcum* (Becking, 1962) and *Clostridium butyricum* (Jensen and Spencer, 1947) has already been reported since long, but it has still to be traced in any of the cyanobacterial members tested so far. Benemann *et al.* (1972) have suggested that the presence of Mo, in traces, cannot be ruled out in these systems even. Particularly, in cyanobacteria, V is incapable of functionally replacing Mo, W or Cr (in the Mo-requiring parent, W-requiring mutant or Cr-requiring mutant, respectively), though it may increase Mo- protein binding in nitrogenase and nitrate reductase resulting in an increased molecular nitrogen and nitrate metabolism.

2.1.2. *Fe-Cu Interaction*

Heavy metals are widely known for their toxicity in plant (Singh and Sharma, 1980), animal (Nriagu, 1978, 1979), microbial (Kumar, 1964; Singh and Kumar, 1972; Sadler and Trudinger, 1967; Blundell and Wild, 1969), and mammalian (Shephard, 1975; Sanyal *et al.*, 1980) systems. Of these, copper and copper compounds have been popularly used as fungicidal and biocidal agents (Roth *et al.*, 1971). However, successful studies on the tolerance of fungal strains to many heavy metals (Ashida, 1965; Ross, 1975; Sadler and Trudinger, 1967) and a saturated solution of copper sulphate (Basu *et al.*, 1955; Gadd and Griffiths, 1978; Starkey, 1973; Starkey and Wakesman, 1943) have been made. Such heavy metal tolerance studies have been made in the cyanobacterium *N. muscorum* as well with regard to copper (Vaishampayan and Prasad, 1982) and mercury (Prasad *et al.*, 1986). In these studies, while Hg-inhibition has been found to be at the level of photosynthetic inhibition of inorganic carbon, and hence exogenous glucose-reversible, results obtained with Cu-toxicity have been found Fe-reversible, and therefore, unfolding the physiological relation of the Cu-Fe interactions in cyanobacteria (Vaishampayan, 1982b, 1984i).

Cu (applied as copper sulphate or chloride) was found to exert its toxic action in cyanobacteria in two ways. At lower concentrations, i.e., 25 and 50 ppm, it inhibited the growth and nitrogenase activity of the organism in N-free medium, but such inhibitory effects were reversed to a great extent on addition of 50 ppm concentration of Fe (applied as a sulphate or chloride salt). Cu did not inhibit ammonia-mediated growth at these concentrations, suggesting that the preliminary effect of Cu was related to the nitrogen-fixing condition and that it was reversible with either of the two Fe salts. At higher concentrations (75 and 100 ppm), Cu inhibited growth of the organism in both N-free and ammonium media besides inhibiting nitrogenase activity, and these effects could not be reversed to any extent of supplementation with exogenous Fe. Microscopic examination of such cultures revealed Cu-induced massive fragmentation and lysis of the cultures. The reason why the inhibitory effect of Cu on nitrogenase activity and consequently growth of the organism in N-free medium is reversed on exogenous Fe supplementation, is explicable. Nitrogen activity is due primarily to the Fe-Mo co-factor (Shah and Brill, 1977; Pienkos *et al.*, 1977) which is an active site of component-I of nitrogenase and the site to which molecular dinitrogen binds and is subsequently reduced during nitrogen fixation (Shah *et al.*, 1978; Rawlings *et al.*, 1978). Naturally then, any hindrance in Fe-uptake is expected to block the formation of the nitrogenase enzyme precursor. Since the role of Cu in causing Fe-deficiency in plants is established (Brown and Ambler, 1973; Brown and Jones, 1977; Lingle *et al.*, 1963; Wallace and De Kock, 1966), it is suggested that the Cu salt interfered with Fe-uptake in cyanobacteria leading to inactivation of the Fe-Mo co-factor. Further, the amount of Fe and Mo prescribed in modified Chu 10 medium (Gerloff *et al.*, 1950) has been tested to be such to balance the toxic effect of Cu, if any, on nitrogenase in cyanobacteria (Vaishampayan, 1982b, 1984i).

2.1.3. *Amino Acid Nutrition*

Amino acids in general are inhibitory to the growth of methylotrophs (Eccleston and Kelly, 1972; Smith and Hoare, 1977) and poor sources of nitrogen for cyanobacteria (Neilson and Doudroff, 1973). However, systematic studies on the nutritive role of amino acids have been done with the help of DCMU [3-(3,4-dicholorophenyl)-1, 1-dimethylurea] in parent *Nostoc muscorum* and its various nitrogen and carbon auxotrophs, based on which the L-isomers of 21 amino acids have been classified for their abilities to serve as carbon and/or nitrogen source(s)/as a source of no nutrition/as a toxic organic substance (Vaishampayan *et al.*, 1978; Vaishampayan, 1981, 1982c, 1984j). The amino acids glutamate, alanine, tyrosine and cysteine were toxic; glutamine, histidine, asparagine, tryptophan and serine were utilized as nitrogen sources; arginine, proline and phenylalanine were utilized as carbon sources; leucine, isoleucine, lysine, methionine, valine and citrulline were utilized as sources of both carbon and nitrogen; aspartate, threonine and glycine served neither as carbon nor nitrogen sources.

It is remarkable to note in these experiments that the organism could apparently utilize carbon from the amino acids arginine, proline and phenylalanine without also utilizing nitrogen which should have been released in the form of ammonium. It is difficult to explain how this could be so and further study of the mechanisms of utilization of these amino acids is necessary. It can be, but in any event, inferred that several amino acids can serve as good carbon and/or nitrogen source(s) in cyanobacteria. However, a few amino acids have shown a poor support to the cyanobacterial growth which may be due to the incorporation of their relatively small amounts into the cyanobacterial cells. The reason for this has been attributed, for methylotrophs, to a limited capacity of the cells to utilize the compounds which are for the most part incorporated unchanged into protein (Eccleston and Kelly, 1972). It is thus essential to characterize the reactions which make the amino acids so specific in their nutritive behaviour.

2.2. Lichens

Owing to their sensitivity to pollution and environmental changes, lichen samples are generally maintained for experimental purposes in growth chambers. Through 0.01% (w/v) digitonin treatment the cyanobacterial symbionts (also known as cyanobionts) are freed from the fungal cells. The lichens are surface sterilized with 3% sodium hypochlorite solution for 30 s (Rai, 1990). The cephalodia, excized and washed with sterilized distilled water, are squashed with sterilized needle and inoculated on BG-11o agar plates (Rippka *et al.*, 1979). The inoculated petri dishes are incubated at 22°C and under 1 klx continuous light. When colonies develop, a single cyanobacterial colony is picked up and spread on another set of agar dish. At this stage, a procedure earlier described by Bunt (1961) is followed to make the cyanobacterial culture bacteria-free. In this method, a 3-cm block of BG-11o agar (BG-11 medium without nitrate) is poured into 15 x 1.5 cm test tube and cut out aseptically from the petri plate and placed on the top of agar earlier inserted inside the test tube. It is overlayered with the same sterilized agar medium (nearly 2-cm deep) containing polymixin-B sulphate (0.01 mg/ml) and cycloheximide (0.1 mg/ml), followed by dropping nearly 2 cm of liquid medium in the test tube before it is plugged with sterilized cotton wool. These samples are placed in dark for 24 h and then transferred to light. This is done in order to stimulate hormogonia production which migrate through the antibiotic containing agar and reach the top liquid layer during light incubation. Polymixin-B and cycloheximide are used to kill the bacterial and eukaryotic contaminants, respectively. The Cu, Mn, Mo, Co, Zn and Fe, as well as EDTA concentrations of the BG-11o medium are so specific that these not only help isolation of cyanobionts from the lichens, but also their mass culture (Rippka *et al.*, 1979).

2.3. Liverworts and Hornworts

Anthoceros and *Blasia* are the two bryophyte species, enlodging cyanobacterial cells as microsymbiotic partners, on which most of the physiological and biochemical investigations are centred around. A special Proskauer solution with a variety of micronutrients, as detailed under Table 3 (Rodgers, 1974) is used to culture these organisms. Enderlin and Meeks (1983) have developed quite a workable method to raise axenic thalli of these bryophytes containing the cyanobionts. In fact these have been raised (*Anthoceros* gametophytes) by germinating surface-sterilized spores. For this, mature sporophytes are excized, treated with 0.5% sodium hypochlorite and 0.1% detergent for 1/2 to 2 min, and then washed with sterilized distilled water. A lengthwise splitting of sporophytes is observed and the spores are spread over the basal agar medium, containing 0.2 g/l ammonium nitrate, 0.1 g/l each of calcium chloride, magnessium sulphate and potassium dihydrogen phosphate, 1 ml/l micronutrient solution (Stein, 1958) and 1.2% agar (pH adjusted to 6.4 before autoclaving). The petri dishes are incubated at 20°C under a 14/10 h light-dark cycle. As germlings appear, they are transferred to liquid basal medium containing 1% glucose buffered with 5 mM 2-N-morpholino ethanesulfonic acid (MES)-NaOH buffer (pH:6.4). This way, the axenic *Anthoceros-Nostoc* associations are raised.

TABLE 3. Composition of Proskauer solution used for the culture of macrosymbiotic bryophytes free from the endophytic (microsymbiotic) cyanobionts (with KNO_3) and the whole association (without KNO_3)

Macronutrients	Concentration	Micronutrients	Concentration
$MgSO_4.7H_2O$	100 mg/L	$LiCl_2$	28 µg/L
$CaCl_2$	100 mg/L	$CuSO_4$	56 µg/L
KH_2PO_4	100 mg/L	$ZnSO_4$	56 µg/L
$FeSO_4$	50 mg/L	H_3BO_3	611 µg/L
KNO_3	250 mg/L	Al_2SO_4	56 µg/L
		$SnCl_2.2H_2O$	28 µg/L
		KBr	28 µg/L
		$MnCl_2.4H_2O$	389 µg/L
		$NiSO_4.6H_2O$	56 µg/L
		$Co(SO_4)_2 6H2O$	56 µg/L
		TiO_2	56 µg/L
		KI	28 µg/L

Following a brief method of Rodgers (1974) the cyanobionts from *Blasia* and *Anthoceros* have been successfully isolated. The N-free ASM medium (Gorham *et al.*, 1964), as detailed in Table 1, has been used in such isolations. But these isolations take nearly three weeks. For an immediate study of the bryophytic cyanobionts, another method has been used by Meeks *et al.* (1985) which takes approximately 3-4 h to isolate just a few cyanobiont cells. In this, very useful method known so far, *Anthoceros* thalli containing *Nostoc* colonies are suspended in 5 mM 3-N-morpholino propane sulphonic acid (MOPS)-NaOH buffer (pH:7.8), containing polyvinyl pyrrolidone-40 (1%) in 1:1 ratio (w/v). For 20 s this suspension is blended at high speed. The resulting slurry (1000 g) is centrifuged twice in the same buffer and incubated for 15 min at 30°C in the presence of 0.05% (w/v) digitonin. The same is then supplemented with macerase, cellulase and B-glucuronidase. The sample is digested for 1 h and centrifuged twice in 5 mM MOPS buffer (pH:7.8) with 2 min vigorous mixing between the centrifugations. The resulting pellet is resuspended in MOPS buffer and loaded on a discontinuous gradient composed of 5% Ficoll-400 (w/v) in addition to 5% neutralized Ludox AM [colloidal silica (v/v)], 7% Ficoll + 4% Ludox, 9% Ficoll + 3% Ludox, and 11% Ficoll + 2% Ludox. This is followed by centrifugation at 16,000 g for 10 min at 10°C in a horizontal rotor. A band of most of the cyanobacterial colonies is found in 11% Ficoll + 2% Ludox region of the gradient which is carefully removed and the gradient centrifugation repeated twice. Still small amounts of contaminating undigested material is generally removed with the aid of stereomicroscope (in modifications to Meek's method) and small inoculum of cyanobiont from the bryophytic associations are obtained.

2.4. Azolla

As a general, the medium devised by Ito and Watanabe (1983) has been in use for *Azolla* culture in the laboratory. But modifications of the same has proved to be even better (Rai, 1990; Vaishampayan *et al.*, 1992b; Vaishampayan, 1993a), detailed in Table 4. Basal medium with nitrate is used for growing cyanobiont-free *Azolla* and basal medium unsupplemented with any combined inorganic nitrogen source for *Azolla-Anabaena* symbiotic nitrogen-fixing complex. Field samples of *Azolla* are generally contaminated with bacteria. For growing them in laboratory, these are thoroughly washed with distilled water and then surface-sterilized with 2% Chlorox (or 0.1% sodium hypochlorite) and 0.01% Triton X-100 (as wetting agent) for 20 min. Some times mere treatment with 0.2% mercuric chloride serves the purpose. The vegetative fronds of *Azolla* are then again washed with sterilized distilled water and transferred to pre-autoclaved growth medium with or without nitrate. After a week or so, healthy fronds are subcultured in fresh medium. Peters and Mayne (1974) have described methods to obtain microsymbiont-free pure macrosymbiont *Azolla*. The same has been slightly modified (Reddy, 1988; Vaishampayan *et al.*, 1992b). In this useful method, surface- sterilized *Azolla* fronds are transferred to antibiotic-containing *Azolla* growth medium (+ nitrate) and maintained for a week. These are then transferred to fresh medium containing a different antibiotic. This process is continued until the four different combinations, i.e., A to D (as given in Table 5), have been employed. The variously antibiotic-treated fronds are transferred to basal medium (+ nitrate) unsupplemented with any antibiotic, and the same appear to be pure cultures of non-nitrogen-fixing cyanobiont-free *Azolla* macrosymbiont.

Methods are also available for the isolation of cyanobacterial packets from the *Azolla-Anabaena* symbiotic complex. Earlier, the same was used to be achieved through causing disintegration of *Azolla* tissue by enzymatic treatments. However, the method used by Peters and Mayne (1974) and Reddy (1988) is simple and does not require enzymic digestion. In this method, fresh sterile *Azolla* fronds are gently crushed with rolling pin in BG-11o medium containing 1% polyvinyl pyrrolidon (PVP). This is then filtered through a coarse nylon mesh followed by filtration through 2-12 layers of cheesecloth. The filtrate is repeatedly centrifuged at low speed to yield *Anabaena azollae* in the pellet. Through this method remarkably clean cyanobiont cells are isolated for immediate studies. However, regarding culture of *ex-planta Anabaena azollae* there has been, till date, only one authentic procedure proposed by Newton and Herman (1979). In this method, 40 g (fresh weight) of *Azolla* is suspended in 100 ml liquid growth medium (Table 4) containing 10 mM dithiotreitol (DTT) and 500 mg each of pectinase and cellulase inside a 500 ml suction flask. The flask is evacuated and filled with Argon to help infiltration of enzymes into the *Azolla* tissue and incubated for 16 h at 25°C.

Its volume is doubled by dilution with similar volume of sterilized distilled water and the same is filtered through a coarse mesh into a sterile 250-ml graduated cylinder.

TABLE 4. Composition of nutrient medium for *Azolla-Anabaena* symbiotic nitrogen fixing complex, as well as the cyanobiont-free non-nitrogen-fixing *Azolla* macro-symbiont

Compounds	$-NO_3^-$ medium*	$+NO_3^-$ medium**
KCl	2.00 mM	–
$CaCl_2$	2.00 mM	–
$Ca(NO_3)_2$	–	2.00 mM
KNO_3	–	2.00 mM
KH_2PO_4	0.40 mM	0.40 mM
$MgSO_4$	0.80 mM	0.80 mM
FeEDTA	6.58 mM	6.58 mM
$Co(NO_3)_2$	0.17 μM	0.17 μM
NH_4VO_3	0.10 μM	0.10 μM
Na_2WO_4	0.05 μM	0.05 μM
$TiO(Cr_2O_4)_x$	0.05 μM	0.05 μM
$Cr_2(SO_4)_3\,K_2SO_4$	0.10 μM	0.10 μM
H_3BO_3	46.00 μM	46.00 μM
$MnSO_4$	9.20 μM	9.20 μM
MoO_3	1.00 μM	1.00 μM
$ZnSO_4$	0.77 μM	0.77 μM
$CuSO_4$	0.32 μM	0.32 μM
$NiSO_4$	0.10 μM	0.10 μM

* Used for growing *Azolla-Anabaena* association;

** Used for growing *Anabaena*-free *Azolla*.

The cyanobiont packets are allowed to settle which are repeatedly washed by decantation with sterile distilled water. The resulting pellet of *Anabaena azollae* is collected in 10 ml sterilized distilled water, and 1 ml from this is inoculated into 15 ml of BG-11o medium. This is incubated for two continuous weeks under light. Cyanobacterial pigments are apparently lost and the inoculum sediments to the bottom as a brown residue. Thereafter, the pigments gradually reappear. At this stage, pure cultures are obtained by picking up short filaments of the cyanobiont and placing them on BG-11o agar medium, using a micromanipulator. The micronutrients of this synthetic medium play specific inevitable roles in the isolation and development of pure cultures of the nitrogen-fixing microsymbiont (Newton & Herman, 1979).

TABLE 5. Different sets of antibiotics used to raise a cyanobiont-free culture of *Azolla* macrosymbiont

Set No.	Combination code	Antibiotic combination
1	A	Aureomycin 2 μg/ml
	B	Penicillin G 50 μg/ml
	C	Streptomycin sulphate 10 μg/ml
	D	Bacitracin 10 μg/ml + polymixin - B-sulphate 12.5 μg/ml
2	A	Polymixin-B-sulphate 25 μg/ml
	B	Tetracycline 100 μg/ml
	C	Bacitracin 20 μg/ml
	D	Penicillin G 100 μg/ml + Streptomycin sulphate 20 μg/ml
3	A	Tetracylcine 10 μg/ml + polymixin-B-sulphate 25 μg/ml + bacitracin 20 μg/ml
	B	Penicillin G 100 μg/ml + Streptomycin sulphate 20 μg/ml + aureomycin 4 μg/ml
	C	Tetracycline 10 μg/ml + polymixin-B-sulphate 25 μg/ml + bacitracin 20 μg/ml
	D	Penicillin G 100 μg/ml + Streptomycin sulphate 20 μg/ml + aureomycin 4 μg/ml

2.5. Cycads and Gunnera

Cycads and *Gunnera* are higher plants, far larger than rest of the symbiotic systems involving nitrogen-fixing cyanobacteria. Either these are maintained in pots in glass houses for experimental purposes, or taken directly from fields. *Gunnera* can be raised through seeds as well. For the isolation of cyanobionts from these systems, the starting materials are *Gunnera* glands and the thin slices of cycad root nodules.

Cyanobiont cells can be obtained from the cycad root nodules for a direct study. For this, the root nodules are externally cleaned and cut into slices which are put in 4-(2-hydroxyethyl)-1-piperazine ethanesulphonic acid (HEPES)-NaOH buffer (pH:7) for 1 h. The cyanobiont filaments move to the buffer which are taken out with a syringe and washed by centrifugation. Although high yields of cycad cyanobionts are observed through this method, yet synthetic media with a proper combinations/concentration of the different micronutrients are still to be worked out. Moreover, success has not yet been achieved even up to this stage in the case of *Gunnera*- cyanobiont associations.

3. Cyanobacteria As A Phototrophic Component In Artificial Associations With Plant Cells

MS medium (Murashige and Skoog, 1962), which is generally used for plant cell cultures, was chosen to check the ability for joint growth of plant cells (isolated protoplasts) and cyanobacteria in suspension cultures, in attempts to make an artificial plant-cyanobacterial nitrogen-fixing symbiosis (Gusev and Korzhenevskaya, 1990). The MS medium contains basic mineral compounds, many of which are incorporated in BG-11 medium (Table 1) for culturing cyanobacteria (Rippka *et al.*, 1979). In addition, it contains phytohormones necessary for the division of plant cells (and some other components) as well as 3% sucrose. However, the concentration of mineral salts, in general, and micronutrients, in particular, in MS medium, exceeds that in BG-11 medium. Therefore, the concentration of all the components in the MS medium for producing mixed cultures/artificial associations has been reduced to half (Butenko *et al.*, 1979; Gusev *et al.*, 1982) or 1/4 (Gusev *et al.*, 1980; Butenko *et al.*, 1982; Korzhenevskaya *et al.*, 1985).

In order to establish the interaction of the partners on the basis of photosynthetic products of cyanobacteria, the mixed cultures were grown in the light at a low concentration of sucrose (1.875 g/l). This modified MS medium was not optimal for the growth of either plant cells or cyanobacteria. The growth of different plant cell cultures on this medium without cyanobacteria (i.e., in monocultures) was 1/2 to 1/4 or even lower than their growth on a full MS medium. Cyanobacteria in monocultures on this medium exhibited very slight growth or no growth at all. However, cyanobacteria were found to stimulate the growth of plant cells in tobacco (Butenko *et al.*, 1979; Gusev *et al.*, 1982), one of the ginseng strains (Gusev *et al.*, 1980; *Butenko et al.*, 1982) and *Dioscorea deltoidea* cells (Gusev & Korzhenevskaya, 1990) in diculture. This is apparently connected with conditioning the medium by one partner and creation of more favourable conditions for the growth of the other partner. Apart from this, the formation of associative interactions are influenced by secondary products of plant cells. For example, cells of *Solanum laciniatum* and *Papaver somniferum* produce alkaloids, but the former accumulate them intracellularly (Rerabek and Gorelova, 1986) while the latter excrete them into the medium (Shamina *et al.*, 1979). This results in an inhibitory effect of *P. somniferum* on the probable cyanobacterial partner *Anabaena variabilis*, while *S. laciniatum* cells do not exert such an inhibitory effect. In general, the systems in which cyanobacteria were found to have a stimulating effect, or no inhibiting influence, on the growth of plant cells have been of great interest, and to facilitate this a number of global efforts are going on to synthesize one or the other suitable medium with proper assembly of micronutrients to make an artificial association between crop/vegetable/fruit plants and nitrogen-fixing cyanobacteria, a success.

4. Conclusions And Future Prospects

Of the various synthetic nutrient media discovered so far and modified from time to time, none can be universally applied as a single suitable medium for the cultivation of free-living and symbiotic cyanobacteria or for raising an artificial plant-cyanobacterial symbiosis. While modifying any standard medium for any new species/genetically manipulated/engineered strain it has been found more important to fix up a suitable concentration of a particular macro- or micro-element/organic constituent, than to delete, add, substitute or superscribe any component of the medium, as the latter has given erratic results in most of the cases.

Although an "*Azotobacter*-like" V-nitrogenase has not yet been traced so far in cyanobacteria in spite of a significant increase in the rate of nitrogen fixation with a threshold concentration of vanadium in Mo-containing medium, we are not yet in a position to say with conformity about the existence/co-existence/non- existence of the same since quite a large number of sub-groups remain to be screened for fixing nitrogen exclusively with this trace element. Probably, the non-photosynthetic (heterotrophic) mutants (carbon auxotrophs) of a few cyanobacteria, isolated so far, may act as good materials for such a screening as, alike to the situation in *Azotobacter*, their nitrogen-fixing potential is dependent upon energy derived from the exogenous metabolizable carbon substrates. May be that a pleiotropy for the appearance of V-nitrogenase occurs in these mutants as a genetic boost, and it has got to be evaluated critically.

Iron-copper interactions, although established in vegetables and other higher plants, are yet to be studied in detail at the biochemical level in cyanobacteria. So far, it has been studied in cyanobacteria only with regard to nitrogen fixation, while Fe has a role in a number of physiological processes apart from its being the most limiting element, after Mo, for *in vitro* and *in vivo* nitrogenase activities.

In established natural and artificial associations, cyanobacteria can function as a source of at least some important substrates for the growth as well as for the biosynthesis of species-specific products by plant cells. Plant cells have been shown to utilize photosynthetic products of cyanobacteria. In all other cases, using phototrophic microorganisms, including numerous experiments with cyanobacteria, no objective for applying their photosynthetic potential in artificial associations has been set. The growth of plant cells can also be supported by the uptake of cyanobacterial cell lysates. In the presence of cyanobacteria, in a suitable synthetic nutrient medium, the biosynthetic activity of plant cells is preserved or even increased. Rough estimates made so far, have revealed the efficiency of such cyanobacterial systems in obtaining valuable natural products. All this is of much interest in terms of prospects of using the method of mix culturing, based on proper manipulation of macro- and micro-nutrients for raising plant-cyanobacterial co-cultures, in biotechnology.

ACKNOWLEDGEMENTS

The author is grateful to the Head, Department of Genetics and Plant Breeding, Institute of Agricultural Sciences, Banaras Hindu University for providing facilities for the work by author's team, a part of which is included in this article.

REFERENCES

Allen, M.B. and Arnon, D.I. 1955. Studies on nitrogen-fixing blue-green algae. I. Growth and nitrogen fixation by *Anabaena cylindrica Lemm. Pl. Physiol.*, **30** : 366-371.

Ashida, J. 1965. Adaptation of fungi to metal toxicants. *Ann. Rev. Phytopathol.*, **3** : 153-174.

Basu, S.N., Bose, R.G. and Bhattachharya, J.P. 1955. Some physiological studies on a copper tolernat *Penicillium* species. *J. Sci. Ind. Res.*, **14** : 46-53.

Becking, J.H. 1962. Species differences in molybdenum and vanadium requirements and combined nitrogen utilization by *Azotobacteriaceae. Pl. & Soil,* **16** : 171-201.

Benemann, J.R., McKenna, C.E., Lie, R.F., Taylor, T.G. and Kamen, M.D. 1972. The vanadium effect in nitrogen fixation by *Azotobacter. Biochim. Biophys. Acta,* **264**: 25-38.

Blundell, M.R. and Wild, D.G. 1969. Inhibition of bacterial growth by metal salts : A survey of effects on the synthesis of RNA and protein. *Biochem. J.,* **115** : 207-212.

Brill, W.J. 1980. Biochemical genetics of nitrogen fixation. *Microbiol. Rev.,* **44** : 449-467.

Brill, W.J., Steiner, A.L. and Shah, V.K. 1974. Effect of molybdenum starvation and tungsten on the synthesis of nitrogenase components in *Klebsiella pneumoniae. J.Bacteriol.*, **118** : 986-989.

Brown, J.C. and Ambler, J.C. 1973. 'Reductants' released by roots of Fe-deficient soybeans. *Agron. J.*, **65** : 311-314.

Brown, J.C. and Jones, W.E. 1977. Fitting plants nutritionally to soils. I. Soybeans. *Agron. J.*, **69** : 399-404.

Burns, R.C., Fuchsman, W.H. and Hardy, R.W.F. 1971. Nitrogenase from vanadium-grown *Azotobacter* : isolation, characteristics and mechanistic implications. *Biochem. Biophys. Res. Commun.*, **42** : 353- 358.

Bunt, J.S. 1961. Isolation of bacteria-free cultures from hormogone producing blue-green algae. *Nature (Lond.)*, **192** : 1278-1279.

Butenko, R.G., Gusev, M.V., Korzhenevskaya, T.G. and Yagodina, I.B. 1979. Cultivation of the tabacco cells and cyanobacterium association. In : *Proc. III All-Union Cong. Pl. Cell Culture* (Ed. Butenko, R.G.), Academia Nauk Armenian SSR, Abovyan, pp. 40-45.

Butenko, R.G., Gusev, M.V., Korzhenevskaya, T.G. and Lobakova, E.S. 1982. Growth of the associative culture of ginseng (*Panax ginseng*) cells and cyanobacterium (*Chlorogloea fritschii*) in the light and darkness. *Fiziol. Rast.*, **29** : 995-999.

Callis, G.E. and Wentworth, R.A.D. 1977. Tungsten vs. molybdenum in models for biological systems. *Bioorgan. Chem.*, **7** : 57-70.

Cardenas, J. and Mortenson, L.E. 1975. Role of molybdenum in dinitrogen fixation by *Clostridium pasteurianum*. *J. Bacteriol.*, **123** : 978-984.

Cardenas, J., Paneque, A., Vega, J.M., Herrera, J., Aparicio, P.J. and Losada, M. 1972. Molybdenum in cyanobacterial nitrogen fixation. *Pl. Cell Physiol.*, **13** : 175-178.

Chu, S.P. 1942. The influence of the mineral composition of the medium on the growth of planktonic algae. I. Methods and culture media. *J. Ecol.*, **30** : 284-287.

Eccleston, M. and Kelly, D.P. 1972. Assimilation and toxicity of exogenous amino acids in the methane-oxidizing bacterium *Methylococcus capsulatus*. *J. Gen. Microbiol.*, **71** : 541-554.

Enderlin, C.S. and Meeks, J.C. 1983. Pure cultures and reconstitution of the *Anthoceros-Nostoc* symbiotic association. *Planta*, **158** : 157-163.

Fogg, G.E., Stewart, W.D.P., Fay, P. and Walsby, A.E. 1973. *The blue-green Algae*, Academic Press, London.

Gadd, G.M. and Griffiths, A.J. 1978. Copper tolerance of *Aureobasidium pullulans*. *Microbios Lett.*, **6** : 117-124.

Gerloff, G.C., Fitzgerald, G.P. and Skoog, F. 1950. The isolation, purification and culture of blue-green algae. *Am. J. Bot.*, **37** : 216-218.

Gorham, P.R., McLachlan, J.S., Hammer, U.T. and Kim, W.K. 1964. Isolation and culture of toxic strains of *Anabaena flos-aquae* (Lyngb.) de Breb. *Verh. Int. Verein. Theor. Angew., Limnol.*, **15** : 796-800.

Gusev, M.V. and Korzhenevskaya, T.G. 1990. Artificial associations. In : *Handbook of Symbiotic Cyanobacteria* (Ed. A.N. Rai), CRC Press, Boca Raton, Florida, USA, pp. 173-230.

Gusev. M.V., Butenko, R.G., Korzhenevskaya, T.G., Lobacova, E.S. and Baulina, O.I. 1980. Intercellular symbiosis of suspension ginseng culture cells and cyanobacteria. *Eur. J. Cell Biol.*, **22** : 503-505.

Gusev, M.V., Butenko, R.G., Korzhenevskaya, T.G. and Yagodina, I.B. 1982. Associative culture of tobacco (*Nicotiana tabacum*) and cyanobacterium (*Anacystis nidulans*) cells. *Fiziol. Rast.*, **29** : 550-557.

Hewitt, E.J. 1974. Nitrogen fixation by blue-green algae. *MTP Int. Rev. Sci. Biochem. Ser.*, **1** : 199-245.

Ito, O. and Watanabe, I. 1983. The relationship between combined nitrogen uptakes and nitrogen fixation in *Azolla-Anabaena* symbiosis. *New Phytol.*, **95** : 647-654.

Jensen, H.L. and Spencer, D. 1947. The influence of molybdenum and vanadium on nitrogen fixation by *Clostridium butyricum* and related organisms. *Proc. Linn. Soc. N.S.W.*, **72** : 73-86.

Korzhenevskaya, T.G., Gorelova, O.A., Gusev, M.V. and Butenko, R.G. 1985. Cocultivation of higher plant cells and nitrogen-fixing cyanobacteria. *Fiziol. Rast.*, **32** : 88-93.

Kratz, W.A. and Myers, J. 1955. Nutrition and growth of several blue-green algae. *Am. J. Bot.*, **42** : 282-284.

Kula, R.J. 1966. Solution equilibriums and structures of molybdenum (VI) chelates. N-methyliminodiacetic acid. *Anal.Chem.*, **38** : 1382-1388.

Kula, R.J. 1967. Structural and bonding characteristics in the molybdenum (VI)-iminodiacetate system determined by infra red and proton nuclear magnetic resonance techniques. *Anal. Chem.*, **39** : 1171-1175.

Kumar, H.D. 1964. Adaptation of a blue-green alga to sodium selenate and chloramphenicol. *Pl. Cell Physiol.*, **5** : 465-472.

Lee, K.Y., Pan, S.S., Erickson, R. and Nason, A. 1974. Molybdenum biology of nitrogenase. *J. Biol. Chem.*, **249** : 3941-3952.

Lingle, J.C., Tiffin, L.O. and Brown, J.C. 1963. Iron uptake- translocation of soybeans as influenced by other cations. *Pl. Physiol.*, **38** : 71-76.

Meeks, J.C., Enderlin, C.S., Joseph, C.K., Chapman, J.S. and Lollar, M.W.L. 1985. Fixation of (13-N)N_2 and transfer of fixed nitrogen in the *Anthoceros-Nostoc* symbiotic association, *Planta*, **164** : 406-409.

Mishra, A.K., Vaishampayan, A. and Singh, V.P. 1991a. Isolation and physiological characterization of a herbicide-resistant mutant of a nitrogen-fixing cyanobacterium. In : *Physiological Strategies for Crop Improvement* (Eds. Tyagi, D.N., Bose, B., Hemantaranjan, A. & Devi, T.M.), Varanasi, India, pp. 63-68.

Mishra, A.K., Vaishampayan, A. and Singh, V.P. 1991b. A herbicide-resistant cyanobacterial mutant derepressed for N_2 fixation under NH^+_4 mediated growth condition. *Proc. Golden J. Symp. Genetic Res. & Educa. Current Trends & Next Fifty Years,* Ind. Soc. Genet. Pl. Breed., New Delhi, India, pp. 98-100.

Murashige, T. and Skoog, F. 1962. A revised medium for rapid growth and bioassay with tabacco tissue cultures. *Physiol. Plant.*, **15** : 473-479.

Nagatani, H.H. and Brill, W.J. 1974. Nitrogenase. V. The effect of Mo, W and V on the synthesis of nitrogenase components in *Azotobacter vinelandii. Biochim. Biophys. Acta.*, **362** : 160-166.

Nason, A., Lee, K.Y., Pan, S.S., Ketchum, P.A., Lamberti, A. and DeVries, J. 1971. *In vitro* formation of assimilatory reduced nicotinamide adenine dinucleotide phosphate : nitrate reductase from a *Neurospora* mutant and a component of molybdo-enzymes. *Proc. Nat. Acad. Sci. USA,* **68** : 3242-3246.

Neilson, A.H. and Doudoroff. M. 1973. Ammonia assimilation in blue-green algae. *Arch. Microbiol.,* **89** : 15-22.

Newton, J.W. and Herman, A.I. 1979. Isolation of cyanobacteria from aquatic fern *Azolla. Arch. Microbiol.*, **120** : 161-165.

Notton, B.A. and Hewitt, E.J. 1971. The role of tungesten in the inhibition of nitrate reductase activity in spinach (*Spinacea oleraceae*) leaves. *Biochem. Biophys. Res. Commun.*, **44** : 702-710.

Nriagu, J.O. 1978. The biochemistry of lead in the environment. Elsevier, Amsterdam.

Nriagu, J.O. 1979. The biochemistry of mercury in the environment. Elsvier, Amsterdam.

Peters, G.A. and Mayne, B.C. 1974. The *Azolla-Anabaena azollae* relationship. II. Localization of nitrogenase activity as assayed by acetylene reduction. *Pl. Physiol.,* **53** : 280.

Prasad, A.B., Samanta, R., Vishwakarma, M.L. and Vaishampayan, A. 1986. Biological effects of a mercury fungicide on a nitrogen-fixing blue-green alga *Nostoc muscorum* : Isolation and preliminary characterization of a Hg-resistant mutant. *New Phytol.*, **102** : 45- 50.

Prasad, A.B., Vaishampayan, A., Vishwakarma, M.L. and Shrivastava, D.K. 1991. Development of fungicide-resistance coupled with N_2-fixing ability during NH_4^+ mediated growth, as a mutational event, in a cyanobacterium *Nostoc muscorum. Biomed. Lett. (Camb.)* , **46** : 245- 251.

Provasoli, L., McLaughlin, J.J.A. and Droop, M.R. 1957. A cyanobacterial growth medium. *Arch. Microbiol.*, **25** : 392-396.

Rai, A.N. 1990. General methods. In : *Handbook of Symbiotic Cyanobacteria* (Ed. Rai, A.N.), CRC Press, Boca Raton, Florida, USA, pp. 231-239.

Rawlings, J., Shah, V.K., Chisnell, J.R., Brill, W.J., Zimmermann, R., Munck, E. and Orme-Johnson, W.H. 1978. Novel metal cluster in the iron- molybdenum co-factor of nitrogenase. *J. Biol. Chem.,* **253** : 1001-1004.

Reddy, Y.R. 1988. Physiological and mutational studies in *Azolla-Anabaena* nitrogen-fixing complex. M.Sc.(Ag.) Thesis in Genetics and Plant Breeding, Banaras Hindu University, India.

Rerabek, J. and Gorelova, O.A. 1986. The dependence of the growth and steroid accumulation on the type and amount of nitrogen compounds in the suspension culture of *Solanum laciniatum.* In : *Plant Cell Culture and Biotechnology* (Ed. Butenko, R.G.), Nauka, Moscow, pp. 76-97.

Rippka, R., Deruelles, J., Waterbury, J.B., Herdman, M. and Stanier, R.Y. 1979. Generic assignments, strain histories and properties of pure cultures of cyanobacteria *J. Gen. Microbiol* , **111** : 1-9.

Rodgers, G.A. 1974. the hepatic-cyanophyte symbiosis, Ph.D. thesis in Biological Sciences, University of Dundee, U.K.

Ross, I.S. 1975. Some effects of heavy metals on fungal cells. *Trans. Brit. Mycol. Soc.*, **64** : 175-193.

Roth, J.A., Wallihan, E.F. and Sharpless, R.G. 1971. Uptake by oats and soybeans of copper and nickel added to a peat soil. *Soil Sci.*, **112** : 338-342.

Sadler, W.R. and Trudinger, P.A. 1967. The inhibition of micro-organisms by heavy metals. *Mineralium Deposita.*, **2** : 158-168.

Sanyal, R., Giri, A.K.,Talukdar, G. and Sharma, A. 1980. Effects of cobalt and nickel on mammalian cellular systems. *Biol. Bull. India* ,**2** : 85-90.

Shah, V.K. and Brill, W.J. 1977. Isolation of an iron-molybdenum cofactor from nitrogenase (nitrogen fixation). *Proc. Nat. Acad. Sci. USA*, **74** : 3249-3253.

Shah, V.K., Chisnell, J.R. and Brill, W.J. 1978. Acetylene redution by the iron-molybdenum co-factor from nitrogenase. *Biochem. Biophys. Res. Commun.*, **81** : 232-236.

Shephard, D.A.E. 1975. Genetic hazards to man from environmental agents. *J. Canad. Med. Assoc.,* **112** : 460-470.

Shamina, Z.B., Frolova, L.V., Paukov, V.N. and Sovalova, L.A. 1979. The study of the poppy cell culture. *Proc. III All-Union Cong. Pl. Cell Culture* (Ed. R.G. Butenko), Akademia Nauk Armenian SSR, Abovyan, pp. 72-77.

Singh, H.N. and Kumar, H.D. 1972. A study of sulphur-selenium antagonism in the blue-green alga *Anabaena doliolum. Proc. Internat. Symp. Taxon. Biol. Blue-green Algae,* University of Madras, India, pp. 545-551.

Singh, H.N., Ladha, J.K. and Kumar, H.D. 1977a. Genetic control of heterocyst formation in the blue-green algae *Nostoc muscorum* and *Nostoc linckia. Arch. Microbiol.,* **114** : 155-159.

Singh, H.N., Sonie, K.C. and Singh, H.R. 1977b. Nitrate regulation of heterocyst differentiation and nitrogen fixation in a chlorate-resistant mutant of the blue-green alga *Nostoc muscorum. Mut. Res.*, **42** : 447-452.

Singh, H.N. and Singh, H.N.J. 1978. An azide-resistant mutant of the blue-green alga *Nostoc muscorum* producing heterocysts and nitrogenase in the presence of fixed inorganic nitrogen source. *Arch. Microbiol.*, **119** : 197-201.

Singh, H.N., Singh, H.R. and Vaishampayan, A. 1979. Toxic and mutagenic action of the herbicide alachlor (*lasso*) on various strains of a nitrogen-fixing blue-green alga *Nostoc muscorum* and characterization of herbicide-induced mutants resistant to methylamine and L-methionine-DL-sulfoximine. *Environ. Exp. Bot.*, **19** : 5-12.

Singh, H.N. and Vaishampayan, A. 1978. Biological effects of rice-field herbicide *machete* on various strains of the nitrogen-fixing blue-green alga *Nostoc muscorum. Environ. Exp. Bot.*, **18** : 87-94.

Singh, H.N., Vaishampayan, A. and Singh, R.K. 1978a. Evidence for the involvement of a genetic determinant controlling functional specificity of group VIB elements in the metabolism of N_2 and NO_3^- in a blue- green alga *Nostoc muscorum. Biochem. Biophys. Res. Commun.,* **81** : 67- 74.

Singh, H.N., Vaishampayan, A. and Sonie, K.C. 1978b. Mutation from molybdenum-dependent growth to tungsten-dependent growth and further evidence for a genetic determinant common to nitrogenase and nitrate reductase in a blue-green alga *Nostoc muscorum*. *Mut. Res.*, **50** : 427-432.

Singh, O.P. and Sharma, A. 1980. Effects of certain metallic pollutants on plant genetic systems. *Nucleus*, **23** : 15-29.

Smith, A.J. and Hoare, D.S. 1977. Specialist phototrophs, lithotrophs and methylotrophs : a unity among a diversity of prokaryotes. *Bacteriol. Rev.*, **41** : 419-448.

Starkey, R.L. 1973. Effect of pH on toxicity of copper to *Scytalidium* species, a copper tolerant fungus, and some other fungi. *J. Gen. Microbiol.*,**78** : 217-225.

Starkey, R.L. and Wakesman, S.A. 1943. Fungi tolerant to extreme acidity and high concentration of copper sulphate. *J. Bacteriol.*, **45** : 505-519.

Stein, J.R. 1958. A morphologic and genetic study of *Gonium pectorales*. *Am. J. Bot.*, **45** : 665-667.

Vaishampayan, A. 1981. Nutritional screening of amino acids in an auxotroph mutant of the cyanobacterium *Nostoc muscorum*. *Microbios Lett.*, **18** : 111-116.

Vaishampayan, A. 1982a. A methylamine-resistant mutant of the blue-green alga, *Nostoc muscorum* : possible involvement of glutamine synthetase in methylamine metabolism. *New Phytol.*, **91** : 607-613.

Vaishampayan, A. 1982b. Cu-Fe interactions in the N_2-fixing cyanobacterium *Nostoc muscorum*. *Microbios Lett.*, **21** : 17-23.

Vaishampayan, A. 1982c. Amino acid nutrition in the blue-green alga *Nostoc muscorum*. *New Phytol.*, **90** : 545-549.

Vaishampayan, A. 1983a. UV-induced wide-range mutations in a cyano-bacterium *Nostoc muscorum*. *Proc. VII I nternat. Cong. Radiation Res.*, Rijswijk, The Netherlands **B** : 4-41.

Vaishampayan, A. 1983b. Mo-V interactions during N_2 and NO_3^- metabolism in a N_2 fixing blue-green alga *Nostoc muscorum*. *Experientia*, **39** : 358-360.

Vaishampayan, A. 1983c. Vanadium as a trace element in the blue-green alga, *Nostoc muscorum* : Influence on nitrogenase and nitrate reductase. *New Phytol.*, **95** : 55-60.

Vaishampayan, A. 1984a. Biological effects of the rice-field herbicide monuron on a N_2-fixing cyanobacterium *Nostoc muscorum*. *Microbios Lett.*, **28** : 105-111.

Vaishampayan, A. 1984b. Collection of a carbon-requiring strain of a nitrogen-fixing cyanobacterium. *Proc. V Internat. Cong. Culture Collections*, Bangkok (Thailand), pp. 63-72.

Vaishampayan, A. 1984c. Studies on diuron uptake in a blue-green alga *Nostoc muscorum*. *J. Exp. Bot. (Oxford)*, **35** : 897-904.

Vaishampayan, A. 1984d. Strong mutagenicity of a bipyridylium herbicide in a N_2-fixing cyanobacterium *Nostoc muscorum*. *Experientia*, **40** : 1016-1019.

Vaishampayan, A. 1984e. Powerful mutagenicity of a bipyridylium herbicide in a nitrogen-fixing blue-green alga *Nostoc muscroum. Mut. Res.*, **138** : 39-46.

Vaishampayan, A. 1984f. Mutagenicity of a bypyridylium herbicide in a N_2-fixing cyanobacterium *Nostoc muscorum. Environ. Internat.*, **10** : 18-27.

Vaishampayan, A. 1984g. Biological effects of a herbicide in a nitrogen-fixing cyanobacterium (blue-green alga) : an attempt for introducing herbicide-resistance. *New Phytol.*, **96** : 7-11.

Vaishampayan, A. 1984h. Genetic manipulation and recombination leading to pesticide-resistance in a N_2-fixing cyanobacterium. *Proc. III Europ. Cong. Biotechnol.*, Munich (Germany) **3** : 353-358.

Vaishampayan, A. 1984i. Copper-iron interactions in a N_2-fixing cyanobacterium *Nostoc muscorum. J. Pl. Nut.*, **7** : 567-573.

Vaishampayan, A 1984j. Screening of amino acids for carbon and nitrogen enrichments in different strains of a cyanobacterium *Nostoc muscorum. Biochem. Physiol. P flanzen*, **179** : 411- 417.

Vaishampayan, A. 1985a. Potent mutagenicity of a bipyridylium herbicide in a nitrogen-fixing blue-green alga, *Nostoc muscorum. Biochem. Physiol. Pflanzen,* **180** : 327-331.

Vaishampayan, A. 1985b. Mutagenicity of bipyridylium salts in a N_2- fixing cyanobacterium. *Microbios,* **43** : 53-65.

Vaishampayan, A. 1985c. Mutagenic activity of alachlor, butachlor and carbaryl to a N_2-fixing cyanobacterium *Nostoc muscorum. J. Agric. Sci. (Camb.),* **104** : 517-526.

Vaishampayan, A. 1985d. Growth of *Nostoc muscorum* mutants with diuron (DCMU) and L-methionine-DL-sulfoximine. *Experientia,* **41** : 137-139.

Vaishampayan, A. 1988. Collection and reconstruction of a non- heterocystous filamentous cyanobacterial culture from a sugar factory effluent, fixing nitrogen with a remarkably higher rate under a simple microaerophilic condition. *Proc. VI Internat. Cong. Culture Collections : Microbial Gene Pool,* University of Maryland, USA, pp. 71-85.

Vaishampayan, A. 1994a. Recent advances in the molecular biology of *Azolla-Anabaena* symbiotic N_2-fixing complex and its use in agriculture. In : *Microbes and Environment* (Eds. A.B. Prasad & R.S. Bilgrami), Narendra Publishing Co., India, pp. 121-144.

Vaishampayan, A. 1994b. Molecular genetics of nitrogen fixation in cyanobacteria. In : *Biology and application of nitrogen-fixing organisms : Problems and prospects* (Eds. A.B. Prasad & A. Vaishampayan), Sci. Pub., India, pp. 1-18.

Vaishampayan, A. and Hemantaranjan, A. 1984. Physiological significance of vanadium uptake during N_2 and NO_3- metabolism in various strains of a N_2-fixing cyanobacterium *Nostoc muscorum. Pl. Cell Physiol.*, **25** : 845-850.

Vaishampayan, A. and Mishra, A.K. 1990. Cyanobacteria : A gene bank for the pesticide-resistance markers. *Proc. III World Cong. Environ. Hazards Pesticides,* Cairo (Egypt), pp. 1-51.

Vaishampayan, A. and Prasad, A.B. 1981. A pesticide-resistant mutants of the N_2-fixing blue-green alga *Nostoc muscorum. Experientia*, **37** : 1285-1286.

Vaishampayan, A. and Prasad, A.B. 1982. Blitox-resistant mutant of the N_2-fixing blue-green algae *Nostoc linckia* and *Nostoc muscorum. Environ. Exp. Bot.*, **22** : 427-435.

Vaishampayan, A. and Prasad, A.B. 1984. Inter-strain transfer of a pesticide-resistant marker in the N_2-fixing cyanobacterium *Nostoc moscorum. Molec. Gen. Genet.*, **193** : 195-197.

Vaishampayan, A. and Singh, H.N. 1981a. Mutation in the blue-green alga *Nostoc muscoum*. I. Methylamine-resistant mutants. *Biochem. Physiol. Pflanzen*, **176** : 625-630.

Vaishampayan, A. and Singh, H.N. 1981b. Mutation in the blue-green alga *Nostoc muscorum*. II. L-methionine-DL-sulfoximine- resistant mutants. *Biochem. Physiol. Pflanzen*, **176** : 631-637.

Vaishampayan, A., Singh, H.R. and Singh, H.N. 1978. Biological effects of rice-field herbicide *stam f-34* on various strains of the nitrogen-fixing blue-green alga *Nostoc muscorum. Biochem. Physiol. Pflanzen*, **173** : 410-419.

Vaishampayan, A., Mishra, A.K. and Singh, V.P. 1992a. Uptake of monuron in a cyanobacterial mutant. *Biomed. Lett.(Cambridge)*, **47** : 39-46.

Vaishamapayan, A., Reddy, Y.R., Singh, B.D. and Singh, R.M. 1992b. Reduced phosphorus requirement of a mutant *Azolla-Anabaena* symbiotic nitrogen-fixing complex. *J. Exp. Bot. (Oxford)*, **43** : 851-856.

Verma, S.K., Singh, A.K., Katiyar, S. and Singh, H.N. 1990. Genetic transformation of glutamine auxotrophy to prototrophy in the cyanobacterium *Nostoc muscorum. Arch. Microbiol.*, **154** : 414-416.

Wallace, A. and DeKock, P.C. 1966. Translocation of iron in tabacco, sunflower and bush bean plants. In : *Current topics in Plants Nutrition* (Ed. A. Wallace), Edward Bros. Publ., pp. 3-9.

Advancements In Micronutrient Research
Editor : A. Hemantaranjan
1996, Scientific Publishers, Jodhpur, pp. 127–150.

6

LEGUME/RHIZOBIAL INTERACTIONS WITHIN THE LEGUME ROOT NODULE : ROLE OF THE PLASTID IN METABOLIC REACTIONS INVOLVING IRON NUTRITION

JAMES A. GUIKEMA, WILLIAM R. ODOM, AND PETER P. WONG
Division of Biology
Kansas State University
231 Ackert Hall
Manhattan, KS 66506-4901
U.S.A.

CONTENTS

1. INTRODUCTION

Iron plays a critical role in the metabolic reactions of plant and microbial systems (Terry and Abadia, 1986; Abadia and Abadia, 1993; Johnson and Barton, 1993; Abadia, 1992). It is an essential feature in many oxidation/reduction components, including cytochromes, iron sulfur proteins, and peptides which function either as monitors of (Morrett *et al.*, 1991; Lois *et al.*, 1993) or carriers for (Appleby, 1984; Bergersen *et al.*, 1973; Dilworth, 1980) available oxygen. As such, iron is particularly important in the processes of photosynthesis (Terry and Abadia, 1986; Abadia, 1992), respiration, and nitrogen fixation (Hirsch, 1992; Tang *et al.*, 1992; Guerinot, 1993). This importance is magnified by the limited availability of iron in aerobic environments. Even though iron is the fourth most abundant element on earth, in aerobic environments Fe(II) is readily oxidized to Fe(III) and precipitates as ferric hydroxide (Biedermann and Schindler, 1957; Lankford, 1973). It is apparent, therefore, that the acquisition of this scarce resource, and its partitioning among the various iron-requiring processes, will play a determining role in the success of interactions between species.

The cellular mechanisms which guide the interactions between species are difficult to define by conclusive experimentation. This is especially true in both pathogenic and symbiotic relationships, where extensive chemical signalling occurs between partners. It has become increasingly apparent that iron partitioning in these systems is one controlling feature (Expert and Gill, 1991). An excellent example of such can be found in the interaction of mammals with microbial pathogens (Emery, 1982; Gadd, 1992; Gong *et al.*, 1993), where the ability of the microbe to acquire iron within its mammalian host is thought to be a factor in virulence. Similar examples are evident in plant/microbial interactions. For example, systemic virulence in *Erwinia chrysanthemi* also requires an efficient iron acquisition mechanism (Enard *et al.*, 1988), brought about by the production of siderophores (Expert *et al.*, 1992) and the biosynthesis of siderophore recognition proteins on the bacterial outer membrane (Neilands, 1982; Page, 1993).

An equally intriguing example of plant/microbial interactions is the symbiotic interaction between legume hosts and the rhizobia in the legume root nodule (Expert and Gill, 1991). Owing to its activity in nitrogen fixation, the bacterial partner has a high iron demand for electron transport components and nitrogenase. This involves the development of the bacteroid state (McDermott *et al.*, 1987) within a microaerophilic environment (cf. (Lois *et al.*, 1993; Langford and Moxon, 1993)), and attending genetic controls (Lois *et al.*, 1993; Gabel and Maier, 1993). The plant partner also has a high respiratory activity, necessitating the biosynthesis of mitochondrial iron-containing electron transfer proteins. Furthermore, root nodule cells have leghemoglobin as nearly 15% of the total soluble protein. Not only does this impose a strong demand for iron, the high leghemoglobin content implies that the root nodule is second only to the developing leaf for rates of tetrapyrrole biosynthesis.

There are ample studies which document the linkage between iron nutrition and the ability of the root nodule symbiosis to provide fixed nitrogen to the plant (Guerinot, 1993). In general, during periods of low iron availability, fewer root nodules are formed. This has been shown in a number of legumes, including lupine (Tang *et al.*, 1990a; 1990b), french bean (Hemantaranjan and Garg, 1986; Hemantaranjan, 1988), lentil (Rai *et al.*, 1984), chickpea (Rai *et al.*, 1982), pigeonpea (Herridge and Holland, 1993) and peanut (O'Hara *et al.*, 1988; Hartzook, 1982). The mechanism of this linkage, and the signal transduction mechanism by which iron unavailability causes a loss in nodule mass, are not at all clear. Split root experiments of Tang *et al.* (1990a) demonstrated that iron limitation of nodule formation was independent of host plant growth. Nodule initials and nodules were only found on the half of the root system which was supplied with iron. Iron was not translocated from that half to the other half of the root which received a low iron concentration, even though, in peanut, foliar-applied iron is rapidly translocated to the root (O'Hara *et al.*, 1988). Furthermore, there was a window of developmental sensitivity (Tang and Dilworth, 1991), in that lupine nodulation is most sensitive to iron deficiency about 4 days after inoculation and just before nodule initials were observed.

That the root nodule is a strong sink for plant iron is also seen in the stress responses of other tissues. For example, lupine which rely on root nodules for fixed nitrogen require more iron than do lupine which are grown in the presence of a fixed nitrogen source (Tang *et al.*, 1990b). Work by Terry and colleagues (Terry *et al.*, 1991) demonstrates that nodulated soybeans show a stronger iron deficiency response in leaf tissue than to non-nodulated plants. That the root nodule is a strong sink for iron is easily understood from the roles which iron-containing proteins play in symbiotic nitrogen fixation.

In this review, we examine the intricacies of iron partitioning in the legume root nodule/*Rhizobium* symbiosis. We will focus on two specific aspects of this problem, since they imply a role for the plastids in root nodule cells in the iron partitioning mechanisms. First, we will examine the roles which iron acquisition and storage may play in this interaction. In plants, ferritin storage is the provence of the plastid. Second, we will focus on the biosynthesis of the tetrapyrrole component of leghemoglobin. Experiments during the past decade suggest that one fascinating difference between plant and mammalian metabolic pathways is the source of carbon in the pyrrole ring of cytochromes and chlorophylls.

2. Iron Acquisition and Storage

2.1. Compartmentation within the Legume Root Nodule

The partitioning of iron within the legume root nodule implies a rather fastidious subcellular compartmentation. Subcellular compartments within root nodules, in the context of developmental biology, have been the focus of recent reviews by Sanchez *et al.*, (1991), Hirsch (1992) and Mellor (1989). In brief, there are two types of nodules: determinate and indeterminate. These are shown diagrammatically in Fig. 1. They differ from each other in several ways. The indeterminant type has a persistent meristem, and therefore an increase in size is accompanied by an increase in cell number. Determinant nodules lack such a meristem. Cell division ceases early in nodule development, and an increase in size is due only to cell expansion. These two types of nodules offer a useful contrast to the developmental expression of mRNA's and proteins important to nodule activity. mRNA's for leghemoglobin, for example, and of interest owing to our focus on iron partitioning, are first observed in the pre-fixing portion of zone II. Interesting developmental changes are also occuring with plastid structure/function in this zone. Within the zone II-III interface, plastids accumulate carbohydrate to become amyloplasts (de Billy *et al.,* 1991). This suggests a control on transport across the plastid envelope. The nitrogen fixing regions of both nodule types contain a mixture of infected and uninfected plant cells. Similar developmental changes are evident in determinate nodules as well. However, these are observed during the temporal development of the nodule and are not observed as developmental zones shown for indeterminate nodules in Fig.1.

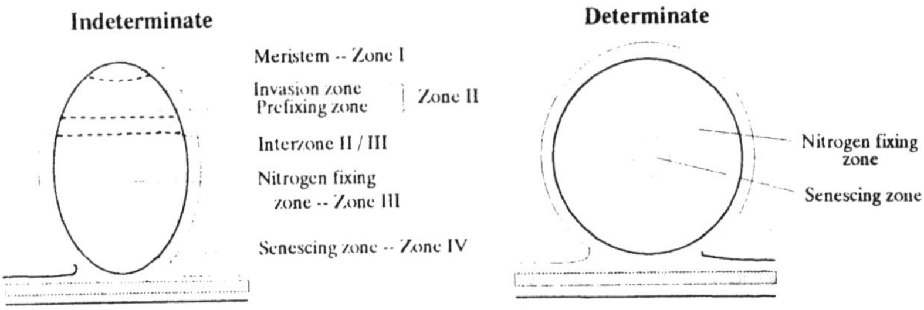

Fig. 1. Structural models of indeterminate and determinate legume root nodules.

Within infected cells, rhizobia, which have developed into a nonterminal bacteroid form (McDermott *et al.*, 1987), are enclosed with a peribacteroid vesicle. This vesicle has functional similarities to lytic compartments within a plant cell (Mellor, 1989), with the membrane of this vesicle largely provided by a process of membrane flow from the golgi (Robertson *et al.*, 1978a). This membrane has a high protein content (Robertson *et al.*, 1978b) and is particularly rich in glycoprotein (Mellor, 1989). Exchanges between the plant host cytoplasm and the bacteroid are regulated by the peribacteroid membrane. A carrier for malate, which also transports succinate, has been described (Udvardi *et al.*, 1988). The ability of exogenous δ-aminolevulinic acid to rescue some but not all *hem A* mutants suggests a species diversity in peribacteroid membrane transport mechanisms (Guerinot and Chelm, 1986; Leong *et al.*, 1982). The suggestions that heme for leghemoglobin biosynthesis is a bacterial contribution to the symbiosis (c.f.(Guerinot, 1993; Guerinot and Chelm, 1986)) is especially interesting, since it implies an interesting and complicated transport system across the bacterial membrane and the peribacteroid vesicle.

2.2. Iron Storage

In higher plants, iron storage is the provence of the plastid, with the iron storage protein, ferritin, as a plastid-targeted polypeptide having significant identity with its mammalian counterpart (Theil and Hase, 1993; Laulhere *et al.*, 1992; Lobreaux and Briat, 1991). The 20 - 28 kD ferritin monomers assemble in the plastid as a 24-mer, forming a hollow protein sphere (Crichton *et al.*, 1978; Lobreaux *et al.*, 1992). Within the hollow core of this sphere, up to 4,500 iron atoms can be sequestered, either as crystalline mineral ferrihydrite or as amorphous hydrous ferric oxyphosphate (Theil and Hase, 1993). The presence of phosphate suggests a storage of that nutrient as well (Fe:P = about 8:1 in phytoferritin and about 1:0.8 in bacteroferritin (Theil and Hase, 1993)). Although plant and animal ferritin are immunologically distinct, they share a conserved predicted three-dimensional structure (Lobreaux *et al.*, 1992), with a 4-helix bundle and a seven residue putative metal binding/ferroxidase site in the center.

Plants and animals also differ in the mechanisms of ferritin developmental regulation. In mammalian systems, ferritin [and likely several enzymes of the heme biosynthetic pathway (Melefors *et al.*, 1993)] is regulated at the level of translation by intracellular iron concentrations (Emery-Goodman *et al.*, 1993; Melefors and Hentze, 1993). A 90 kD proteinaceous *iron regulatory factor* is responsible for this regulation. During times of iron-sufficiency, *iron regulatory factor* is thought to be an iron-sulfur protein (Emery-Goodman *et al.*, 1993), having aconitase activity and no affinity for RNA-binding. In the absence of iron, however, this polypeptide can bind with an *iron regulatory element* on the ferritin mRNA, both prohibiting the translation of the transcript and protecting it from degradation. It is possible, therefore, that the Km for iron assimilation into the *iron regulatory factor* Fe/S cluster may play a critical role in this biosynthetic control. Furthermore, recent evidence suggests that an unbound iron regulatory element is a positive regulator, enhancing translational rates (Dix *et al.*, 1992).

In plant systems, however, the regulation of ferritin levels is not as well understood. Certainly, iron supply is a major factor regulating the synthesis of the peptide (Seckback, 1987), at least in part by regulation at the level of transcription (Van der Mark *et al.*, 1983; Leascure *et al.*, 1991), since the transition from iron-deficiency to iron-abundance is generally accompanied by *de novo* ferritin mRNA and protein biosynthesis. However, the controls on ferritin biosynthesis and stability during periods of iron removal from the storage molecule are less clear. Ferritin concentrations decrease during leaf development in maize (Theil and Hase, 1993), consistent with the partitioning of iron from ferritin into the iron-containing cofactors of the developing photosynthetic apparatus. However, concentrations of the ferritin protein are not well linked with decreases in the ferritin mRNA levels (Theil and Hase, 1993). It may be the case, therefore, that protein stability is regulated by iron trafficking to and from the protein sphere, indicating a role for iron trafficking in protein turnover and accumulation. A similar argument can be made for ferritin concentrations in the developing root nodule (Lobreaux and Briat, 1991; Ragland and Theil, 1993).

The regulation of iron trafficking to and from the plastid ferritin storage pool is not understood. Iron release is certainly dependent upon pH and the prevalent oxidation/reduction conditions (Laulhere and Briat, 1993). This may rely in part on a stacked aromatic amino acid network to the ferritin core, which may provide an electron transport network to the protein core (Lobreaux *et al.*, 1992). Reducing equivalents would be especially available during the onset of photosynthesis in greening leaf tissue, which may explain the low ferritin levels found in leaves under normal conditions (Lobreaux and Briat, 1991; Lobreaux *et al.*, 1992).

Ragland and Theil, (1993) recently followed the ferritin and iron concentrations during the development of soybean (determinate) nodules. Ferritin mRNA levels were observable in non-inoculated root tissues, and steadily increased during root nodule development to a peak value at about 21 d. Ferritin protein concentrations, however, peaked at 12 d and declined thereafter during periods of leghemoglobin and nitrogenase biosynthesis. The microaerophilic environment of the nodule during nitrogen fixation is also conducive to a more negative oxidation/reduction potential, and the increased reducing power may facilitate iron-unloading from the nodule plastid in a manner similar to that suggested for greening leaves (Lobreaux and Briat, 1991). The inverse relationship between ferritin accumulation and iron utilization, coupled with the observations of ferritin accumulation when such environmental conditions as nematode infection prohibit iron use (Ko *et al.*, 1985), reinforce the notion that cellular levels of iron regulate accumulation of the storage apparatus.

Concomitant iron storage by the bacterial partner in the legume symbiosis, compartmentalized within the peribacteroid vesicle, is not as well understood. Bacterial ferritin is a distant cousin of the plant form; but unlike phytoferritin, the bacterial form is a *b-type* heme containing protein (Theil and Hase, 1993). Furthermore, there is little sequence similarity between them, such that sophisticated analyses were required to document similarities, particularly in the ferroxidase residues (Andrews *et al.*, 1991). It is certain, however, that bacterioferritin is upregulated at the level of transcription by the presence of iron. Whether such a class of proteins is expressed in bacteroids during symbiotic nitrogen fixation, or if levels of iron within the peribacteroid vesicle warrant such expression, remains to be determined.

2.3. Role of the Bacteroid in Iron Partitioning

Since the acquisition mechanisms of the host plant (Takagi, 1993; Welkie and Miller, 1993) are responsible for providing iron to the nodule, it is of interest to examine the potential role that the bacterial partner may play in iron partitioning. Free living rhizobia, under the appropriate culture conditions, are known to excrete low molecular weight ferric iron binding molecules called siderophores. In this, they are similar to other microbial systems (Silver and Walderhaug, 1992), from *E. coli* to cyanobacteria (Armstrong and Van Baalen, 1979; McKnight and Morel, 1980). The best studied example of the rhizobial siderophores is a novel molecule called rhizobactin (Smith and Neilands 1984), containing an ethylenediamine nucleus (Neilands, 1983) and observed in a number of *Rhizobium meliloti* isolates. Other examples include an hydroxamate-type siderophore from *R. leguminosarum* (Miller *et al.*, 1984), a catechol-like molecule from the cowpea *Rhizobium* (Modi *et al.*, 1985), a phenolate compound from *R. leguminosarum* (Patel *et al.*, 1988), and citrate as a chelator excreted by *Bradyrhizobium japonicum* (Guerinot *et al.*, 1990). Interestingly, in a number of strains so far examined, there is a correlation of siderophore production with new outer membrane protein production during the synthesis of the iron acquisition apparatus (Reight and O'Connel, 1993), suggesting the assembly of a siderophore receptor system as well.

The genetic control of microbial siderophore production is maintained at the transcriptional level, mediated by the availability of iron (Neilands, 1993). During periods of iron availability, metal binding to the product of the *fur* gene promotes its binding to 'iron box' regions of DNA as a repressor of transcription. Iron deprivation causes the loss of iron/protein interaction, the subsequent removal of the *Fur* protein from promotor regions derepresses the iron acquisition genes, and causes the upregulation of iron acquisition activity. Indications to date (Neilands, 1993; Gill and Neilands, 1989) suggest that the regulation of iron acquisition genes in the rhizobia is very similar to that observed in the more extensively studied model systems.

Is this iron acquisition system important in the trafficking of iron from storage within the plastid to use in the biosynthesis of bacteroid iron-proteins? Fascinating evidence during the past several years suggest that a high affinity iron acquisition system is operative in bacteroids (Nadler *et al.*, 1990), and may be required for bacteroid development (Gill *et al.*, 1991). There is a correlation between the presence of a functioning iron acquisition system *ex planta* (as modified by Tn5 mutagenesis) and the *in planta* nitrogen fixation activity of *Rhizobium meliloti* (Nadler *et al.*, 1990). This phenomenon was not attributable to losses in ferric reductase activity, or to a reduced number of bacteroids per nodule (Barton *et al.*, 1992). Rather, rates of nitrogenase activity per bacteroid were diminished (Barton *et al.*, 1992), suggesting an iron-limited decreased biosynthesis of a stable nitrogen-fixation apparatus. Interestingly, in cyanobacteria, nitrogen fixation and siderophore production are thought to be mutually exclusive strategies (Godfrey and Dilworth, 1971). If this were also the case in the rhizobia, it would suggest a developmental or temporal profile, in which iron acquisition would precede the differentiation into the bacteroid form. Changes in outer membrane composition during bacteroid differentiation have been documented at the electron microscopic level by immunogold labelling (Goosen-de Roo *et al.*, 1993), and it would be of interest to examine similar nodules for an iron-deficiency outer membrane protein epitope (Reigh and O'Connell, 1993). In indeterminate nodules, this would correspond to the prefixing zone II, in which plastids have accumulated large amounts of starch and leghemoglobin biosynthesis has begun (de Billy *et al.*, 1991), and suggests that a bacterial chelator in part aids the unloading of iron from ferritin-bound stores.

3. LEGHEMOGLOBIN BIOSYNTHESIS

The microaerophilic environment of the legume root nodule requires special considerations in terms of mitochondrial and bacteroid respiration (Rawsthorne and La Rue, 1986), and of oxygen diffusion to the nodule core (Bergersen *et al.*, 1973). The accumulation of leghemoglobin during development, immediately prior to the onset of nitrogenase biosynthesis, serves both as an oxygen buffer and as a facilitated diffusion mechanism to afford a uniform, albeit low, oxygen concentration throughout the nodule (Appleby, 1984; Bergersen *et al.*, 1973; Dilworth, 1980). Both uninfected and infected nodule cells contain leghemoglobin. The former, however, have only about 25% of the leghemoglobin found in infected cells. In mature nodules grown in soil, leghemoglobin concentrations can reach as high as 15-20% of the total soluble protein (Dilworth, 1980), although this is somewhat under the influence of environmental conditions. Nodules from hydroponically-grown plants, for example seem to have a lower concentration than those grown in soil (Klucas *et al.*, 1985). Nevertheless, the high leghemoglobin concentration implies a high requirement for both the iron and the tetrapyrrole ring of the heme, and makes the developing root nodule nearly equal to the greening leaf for rates of tetrapyrrole biosynthesis.

The leghemoglobin apoproteins are products of a family of genes found in the nucleus of legumes. Therefore, the globin is contributed by the host, with transcriptional regulation which turns the gene on during the appropriate time during nodule development (Hirsch, 1992; Sanchez *et al.*, 1991). The promoters responsible for this transcriptional upregulation are not understood, and this is the focus of interest from a number of laboratories. It is known, however, that globin synthesis and heme availability are coupled (May *et al.*, 1987) and that heme may regulate leghemoglobin expression at a postranscriptional level (Jensen *et al.*, 1986), such that neither free globin nor free heme are observed in nodules.

3.1. Role of the Plastid in Tetrapyrrole Biosynthesis

One of the most interesting differences between plant and mammalian metabolic pathways to emerge during the past decade is the initial phase of tetrapyrrole biosynthesis. In mammals and in many bacteria (including the rhizobia (Guerinot, 1993; Avissar and Beale, 1989; Frustaci *et al.*, 1991)), the first committed step in heme biosynthesis is the condensation of glycine and succinyl-CoA, forming δ-aminolevulinic acid (ALA), by the enzyme-aminolevulinic acid synthase.

In plant systems, δ-aminolevulinic acid biosynthesis occurs in the plastid (Beale, 1990; Kannangara and Schouboe, 1985), mediated by enzymes found in the stroma (Gough and Kannangara, 1977), with glutamate in place of glycine/succinyl-CoA as the starting material. Conversion of glutamate to δ-aminolevulinic acid is a three-step process, involving first an ATP-dependent formation of a glutamyl-tRNA (Avissar and Beale, 1988; Schon *et al.*, 1986), an NADPH-dependent reduction forming glutamate-1-semialdehyde (Kannangara and Schouboe, 1985; Weinsein *et al.*, 1986), and followed by a pyridoxyl phosphate-dependent transamination (Weinsein *et al.*, 1986) of the semialdehyde to δ-aminolevulinic acid. This pathway, one of the few having a tRNA-dependent step in the absence of protein synthesis (Avissar and Beale, 1988; Schon *et al.*, 1986), is responsible for all chlorophyll biosynthesis in higher plants. Evidence suggests that mitochondrial heme has its origins in this pathway as well (Schneegurt and Beale, 1986), with the only exception so far observed being the phytoflagellate, *Euglena*, which maintains both pathways (Corriveau and Beale, 1986a; 1986b). Smith (1987), in a cell fractionation study of etiolated pea leaves, concluded that δ-aminolevulinic acid dehydratase and porphobilinogen deaminase were not localized in the mitochondrion. Rather, they were partitioned between cytosolic and etioplast fractions. If these findings are correct, they imply a coordinated assembly model for plant mitochondrial and cytosolic heme. The responsibility for heme biosynthesis is shared between the mitochondrion, cytoplasm, and plastid.

Owing to this recently observed role of the plastid in the synthesis of tetrapyrrole precursors, the source of heme for root nodule leghemoglobin is of great interest. This heme has long been thought to be a rhizobial contribution to the nodule symbiosis (c.f. (Guerinot, 1993; Leong *et al.*, 1982; May *et al.*, 1987; O'Brian and Maier, 1989; Nadler and Avissar, 1977)). The origins of this concept lie in two interrelated observations. First, early studies (Godfrey and Dilworth, 1971; Cutting and Schulman, 1969) showed that the bacteroid fractions from root nodules were very active in heme biosynthesis. The bacteroid tissue maintained high rates of δ-aminolevulinic acid synthase, δ-aminolevulinic acid dehydratase, and porphobilinogen deaminase. Furthermore, culture conditions which mimic those within the peribacteroid vesicle-- such as low oxygen tensions and high iron -- induced high rates of both δ-aminolevulinic acid synthase and δ-aminolevulinic acid dehydratase in free-living *Rhizobium japonicum* (Nadler and Avissar, 1977; Gollop *et al.*, 1984).

Second, these early experiments could find no δ-aminolevulinic acid synthase activity associated with the plant fractions. Given our current understanding of δ-aminolevulinic acid synthesis in plants, however, this finding is not surprising. At the time, however, these results were viewed as indicating that the plant fractions were incapable of contributing the δ-aminolevulinic acid required for leghemoglobin heme. Recent studies have documented high rates of δ-aminolevulinic acid biosynthesis (with labeled glutamate as the carbon source) from plant-derived legume root nodule cytoplasmic fractions, which suggests that the plastid-contained pathway is active in these tissues (Sangwan and O' Brian, 1991).

3.2. Studies with Mutant Rhizobial Strains

During the past few years, a number of approaches have been used to examine the mechanism of δ-aminolevulinic acid biosynthesis in the rhizobia, and its implications on rhizobial respiratory activity and leghemoglobin formation. The rhizobial gene for δ-aminolevulinic acid synthase, *hemA*, has been a target of several such studies. For example, *hemA*-less mutants of *Azorhizobium caulinodans* incite nodules of *Sesbania rostrata* which are lacking leghemoglobin mRNA transcripts (Pawlowski *et al.*, 1988). Similarly, *hemA*-less mutants of *R. meliloti* form fix-minus nodules on alfalfa (Leong *et al.*, 1982). Stanley and coworkers (Stanley *et al.*, 1988) found that a broad host *Rhizobium* NGR234 formed ineffective nodules on five genera of tropical legumes, independent of whether the nodules were determinate or indeterminate. In all of these cases, leghemoglobin expression was minimal, reinforcing the suggestion that the heme-ligand for this protein is bacterially derived.

These results contrast with studies initiated by Guerinot and Chelm

(1986) using *Bradyrhizobium japonicum*. A *hemA*-less mutant formed fully effective nodules on soybean, suggesting that bacterial δ-aminolevulinic acid is not the source for leghemoglobin heme and indicating that plant derived δ-aminolevulinic acid could rescue the mutant *in planta* (Sangwan and O'-Brian, 1991). In some rather controversial (Bott *et al.,* 1992) observations, O'Brian and colleagues (Frustaci *et al.,* 1991) showed that this mutant (if provided yeast extract) was capable of nearly-normal growth rates, with intracellular heme levels approaching zero. O'Brian's group has recently isolated a *hemH*-less mutant of *Bradyrhizobium japonicum*, defective in ferrochelase which is the last step in heme biosynthesis (Frustaci and O'Brian, 1992). Interestingly, unlike the *hemA* mutant, the *hemH* mutant could not be rescued *in planta*. These nodules do not fix nitrogen, contain very few bacteria, and do not express leghemoglobin. Apparently, free heme is not available to this mutant from the plant, a problem of heme synthesis, of transport across the peribacteroid membrane, or of transport into the bacteroid. Furthermore, this mutant does not provide the correct signalling to cause the onset of leghemoglobin biosynthesis. It is uncertain, however, that heme for leghemoglobin biosynthesis is such a bacterially-derived signal.

3.3. Studies on Plant δ-aminolevulinic Acid Synthesis in Root Nodules

The *hemA* mutation has been extremely useful in studies of the rhizobium-legume symbiosis. No such genetic system is available to examine the plant partner in the symbiosis. Fortunately, a series of powerful inhibitors are available which block the pathway from glutamate to δ-aminolevulinic acid and, therefore, inhibit the plant, but not the rhizobial, capacity to make δ-aminolevulinic acid.

Gabaculine (3-amino-2,3-dihydrobenzoic acid), shown by Flint (1984) to be a potent inhibitor of chlorophyll biosynthesis, was first isolated from the culture filtrates of *Streptomyces toyocaensis* subsp. 1039 (Kobayashi *et al.*, 1976). This compound (and a class of molecules developed subsequently) gained medical importance because it strongly inhibits δ-aminobutyric acid aminotransferase in mammalian systems (Kobayashi *et al.*, 1976; Rando, 1977; Rando and Bangerter, 1993) and, thus, inhibits the breakdown of an important mammalian neurotransmitter. Its mechanism of action is fairly well established, as a mechanism-based 'suicide inhibitor' of a pyrixodal-phosphate linked enzyme (Rando, 1977; Soper and Manning, 1982). At the active site, a beta-proton elimination of gabaculine produces an aromatic derivative (Soper and Manning, 1982) which binds so tightly to pyridoxal phosphate that it is only liberated by denaturation (Gehring *et al.*, 1976). The mechanism-based mode of inhibition makes gabaculine extremely specific, unlike inhibitors such as aminooxyacetic acid.

In higher plant δ-aminolevulinic acid biosynthesis, a pyridoxyl phosphate-linked transaminase catalyses the conversion of glutamate-1-semialdehyde to δ-aminolevulinic acid. This enzyme is exceedingly sensitive to inhibition by gabaculine (May *et al.*, 1987; Schneegurt and Beale, 1986; Corriveau and Beale, 1986a; Caiger *et al.*, 1986; Granick and Beale, 1978; Hill *et al.*, 1985; Hoult *et al.*, 1986a; 1986b). This inhibitor has been useful in examining chlorophyll synthesis (Caiger *et al.*, 1986), phytochrome assembly (Gardner and Gorton, 1985; Konomi and Furuya, 1986), and the turnover of tetrapyrroles in plant tissues (May *et al.*, 1987; Hendry, 1987). Gabaculine readily enters plant tissues and is transported to sites of action in the leaves and roots (May *et al.*, 1987). Thus, since it blocks δ-aminolevulinic acid biosynthesis at micromolar concentrations while it does not inhibit rhizobial growth rates at 500 times higher concentrations (May *et al.*, 1987), gabaculine presents a useful counterpoint, to examine the plant partner in a manner analogous to the *hemA* mutant rhizobial strains presented above.

May *et al.* (1987) applied gabaculine to the roots of 9 day *Phaseolus lunatus*, 7 day after inoculation by *Rhizobium* sp. 127E15. The inhibitor was applied immediately prior to the onset of leghemoglobin biosynthesis, and well after the establishment of nodule initials. Thus, unlike the experiments using mutant strains, inhibition was targeted to the window of development in which demand for both tetrapyrrole and globin would be the greatest. They observed a parallel decrease in globin and heme, confirming the linkage in synthesis between these two components and suggesting that a block in plant δ-aminolevulinic acid biosynthesis influences the nodule in much the same way as a blockage of δ-aminolevulinic acid synthase in the bacteroid. Interestingly, however, acetylene reduction was much less sensitive.

Figure 2 presents similar experiments performed in soil, in place of the hydroponic conditions employed by May *et al.* (1987). Inhibitor concentrations between 20 - 50 μM resulted in a dramtic reduction of leghemoglobin biosynthesis, but not nitrogen fixation. To some extent, exogenously supplied δ-aminolevulinic acid is able to overcome the inhibitor effects of gabaculine. Figure 3 shows that nodule phosphoenolpyruvate carboxylase (PEP Case) and aspartate aminotransferase (Asp AT) are only marginally influenced by gabaculine inhibition. Exogenous δ-aminolevulinic acid, however, alleviates this partial inhibition. Nodule heme content, in contrast, is severely inhibited by gabaculine, and δ-aminolevulinic acid addition has a dramatic effect.

Thus, the findings of May *et al.* (1987) and the results shown in Figures 2 and 3 suggest that an inhibitor of the plant δ-aminolevulinic acid biosynthesis pathway blocks the accumulation of leghemoglobin. Bacteroid viability is maintained, as evidenced by the high rates of acetylene reduction in gabaculine-treated nodules. These results, coupled with the insensitivity of rhizobia to δ-aminolevulinic acid and the mutant studies of Guerinot and Chelm (1986), make a strong case for the importance of a plant contribution to the synthesis of heme for leghemoglobin.

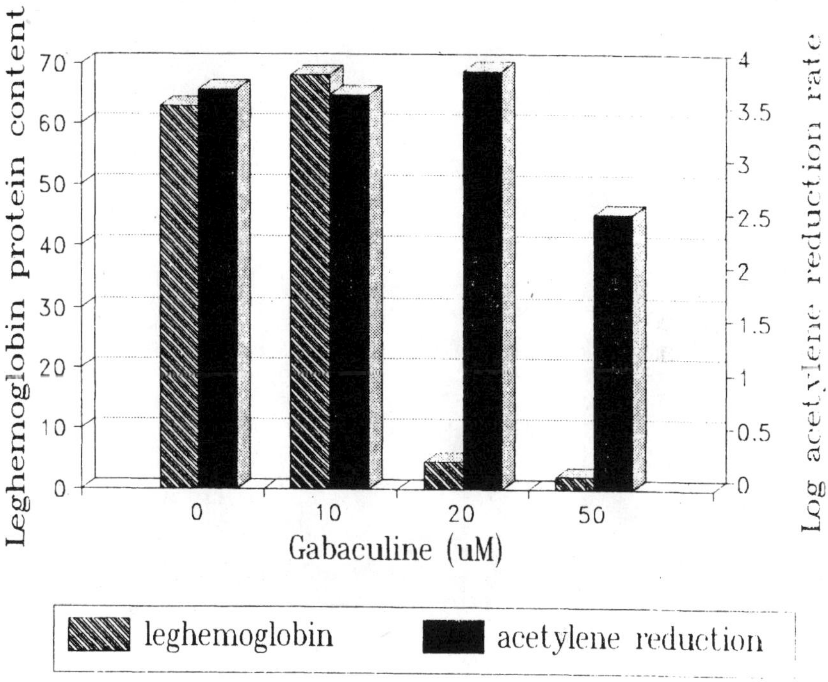

Fig. 2. Acetylene reduction rates and the leghemoglobin protein content of lima bean root nodules after treatment with gabaculine. Lima beans were inoculated at 2 d, and were treated at 7 d with 150 mL aqueous gabaculine. Plants were harvested 21 - 28 d, examined for acetylene reduction rates (expressed as log nmol acetylene reduced/h/g nodule fresh weight), and the harvested root nodules were examined by Western blotting for the presence of leghmoglobin (expressed as %g leghemoglobin/mg nodule protein). These results were averaged from 4 separate experiments. Notice that, at 20 %M gabaculine, normal rates of acetylene reduction were obtained with nodules which had about 3% normal leghemoglobin levels.

4. Leghemoglobin Oxidation And Reduction

The role of leghemoglobin as an oxygen buffer and in the facilitated diffusion of oxygen (Appleby, 1984; Bergersen *et al.*, 1973; Dilworth, 1980) implies an oxygen binding mechanism which has a high affinity and is reversible. Interestingly, only the ferrous (Fe(II)) leghemoglobin form can bind oxygen while the ferric (Fe(III)) form does not (Becana and Klucas, 1992). From the chemical properties of leghemoglobin, ferric leghemoglobin

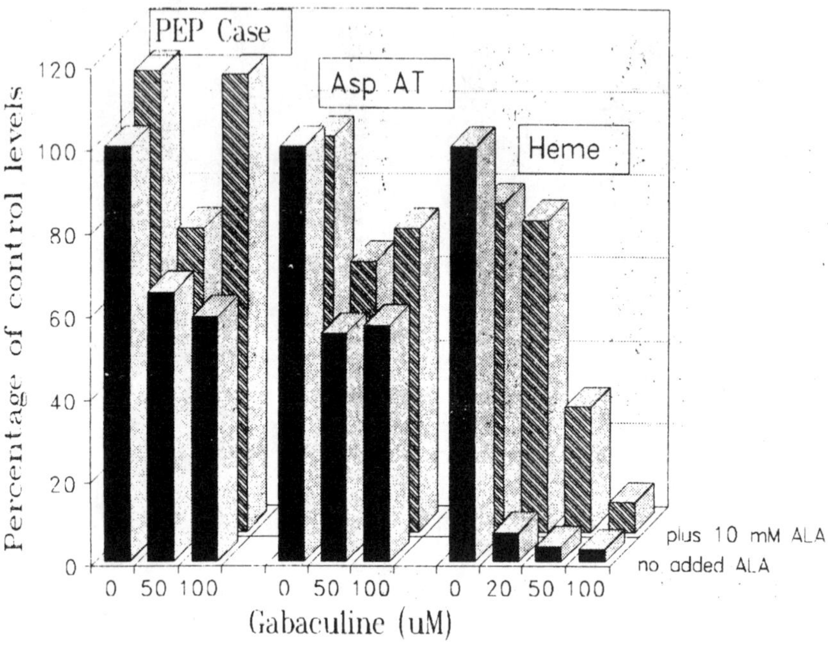

Fig. 3. The effects of exogenously supplied δ-aminolevulinic acid on gabaculine inhibition of PEP carboxylase (PEP Case), aspartate aminotransferase (Asp AT), and nodule heme content (Heme) in soybean nodules. Soybean was inoculated at 2 d, and treated with 150 mL aqueous gabaculine (+/- 10 mM δ-aminolevulinic acid (ALA)) at 7 d. At 28 d, nodules were harvested, and soluble extracts examined for PEP carboxylase activity, aspartate aminotransferase activity, and heme content. The data are expressed as percentage compared with samples untreated with either gabaculine or ALA. Notice that, at 20 - 50 %M gabaculine, there is a significant recovery of nodule heme after exogenous δ-aminolevulinic acid supply.

should be formed continually *in vivo* by oxidation of the ferrous form. For example, ferric leghemoglobin can be formed by a typical electron oxidation of the ferrous form, with a redox potential of around +0.22 volt at pH 7 (Becana and Klucas, 1992). In order to maintain leghemoglobin in a functioning state, the ferrous form must be maintained by reduction, and mechanisms to provide reducing equivalents to leghemoglobin have been described nodules (Klucas *et al.*, 1985).

4.1. Reduction of Leghemoglobin by Enzymes

Kretovich and coworkers reported a protein with leghemoglobin reductase activity isolated from lupine nodules (Kretovich *et al.*, 1982). This enzyme requires NADH as the electron donor and an intermediate electron carrier (such as methylene blue) between NADH and leghemoglobin. This enzyme contains FAD and active thiol groups but no metals (Golubeva *et al.*, 1987; Golubeva *et al.*, 1988). It has a molecular mass of 60 kD, a pI value of 5.8. It exhibits Km values of 8.7 μM for NADH and 10 μM for leghemoglobin.

A protein with leghemoglobin reductase activity has been purified to homogeneity from soybean nodules (Ji *et al.*, 1991; Saari and Klucas, 1984). The enzyme has a molecular mass of 110 kD and is composed of two subunits. Isoelectric focusing of the native protein revealed three equal proteins bands with pI values of 5.5, 5.6, and 5.7, which suggest slightly modified or different proteins. The enzyme contains FAD and active thiol groups. It requires NADH as the electron donor but NADPH is 30% as effective. Unlike the lupine enzyme, the soybean enzyme does not require an intermediate electron carrier. The soybean enzyme has Km values of 9.5 μM for leghemoglobin and 37 μM for NADH.

4.2. Reduction of Leghemoglobin by Small Molecules

Leghemoglobin also can be reduced nonenzymatically by reductants such as NAD(P)H, ascorbate, reduced glutathione, and cysteine (Becana and Klucas, 1990). The physiological levels of these reductants were estimated in the order of 150 to 250 μM for NADH + NADPH, 200 μM for cysteine (Becana and Klucas, 1990), 40 to 150 μM for reduced glutathione, and 1 to 2 mM for ascorbate (Dalton *et al.*, 1986). At these levels, only ascorbate and cysteine significantly reduce leghemoglobin (Becana and Klucas, 1990). In the presence of NAD(P)H, free flavins efficiently reduce leghemoglobin (Becana and Klucas, 1990), stimulated by the microaerophilic environment found within the nodule. As a result, free flavins may play a role in reducting leghemoglobin. Furthermore, nodules apparently have abundance of free flavins, especially riboflavin (Becana and Klucas, 1990).

5. SUMMARY

In this discussion, we have focused our attention on the various roles which the plastid may play in events which impact on the successful partitioning of iron in the legume root nodule. These include the storage of iron by ferritin and the synthesis of molecules which are the precursors for tetrapyrroles in leghemoglobin and the plant-derived respiratory components. As shown in Figure 4, trafficking at the levels of the plastid envelope and the peribacteroid membrane must play critical roles in directing these processes.

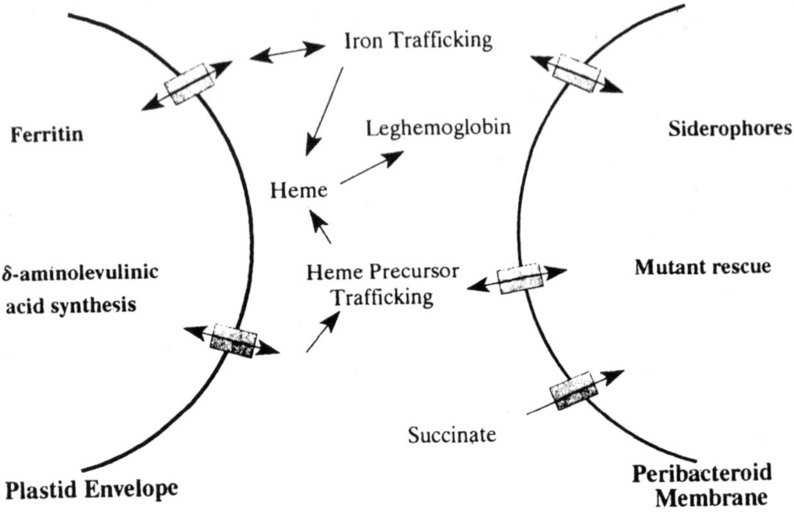

Fig. 4. Interaction of the plastid and the peribacteroid vesicle in the regulation of iron trafficking in the legume root nodule.

Future research will elucidate the molecular mechanisms of these transport processes, and how developmental changes in transport proteins within these membranes facilitate iron partitioning in the legume/rhizobial nitrogen fixation symbiosis.

ACKNOWLEDGEMENTS

This contribution was funded in part by the NASA Specialized Center of Research and Training in Gravitational Biology (NAGW 2328; JAG and WRO), by BioServe Space Technologies (NAGW 1197; JAG and PPW), and the USDA National Research Initiative Competitive Grants Program (JAG and PPW). This represents contribution 93-538-B of the Kansas Agricultural Experiment Station.

REFERENCES

Abadia, J. 1992. Leaf responses to Fe deficiency: A review. *J. Plant Nutr.*, **15**:1699-1713.

Abadia, J. and Abadia, A. 1993. Iron and plant pigments. In: *Iron chelation in plants and soil microorganisms,* Barton, L.L. and Hemming, B.C. (Eds.) Academic Press, Inc., New York, pp. 327-343.

Andrews, S.C., Smith, J.M.A., Yewdall, S.J., Guest, J.R. and Harrison, P.M. 1991. Bacterioferritins and ferritins are distantly related in evolution. *FEBS Lett.* **293**:164-168.

Appleby, C. 1984. Leghemoglobin and *Rhizobium* respiration. *Annu. Rev. Plant Physiol.*, **35**:443-478.

Armstrong, J.E. and Van Baalen, C. 1979. Iron transport in microalgae: The isolation and biological activity of a hydroxamate siderophore from the blue-green alga, *Agmenellum quadruplicatum. J. Gen. Microbiol.*, **111**:253-262.

Avissar, Y.J. and Beale, S.I. 1988. Biosynthesis of tetrapyrrole pigment precursors. Formation and utilization of glutamyl-tRNA for delta-aminolevulinic acid synthesis by isolated enzyme fractions from *Chlorella vulgaris. Plant Physiol.*, **88**:879-886.

Avissar, Y.J. and Beale, S.I. 1989. Distribution of the enzymatic basis for delta-aminolevulinic acid biosynthesis pathways among phototrophic bacterial groups. *Arch. Microbiol.*, **151**:513-519.

Barton, L.L., Fekete, F.A., Vester, C.R., Gill, Jr.,P.R. and Neilands, J.B. 1992. Physiological characteristics of *Rhizobium meliloti* 1021 Tn5 mutants with altered rhizobactin activities. *J. Plant Nutr.*, **15**:2145-2156.

Beale, S.I. 1990. Biosynthesis of the tetrapyrrole pigment precursor, delta-aminolevulinic acid, from glutamate. *Plant Physiol.,* **93**:1273-1279.

Becana, M. and Klucas, R.V. 1990. Enzymatic and nonenzymatic mechansims for ferric leghemoglobin reduction in legume root nodules. *Proc. Natl. Acad. Sci. USA,* **87**:7295-7299.

Becana, M. and Klucas, R.V. 1992. Oxidation and reduction of leghemoglobin in root nodules of leguminous plants. *Plant Physiol.,* **98**:1217-1221.

Bergersen, F.J., Turner, G.L. and Appleby, C.A. 1973. Studies of the physiological role of leghemoglobin in soybean root nodules. *Biochim. Biophys. Acta*, **292**:271-282.

Biedermann, G. and Schindler, P. 1957. On the solubility of precipitated iron (III) hydroxide. *Acta Chem. Scand.*, **11**:731-740.

Bott, M., Preisig, O. and Hennecke, H. 1992. Genes for a second terminal oxidase in *Bradyrhizobium japonicum. Arch. Microbiol.*, **158**:335-343.

Caiger, D.P., Pearson, S.A., Smith, A.J. and Rogers, L.J. 1986. Differential effects of gabaculine and levulinic acid on protochlorophyllide regeneration. *Plant Cell Environ.*, **9**:495-499.

Corriveau, J. and Beale, S.I. 1986a. Influence of gabaculine on growth, chlorophyll synthesis, and delta-aminolevulinic acid synthetase in *Euglena gracilis. Plant Sci.*, **45**:9-17.

Corriveau, J.L. and Beale, S.I. 1986b. Regulatory influence of the chloroplast tetrapyrrole pathway over delta-aminolevulinic acid synthetase levels in greening

Euglena gracilis. Plant Physiol., **80S**:52.

Crichton, R.R., Ponce-Ortiz, Y., Koch, M.H.J., Parfait, R. and Stuhrmann, H.B. 1978. Isolation and characterization of phytoferritin from pea (*Pisum sativum*) and lentil (*Lens esculenta*). *Biochem. J.*, **171**:349-356.

Cutting, J. and Schulman, H. 1969. The site of heme synthesis in soybean root nodules. *Biochim. Biophys. Acta*, **192**:486-493.

Dalton, D.A., Russell, S.A., Hanus, F.J., Pascoe, G.A. and Evans, H.J. 1986. Enzymatic reactions of ascorbate and glutathione that prevent peroxidase damage in soybean root nodules. *Proc. Natl. Acad. Sci. USA*, 83:3811-3815.

de Billy, F., Barker, D.G., Gallusci, P. and Truchet, G. 1991. Leghaemoglobin gene transcription is triggered in a single cell layer in the indeterminate nitrogen-fixing nodule of alfalfa. *Plant J.*, **1**:27-35.

Dilworth, M.J. 1980. Leghemoglobins. *Methods Enzymol.*, **69c**:812-823.

Dix, D.J., Lin, P.-N., Kimiata, Y. and Theil, E.C. 1992. The iron regulatory region of ferritin mRNA is also a positive control element for iron-independent translation. *Biochemistry*, **31**:2818-2822.

Emery, T. 1982. Iron metabolism in humans and plants. *Amer. Sci.*, **70**:626-632.

Emery-Goodman, A., Hirling, H., Scarpellino, L., Henderson, B. and Kühn, L.C. 1993. Iron regulatory factor expressed from recombinant baculovirus: Conversion between the RNA-binding apoprotein and Fe-S cluster containing aconitase. *Nucleic Acids Res.*, **21**:1457-1461.

Enard, C., Diolez, A. and Expert, D. 1988. Systemic virulence of *Erwinia chrysanthemi* 3937 requires a functional iron assimilation system *J. Bacteriol*, **170**:2419-2426.

Expert, D., Sauvage, C. and Neilands, J.B. 1992. Negative transcriptional control of iron transport in *Erwinia chrysanthemi* involves an iron-responsive two-factor system. *Molec. Microbiol.*, **6**:2009-2017.

Expert, D. and Gill, Jr.,P.R. 1991. Iron: A modulator in bacterial virulence and symbiotic nitrogen fixation. In: *Signals in Plant-Microbe Communication*, Verma, D.-P. (Ed.)

Flint, D. 1984. Gabaculine inhibits delta-ALA synthesis in chloroplasts. *Plant Physiol.*, 75S:170.

Frustaci, J.M., Sangwan, I. and O'Brian, M.R. 1991. Aerobic growth and respiration of delta-aminolevulinic acid dehydratase (*hemA*) mutant of *Bradyrhizobium japonicum. J. Bacteriol.*, **173**:1145-1150.

Frustaci, J.M. and O'Brian, M.R. 1992. Characterization of a Bradyrhizobium japonicum ferrochelatase mutant and isolation of the *hem H* gene. *J. Bacteriol.*, **174**:4223-4229.

Gabel, C. and Maier, R.J. 1993. Oxygen-dependent transcriptional regulation of cytochrome *aa3* in *Bradyrhizobium japonicum. J. Bacteriol.*, **175**:128-132.

Gadd, G.M. 1992. Metals and microorganisms: A problem of definition. *FEMS Microbiol. Lett.*, **100**:197-204.

Gardner, G. and Gorton, H. 1985. Inhibition of phytochrome synthesis by gabaculine.

Plant Physiol., **77**:540-543.

Gehring, H., Rando, R.R. and Christen, P. 1976. Active site labeling of aspartate aminotransferases by the a,b-unsaturated amino acid vinyl glycine. *Biochemistry*, **16**:4832-4836.

Gill, P. R., Barton, L. L., Scoble, M. D. and Neilands, J. B. 1991. A high affinity iron transport system of *Rhizobium meliloti* may be required for efficient nitrogen fixation *in planta*. In: *Iron Nutrition and Interactions in Plants*, Chen, Y. & Hadar, Y. (Eds.) Kluwer Academic Publishers, The Netherlands, pp. 251-257.

Gill, P.R. and Neilands, J.B. 1989. Cloning a genomic region required for a high-affinity iron-uptake system in *Rhizobium meliloti* 1021. *Mol. Microbiol.*, 3:1183-1189.

Godfrey, C. and Dilworth, M. 1971. Haem biosynthesis from [14C]-delta-aminolevulinic acid in laboratory grown and root nodule *Rhizobium lupini*. *J. Gen. Microbiol.*, **69**:385-390.

Gollop, R., Santhaguru, K. and Avissar, Y.J. 1984. Control of the initial steps in heme biosynthesis in free-living *Rhizobium* sp. by culture conditions. *Curr. Microbiol.*, **11**:101-106.

Golubeva, L.I., Topunov, A.F., Goncharova, S.S., Aseeva, K.B. and Kretovich, V.L. 1988. Production and properties of a homogenous preparation of metleghemoglobin reductase of lupine root nodule cytosol. *Biokhimiva (English Translation)*, **53**:1478-1482.

Golubeva, L.I., Topunov, A.F., Talyzin, V.V. and Kretovich, V.L. 1987. Interaction of metleghemoglobin reductase of lupine nodule cytosol with oxygen-transporting hemoproteins. *Dokl. Akad. Nauk. SSSR (English Translation)*, **294**:178-181.

Gong, H., Nilsen, S. and Allen, J.F. 1993. Photoinhibition of photosynthesis *in vivo:* Involvement of multiple sites in a photodamage process under CO_2- and O_2-free conditions. *Biochim. Biophys. Acta Bio-Energetics*, **1142**:115-122.

Goosen-de Roo, L., De Maagd, R.A. and Lugtenberg, B.J.J. 1993. Outer membrane protein gene expression traced by immunogold labelling in *Rhizobium leguminosarum* bv. viciae strain 248 in *Vicia sativa* ssp *Nigra*. *Cell Biol. Int. Rep.*, **17**:221-226.

Gough, S.P. and Kannangara, C.G. 1977. Synthesis of delta-aminolevulinate by chloroplast stroma preparations from greening barley leaves. *Carlsberg Res. Commun.*, **42**:459-464.

Granick, S. and Beale, S.I. 1978. Hemes, chlorophylls, and related compounds: Biosynthesis and metabolic regulation. *Adv. Enzymol.*, **46**:33-203.

Guerinot, M. L. 1993. Iron and the nodule. In: *Iron chelation in plants and soil microorganisms*, Barton, L.L. & Hemming, B.C. (Eds.) Academic Press, Inc., New York, pp. 197-217.

Guerinot, M.L., Meidl, E.J. and Plessner, O. 1990. Citrate as a siderophore in *Bradyrhizobium japonicum*. *J. Bacteriol.*, **172**:3298-3303.

Guerinot, M.L. and Chelm, B. 1986. Bacterial delta-aminolevulinic acid synthase activity is not essential for leghemoglobin formation in the soybean/*Bradyrhizobium japonicum* symbiosis. *Proc. Natl. Acad. Sci. USA*, **83**:1837-1841.

Hartzook, A. 1982. The problem of iron deficiency in peanuts (*Arachis hypogaea* L.) on basic and calcareous soils of Israel. *J. Plant Nutr.*, **5**:923-926.

Hemantaranjan, A. 1986. Iron fertilization in relation to nodulation and nitrogen fixation in french bean (*Phaseolus vulgaris* L.). *J. Plant Nutr.*, **11**:829-873.

Hemantaranjan, A. and Garg, O.K. 1986. Introduction of nitrogen-fixing nodules through iron and zinc fertilization in the non-nodule forming French bean (*Phaseolus vulgaris* L.). *J. Plant Nutr.* **9** :281-288.

Hendry, G. 1987. Tetrapyrrole turnover in plants. In: Progress in Photosynthesis Research, Biggins, J. (Ed.) Martinus Nijhoff Publishers, Dordrecht, pp. 499-502.

Herridge, D.F. and Holland, J.F. 1993. Low nodulation and nitrogen fixation limits yield of pigeonpea on alkaline vertisels of Northern NSW -- effect of iron, rhizobia, and plant genotype. *Aust. J. Agr.*, **44**:137-149.

Hill, C., Pearson, S.A., Smith, A.J. and Rogers, L.J. 1985. Inhibition of chlorophyll synthesis in *Hordeum vulgare* by 3-amino-2,3-dihydrobenzoic acid (gabaculin). *Biosci. Rep.*, **5**:775-781.

Hirsch, A.M. 1992. Developmental biology of legume nodulation. *New Phytol.* **122**:221-237.

Hoult, R.C., Rees, D., Rogers, L.J. and Smith, A.J. 1986a. Specific inhibition of tetrapyrrole biosynthesis by 3-amino-2,3-dihydrobenzoic acid (gabaculin) in the cyanobacterium *Synechococcus* 6301. *Arch. Microbiol.*, **146**:57-62.

Hoult, R.C., Richardson, K., Rogers, L.J. and Smith, A.J. 1986b. Effect of an inhibitor of chlorophyll biosynthesis on the cyanobacterium *Synechococcus* 6301. *Biochem. Soc. Trans.*, **14**:37-38.

Jensen, E.O., Marcker, K.A. and Villadsen, I.S. 1986. Heme regulates the expression in *Saccharomyces cervesia* of chimeric genes containing 5'-flanking soybean leghemoglobin sequences. *EMBO J.*, **5**:843-847.

Ji, L., Wood, S., Becana, M. and Klucas, R.V. 1991. Purification and characterization of soybean root nodule ferric leghemoglobin reductase. *Plant Physiol.*, **96**:32-37.

Johnson, G. V. and Barton, L. L. 1993. Selected physiological responses associated with Fe(III) and Fe(II) metabolism. In: *Iron chelation in plants and soil microorganisms,* Barton, L.L. and Hemming, B.C. (Eds.) Academic Press Inc; New York pp. 371-386.

Kannangara, C.G. and Schouboe, A. 1985. Biosynthesis of delta-aminolevulinate in greening barley leaves. VII. Glutamate 1-semialdehyde accumulation in gabaculine treated leaves. *Carlsberg Res. Commun.*, **50**:179-191.

Klucas, R. V., Lee, K. K., Saari, L. and Erickson, B. K. 1985. Factors affecting functional leghemoglobin in legume root nodules. In: *Nitrogen Fixation and CO_2 Metabolism,* Ludden, P.W. & Burris, J.E. (Eds.) Elsevier, New York, pp. 13-20.

Ko, M.P., Huang, P.Y., Huang, J.S. and Barker, K.R. 1985. Accumulation of phytoferritin and starch granules in developing nodules of soybean roots infected with *Heterodera glycines. Cytol. Histol.*, **75**:159-164.

Kobayashi, K., Miyazawa, S. and Terahara, A. 1976. Gabaculine: gamma-aminobutyrate aminotrasnferase inhibitor of microbial origin. *Tetrahedron Lett.*, **7**:537-540.

Konomi, K. and Furuya, M. 1986. Effects of gabaculine on phytochrome synthesis during imbibition in embryonic axes of *Pisum sativum* L. *Plant Cell Physiol.*, **27**:1507-1512.

Kretovich, V.L., Melik-Sarkissyan, S.S., Bashirova, N.F. and Topunov, A.F. 1982. Enzymatic reduction of leghemoglobin in lupine nodules. *J. Appl. Biochem.*, **4**:209-217.

Langford, P.R. and Moxon, E.R. 1993. The dilution rate affects the outer membrane protein and lipopolysaccharide composition of *Haemophilus influenzae* type b grown under iron limitation. *J. Bacteriol.*, **175**:2462-2464.

Lankford, C. 1973. Bacterial assimilation of iron. *CRC Crit. Rev. Microbiol.*, **2**:273-331.

Laulhere, J.-P., Laboure, A.M., Van Wuytswinkel, O., Gagnon, J. and Briat, J.-F. 1992. Purification, characterization and function of bacterioferritin from the cyanobacterium *Synechocystis* P.C.C. 6803. *Biochem. J.*, **281**:785-793.

Laulhere, J.-P. and Briat, J.-F. 1993. Iron release and uptake by plant ferritin: Effects of pH, reduction and chelation. *Biochem. J.*, **290**:693-699.

Leong, S.A., Ditta, G. and Helinski, D.A. 1982. Heme biosynthesis in *Rhizobium*. Identification of a cloned gene coding for delta-aminolevulinic acid synthetase from *Rhizobium meliloti*. *J. Biol. Chem.*, **257**:8724-8730.

Lescure, A.M., Proudhon, D., Pesey, H., Ragland, M., Theil, E.C. and Briat, J.F. 1991. Ferritin gene transcription is regulated by iron in soybean cell cultures. *Proc. Natl. Acad. Sci. USA*, **88**:8222-8226.

Lobreaux, S., Yewdall, S.J., Briat, J.-F. and Harrison, P.M. 1992. Amino-acid sequence and predicted three-dimensional structure of pea seed (*Pisum sativum*) ferritin. *Biochem. J.*, **288**:931-939.

Lobreaux, S. and Briat, J.-F. 1991. Ferritin accumulation and degradation in different organs of pea (*Pisum sativum*) during development. *Biochem. J.*, **274**:601-606.

Lois, A.F., Ditta, G.S. and Helinski, D.R. 1993. The oxygen sensor FixL of *Rhizobium meliloti* is a membrane protein containing four possible transmembrane segments. *J. Bacteriol.*, **175**:1103-1109.

May, T.B., Guikema, J.A., Henry, R.L., Schuler, M.K. and Wong, P.P. 1987. Gabaculine inhibition of chlorophyll biosynthesis and nodulation in *Phaseolus lunatus* L. *Plant Physiol*, **84**:1309-1313.

McDermott, T.R., Graham, P.H. and Brandwein, D.H. 1987. Viability of *Bradyrhizobium japonicum* bacteroids. *Arch. Microbiol.*, **148**:100-106.

McKnight, D.M. and Morel, F.M.M. 1980. Copper complexation by siderophores from filamentous blue-green algae. *Limnol. Oceanogr.*, **25**:62-71.

Melefors, O., Goossen, B., Johansson, H.E., Stripecke, R., Gray, N.K. and Hentze, M.W. 1993. Translational control of 5-aminolevulinate synthase mRNA by iron-responsive elements in erythroid cells. *J. Biol. Chem.*, **268**:5974-5978.

Melefors, O. and Hentze, M.W. 1993. Translational regulation by mRNA/protein interactions in eukaryotic cells: Ferritin and beyond. *BioEssays,* **15**:85-90.

Mellor, R.B. 1989. Bacteroids in the *Rhizobium*-legume symbiosis inhabit a plant internal lytic compartment: Implications for other microbial endosymbioses. *J. Exp. Bot.*, **40**:831-839.

Miller, G.W., Pushnik, J.C. and Welkie, G.W. 1984. Iron chlorosis, a world-wide problem: The relation of chlorophyll biosynthesis to iron. *J. Plant Nutr.*, **7**:1-22.

Modi, M., Shah, K.S. and Modi, V.V. 1985. Isolation and characterization of a catechol-like siderophore from cowpea *Rhizobium* RA-1. *Arch. Microbiol.*, **141**:156-158.

Morrett, E., Fischer, H.-M. and Hennecke, H. 1991. Influence of oxygen on DNA binding, positive control, and stability of the *Bradyrhizobium japonicum* NifA regulatory protein. *J. Bacteriol.*, **173**:3478.

Nadler, K.D., Johnson, A.W.B., Chen, J.-W. and John, T.R. 1990. A *Rhizobium leguminosarum* mutant defective in symbiotic iron acquisition. *J. Bacteriol.*, **172**:670-677.

Nadler, K.D. and Avissar, Y.J. 1977. Heme synthesis in soybean root nodules. I. On the role of bacteroid delta-amino levulinic acid synthase and delta-amino levulinic acid dehydrase in the synthesis of the heme of leghemoglobin. *Plant Physiol.*, **60**:433-436.

Neilands, J. B. 1993. Overview of bacterial iron transport and siderophore systems in Rhizobia. In: *Iron chelation in plants and soil microorganisms*, Barton, L.L. & Hemming, B.C. (Eds.) Academic Press, Inc., New York, pp. 179-195.

Neilands, J.B. 1982. Microbial envelope proteins related to iron. *Annu. Rev. Microbiol.*, **36**:285-309.

O'Brian, M.R. and Maier, R.J. 1989. Molecular aspects of the energetics of nitrogen fixation in Rhizobium-legume symbioses. *Biochim. Biophys. Acta*, **974**:229-246.

O'Hara, G.W., Dilworth, M.J., Boonkerd, N. and Parkian, P. 1988. Iron deficiency specifically limits nodule development in peanut inoculated with *Bradyrhizobium* sp. *New Phytol.*, **108**:51-57.

Page, W. J. 1993. Growth conditions for the demonstration of siderophores and iron-repressible outer membrane proteins in soil bacteria, with an emphasis on free-living diazotrophs. In: *Iron chelation in plants and soil microogranisms,* Barton, L.L. & Hemming, B.C. (Eds.) Academic Press, Inc., New York, pp. 75-110.

Patel, H.N., Chakraborty, R.N. and Desai, S.B. 1988. Isolation and partial characterization of phenolate siderophore from *Rhizobium leguminosarum. FEMS Microbiol. Lett.*, **56**:131-134.

Pawlowski, K., Schell, J. and deBruijn, F. J. 1988. Construction and characterization of heme biosynthesis (*hemA*) mutants of *Azorhizobium caulinodans* ORS571; *hem A* is essential for nodulation and symbiotic nitrogen fixation. In : *Nitrogen Fixation : Hundred Years* After deBruijn, F.J. & Newton, W.E. (Eds.) Gustav Fischer, Stutgart, PP. 380.

Ragland, M. and Theil, E.C. 1993. Ferritin (mRNA, protein) and iron concentrations during soybean nodule development. *Plant Mol. Biol.*, **21**:555-560.

Rai, R., Prasad, V., Choudrhury, S.K. and Sinha, N.P. 1984. Iron nutrition and symbiotic N_2-fixation of lentil (*Lens culinaris*) genotypes in calcareous soils. *J. Plant Nutr.*, 7:399-405.

Rai, R., Singh, S.N. and Prasad, V. 1982. Effect of pressmud amend pyrite on symbiotic N_2 fixation , active iron contents of nodules, grain yield and quality of chickpea (*Cicer arietinum* Linn.) genotypes in calcareous soils. *J. Plant Nutr.*, 5:905-913.

Rando, R. 1977. Mechanism of the irreversible inhibition of the GABA-alpha-ketoglutaric acid transaminase by the neurotoxin gabaculine. *Biochemistry*, **16**:4604-4610.

Rando, R. and Bangerter, F. 9893. The irreversible inhibition of mouse brain GABA-alpha-ketogutaric acid transaminase by gabaculine. *J. Am. Chem. Soc.*, 6762:6766.

Rawsthorne, S. and LaRue, T.A. 1986. Metabolism under microaerobic conditions of mitochondria from cowpea nodules. *Plant Physiol.*, **81**:1097-1102.

Reigh, G. and O'Connell, M. 1993. Siderophore-mediated iron transport correlates with the presence of specific iron-regulated proteins in the outer membrane of *Rhizobium meliloti*. *J. Bacteriol.*, **175**:94-102.

Robertson, J.G., Lyttleton, P., Bullivant, S. and Grayston, G.F. 1978a. Membranes of lupin root nodules. I. The role of golgi bodies in the biogenesis of infection threads and peribacteroid membranes. *J. Cell Sci.*, **30**:129-149.

Robertson, J.G., Warburton, M.P., Fordyce, A.M. and Bullivant, S. 1978b. Membranes in lupin root nodules. II. Preparation and properties of peribacteroid membranes and bacteroid envelope inner membranes from developing lupin nodules. *J. Cell Sci.*, **30**:151-174.

Saari, L.L. and Klucas, R.V. 1984. Ferric leghemoglobin reductase from soybean root nodules. *Arch. Biochem. Biophys.*, **231**:102-113.

Sanchez, F., Padilla, J.E., Perez, H. and Lara, M. 1991. Control of nodulin genes in root-nodule development and metabolism. *Annu. Rev. Plant Physiol. Plant Mol. Biol*, **42**:507-538.

Sangwan, I. and O'Brian, M.R. 1991. Evidence for an interorganismic heme biosynthetic pathway in symbiotic soybean root nodules. *Science*, **251**:1220-1222.

Schneegurt, M.A. and Beale, S.I. 1986. Biosynthesis of protoheme and heme a from glutamate in maize. *Plant Physiol.*, **81**:965-971.

Schon, A., Krupp, G., Gough, S., Berry-Lowe, S., Kannangara, C.G. and Soll, D. 1986. The RNA required in the first step of chlorophyll biosynthesis is a chloroplast glutamate tRNA. *Nature*, **322**:281-284.

Seckback, J. 1987. Studies on the detection of plant ferritin as influenced by iron supply to iron-deficient beans. *J. Ultrastruct. Res.*, **22**:413-423.

Silver, S. and Walderhaug, M. 1992. Gene regulation of plasmid- and chromosome-determined inorganic ion transport in bacteria. *Microbiol. Rev.*, **56**:195-228.

Smith, A. G. 1987. Subcellular localization of porphyrin synthesis enzymes in pea and arum. In: *Progress in Photosynthesis Research, vol. IV*, Biggins, J. (Ed.) Martinus Nijhoff Publishers, Dordtrecht, pp. 453-456.

Smith, M.J. and Neilands, J.B. 1984. Rizobactin, a siderophore from *Rhizobium meliloti. J. Plant Nutr.*, **7**:449-458.

Soper, T. and Manning, J. 1982. Inactivation of pyridoxal phosphate enzymes by gabaculine. *J. Biol. Chem.*, 13920:13936.

Stanley, J., Dowling, D.N. and Broughton, W.J. 1988. Cloning of hem A from *Rhizobium* sp. NGR234 and symbiotic phenotype of a gene-directed mutant in diverse legume genera. *Mol. Gen. Genet.*, **215**:32-37.

Takagi, S. 1993. Production of phytosiderophores. In: Iron Chelation in Plants and Soil Microorganisms, Barton, L.L. and Hemming, B.C. (Eds.) Academic Press, Inc., New York, pp. 111-133.

Tang, C., Robson, A.D. and Dilworth, M.J. 1990a. A split-root experiment shows that iron is required for nodule initiation in *Lupinus augustifolius* L. *New Phytol.*, **115**:61-67.

Tang, C., Robson, A.D. and Dilworth, M.J. 1990b. The role of iron in nodulation and nitrogen fixation in *Lupinus augustifolius* L. *New Phytol.*, **114**:173-182.

Tang, C., Robson, A.D. and Dilworth, M.J. 1992. The role of iron in the (Brady) *Rhizobium* legume symbiosis. *J. Plant Nutr.*, **15**:2235-2252.

Tang, C. and Dilworth, M.J. 1991. Which stage of nodule initiation in *Lupinus angustifolius* L. is sensitive to iron deficiency? *New Phytol.*, **117**:243-250.

Terry, N. and Abadia, J. 1986. Function of iron in chloroplasts. *J. Plant Nutr.*, **9**:609-646.

Terry, R.E., Sorensen, K.U., Jolley, V.D. and Brown, J.C. 1991. The role of active *Bradyrhizobium japonicum* in iron stress response of soybean. *Plant Soil*, **130**:225-230.

Theil, E. C. and Hase, T. 1993. Plant and microbial ferritins. In: *Iron chelation in plants and soil microorganisms*, Barton, L.L. & Hemming, B.C. (Eds.) Academic Press, Inc., New York, pp. 133-155.

Udvardi, M.K., Price, G.D., Gresshoff, P.M. and Day, D.A. 1988. A dicarboxylate transporter on the peribacteroid membrane of soybean nodules. *FEBS Lett.*, **231**:36-40.

van der Mark, F., Bientait, F. and van den Ende, H. 1983. Variable amounts of translatable ferritin mRNA in bean leaves with various iron contents. *Biochem. Biophys. Res. Commun.*, **115**:463-469.

Weinsein, J.D., Mayer, S.M. and Beale, S.I. 1986. Heterologous reconstitution of tetrapyrrole precursor formation by algal extracts. *Plant Physiol.*, **80**S:52.

Welkie, G. W. and Miller, G. W. 1993. Plant iron uptake physiology by nonsiderophore systems. In: *Iron chelation in plants and soil microorganisms*, Barton, L.L. & Hemming, B.C. (Eds.) Academic Press, Inc., New York, pp. 345-369.

Advancements In Micronutrient Research
Editor : A. Hemantaranjan
1996, Scientific Publishers, Jodhpur, pp. 151–178.

7

PHYSIOLOGICAL AND BIOCHEMICAL SIGNIFICANCE OF ZINC NUTRITION IN PLANTS

A. HEMANTARANJAN
Department of Plant Physiology,
Institute of Agricultural Sciences,
Banaras Hindu University,
Varanasi 221 005, INDIA

CONTENTS

1. INTRODUCTION

Recent achievements in global agricultural productivity have had no historical parallel. We are in the midst of the greatest food production revolution the world has known. A glance at the history of agriculture reveals that man has always depended on plants for his existence on this planet. The period in the development of human race during which man began cultivation of plants marks the dawn of agriculture. Since then, man has been preoccupied with the question of improving the productivity efficiency of plants through materialistic adjustments. This might be responsible for the gradual exploration of the mystery behind the growth changes. The revelation of the existence, activity, and benefits of mineral nutrients is indeed an important milestone towards modern agriculture, and has created a subject of great interest to plant researchers. Contributions in this direction have unfolded many facts about the physiology of nutrition. The growing knowledge of mineral nutrition of plants with special reference to micronutrients has made us realize that great changes are possible through judicious supply of certain inorganic substances.

The essentiality of micronutrients was proved long before their role in crop production was recognized. The essential microelements are referred to as micronutrients and are required in quantities often less than 1 ppm in solution and not in plants. A more acceptable definition of micronutrients is 'those essential elements which are required or are present in plant and soils equal to or less than iron.' Many elements, as a result of critical investigations and refined techniques, viz., zinc, boron, copper, iron, molybdenum, manganese, cobalt, sodium, chlorine, silicon, vanadium, nickel, aluminium, selenium, florine, aluminium, tungsten, beryllium, etc. were added to the list of microelements, required by plants only in extremely small quantities.

Zinc is one of the micronutrients which, even in minute quantities in soil, is indispensable for plant growth. Zinc deficiency in crop plants is one of the most common features observed in agricultural fields. Many agricultural soils are incapable of meeting even a small demand for zinc. Consequently, the standard fertilizer recommendation of N, P and K is sometimes inadequate in today's quest for higher crop production. Researchers have established that in many such instances zinc treatment could bring about a striking improvement in the yield. In fact, zinc deficiency has the distinction of being the most widespread of all the micronutrient problems.

It is felt widely that modern agriculture has brought about zinc deficiency disorders in soils to a considerable extent (Safaya *et al.*, 1975). Introduction of high yielding varieties has intensified the removal of all essential nutrients, including zinc, from soils. Neglect of bulky organic manures and increased dependence on strong inorganic fertilizers have led to less incidental additions of zinc, either as a constituent or as a contaminant. These have caused development of zinc deficiency in soils in a relatively shorter time. Most of the agricultural soils are well supplied with total zinc which can sustain crops for decades. However, total zinc is immaterial as far as plant productivity is concerned. It is the available zinc which regulates the crop performance. Alternatively, these can be recognized by soil analysis or plant tests. Soils that contain less than 0.6 ppm DTPA (diethylenetriamine penta acetic acid) –extractable zinc are supposed to be zinc deficient.

Most rocks of earth's crust contain zinc. Relatively higher concentrations are found in eruptive rocks. The carbonated rocks and sandstones have relatively low values – 20 and 16 ppm Zn respectively (Aubert and Pinta, 1977). Zinc in soils is mostly present in the ferro-magnesian minerals, augite, horn-blende and biotite. Zinc salts like ZnS, sphelerite (ZnFeS), zincite (ZnO), smithsonite ($ZnCO_3$) and the two zinc-silicates $ZnSiO_3$ and $ZnSiO_4$ constitute other important sources for soil zinc. Zinc is also found in exchange sites of clay minerals and organic matter or adsorbed on solid surfaces in the form of Zn^{2+}, ZnOH or ZnCl. The availability of zinc to the plants is affected by several factors: edaphic, environmental and genotypic. From the studies of zinc distribution in the principal soil profiles of Uttar Pradesh (Chatterjee *et al.*, 1976), it is quite obvious that availability of zinc in a particular soil is related not only to the zinc content of the soil but also to the type of the soil. Another important soil property that affects availability of zinc is soil pH because the solubility on Zn^{++} is highly dependent on pH.

In view of the fact, a deficient supply of a nutrient is reflected in visual changes in that the plant appearance, particularly on the foliage, a study of foliar symptoms of nutrient deficiencies is desirable (Wallace, 1961; Sprague, 1964). This is specially so in some particular plant species which develop characteristic and specific symptoms of the deficiency, and thus, serve as 'indicator' plants Another important aspect in which zinc seems to be involved is auxin metabolism of plants and as biocatalyst in cellular metabolism (Vallee and Wacker, 1970; Agarwala, 1979).

In view of the physiological significance of zinc and its poor presence in soils all over the world, in general, and Indo-Gangetic soils, in particular, it appears that there is enough scope of improvement of crops through zinc fertilization. In this review various physiological aspects related to zinc nutrition with reference to plant metabolism, chlorophyll pigments, zinc uptake and distribution in plants, and interactions with other nutrients are described and discussed in the light of recent advances to gain a better understanding of the biochemistry and physiology in plant zinc nutrition research.

2. Zinc Uptake Mechanism and Distribution in Plants

The first mention of the utility of zinc in the growth of plants was made by Javellier (1912), who considered that this element was being used by plant growers as far back as the year 1900. However, the recognition of zinc as an essential element for plants was made by Sommer and Lipman (1928) in California, U.S.A. Since then considerable interest was aroused among the plant scientists and led them to use zinc salts as a plant nutrient to bring about improvement in the growth and development of crops.

Zinc is taken up predominantly as a divalent cation (Zn^{2+}) and presumably at higher pH it is also taken up as a monovalent cation ($ZnOH^{+}$). The long distance transport of zinc takes place mainly in the xylem, where zinc is either bound to organic acids or exists as the free divalent cation (White *et al.*, 1981 a,b). Diffusion is a critical phenomenon in regulating zinc uptake by plants as well as in determining the effectiveness of various zinc fertilizers. The intensity-capacity factor is also involved since both the labile solid phase and soil solution contribute to the zinc supply (Lindsay, 1972). Chelated zinc is often more effective than inorganic zinc for many upland crops, but this was claimed opposite for flooded rice by Mikkelsen and Kuo (1977) and Giordano (1977). A possible explanation is greater diffusibility of chelates, which transport zinc downward and beyond the effective root zone of small rice plants. In addition, exchange reactions of zinc in the chelate molecule with iron and calcium could lessen chelate effectiveness under reduced soil conditions. Giordano *et al.* (1974) reported that uptake of zinc by rice was retarded by metabolic inhibitors when substrate zinc concentrations approximated to those found in soil solution. The presence of 10^{-4} M $FeCl_2$ or $MnCl_2$ in absorption solutions depressed zinc uptake significantly, while $MgCl_2$ reduced the concentrations of zinc in roots but enhanced translocation to shoots. Most evidence suggests that uptake of zinc is predominantly an active process.

Apart from naturally occurring reactions in soils which limit plant uptake of zinc, other parameters interact and influence availability. The most influential single factor is soil pH. Since the solubility of zinc decreases as pH increases, deficiency is most prevalent on calcareous soils of pH 7.4 or higher. Chapman (1966) indicates that certain acid peats and acidification of some soils through continued use of acidforming fertilizers can lead to zinc toxicity of sensitive crops. Zinc deficiency has been associated with irrigated soils from which topsoil has been removed for land leveling. The exposed subsoil is generally of lower zinc content, is higher in pH, and may contain free $CaCO_3$. Supplying organic matter by growing corn or green manure crops is often beneficial to supplement the available zinc supply. In contrast to this, adding organic manure to flooded neutral or alkaline rice soils is detrimental in that resulting bicarbonate aggravates zinc deficiency (Forno *et al.*, 1975).

Since organic matter is a source of zinc for plants, cool soils may reduce the rate of microbial release of available zinc. A second factor is delayed root growth at low temperature, which would limit zinc uptake. Sharma *et al.* (1968) and Mikkelsen and Kuo (1977) found low temperature (16°C) to inhibit zinc uptake by rice, and neither phosphorus nor zinc additions appreciably increased zinc concentration of uptake. Giordano and Mortvedt (1978) reported the uptake of zinc to be greater with triammonium pyrophosphate than monoammonium phosphate. The fact that uptake of zinc was greater at 32 than at 16°C appeared to reflect increased growth.

Peribock (1959) observed that soil application of Zn increased the concentration and uptake of zinc in young plants of wheat and flax. A similar increase in Zn absorption in *Phaseolus vulgaris* was reported by Burlson *et al.* (1961). An application of trace element mixture at commercial rate, increased zinc concentration by 58% above the control plants while the uptake of applied zinc was more efficient at the lower rate of zinc application (Gladstones and Loneragan, 1966). A linear increase in zinc uptake was noted by Caroll and Loneragan (1968) from 2 to 400 ng atoms Zn g^{-1} fresh roots per day, as zinc concentration in solution culture increased from 0.01 to 0.25 μM. In concentration dependent uptake experiments with intact barley roots, three phases in the range upto 1.38 μ M zinc could be detected independent of the tested uptake period. The kinetic constants increased with higher phases and the transition points were lowered with increasing time. It is concluded that zinc and copper are taken up by separate mechanisms (Werner, 1978). Emine Polar (1970) reported the uptake of ^{65}Zn by seeds of *Vicia faba* soaked in a $^{65}ZnCl_2$ solution. It was found to be constant and did not depend on the amount of radionucleide given. However, the percentage of the total absorbed ^{65}Zn retained by seed coats varied according to the cultural variety of *Vicia faba.* Most of the ^{65}Zn entering the seed remained bound to the seed coat and the remainder entered the embryo and cotyledons which was concentrated around the abaxial part of this latter, lessening towards the adaxial region. The fate of the isotope was followed in all the stages of plant development. Emine Polar (1970) calculated ^{65}Zn distribution on percentage basis increased in that order: roots, stem, and leaves in one month old plants and older :

^{65}Zn content and uptake in roots, sheaths and blades of maize and barley plants increased significantly with increased levels of zinc application. The sheaths contain highest ^{65}Zn content followed by roots and blades. The distribution of total zinc, however, differed from that of radioactive zinc. The roots had the highest content, followed by sheaths and blades. The two species differed very little in zinc distribution patterns. The autoradiograph of intact maize and barley plants showed that ^{65}Zn was fairly evenly distributed in the main and auxiliary roots, but, there was relatively higher concentration at the root-stem junction. The ^{65}Zn concentration was higher in node than in internodes and in young emerging leaves (Singh and Steenberg, 1974; Singh and Singh, 1973). Navrot *et al.* (1967) found that Zn fertilization of tomato plants in calcareous soil increased yield as well as ^{65}Zn content and uptake. Brown and Taylor (1966) found ^{65}Zn concentrations to rank : roots > stem > leaves. On the other hand Makhonina *et al.* (1960), found ^{65}Zn to be concentrated in the above ground portions of several plant species. Langston (1956) reported that ^{65}Zn accumulated to a marked degree in the stem and in veins and petioles of older leaves of tomato. In subterrenean clover zinc is always present in high concentration in the terminal portions of the main axis, runners and lateral, young and expanding leaves, petioles and root tips (Riceman and Jones, 1958, 1960). Turner and Gregory (1967) also reported high concentration of zinc throughout the root systems as well as in the nodes and basal areas of the sheaths. Singh and Steenberg (1974) concluded on the above ground that zinc concentration is higher in young parts and especially in veins, and is mobile through the sieve tubes in the plants.

Zinc uptake by *chlorella* and barley (*Hordeum vulgare* L.) root is predominantly passive process resulting from ion exchange (Broda, 1965). Zinc was present in the absorption solution at concentrations of 10^{-3} and 10^{-5}M. Rathmore *et al.* (1970) and Joseph *et al.* (1971) measured Zn uptake by intact leaves and plants, isolated leaf cells, leaf disks, excised roots, and stem callus tissue of two field bean (*Phaseolus vulgaris* L.) cultivars from Zn solutions varying in concentration from 2×10^{-6} to 2×10^{-4}M. At these concentrations, respiratory inhibitors and low temperature did not impede Zn uptake.

Absorption of Zn by wheat (*Triticum aestivum* L.) from a 10^{-6}M $ZnCl_2$ solution was strongly temperature-dependent, especially over a range of 2 to 10°C (Chaudhary and Loneragan, 1972). However, Schmidt *et al.* (1965) and Schmidt and Hawf (1966) have reported conflicting results regarding the effects of metabolic inhibitors on Zn uptake over an absorption range of 1×10^{-6} to 5×10^{-6}M Zn. They suggested that perhaps an 'unclarified mechanism' may be involved. A two component system for ion uptake may exist (Hiatt, 1968). One component could account for uptake from dilute solutions, whereas a second component might result from diffusion according to Donnan phenomena at higher salt concentrations (Brouwer, 1965). Kinetics of ion uptake at low salt concentrations indicate that a multiphasic mechanism may be present in the plasmalemma and the tonoplast. Palsmalemma may regulate ion uptake at low salt concentrations, whereas the tonoplast may limit uptake at higher concentrations (Nissen, 1973). Zinc uptake by an aquatic moss *Fontinalis antipyretica* is a step-wise process. Uptake occurred by exchange absorption, presumably on Donnan sites during the first 20 min. of absorption. Thereafter, uptake was probably active rather than passive since DNP effectively suppressed absorption (Pickering and Puia, 1969). The presence of other cations in solution also influenced the uptake of Zn; anions had little effect on Zn absorption (Chaudhary and Loneragan, 1972 b).

In recent years studies have been made on the influence of low molecular weight organic acids on zinc distribution within micronutrient pools and zinc uptake. One such study indicates that organic acid may be important in influencing the distribution of Zn between the various soil fractions and thus affecting its availability to the plant (Andrew, 1991).

3. Zinc and Plant Metabolism

The relationship of zinc in soil and plant nutrition cannot be accomplished until we know the exact role that zinc plays in the metabolism of plants. A few decades back carbonic anhydrase was the only authenticated zinc metalloenzyme in living organisms (Price *et al.*, 1972).

3.1. Zinc Containing Enzymes

Presently, zinc is recognized as an essential component in several dehydrogenases, proteinases and peptidases (Vallee and Wacker, 1970). For plants, these include carbonic anhydrase, alcohol dehydrogenase, glutamic dehydrogenase, L-lactic dehydrogenase, malic dehydrogenase, D-glyceraldehyde-3-phosphate dehydrogenase, D-lactic cytochrome c reductase, Cu-Zn superoxide dismutase, aldolase, and RNA polymerase. Several cellular proteins in plants (and more so in animals) are known to contain zinc as a constituent part and many of these function as biocatalysts, mediating rate limiting reactions in cellular metabolism. Many of these enzymes, in which zinc is tightly bound to the enzyme, have been well characterized. The zinc atom of these enzymes has no unpaired electrons, and because of this it has only one stable state in solution. It forms strong complexes with radicals of polar groups containing oxygen, nitrogen and sulphur (Agarwala, 1979). The zinc containing enzymes from the different sources (micro-organism, higher plants, animals and human beings) catalyse reactions characteristic of all the six classes of enzymes oxido-reductases, transferases, lyases, isomerases, hydrolases and lagases and thus illustrate a multiplicity of the functions of zinc.

The list of metalloenzymes has grown over the last four decades. As reported by Marschner in his book, at least four plant enzymes contain bound zinc; alcohol dehydrogenase, Cu-Zn superoxide dismutase, carbonic anhydrase, and RNA polymerase. Alcohol dehydrogenase catalyzes the reduction of acetaldehyde to ethanol. Ethanol formation takes place mainly in meristematic zones (root apices) under aerobic conditions in higher plants. Zinc deficiency in higher plants results into decreased alcoholic dehydrogenase activity. In lower plants (e.g., yeast), this decline in activity is responsible for gross metabolic changes.

Superoxide dismutase (SOD) has been found in chloroplast stroma (Asada *et al.*, 1973). In this isoenzyme zinc is associated with copper (Cu-Zn-SOD). Although the functions of zinc in the enzyme are not yet clear (Walker and Webb, 1981), the activity of the Cu-Zn-SOD is much lower than normal under conditions of zinc deficiency. It can be restored *in vitro* by zinc supply to the assay medium (Vaughan *et al.*, 1982). The copper-zinc SOD has a molecular weight of ~ 32,000, and at the active site one copper and one zinc atom are possibly closely connected by a common histidine nitrogen (Sandmann and Boger, 1983). The copper atom is involved in the mechanism of detoxification of $O^{\cdot}_2$ generated in photorespiration (Halliwell, 1978). The inactivation of superoxide is catalyzed by SOD, and the subsequent dismutation of H_2O_2 is facilitated by catalase (Elstner, 1982). In illuminated chloroplast thylakoids, the superoxide anion radical ($O^{\cdot}_2$) is generated through autoxidation of the primary electron acceptor in photosystem I (Asada *et al.*, 1974). The $O^{\cdot}_2$ generated in the thylakoid membranes is thought to diffuse to both sides of membranes : stroma and intrathylakoid space (lumen) (Asada *et al.*, 1983). Thylakoids obtained from intact spinach chloroplasts showed no superoxide dismutase (SOD) activity, but Cu, Zn- and Mn-SOD activities were detected in the presence of Triton X-100. Thylakoid membranes and the lumen fraction were separated by centrifugation after treatment of the thylakoids with a Yeda pressure cell. Cu-Zn-SOD was found in the lumen fraction. Antibody against spinach Cu-Zn-SOD did not interact with the latent Cu-Zn-SOD in the thylakoids unless triton was added. These results indicate that Cu-Zn-SOD occurs in the lumen in addition to the stroma of spinach chloroplasts, and Mn-SOD binds to the thylakoid membranes (Hayakawa *et al.*, 1984).

Carbonic anhydrase activity in plants in relation to zinc content was previously reported by Wood and Silby (1952). It is zinc metallo-enzyme and further it has been shown to be largely localized in the stroma of the chloroplasts (Jacobson *et al.*, 1975). It is suggested that the enzyme exerts a buffering action by mediating pH changes in the stroma and thus prevents the chloroplast-proteins from being denatured due to pH changes resulting from H^+ pumps or incorporation of CO_2 into ribulose-1-5 biphosphate. Zinc deficiency reduces photosynthesis and carbonic anhydrase activity in soybean (Edward and Mohamad, 1973; Randall and Bouma, 1973; Ohki, 1978).

Carbonic anhydrase from dicotyledons consists of six subunits and has a molecular weight of 180,000 and six zinc atoms per molecule (Sandmann and Boger, 1983). Marschner (1986) indicates the presence of carbonic anhydrase in cytoplasm too. In C_3 plants the carbonic anhydrase (CA) in the cytoplasm could increase the pool size of dissolved CO_2 in the cytosol of mesophyll cells. It is not HCO^-_3, however, but CO_2 that moves freely and rapidly across the chloroplast envelope, and CA in the stroma would again increase the pool size of stored CO_2. This also concludes that the substrate for RuBP carboxylase is CO_2 rather than HCO^-_3.

Agarwala *et al.* (1976) observed that in zinc deficient radish plants, aldolase activity was markedly depressed. It is presumably so because of its dimetric dissociation in the absence of sufficient zinc. Supply of zinc to the deficient plants resulted in reconstitution of the enzyme activity. Study of isoenzyme patterns associated with zinc deficiency and recovery therefrom may provide valuable information about the role of zinc in enzymes involving allostering mechanisms.

Zinc is also concerned with the synthesis of starch in higher plants (Jyung *et al.*, 1975). A comparative study of the starch metabolism of two varieties of navy beans, one susceptible to zinc deficiency and the other tolerant of the disorder. Under zinc stress the susceptible variety showed a marked decrease in the starch concentration and the activity of starch synthetase. This observation may have a great significance in cereal crop production under zinc sress. The precise role of zinc in starch synthesis has yet to be worked out.

Since zinc is reported to be involved in a number of enzymatic processes directly or indirectly, it was observed by Reed (1946) that dehydrogenase activity decreased in zinc deficient plants, while the quinones arising by the action of phenol oxidase were more abundant in them. Accumulation of inorganic phosphate in zinc deficient plant suggests a possible role of zinc in the activation of a phosphate transferring enzymes, possibly hexokinase (McElroy and Nason, 1954). Similarly, zinc exerts a beneficial effect on the N-assimilation indirectly through its influence on the nitrate reductase activity (Rossel and Ulrich, 1964; Garg *et al.*, 1986). Sharma *et al.* (1979) reported that zinc deficiency caused a marked decrease in the activity of catalase in the mature anthers of wheat. Marked reduction in nitrate reductase and catalase activities also suggest a role of zinc in the metabolism and development of wheat plants. The findings of Clarkson and Hanson (1980) strengthen this concept and provides important information that plant cells need stable metallo-enzyme complexes in the tetrahedral coordination. Zinc can partially fulfil this need. Zinc has saturated d orbitals and a small molecular size which favours tetrahedral complexes. This might indicate the importance of Zn in the various processes of plant metabolism.

But the significance of the functions of zinc containing enzymes in plant metabolism is still not well understood. A large number of growth and metabolic effects of zinc deficiency still remain unexplained in the light of the known role of zinc as a biocatalyst. While it is logical to find a decrease in the activity of zinc enzymes in the living systems subjected to zinc stress, explanations have yet to be found for the observed effect of zinc deficiency on the activities of the enzymes of which it is not the constituent part.

3.2. Tryptophan and Indole Acetic Acid Synthesis

Another important aspect in which zinc seems to be involved is tryptophan and indole-3-acetic acid synthesis. Tsui (1948) suggested that zinc deficiency affects the auxin activities of plants. The most distinct zinc deficiency symptoms - stunted growth and "little leaf"- are related to disturbances in the metabolism of auxins, in general, and indole-3-acetic acid, in particular. Zinc-deficient tomato plants showed low tryptophan content, low rate of stem elongation and low auxin activities (Tsui, 1948). Low auxin levels in zinc-deficient plants may be the result of high IAA oxidase activity (Skoog, 1940). The auxin content of the shoot apices of zinc-deficient plants is extremely low, and after the supply of zinc is restored, the level rapidly increases before the resumption of growth. Experimental evidence, however, supports the view that zinc is required for the synthesis of tryptophan, a precursor for the synthesis of indole-3-acetic acid (IAA) according to the following scheme:

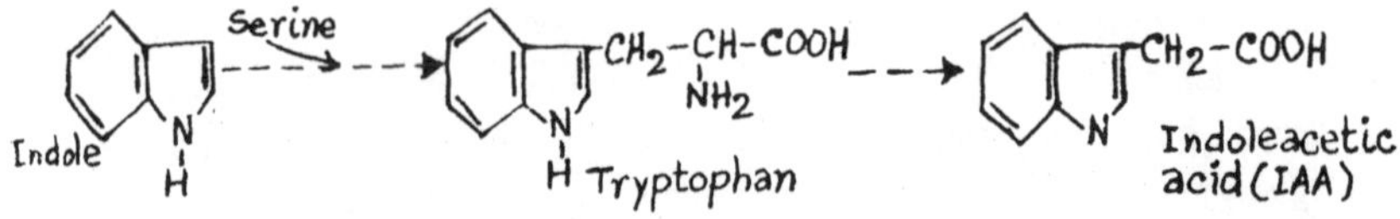

Furthermore, Salami and Kenefic (1970) observed that addition of either zinc or tryptophan to zinc-deficient solution cultures ameliorated mild zinc deficiency symptoms in maize. In agreement with this an increasing zinc supply results in an increase in the tryptophan content in the grains of rice plants growing in zinc-deficient calcareous soil (Singh, 1981). On the contrary, Takaki and Kushizaki (1970) found accumulation of tryptophan in zinc-deficient maize plants which is difficult to interpret. However, it may be suggested that in zinc-deficient maize plants accumulation of tryptophan indicates an involvement of zinc in the synthesis of auxin from tryptophan. These results confirm the findings of Chesters and Robinson (1951) and McElroy and Nason (1954). Zinc supply to wheat plants improved the seed quality with increased levels of indole-3-acetic acid in addition to other physiological attributes in leaves (Hemantaranjan, 1982; Hemantaranjan and Garg, 1988), as well as in case of french bean leaves (Hemantaranjan, 1991) zinc, either individually or in combination with iron and/or vanadium, increased IAA levels.

In a recent report, free tryptophan was detected in higher amounts of zinc-deficient radish (*Raphanus sativus*) shoots than in control shoots with the use of HPLC. Two dimensional TLC and GC-MS analysis revealed the presence of IAA in zinc-deficient radish shoots. Quantitative estimation of IAA by gas chromatography showed that zinc-deficient and control radish shoots contained 0.32 and 0.36 µg/100 g fresh weight, respectively. It could be deduced from these results that zinc nutrition did not affect significantly the amount of IAA in radish shoots (Domingo *et al.*, 1992). In recent years, not much work has been done in this direction and it needs a gross study to correlate the actual role of zinc with the growth hormone auxin.

3.3. Functions of Zinc in Chloroplast

3.3.1. *Chlorophyll and Leaf Senescence*

Zinc in relation to chlorophyll pigments and senescence has been reported by many workers. It is of common observation that chlorophyll content per unit area of leaf was decreased in zinc-deficient plants. Godnev and Lipskaya (1965) observed that there is an increase in the chlorophyll content per unit leaf area with zinc application. Previously, it was opined that adequate supply of zinc to plants within optimum limit increases not only the chlorophyll content of the leaves but also elevates the efficiency of the chloroplasts to capture solar energy (Lonstalot *et al.*, 1945). It was further suggested that these two factors, due to the involvement of zinc in metabolism, improved the photosynthetic efficiency of the plants.

One of the striking phenomena of senescence is the loss of chlorophyll (Patterson and Moss, 1979). Zinc deficiency reduces photosynthesis and carbonic anhydrase activity in soybean (Edward and Mohamad, 1973; Randall and Bouma, 1973; Ohki, 1978). It is likely that in control (+Zn) plants there was an increase in the duration of photosynthesis which possibly delayed the senescence and thereby increased the yield of wheat crop (Hemantaranjan, 1982; Hemantaranjan and Garg, 1988). The ratio of chlorophyll 'a' and chlorophyll 'b' has been suggested as a good measurement of senescence (Woolhouse, 1974; Hemantaranjan and Garg, 1984; Garg *et al.*, 1986) because, as leaves senesce, there is a more rapid destruction of chlorophyll 'a' than of chlorophyll 'b'. Leaf senescence in wheat occurs during the post anthesis period (Puckridge, 1971; Moss, 1975), which causes the canopy photosynthetic rate to decline rapidly for a maximum at anthesis. The drop in photosynthetic activity, which occurs while the grain dry weight is increasing rapidly, may be a limiting factor to yield. If leaf senescence could be delayed, then the photosynthetic activity of the canopy could be maintained. A positive response of grain yield to canopy CO_2 fertilization during the post anthesis period has been shown by Krenzer and Moss (1975) and Fischer and Aguilar (1976), this being presumably due to increase in the photosynthetic rate. Delaying senescence in plants should depend on either finding effective chemical growth regulators or identifying genetic variability for leaf senescence (Patterson *et al.*, 1980). Leaf area duration (LAD), an integration of the green leaf area index over time, is positively correlated to grain yield (Evans *et al.*, 1969; Spiertz *et al.*, 1971). IAA is also known to inhibit chlorophyll loss during aging of wheat (Mishra and Biswal, 1980; Hemantaranjan and Garg, 1988). It may be explained that IAA, photosynthetic pigments chlorophyll 'a' and 'b' and the leaf area duration, can very well be helpful in delaying the senescence of leaf for a higher yield of the crop. Obviously, that zinc has the ability to readily incorporate in the organic complexes of cells, tends to be more abundant in chloroplasts, which increases the duration of photosynthesis. This may be because of the enzyme carbonic anhydrase, a zinc metallo-enzyme, and exerts a buffering action by mediating pH changes in the stroma, and thus prevents the chloroplasts proteins from being denatured, because of pH changes (Jacobson *et al.*, 1975). Zinc deficiency very likely hampers the formation of cellular proteins and the enzyme carbonic anhydrase.

3.3.2. *Carbohydrate Metabolism*

Zinc brought about a decrease in the carbohydrate content of the leaves and stems during ear head formation which apparently facilitated the flow of carbohydrates to the reproductive organs (Shkol'nik and Chirkova, 1958; Hemantaranjan, 1982) and contributed to improved grain yield. This may be due to the influence of zinc on various redox processes and hydrolysing enzymes like carbonic anhydrase, carboxylase and dehydrogenases (Jacobson *et al.*, 1975). A sharp decline in the activities of various photosynthetic enzymes were noted in leaves suffering from increasing zinc deficiency (Shortri *et al.*, 1983). In CO_2 assimilation pathway, the activity of fructose-1, 6-biphosphatase declines, fairly rapidly, which is a key enzyme in the partitioning of C_6 sugars in the chloroplasts and the cytoplasm according to the following scheme :

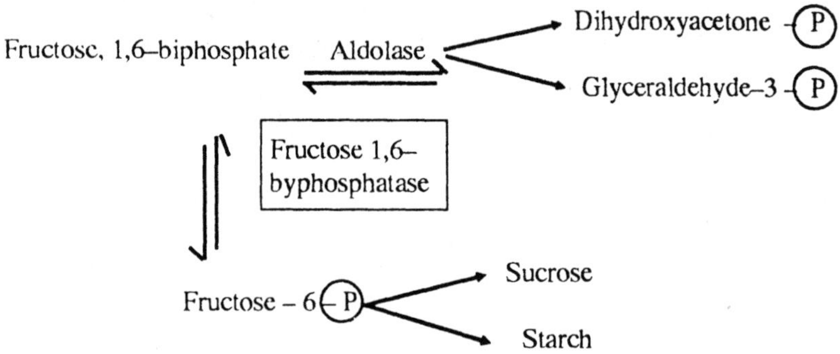

Aldolase isoenzymes, another group of key enzymes in carbohydrate metabolism in the chloroplast, regulate the transfer of C_3 photosynthates to the cytoplasm (see diagram), and in the cytoplasm they regulate the transfer of C_6 sugars to the glycolytic pathway. A direct connection between zinc nutritional status and starch formation has been found only in bean (Jyung *et al.*, 1975), where zinc deficiency leads to a decrease in both starch contents and starch synthase activity. According to O'Sullivan (1971), aldolase activity is drastically and specifically reduced in various plant species and may serve as an indicator of the zinc nutritional status of plants.

A study on the effect of zinc deficiency and resumption of zinc supply on zinc and carbohydrate contents of cabbage leaves indicate sugar and starch accumulation in zinc-deficient leaves despite a drastic reduction in the rate of photosynthesis (as indicated by Hill reaction activity) (Sharma *et al.*, 1982). As early as twenty four hours after the zinc supply is restored, the sugar levels and the Hill reaction activity are again comparable to that of the adequately supplied control plants continuously receiving 1.0 μ Møzinc

Marschner (1986) concluded that most experimental evidence obtained with green plants supports the view that zinc deficiency-induced changes in carbohydrate metabolism are not primarily responsible for either growth retardation or the visible symptoms of zinc deficiency.

TABLE 1 : Effect of zinc deficiency on the total carbohydrate (%) in different parts of wheat plants at different stages of growth*

Treatment	Age of plants (days after sowing)					
	35	50	65	80	95	120
			Stems			
Control + Zn	14.50	16.80	21.00	20.00	18.80	14.20
- Zn	8.50	11.00	15.00	12.00	10.80	8.20
			Leaves			
Control + Zn	17.10	21.00	23.00	20.80	17.90	13.50
- Zn	13.00	16.10	19.00	15.50	14.00	10.20
			Grains (recorded after harvest, i.e. 120 days after sowing)			
Control + Zn			42.10			
- Zn			30.50			

* Hemantaranjan (1982)

It is apparent from table 1 that there is a positive relationship between zinc and acid hydrolyzable carbohydrate content in stem, leaves and grains of the wheat plants. It reveals well that zinc deficiency reduced carbohydrate in plant organs. At 65 days stage, carbohydrate was maximum in stems and leaves whereafter a gradual decrease appeared upto the harvest stage. In grains also, the control plants exhibited a higher percentage of carbohydrates, i.e., 42.10 per cent when it was found only 30.50 per cent in zinc-deficient plants. This proves the incorporation of zinc towards better production of carbohydrates in grains (Hemantaranjan, 1982).

3.4. Protein synthesis

The rate of protein synthesis and the protein content of zinc-deficient plants are drastically reduced along with the accumulation of amino acids and amides in tomato plants, which indicates the importance of zinc for protein synthesis (Possingham, 1957). Improved translocation of proteins to grains in the zinc applied plants contributed to the improved grain yield (Shkol'nik and Chrikova, 1958). Zinc exerts a beneficial effect on the N-assimilation indirectly through its influence on the nitrate reductase activity (Russel and Ulrich, 1964). The role of zinc in protein synthesis is further strengthened through its involvement in the nucleic acid metabolism (Wacker and Vallee, 1959; Tal, 1969). Enzymes concerned with the synthesis of DNA and RNA have been shown to contain zinc. Scrutton *et al.* (1971) showed that DNA-dependent RNA polymerase from *Escherichia coli*, contains two tightly bound zinc atoms per molecule of the enzyme. The concentration of cellular protein as well as nucleoproteins is also reported to be decreased by zinc deficiency (Praske and Plocke, 1971). Furthermore, ribosomes become very unstable or even disappear under conditions of zinc deficiency, thus causing impairment in protein synthesis. The decreased levels of RNA and accumulation of RNA nucleosides and their di- and triphosphatases could be an effect of increased hydrolysis of RNA under conditions of zinc deficiency (Agarwala, 1979).

TABLE 2 : Effect of zinc deficiency on the ascorbic acid (mg/g) content in leaves and stem of wheat (at different stages of growth) Hemantaranjan (1982)

Treatment	Age of plants (days after sowing)			
	35	50	65	80
		Leaves		
Control + Zn	2.40	8.50	16.60	4.40
- Zn	1.60	5.60	9.50	2.50
		Stem		
Control + Zn	2.20	5.80	10.90	4.80
- Zn	1.50	3.80	6.50	2.70

From studies with *Euglena* it is well established that zinc is an essential component of RNA polymerase. As mentioned, there are about two atoms of zinc per molecule, and if the zinc is removed, the enzyme is inactivated (Falchuk *et al.*, 1977). The zinc content of ribosomal RNA in zinc-sufficient cells of *Euglena* is in the range of 300 to 380 µg per g RNA (Prask and Plocke, 1971). The decrease in the protein content of zinc-deficient plants is also the result of rapid rates of RNA degradation. Higher rates of RNase activity are a typical feature of zinc deficiency (Sharma *et al.*, 1982). An inverse correlation exists between zinc supply and RNase activity and also between RNase activity and protein content, as observed through the studies of Johnson and Simons (1979). An increase in RNase activity is observed even before symptoms of zinc deficiency such as stunted growth and changes in leaf anatomy become apparent (Dwivedi and Takkar, 1974).

In addition to the obvious involvement of zinc in enzymatic activities, auxin metabolism, chlorophyll and photosynthesis, carbohydrate metabolism and protein synthesis, zinc also appears to be involved in ascorbic acid synthesis (Hemantaranjan, 1982). Zinc deficiency caused a reduction in the ascorbic acid content in leaves and stems of wheat (Table 2), which appears to be related to photosynthetic efficiency of plants. Vallee (1977) opined that many functions of zinc are similar to those of vitamins. Zinc seems to serve as a cofactor with many roles, since zinc may be essential for activity, regulation, stabilization of structure, or any combination of these.

4. INTERACTIONS WITH OTHER NUTRIENTS

Interactions of zinc with other elements modify plant nutrition by affecting some growth factors. Adequate zinc nutrition in plants depends on various factors. An interaction in plant growth can be defined as a mutual or reciprocal action of one element upon another. In this section some of the main interactions of zinc with other elements are briefly discussed.

4.1. Zinc-Phosphorus

Interaction of zinc and phosporus has been widely studied, with the earliest work reported by Barnette *et al.* (1936). Interaction between Zn and P in plant nutrition is primarily a plant physiological phenomenon rather than a soil-related problem (Burleson *et al.*, 1961; Stuckenholtz *et al.*, 1966; Burleson and Page, 1967). The Zn-P interaction is usually associated with high levels of available P or with application to soil that are low in available zinc. High uptake of P antagonizes or reduces uptake of Zn frequently to the point of zinc deficiency. This disorder, termed a phosphorus-induced zinc deficiency, can usually be prevented or corrected by applying zinc to the soil.

The site of Zn-P interaction has been recognized in the root zone (Loneragan and Asher, 1967; Sharma *et al.*, 1968 a,b) and the zinc deficiency has been attributed to the immobilization of zinc owing to increase in the concentrations of phosphorus in the roots above the threshold values (Boawn and Leggett, 1964). The autoradiographs and radiochemical analysis showed that a high level of P rendered the applied zinc unavailable to the leaves by immobilizing almost 40 per cent or more of the total absorbed zinc in the roots and 20 per cent or more at the nodes of the stem than that with 12 and 6 per cent in the roots and the nodes respectively at the normal dose of P supply (30 ppm P) (Dwivedi *et al.*, 1975).

As summarized by Loneragan *et al.* (1979), three different factors may be responsible for this : (a) "dilution" of zinc in plants by the increase in growth induced by phosphorus fertilizers, (b) inhibition of zinc uptake by the cations (Ca^{2+} in particular) added with phosphorus fertilizers, and (c) phosphorus enhanced zinc adsorption in the soil to hydroxides and oxides of iron and aluminum and to $CaCO_3$.

Results obtained from the effect of phosphorus supply on shoot dry weight and phosphorus and zinc content of young grapevine leaves (Marschner and Schropp, 1977) strongly support the view that the major Zn-P interactions take place not in the plant (Ghoneim and Bussler, 1980) but in the soil, where the availability of zinc and zinc diffusion rates are reduced by a large phosphorus supply (Schropp and Marschner, 1977). This decrease in zinc availability is brought about by enhanced adsorption of zinc on, for example, iron oxides (Loneragan, 1975; Loneragan *et al.*, 1979) rather than by the precipitation of zinc phosphates, which are relatively good sources of zinc for plant growth (Jurinak and Inouye, 1962).

When rate of plant growth exceeds rate of uptake of nutrient, concentrations of that nutrient decrease in the plant tissue. With marginal zinc supply, increased plant growth with adequate phosphorus can result in decreased plant zinc concentrations to the point of zinc deficiency. With adequate zinc supply in the soil, there should be sufficient zinc for optimum growth even when plant zinc concentrations are diluted because of increased growth from added phosphorus. In past, it has been demonstrated that polyphosphate solutions can maintain certain metallic cations in solution through sequestration (Rogers and Reynolds, 1949). This fact prompted speculation that the ammonium polyphosphates, when applied to a soil, may sequester native and applied zinc and thus increase its concentration in the soil solution. Whether such a sequestering effect would be long-lived is questionable, since the rate of hydrolysis of polyphosphate to orthophosphate in soils is often quite rapid (Sutton and Larsen, 1964; Gilliam and Sample, 1968).

Under conditons of zinc deficiency and ample supply of phosphorus the phosphorus uptake increase and toxic concentration of phosphorus may accumulate (Loneragan *et al.,* 1979). Therefore, under these experimental conditions, the symptoms of phosphorus toxicity (e.g., necrosis on mature leaves) in zinc deficient plants may be mistaken for evidence of an accentuation of the zinc deficiency because of the large phosphorus/zinc ratio.

New insight into Zn-P interactions in plants has been provided in the last few years. It was suggested that relatively low phosphorus/zinc ratios in grass shoots had stimulated the production of higher than normal concentrations of auxin, which in turn had increased the calcium requirement of growing tissue. Using selected data, it was possible to show that relationships existed between the critical concentration of calcium in grass and the concentration of phosphorus and zinc, and between the critical concentrations of phosphorus in grass and the concentrations of calcium and zinc (Bailey and Stevens, 1989). The results suggest that the reputed phosphorus-sparing effect of lime on grassland may be due in part to the lower internal phosporus requirements of grass grown on limited soils.

The effect of zinc supply on phosphorus accumulation was studied in wheat (Webb and Loneragan, 1990), which suggests that the enhanced P concentration in Zn deficient plants was mediated primarily through an effect of zinc in the shoots rather than in roots.

Although the connection between zinc deficiency and phosphorus toxicity is not yet fully understood, there is substantial evidence that zinc affects phosphorus metabolism in roots (Loughman *et al.,* 1982) and increase the permeability of the plasma membranes of root cells to phosphorus, and also to chloride (Welch *et al.,* 1982). It is well documented that zinc stabilizes biomembranes, presumably by reacting with the sulphydryl groups (R–SH) of membrane proteins (Chvapil, 1973). Zinc may therefore have specific functions in the structural orientation of macromolecules within membranes and thus in membrane integrity (Welch *et al.*, 1982). These effects of zinc are presumably related also to the protective role of SOD enzymes against lipid peroxidation of membranes, the latter process being closely related to the senescence of tissues (Pauls and Thompson, 1984).

In some studies conducted at Hissar, India (Gupta *et al.*, 1970; Safaya *et al.*, 1974) on several varieties of maize, a general disruption in the normal mode of phosphate (P^{32}) absorption and translocation resulting in heavy accumulation of phosphate in the shoots has been observed in zinc-deficient plants. These findings are suggestive of some involvement of zinc in the absorption and translocation of phosphorus itself.

4.2. Zinc-Nitrogen

Judicious application of nitrogen as a possible cause of zinc deficiency in citrus was recognized as early as 1936. From one of the studies on citrus die back conducted at Hissar during 1967-68, application of more than 3 kg of CAN per tree proved to be highly detrimental in the utilization of Zn as well as Cu and Fe. Increasing the nitrogen supply usually results in increased plant growth, which can dilute the zinc concentration. The effects of nitrogen on zinc concentrations in plants are also related to nitrogen source effects in decreasing soil pH; this would increase zinc uptake (Giordano *et al.*, 1966). The similar pH effect (increasing zinc uptake) was noted in grain sorghum (*Sorghum bicolor* (L.) *Moench*) and potatoes (*Solanum tuberosum* L.); zinc concentrations and uptake were also found to be increased in sugar beets (*Beta vulgaris* L.) (Boawn *et al.*, 1960). Others suggested that fertilizer nitrogen may enhance zinc uptake by increasing protein nitrogen, which retains zinc, or by increasing the exchange capacity of roots, thus stimulating the absorptive capacity for zinc (Lindsay, 1972).

This has also been established in the recent years that zinc has a beneficial effect on the content of nitrogen and nitrogen uptake values (Garg and Hemantaranjan, 1981; Hemantaranjan, 1982; Yadav and Shukla, 1983; Hemantaranjan and Garg, 1986) in cases of wheat, chick pea and french bean. These findings are in close conformity with the views of Rajgopal and Mehta (1969), Yie (1969), and Shkol'nik and Chrikova (1958).

4.3. Zinc-Iron

Zinc and iron are apparently antagonistic in plant nutrition, but not in all cases and rather have a beneficial effect on the growth and yield of crop if supplied in combinations (Hemantaranjan and Garg, 1986; Garg *et al.*, 1986; Hemantaranjan and Garg, 1988). In case of Hawkeye soybeans (*Glycine max* L.), an iron-efficient cultivar when grown in calcareous Hacienda loam soil low in P and available Fe did not respond to P, FeEDDHA, NH^+_4-N or NO^-_3-N when applied singly nor to Fe + P together; but did give yield response for N + P + Fe applied together. Addition of Fe decreased Zn and Mn concentrations in leaves, while addition of both Fe and P decreased Zn concentration even more. Bicarbonate contributed additionally to low Zn in leaves with treatments of Fe and Fe + P (Wallace and Cha, 1986). Lingle *et al.* (1963) found that zinc interfered with the absorption and translocation of iron in soybeans. Brown and Tiffin (1962) showed that added zinc resulted in iron chlorosis in corn and millet (*Setaria italica* L.). Mortvedt and Giordano (1971) demonstrated mutual iron-zinc antagonism in grain sorghum grown on calcareous soil. Zinc concentration and uptake decreased with increased rates of FeEDDHA, and forage yields decreased at the highest iron rates, with tissue zinc concentrations decreasing to 12 ppm. In contrast, soil applications of zinc in the absence of applied iron also resulted in decreased forage yields. Therefore, iron fertilization of soils that are marginal in zinc can result in zinc deficiencies, and vice versa.

A study on the interactions of iron, manganese, and zinc during uptake and transport by maize (*Zea mays*) and sunflower was made by Scherer and Hoefner (1980) in solution culture. The Fe and Zn concentrations in corn above ground parts were lowered, whereas those of the roots were increased by increasing Mn concentrations (0, 0.03 and 0.33 mg Mn/L nutrient solution). The Fe and Zn concentrations of sunflower plants decreased, but their total amounts in the plants increased with increasing Mn concentrations. An increasing Fe concentration (0.1, 0.7 and 4.9 mg/L) reduced the Zn content of all corn organs, whereas the Mn concentrations were not decreased. A negative effect of increasing Fe concentrations of Mn uptake and accumulation by sunflower plants was observed. This study indicates that in the interaction studies the reciprocal effects of the elements not only on the other element concentration but also on its total accumulation should be taken into account (Scherer and Hoefner, 1980).

5. Conclusions

Amongst the remarkable discoveries, the one relating to mineral nutrition of plants is of special importance in plant growth and development. The growing knowledge of mineral nutrition of plants has made us realize that great changes are possible through judicious supply of certain inorganic substances. According to Hewitt (1963), the deficiency of an essential element creates a metabolic disorder followed by catenary events which result eventually in far-reaching disruption of the normal metabolic pattern. The primary biochemcial events are almost inhibition of enzymatic steps caused by the subnormal activity of an enzyme or coenzymes, or by the deficiency of substrates.

Zinc is one of the micronutrients which, even in minute quantities in soil, is indispensable for plant growth. In fact, zinc deficiency has the distinction of being the most widespread of all the micronutrient problems. The availability of zinc to the plant is affected by several factors : edaphic, environmental and genotypic. Availability of zinc is very much dependent on soil pH. Plant uptake is dependent on solubility and movement to roots. Diffusion is a critical phenomenon in regulating zinc uptake by plants as well as in determining the effectiveness of various zinc fertilizers. Chelating or complexing agents of natural or synthetic origin influence zinc diffusion by forming a larger chelated ion, by decreasing the reaction of zinc with soil, and by increasing zinc concentrations in solution. Most evidence suggests that uptake of zinc is predominantly an active process. However, many factors may interact to influence interpretation.

Several factors are known to affect zinc uptake. Since organic matter is a source of zinc for plants, cool soils may reduce the rate of microbial release of available zinc. A second factor is delayed root growth at low temperature, which would limit zinc uptake. Recent evidence indicate that organic acid may be important in influencing the distribution of zinc between the various soil fractions and thus affecting its availability to the plant (Andrew, 1991).

Several cellular proteins in plants are known to contain zinc as a constituent part, and many of these functions as biocatalysts in cellular metabolism. Many of these enzymes, in which zinc is tightly bound to the enzyme, have been well characterized. However, the significance of the functions of zinc containing enzymes in plant metabolism is still not well understood. A large number of growth and metabolic effects of zinc deficiency still remain unexplained in the light of the known role of zinc as a biocatalyst.

Another important aspect in which zinc seems to be involved is tryptophan and indole-3-acetic acid synthesis. In addition, zinc in relation to chlorophyll pigments and senescence has been reported by many workers. Zinc has the ability to readily incorporate in the organic complexes of cells, tends to be more abundant in chloroplasts, which increases the duration of photosynthesis. Zinc is also important for the functions of enzymes involved in carbohydrate metabolism and protein synthesis, thus, directly or indirectly it regulates the two vital physiological processes. Furthermore, zinc appears to be involved in ascorbic acid synthesis.

Zinc also interacts with other inorganic nutrients, viz., phosphorus, nitrogen, iron, manganese copper, etc., which modify plant nutrition by affecting some growth factors, as described in the preceding pages. Some of the recent researches on zinc nutrition have been discussed in the subsequent chapters by Dr. I.C Dave which doubtless accomplishes our knowledge to a great extent.

REFERENCES

Agarwala, S.C. 1979. Yield component compensation - Alternative interpretations. *Crop Sci.*, **11** : 33-35.

Aubert, H. and Pinta, M. 1977. Trace Elements in soils. E sevier Scientific Publishing Company, Amsterdam-Oxford-New York. pp. 85-94.

Agarwala, S.C., Bisht, S.S., Sharma, C.P. and Afzal, A. 1976. Effect of deficiency of certain micronutrients on the activity of aldolase in radish plants grown in sand culture. *Can.J.Bot.*, **54** : 76-78.

Andrew, E. Jr. 1991. Influence of low molecular weight organic acids on zinc distribution within micronutrient pools and zinc uptake by wheat. *J.Plant Nutr.*, **14** : 1307-1318.

Asada, K., Urano, M. and Takahashi, M. 1973. Subcellular location of superoxide in spinach leaves and preparation and properties of crystalline spinach superoxide dismutase. *Eur. J. Biochem.*, **36** : 257-266.

Asada, K., Kiso, K. and Yoshikawa, K. 1974. Univalent reduction of molecular oxygen by spinach chloroplasts on illumination. *J.Biol.Chem.*, **249** : 2175-2181.

Asada, K., Takahashi, M. and Hayakawa, T. 1983. Photoreduction of superoxide in membranes of chloroplast thylakoids and intrachloroplast distribution of superoxide dismutase. *In* Oxy Radicals and Their Scavenger Systems, Vol.I. Edited by G. Cohen and R.A. Greenwald. pp. 240-245. Elsevier Biochemical, N.Y.

Aubert, H. and Pinta, M. 1977. Trace Elements in Soils. Elsevier Scientific Publishing Company, Amsterdam. pp. 85-94.

Bailey, J.S. and Stevens, J. 1989. Factors explaining variable lime responses in pastures : calcium x zinc x phosphorus interactions. *J. Plant Nutr.*, **12** : 387-405.

Barnette, R.M., Camp, J.P., Warner, J.D. and Gall, O.E. 1936. "Use of zinc sulphate under corn and other field crops.". *Fla. Agr. Exp. Bull.*, **293** : 3.

Boawn, L.C. and Leggett, G.E. 1964. Phosphorus and zinc concentration in Russettee Burbank Potato tissue in relation to development of zinc deficiency symptoms. *Soil Sci. Soc. Amer. Proc.*, **28** : 229-232.

Boawn, L.C., Viets, F.G.Jr., Crawford, C.L. and Nelson, J.L. 1960. Effect of nitrogen carrier, nitrogen rate, zinc rate, and soil pH on zinc uptake by sorghum potatoes and sugarbeets. *Soil Sci.*, **90** : 329-337.

Broda, E. 1965. Mechanism of uptake of trace elements by plants (experiments with radiozinc). *Soil Plant Nutr.Studies Proc. Symp.*, Ankara, pp. 207-215.

Brouwer, R. 1965. Ion absorption and transport in plants. *Annu.Rev.Plant Physiol.*, **16**: 241-261.

Brown, G.N. and Taylor, F.G. 1966. $Zinc^{65}$ distribution and radioactivity of growth and Zn^{65} uptake in red oak seedlings. *Radiation Bot.*, **6** : 519-524.

Brown, J.C. and Tiffin, L.O. 1962. Zinc deficiency and iron chlorosis dependent on the plant species and nutrient-element balance in Tulare clay. *Agron.J.*, **54** : 356-363.

Burleson, C.A., Dacus, A.D. and Gerard, C.J. 1961. The effect of P fertilization on the Zn nutrition of several irrigated crops. *Soil Sci. Soc. Am. Proc.*, **25** : 365-368

Burleson, C.A. and Page, N.R. 1967. Phosphorus and zinc interactions in flax. *Soil Sci. Soc. Am. Proc.*, **31** : 510-513.

Caroll, M.D. and Loneragan, J.F. 1968. Response of plant species to concentrations of zinc in solution. I. Growth and zinc content of plants. *Austr.J.Agric.Res.*, **19** : 859-868.

Chapman, H.D. 1966. *In* "Diagnostic Criteria for Plants and Soils" (H.D. Chapman ed.), pp. 484-499. University of California Press, Riverside.

Chatterjee, C., Agarwala, S.C. and Sharma, P.N. 1976. Micronutrient status of principal soil types of Uttar Pradesh. *Geophytology*, **6** : 245-258.

Chaudhary, F.M. and Loneragan, J.F. 1972. Zinc absorption by wheat seedlings : I. Inhibition of macronutrient ions in short term experiments and its relevance to long term zinc nutrition. *Proc.Soil Sci.Soc. Am.*, **36** : 327-331.

Chaudhary, F.M. and Loneragan, J.F. 1972b. Zinc absorption by wheat seedlings : II. Inhibition by hydrogen ions and by micronutrient cations. *Proc.Soil Sci.Soc.Am.*, **36** : 327-331.

Chesters, C.G.C. and Robinson, G.N. 1951. The role of zinc in plant metabolism. *Biol. Rev.*, **26** : 239-252.

Chvapil, M. 1973. New aspects in the biological role of zinc : A stabilizer of macromolecules and biological membranes. *Life Sci.* **13** : 1041-1049.

Clarkson, T.D. and Hanson, B.J. 1980. The mineral nutrition of higher plants. *Annu. Rev. Plant Physiol.*, **31** : 239-298.

Domingo, Nagatomo, A.L.Y., Tamai, M. and Takaki, H. 1992. Free-tryptophan and indole-3-acetic acid in zinc-deficient radish shoots. *Soil Sci. Plant Nutr.*, **38** : 261-267.

Dwivedi, R.S. and Takkar, P.N. 1974. Ribonuclease activity as an index for the hidden hunger of zinc in plants. *Plant Soil,* **40** : 173- 181.

Dwivedi, R.S., Randhawa, N.S and Bansal, R.L. 1975. Phosphorus-zinc interaction : I. Sites of immobilization of zinc in maize at a high level of phosphorus. *Plant Soil.*, **43** : 639-648.

Edward, G.E. and Mohamad, A.K. 1973. Reduction in carbonic anhydrase activity in zinc deficient leaves of *Phaseolus vulgaris* L. *Crop Sci.*, **13** : 351-354.

Elstner, E.F. 1982. Oxygen activation and oxygen toxicity. *Annu. Rev. Plant Physiol.*, **33** : 73-96.

Emine Polar. 1970. The distribution of Zn^{65} in the cotyledons of *Vicia faba* and its translocation during the growth and maturation of the plant. *Plant Soil*, **32** : 1.

Evans, L.T., Wardlaw, I.F. and Fischer, R.A. 1969. Wheat. pp.101-149. In L.T. Evans (ed.) Crop Physiology : Some case histories, Cambridge Univ. Press, New York.

Falchuk, K.H., Ulpino, L., Mazus, B. and Valee, B.L. 1977. *E.gracilis* RNA polymerase. I. A zinc metalloenzyme. *Biochem. Biophys. Res. Commun.*, **74** : 1206-1212.

Fischer, R.A. and Aguilar, I.M. 1976. Yield potential in a dwarf spring wheat and the effect of CO_2 fertilization. *Crop Sci.*, **68** : 749-752.

Forno, D.A., Yoshida, S. and Asher, C.J. 1975. Zinc deficiency in rice I. Soil factors associated with the deficiency. *Plant Soil,* **42** : 537-550.

Garg, O.K. and Hemantaranjan, A. 1981. Interaction of zinc and nitrogen on the yield of *Triticum aestivum* L. *Proc. Indian Sci. Congress,* Section-Botany, No. 319, p.151.

Garg, O.K., Hemantaranjan, A. and Ramesh, C. 1986. Effect of iron and zinc fertilization on senescence in French bean (*Phaseolus vulgaris* L.). *J.Plant Nutr.,* **9** : 257-266.

Ghoneim, M.F. and Bussler, W. 1980. Diagnosis of zinc deficiency in cotton. *Z. Pflanzenenernaehr. Bodenkd.,* **143** : 377-384.

Gilliam, J.W. and Sample, E.C. 1968. Hydrolysis of pyrophosphate in soils : pH and biological effects. *Soil Sci.,* **106** : 352-357.

Giordano, P.M. 1977. Efficiency of zinc fertilization for flooded rice. *Plant Soil,* **48** : 673-684.

Giordano, P.M. and Mortvedt, J.J. 1978. Response of corn to zinc source in ortho- and pyrophosphate fertilizers, as affected by soil temperature and moisture. *Agron. J.,* **70** : 531-534.

Giordano, P.M., Mortvedt, J.J. and Papendick, R.I. 1966. Response of corn (*Zea mays* L.) to zinc as affected by placement and nitrogen source. *Soil Sci. Soc. Am. Proc.,* **30** : 767-770.

Giordano, P.M., Noggle, J.C. and Mortvedt, J.J. 1974. Zinc uptake by rice as affected by metabolic inhibitors and competing cations. *Plant Soil.,* **41** : 637-646.

Gladstones, J.S. and Loneragan, J.F. 1966. Mineral elements in temperate crop and pasture plants. I. Zinc. *Austr. J. Agric. Res.,* **18** : 427-446.

Godnev, T.N. and Lipskaya, G.A. 1965. Accumulation of chlorophyll in suger beet chloroplasts under the action of Co, Mo and Zn. *Fiziol. Rast.,* **12** : 1012-1016.

Gupta, A.P., Safaya, N.M., Virmani, S.M. and Shukla, U.C.1970. Proc. Deptt. Atomic Energy Symposium on Radiation and Radioisotopes in Soil Studies and Plant Nutrition, Bangalore (See Safaya *et al.,* 1974). pp.321-326.

Halliwell, B. 1978. Biochemical mechanisms accounting for the toxic action of oxygen on living organisms : The key role of superoxide dismutase. *Cell Biol.Int.Rep.,* **2**: 113-128.

Hayakawa,T., Kanematsu, S. and Asada, K. 1984. Occurrence of Cu, Zn-superoxide dismutase in the intrathylakoid space of spinach chloroplasts. *Plant Cell Physiol.,* **25** : 883-890.

Hemantaranjan, A. 1982. Physiological signification of zinc fertilization in wheat (*Triticum aestivum* L.). Ph.D. Thesis, Banaras Hindu University, India.

Hemantaranjan, A. 1991. Yield potentiality of *Phaseolus vulgaris* L. with special reference to vanadium, zinc and iron. *In : Physiological Strategies For Crop Improvement* (Eds. D.N. Tyagi, B.Bose, A. Hemantaranjan and T. Medhabati Devi), Indian Society For Plant Physiology and Banaras Hindu University, India. pp. 49-55.

Hemantaranjan, A. and Garg, O.K. 1984. Effect of zinc fertilization on the senescence of wheat varieties. *Indian J. Plant Physiolgy,* **27** : 239- 245.

Hemantaranjan, A. and Garg, O.K.1986. Introduction of nitrogen-fixing nodules through iron and zinc fertilization in non-nodule forming french bean (*Phaseolus vulgaris* L.). *J. Plant Nutr.,* **9** : 281-288.

Hemantaranjan, A. and Garg, O.K. 1988. Effect of iron and zinc fertilization on grain quality of *Triticum aestivum* L. *J. Plant Nutr.,* **11** : 1439-1450.

Hewitt, E.J. 1963. The essential nutrient elements : requirements and interactions in plants. *In* : "Plant Physiology(F.C.Steward ed.) Academic Press, New York, **3** : 137-360.

Hiatt, A.J. 1968. Electrostatic association and Donnan phenomena as mechanism of ion accumulation. Plant Physiol., **43** : 893-901.

Jacobson, B.S., Fong, F. and Heath, R.L. 1975. Carbonic anhydrase of spinach. Studies on its location, inhibition and physiological function. *Plant Physiol.,* **55** : 468-474.

Javellier, M.1912. The use of zinc as catalytic fertilizer. *Int. Congr. Appl. Chem.,* **15** : 143.

Johnson, A.D. and Simons, J.G. 1979. Diagnostic indices of zinc deficiency in tropical legumes. *J. Plant Nutr.,* **1** : 123-149.

Joseph, C., Rathmore, V.S., Bajaj, Y.P.S. and Wittwer, S.H. 1971. Mechanism of zinc absorption by intact bean plants. *Ann. Bot.,* **35** : 683-686.

Jurinak, J.J. and Inouye, T.S. 1962. Some aspects of zinc and copper phosphate formation in aqueous systems. *Soil Sci. Soc. Am. Proc.,* **26** : 144-147.

Jyung, W.H., Ehmann, A., Schlender,K.K. and Scala, J. 1975. Zinc nutrition and starch metabolism in *Phaseolus vulgaris* L. *Plant Physiol.,* **55** : 414-420.

Krenzer, E.G. Jr. and Moss, D.N. 1975. CO_2 enrichment effects upon yield and yield components in wheat. *Crop Sci.,* **15** : 71-74.

Langston, R. 1956. Distribution patterns of radioisotopes in plants. *Proc. Am. Soc. Hort. Sci.,* **68** : 370-376

Lindsay, W.L. 1972. Zinc in soils and Plant Nutrition. *Adv. Agron., 24* : 147-186.

Lingle, J.C., Tiffin, L.O. and Brown, J.C. 1963. Iron uptake transport of soybeans as influenced by other cations. *Plant Physiol.,* **38** : 71-76.

Loneragan, J.F. 1975. The availability and absorption of trace elements in soil-plant systems and their relation to movement and concentrations of trace elements in plants. *In* : "Trace elements in Soil-Plant-Animal Systems"(D.J.D. Nicholas and A.R. Egan, eds.), pp. 109-134. Academic Press, London and Orlando.

Loneragan, J.F. and Asher, C.H. 1967. Response of plants to phosphate concentration in solution culture. II. Role of phosphate absorption and its relation to growth. *Soil Sci.,* **103** : 311-318.

Loneragan, J.F., Grove, T.S., Robson, A.D. and Snowball, K. 1979. Phosphorus toxicity as a factor in zinc-phosphorus interactions in plants. *Soil Sci. Soc. Am. J.,* **43** : 966-972.

Lonstalot, A.J., Burrows, F.W., Gilbert, S.G. and Nason, A. 1945. Effect of Cu and Zn deficiency on the photosynthetic activity of the foliage of young tung trees. *Plant Physiol.*, **20** : 283-288.

Loughman, B.C., Webb, M.J. and Loneragan, J.F. 1982. Zinc and the utilization of phosphate in wheat plants. *In :"Proceedings of the Ninth International Colloguium, Warwick, England"* (A. Scaife, ed), pp. 335-340. *Commonw. Agric. Bur.*, Farnham Royal, Bucks.

Makhonina,G.I., Molchanna, J.V., Subbotina, E.M., Timofeev, R.N.V., Titlyanova, A.A. and Tyuryukanov, A.N. (1960). The experimental investigation of the distribution of radioisotopes in natural biogeocoenoses. *Dokdady. Akad. Nauk. SSSR.*, **133** : 484-487.

Marschner, H. 1986. Mineral Nutrition of Higher Plants. pp. 300-312. Academic Press, London, Orlando, San Diego, New York, Austin, Boston, Sydney, Tokyo, Toronto.

Marschner, H. and Schropp, A. 1977. Vergleichende Untersuchungen uber die Empfindlichkeit von 6 Unterlagensorten der Weinrebe gegenuber Phosphatinduziertem Zink-Mangel. *Vitis,* **16** : 79-88.

McElroy, W.D. and Nason, A. 1954. Mechanism of action of micronutrient elements in enzyme systems. *Annu. Rev. Plant Physiol.*, **5** : 1.

Mikkelson, D.S. and Kuo, S.1977. Zinc fertilization and behavior in flooded soils. *Commonwealth Bureau of Soils, Spec. Publ.*, **5** : 59.

Mishra, A.N. and Biswal, U.C. 1980. Effect of phytohormones on chlorophyll degradation during aging of chloroplasts *in vivo* and *in vitro. Protoplasma,* **105** : 1-8.

Mortvedt, J.J. and Giordano, P.M. 1971. Response of grain sorghum to iron sources applied alone or with fertilizers. *Agron. J.*, **63** : 758-761.

Moss, D.N. 1975. Studies on increasing photosynthesis in crop plants. *In*: R.H. Burris and C.C. Black (eds.) CO_2 Metabolism in plant productivity. pp.31-41. University Park Press, Baltimore.

Navarot, J., Jacoby, B. and Ravikovitch, S. 1967. Fixation of Zn^{65} in some calcareous soils and its availability to tomato plants. *Plant Soil,* **27** : 141-147.

Nissen, P.1973. Kinetics of ion uptake in higher plants. *Physiol. Plantarum,* **28** : 113-120.

Ohki, K. 1978. Zinc concentration in soybean as related to growth, photosynthesis and carbonic anhydrase activity. *Crop. Sci.*, **18** : 77.

O'Sullivan, M. 1971. Aldolase activity in plants as an indicator of zinc deficiency. *J. Sci. Food Agric.*, **21** : 607-609.

Patterson, T.G. and Moss, D.N. 1979. Senescence in field-grown wheat. *Crop Sci.*, **19**: 635-640.

Patterson, T.G., Moss, D.N. and Burn, W.A. 1980. Enzymatic changes during the senescence of field-grown wheat. *Crop Sci.*, **20** : 15-18.

Pauls, K.P. and Thompson, J.E. 1984. Evidence for the accumulation of peroxidized lipids in membranes of senescing cotyledons. *Plant Physiol.*, **75** : 1152-1157.

Peribock .1959.

Pickering, D.C. and Puia, I.L. 1969. Mechanism for the uptake of zinc by *Fontinalis antipyretica. Physiol. Plantarum,* **22** : 653.

Possingham, J.V. 1957. The effect of mineral nutrition on the content of free amino acids and amides in tomato plants. I. A comparison of the effects of deficiencies of copper, zinc, manganese, iron and molybdenum. *Aust. J. Biol. Sci.,* **10** : 539-551.

Praske, J.A. and Plocke, D.J. 1971. A role of zinc in the structural integrity of the cytoplasmic ribosomes of *Euglena Gracilis. Plant Physiol.,* **48** : 150-155.

Price, C.A., Clark, H.E. and Funkhouser, E.A. 1972. In "Micronutrients in Agriculture" (J.J.Mortvedt, P.M. Giordano and W.L.Lindsay, eds.), pp. 231-242. *Soil Sci. Soc. Am.*, Madison, Wisconsin.

Puckridge, D.W. 1971. Photosynthesis of wheat under field conditions. III. Seasonal trends in CO_2 uptake of crop communities. *Aust. J. Agric. Res.,* **22** : 1-9.

Rajgopal, V. and Mehta, B.V. 1969. Effect of Zn and P application on the respiration and chemical composition of hybrid maize grown on torana soil. *Indian J. Agric. Sci.,* **41** : 778-784.

Randall, P.J. and Bouma, D. 1973. Zinc deficiency, carbonic anhydrase and photosynthesis in leaves of spinach. *Plant Physiol.,* **52** : 229-239.

Rathmore, V.S., Wittwer, S.H., Jyung, W.H., Bajaj, Y.P.J. and Adams, M.W. 1970. Mechanism of zinc uptake in bean (*Phaseolus vulgaris*) tissues. *Physiol. Plantarum,* **23** : 908-919.

Reed, H.S. 1946. Effect of zinc deficiency on phosphate metabolism of the tomato plant. *Am. J. Bot.,* **33** : 778-784.

Riceman, D.S. and Jones, G.B. 1958. Distribution of zinc in subterrenean clover (*Trifolium subterrenean* L.) grown at maturity in a culture solution containing zinc labelled with the radioactive isotopes Zn^{65}. *Aust. J. Agric. Res.,* **9** : 730-744.

Riceman, D.S. and Jones, G.B. 1960. Distribution of recently absorbed zinc in subterrenean clover (*Trifolium subterrenean* L.) determined by adding radioactive zinc to culture solution. *Aust. J. Agric. Res.,* **11** : 887-891.

Rogers, L.B. and Reynolds, C.A. 1949. Interaction of pyrophosphate ion with certain multivalent cations in aqueous solutions. *J. Am. Chem. Soc.,* **71** : 2081-2085.

Rossel, R.A. and Ulrich, A. 1964. Critical zinc concentration and leaf minerals of sugar beet plants. *Soil Sci.,* **97** : 152-167.

Safaya, N.M., Shukla, U.C. and Khanna, S.S. 1974. Behaviour of zinc in soils and plants. *Fert. News.,* **19** : 21-27.

Safaya, N.M., Shukla, U.C. and Khanna, S.S. 1975. Nutritional physiology of zinc and symptoms of its deficiency in some field crops. *Fert. News,* **20** : 31-36.

Salami, U.A. and Kenefic, D.G. 1970. Stimulation of growth in zinc deficient corn seedlings by the addition of tryptophan. *Crop Sci.,* **10** : 291-294.

Sandmann, G. and Boger, P. 1983. The enzymatological function of heavy metals and their role in electron transfer processes of plants. *In : "Encyclopedia of Plant Physiology, New Series"* (A.Lauchli and R.L.Bieleski, eds.), *Vol. 15A,* pp. 563-596. Springer-Verlag, Berlin and New York.

Scherer, H.W. and Hoefner, W. 1980. Interactions of iron, manganese and zinc during uptake and transport by maize (*Zea mays*) and sunflower (*Helianthus annuus*). *Z.Pflanzenphysiol.*, **97** : 25-34.

Schmidt, W.E., Haag, H.P. and Epstein, E. 1965. Absorption of zinc by excised barley roots. *Physiol. Plantarum*, **18** : 860-869.

Schmidt, W.E. and Hawf, L. 1966. Effect of carbonyl cyanide phynylhydrazone on cotton uptake in excised barley roots. *Plant Cell Physiol.*, **7** : 145-148.

Schropp, A. and Marschner, H. 1977. Wirkung hoher Phosphatdungung auf die Wachstumsrate, den Zn-Gehalt und das P/Zn-Verhaltnis in Weinreben (*Vitis vinefera*). *Z. Pflanzenernaehr. Bodenkd.*, **140** : 525-529.

Scrutton, M.C., Wu, C.W. and Goldwait, D.A. 1971. The presence and possible role of zinc in the RNA polymerase obtained from *Escherichia coli*. *Proc. Nat. Acad. Sci. U.S.*, **68** : 2497-2501.

Sharma, C.P., Sharma, P.N. Bisht, S.S. and Nautiyal, B.D. 1982. Zinc deficiency induced changes in cabbage. *In : "Proceedings of the Ninth Plant Nutrition Colloguim, Warwick, England"* (A. Scaife, ed.), pp. 601-606. *Commonw. Agric. Bur.*, Farnham Royal, Bucks.

Sharma, P.N., Chatterjee, C., Sharma, C.P. and Agarwala, S.C. 1979. Effect of zinc deficiency on the development and physiology of wheat pollen. *J. Indian Bot. Soc.*, **58** : 330-334.

Sharma, K.C., Krantz, B.A. and Brown, A.L. 1968a. Interaction of Zn and P on two dwarf wheat. *Agron. J.*, **60** : 329-332.

Sharma, K.C., Krantz, B.A. and Brown, A.L. 1968b. Interaction of Zn and P in top and root of corn and tomato. *Agron. J.*, **60** : 453- 456.

Shkol'nik, M.Y. and Chrikova, T.V. 1958. The effect in corn growing boron, zinc and molybdenum on growth, development, carbohydrate metabolism, photosynthesis and redox processes in corn ontogenesis. *Tr. Bot. Inst. Akad. Nauk. SSSR Ser., 4* : 169-192.

Shortri, C.K., Mohanty, P., Rathore, V.C. and Tewari, M.N. 1983. Zinc deficiency limits the photosynthetic enzyme activation in *Zea mays* L. *Biochem. Physiol. Pflanzen.*, **178** : 213-217.

Singh, M. 1981. Effect of zinc, phosphorus and nitrogen on tryptophan concentration in rice grain grown on limed and unlimed soils. *Plant Soil*, **62** : 305-308

Singh, B.R. and Singh, M. 1973. Studies on the supply of zinc to rice seedlings in khaira affected lands. *Indian J. Argic. Sci.* (see Singh, B.R. and Steenberg, K. 1974).

Singh, B.R. and Steenberg, K. 1974. Plant response to micronutrients. I. Uptake, distribution and translocation of zinc in maize and bar plants. *Plant Soil*, **49** : 637-646.

Skoog, F. 1940. Relationship between zinc and auxin in the growth of higher plants. *Amer. J. Bot.*, **27** : 939-951.

Sommer, A.L. and Lipman, C.B. 1928. Evidence for the indispensable nature of zinc and boron for higher green plants. *Plant Physiol.*, **1** : 231-249.

Spiertz, J.H.J., Hag, B.A.T. and Kupers, L.P.J. 1971. Relation between green area duration and grain yield in some varieties of spring wheat. *Neth. J. Agric. Sci.*, **19** : 215-222.

Sprague, H.B. 1964. Hunger signs in crops—*A Treatise : Inorganic Nutrition of Plants.* Third edition (Ed.) David Mckay Company, New York and London, **3** : 99-109.

Stuckenholtz, D.D., Olsen, R.J., Gogan, G. and Olsen, R.A. 1966. On the mechanism of P-Zn interaction in corn nutrition. *Soil Sci. Soc. Am. Proc.*, **30** : 759-763.

Sutton, C.D. and Larsen, S. 1964. Pyrophosphate as a source of phosphorus for plants. *Soil. Sci.*, **97** : 196-201.

Takaki, H. and Kushizaki, M. 1970. Accumulation of free tryptophan and tryptamine in zinc-deficient maize seedlings. *Plant Cell Physiol.* **2** : 793-804.

Tal, M. 1969. Metal ions and ribosomal conformation. *Biochim. Biophys. Acta.*, **195** : 76-86.

Tsui, C. 1948. The role of zinc in auxin synthesis in tomato plant. *Amer. J. Bot.*, **35** : 172-179.

Turner, R.G. and Gregory, R.P.G. 1967. The use of radioisotopes to investigate heavy metals tolerance in plants. Isotopes in plant nutrition and physiology. *Proc. Sym. IAEA, Vienna,* pp. 493-509.

Vallee, B.L. 1977. *In : Biological Aspects of Chemistry,* ed. A.W. Addison, W.R. Cullen, D. Dolphin and B.R. James, pp. 37-70. New York.

Vallee, B.L. and Wacker, W.E.C. 1970. *In : "Proteins"* (H.Neurath ed.) Second edition, **5**, Academic Press, New York.

Vaughan, D., DeKock, P.C. and Ord, B.G. 1982. The nature and localization of superoxide dismutase in fronds of *Lemma gibba* L. and the effect of copper and zinc deficiency on its activity. *Physiol. Plantarum,* **54** : 253-257.

Wacker, W.E.C. and Vallee, B.L. 1959. Nucleic acids and metals. I. Chromium, manganese, nickel, iron, and other metals in ribonucleic acid from diverse biological sources. *J. Biol. Chem.*, **234** : 3257-3262.

Walker, C.D. and Webb, J. 1981. Copper in plants, forms and behaviour. *In : "Copper in Soils and Plants"* (J.F.Loneragan, A.D. Robson, and R.D. Graham, eds.), pp. 189-212. Academic Press, London and Orlando.

Wallace, T. 1961. The diagnosis of mineral deficiencies in plants by visual methods—A color atlas and guide. Third Ed., H.M. Stationary Office, London.

Wallace, A. and Cha, J.W. 1986. Effects of bicarbonate, phosphorus, iron EDDHA, and nitrogen sources on soybeans grown in calcareous soil. *J. Plant Nutr.*, **9** : 251-256.

Webb, M.J. and Loneragan, J.F 1990. Zinc translocation to wheat roots and its implications for a phosphorus/zinc interaction in wheat plants. *J. Plant Nutr.*, **13**: 1499- 1512.

Welch, R.M., Webb, M.J. and Loneragan, J.F. 1982. Zinc in membrane function and its role in phosphorus toxicity. *In : "Proceedings of Ninth Plant Nutrition Colloguium, Warwick, England"* (A. Scaife, ed.) pp. 710-715. *Commonw. Agric Bur.*, Farnham Royal, Bucks.

Werner, V. 1978. Characteristics of zinc uptake. *Physiol. Plantarum,* **42** : 190.

White, M.C., Decker, A.M. and Chaney, R.L. 1981a. Metal complexation in xylem fluid. I. Chemical composition of tomato and soybean stem exudate. *Plant Physiol.,* **67** : 292-300.

White, M.C., Decher, A.M. and Chaney, R.L. 1981b. Metal complexation in xylem fluid. II. Theoretical equilibrium model and computational computer program. *Plant Physiol.,* **67** : 301-310.

Woolhouse, H.W. 1974. Longivity and senescence in plants. *Sci. Prog.,* **61** : 123-147.

Yadav, O.P. and Shukla, U.C. 1983. Effect of zinc on nodulation and nitrogen fixation in chickpea (*Cicer arietinum* L.). *J. Agric. Sci.,* **101** : 559-563.

Yie, S.T. 1969. A study on the relationship of zinc concentration to nitrogen fixation in soybean. *J. Sci. Eng.* (Taiwan), **6** : 1-8.

Advancements In Micronutrient Research
Editor : A. Hemantaranjan
1996, Scientific Publishers, Jodhpur, pp. 179–202.

8

MICRONUTRIENTS-BIOCHEMICAL AND PHYSIOLOGICAL ACTIVITIES

INDRAVADAN CHHELSHANKAR DAVE
Nuclear Agriculture Division
Bhabha Atomic Research Centre,
Bombay 400 085, INDIA

CONTENTS

1. INTRODUCTION

Trend setting patterns in nutritional physiology of plants essentially consist of ion uptake mechanism of individual cells and roots, regulatory aspects of short distance transport across membranes, radial transport across roots and release of ions into xylem as also of long distance transport emphasizing the relative importance of phloem and xylem transport from roots to shoots and ionic mobilization and remobilization according to the need. It also consists of foliar uptake as also uptake by other aerial parts including leachcates from the plant parts. In case of crop and economically important plants, sink-source relationship plays a significant role. All these processes have a sound well programmed biochemical basis which in turn is fully dependent on inorganic nutrients.

Functionally essential micronutrients, viz., boron, copper, chloride, iron, manganese, molybdenum and zinc form a part of physio-biochemical processes involved in germination, growth, development and senescence, thus, having a significant potential during ontogeny, tissue differentiation, organogenesis and even in defence mechanisms against pathogenic agents and also during stress conditions.

A large number of review articles have contributed in understanding the role of micronutrients in relation to rhizosphere (Woldendrop, 1981), growth (Wild and Breeze, 1981), selectivity and kinetics (Lee, 1982; Gerloff, 1963), a general survey (Chapin, 1980; Clarkson and Hanson, 1980), factors affecting the nutrient acquisition (Clarkson, 1985) and foliar uptake and transport (Kannan, 1986; Kannan, 1990), several interesting topics like energy cost (Veen, 1980), regulation of transport (Glass, 1983), role in photosynthesis (Boardaman, 1975; Govindjee *et al.*, 1978) and biochemical functions especially of Fe, Mn, Cu, Mo and Zn (Amberger, 1973; Bowden, 1975; Bannister *et al.*, 1978; Bowler *et al.*, 1992; Dave, 1987) have also been reviewed.

General as well as specific topics, e.g., molecular aspects of plant peroxidases provide a deep insight into mechanistics of micronutrients regulating the metabolic processes at cellular and molecular phases (Mengel and Kirkby, 1982; Marschner, 1986; Greppin *et al.*, 1986).

Biological processes and their inorganic chemistry (Huges, 1972; Ochiai, 1977), role and nature of structural bonding (Silva and Williams, 1976), electron transfer and oxidative energy (Williams, 1973) and gene expression during stress (Sachs and Ho, 1986) are some of the special topics reviewed earlier.

2. BIOINORGANIC ASPECTS OF MICRONUTRIENTS – A BRIEF SURVEY.

2.1. Boron

Under boron deficient conditions, a sharp decrease in nuclear membrane associated DNA polymerase activity and in thymidine incorporation into

DNA, as also a manifold upsurge in acid RNase activity in been plants (Dave and Nair, 1993; Dave and Kannan, 1980) and a drastic reduction in transfer RNA content (Dave, unpublished data) strongly support the suggestion that B may act to create a stable positive charge or a hole in membranes, as also may scavange electrons for charge separation as in solar cells (Tanada, 1978). The primary effect of B seems to be also apoplastic, affording stabilizing support to metal chelates associated with membrane structure and function through possible linkage with cis-diols of glycoproteins (Clarkson and Hanson, 1980).

2.2. Copper

Occurrence of Cu^{2+} in plants is in the range of 3-10 ppm on dry weight basis. It occurs in two oxidation states Cu^{+} and Cu^{2+} , gets readily reduced even by weak reductants, has strong binding affinity to most ligands and can bind to N-terminal of peptides when in ionized states. Its nutritional essentiality seems to be because of its direct involvement in redox changes through redox enzymes like tyrosinase, laccase, ascorbic acid oxidase, mono and diamine oxidase, D-galactose oxidase cyt. C-oxidase etc. Cu/Zn superoxide dismutase reacts with superoxide ions and scavanges them, thus affording protection to a large number of sequential metabolic reactions. Functional differences between Cu^{+} and Cu^{2+} is the ability of the former to form covalent bonding with ligands like I^{-}, CN^{-} etc., while the latter can electrostatically chelate hard ligands of all the metal-micronutrients. only Cu, through enzymic reactions can help formation of lignins, melanin etc. through hydroxylation of monophenols and their subsequent oxidation terminating electron transport chain, can oxidize amines and act as cytoplasmic oxidases. Plastocyanin, an important copper redox protein with no oxidase function acts as terminal substance in PS-I electron transport chain (Boardaman, 1975; Huges, 1972; Ochiai, 1977, Zuberbukler, 1976).

2.3. Iron

Of the two oxidation states of Fe, Fe^{3+} tends to form coördination complexes, gets immobilized through binding with hard ligands in contrast to Fe^{2+} which through chelation by electron donors (N) form stable complexes (Huges, 1972; Buckingham, 1973). Redox role of Fe in heme and nonheme proteins is through their participation as dehydrogenases and reductants, e.g., ferredoxins, electron transfer reactions in b and c type cytochromes, oxidases and peroxidases, oxygenases as well as in the reversible oxygenation of leghemoglobins. Most of the redox reactions occur in chloroplasts, mitochondria and peroxisomes, need Fe^{2+} for porphyrin synthesis as also of similar substances (Price, 1968; Rains, 1976). Thus, in addition to catalytic and structural functions, atomic properties of Fe, it is able to act in redox-reactions with variable redox potential and redox indices. Fe in fact fulfills such a requirement - a π electron system in outer sphere and ligand coordination in the inner sphere (Clarkson and Hanson, 1980).

2.4. Manganese

Mn being a member of Period IV transition metals, preferentially acts as a metallo-protein member instead of acting as labile metal-enzyme complex. Plant tissues contain Mn in the range of 30-500 ppm dry weight basis. It is acquired in biologically active form - Mn^{2+} at the expenditure of energy, it moves as free ion, and partially has phloem mobility (Tiffin, 1972). Photosynthetic oxygen evolution is dependent on the storage of four oxidizing equivalents of Mn needed to transfer four electrons from split-up of two molecules of water per molecule of P680 reaction centre (PS-II) as speculated from water proton relaxation rates to record transition in Mn^{2+} (Govindjee *et al.*, 1978). Its essentiality lies in producing the requisite conformational state of proteins through binding with strong ligands. Of the three known types of superoxide dismutases (SOD), MnSOD is one (Asada *et al.*, 1977) which is resistant to KCN and H_2O_2 as against FeSOD being sensitive to H_2O_2 only and Cu/ZnSOD being sensitive to both KCN and H_2O_2 inhibitors, and is localised in mitochondrial matrix. During light or chilling mediated stress conditions, MnSOD synthesis registers a sharp rise (Clare *et al.*, 1984; Steinitz *et al.*, 1979).

2.5. Zinc

The chemistry of Zn^{2+} differs from that of Cu in its having relatively lower cationic density despite their having almost similar ionic radii. This feature confers to Zn a property to exist more freely in an ionic state, as can be seen from Zn^{2+} having lower rate of binding to anionic chelates in xylem exudates (Tiffin, 1972). The Zn-complexes in plants have a higher stability constants in comparison to Mg^{2+}, Ca^{2+} or Mn^{2+}. Tissue concentration status of Zn relatively is Fe> Zn> Cu (Loneragan *et al.*, 1976), but its relative mobility is the lowest (Dave and Kannan, 1981). The list of Zn-enzymes and other Zn proteins is very big-about more than 300. Zn^{2+} participation has shown four coordinate tetrahedral bonding with certain degrees of distortion, e.g., in case of alcohol dehydrogenase and few other enzymes (Ochiai, 1977). Thus, even though the essentiality of Zn lies both in structural and functional sites, the expert opinion favours a need for enough information to decide its precise properties to implicate in metabolic operations at every lavel (Clarkson and Hanson, 1980).

2.6. Molybdenum

From Mo (IV) form under mineralized state, Mo is available to plants as its oxyanion- MoO^{2-}_4 -Mo (VI). Being a weak acid as molybdate, Mo can form polyanions and also can undergo sequestration in plants under high concentration conditions. Due to this, plants can withstand high levels of Mo, without facing toxic effects. Its content in leaves is ~ 1 ppm dry weight basis, and is associated with enzyme nitrate reductase. Under symbiotic conditions, it is also present in association with nitrogenase of nodular bacteroids on root or shoot. Mostly present as Mo(VI) its lower oxidation states also can occur, e.g., Mo(V) which can be detected through electron paramagnetic resonance signals. The bonding of Mo with enzyme is reversible (Bray and Swann, 1972). It is not yet known which of its chemical properties makes it an essential micro-nutrient. However, its certain properties are unique to Mo only, e.g., coupling of H^+ and e^- transfer reactions causing thereby a simultaneous giving off an electron and also a proton by Mo- flavoproteins, which is exploited in binding and reducing NO^-_3 through a typical reaction- $2e^- + 2H^+ NO^-_3 \rightarrow NO^-_2 + H_2O$ (Stiffel, 1973; Spence, 1976).

2.7. Chloride

Normally the concentration of Cl^- ranges upto 500 µmoles per g dry weight basis except in halophytes, where it is much higher. Below 20 µmoles per dry weight basis, the deficiency symptoms appear like loss of turgor in leaves. Being readily transportable and tolerated over a wide range, it has been found useful as a counter ion during quick fluxes of K^+ aiding in this way the guard cells to maintain turgor. Biochemically nonreactive, Cl^- may exert a control over several reactions indirectly through making some cations neutral, when in excess (Rains, 1972). Otherwise there is nothing much to review about this anion.

3. RECENT ADVANCEMENTS IN BIOCHEMICAL PHYSIOLOGY OF BORON, COPPER, IRON, MANGANESE, MOLYBDENUM AND ZINC

The coverage that can be given to topics under this heading is so vast that one chapter will not be sufficient to be justice for reviewing sizeable amount of data published during last few years. Further, majority of relevant important references have been covered in earlier reviews listed in previous paragraphs in addition to two authentic and authoritative books by Mengel and Kirkby (1982) and Marschner (1986). Hence, in the following, an attempt has been made to update the events and also to highlight new evidences or hypotheses concerning the role of micronutrients functionally.

3.1. Boron

Attempts have been made to interpret the role of boron on the formation of *cis*-diol borate complexes. Cell division and elongation and their nucleic acid metabolism are the primary processes and reactions severely affected by B deficiency (Cohen and Lepper, 1977; Lovatt *et al.*, 1981; Shkolnik, 1974; Moore and Hirsch, 1983). The effects of B deficiency in mustard plants included lowering of water potential, reduced transpiration rate due to partial stomatal opening accompanied by a significant decrease in chlorophyll content and photosynthesis rate, Hill reaction activity and CO_2 concentration, with an upsurge in proline, sugar and starch content and soil-N (Sharma and Ramchandra, 1990). Application of B to foliage of citrus plants caused an increased root exudation of reducing sugars and amino acids, as a result of which, there was an increase in vesicular-arbuscular fungal colonies in the rhizosphere (Dixon *et al.*, 1989). This clearly showed that boron increases the efficiency of nutrient acquisition and uptake by promoting VAM-plant symbiosis. These observations have an indirect bearing on the role of boron in facilitating sugar transport through boron-sugar complexing (Dugger, 1983). It was earlier concluded that B has a role in formation of cell-wall, based on the observed *cis*-borate esters in cell-walls (Thellier *et al.*, 1979), mainly in hemicellulose fractions and lignin precursors (Lewis, 1980). A model has been proposed wherein a critical acceptor molecule being activated by bond- formation with Ca and B at separate sites during cell-wall biosynthesis (Teasdale and Richards, 1990). Boron deficiency also leads to changes in important enzyme activites. Nitrate reductase (NR) activity was strongly inhibited with an increase in NO^-_3 content of both root and shoot of sunflower under B deficient or toxic levels disturbing the N-assimilation processes (Kastori and Petrovic, 1989). A many fold increase in acid RNase activity in trifoliate leaves of B-deficient bean plants suggested ruptures in tonoplast as this enzyme is exclusively localized in vacuolar lumen. This was preceded by an altered protein profile and decreased uptake of Fe and Mn but increased uptake of Zn in cotyledons during germination (Dave and Kannan, 1980; 1981), and a sharp decrease in tRNA content (unpublished data). It has also been shown that the formation and maintenance of bioelectric potential across the membranes require B which has a stabilizing effect on the membrane function (Tanada, 1983). Support to such a membrane-boron relationship has come from two different laboratories. Based on an observed marked decrease in ^{3}H-thymidine incorporation into DNA of M-Band (DNA- membrane complex) as well as a sharp fall in the M-Band associated DNA- polymerase activity in embryos of B-deficient bean seedlings, it has been concluded that there is a direct implication of B on newly synthesized DNA replication attached to nuclear membrane through its control over available binding site proteins and the physical forces underlying the M-Band formation (Dave and Nair, 1993). Electrophysiological data gathered from sunflower root tip cells also support membrane hypothesis of Tanada (1983). An increased hyperpolarization occurred in plasmalemma of root tip cells within 20 minutes of their transfer to basal medium + 50 μm H_3BO_3, as indicated by membrane- electropotential value-

Em. This was explained on the basis of increased K^+ influx induced by B, which in turn enhanced H^+- ATPase (proton-pump), leading to hyperpolarization. Regulatory role of B for molecular and biological processes in *Lilium* has been suggested (Schon *et al.*, 1990). Other observations on effect of B include better yield performance of winter rape (Polster *et al.*, 1992), higher B requirement with increased growth characteristics of *Macadamia integrifolia* (Schaecke *et al.*, 1990) and an exponential relation between dry weight, daily radiation, transpiration and B uptake confirmed through, nutrient film technique (Fox and Hue, 1989).

3.2. Copper

Humic and fulvic acid bind Cu^{2+} (Heinen *et al.*, 1991) and upto 98% of Cu in soil solution is bound to organic compounds (Stevenson and Fitch, 1981). It is present in plants as (i) plastocyanin-useful in single electron transfer, (ii) non-blue proteins needed for monophenol-oxidation and (iii) multicopper proteins with a minimum of 4 *Cu* atoms per molecule, functioning as oxidases (Hodgson *et al.*, 1966). Fe is also needed in Cu-enzyme cytochrome oxidase for terminal oxidation in mitochondria.

Cu at the concentration of 1 ppm influenced the genetic apparatus of *Dunaliella* and caused an accumulation of RNA resulting into decreased protein content (Sandmann and Bogger, 1983). Cu deficiency damaged the formation of chloroplast ultrastructure with a reduction in chlorophyll content, rates of photosynthesis and synthesis of amino acids (Lyashenko *et al.*, 1991; Qunying *et al.*, 1992). Cu is also involved in ethylene biosynthesis of tobacco leaves and *Spirodela oligorhiza* through conversion from methionine and linoleic acid respectively (Mattoo *et al.*, 1992), showing the need of Cu ions for ethylene biosynthesis pathways. Robson *et al.* (1981) have shown a 50% reduction in the cell-wall lignin in Cu deficient wheat leaves and of *Pinus radiata* (Turvey *et al.*, 1992) seedlings. This is related to multicopper enzymes phenolase and laccase activity (Robson *et al.*, 1981). Apart from the metalloenzyme status, Cu is known to complex with polypeptides also, which form part of root exudates from xylem sap of pumpkin, having molecular weight of 1400-5000 dalton rich in cysteine residues (Przemeck and Hasse, 1991). Cu is shown to activate polyphenol oxidase from sweet potato (Jin *et al.*, 1991) and inhibit α-L-arabinofuranosidase from soybean seeds which degrades water soluble polysaccharides (Hatanaka *et al.*, 1991). Toxic levels of Cu causes oxidative stress by depletion of antioxidants, e.g., glutathione (GSH) by inducing the synthesis of phytochelatin. Cu tolerant species of *Silence cucubalus* was able to prevent the depletion of GSH, wherein the phytochelatin content was only 40% of that produced in Cu sensitive plants (DeVos *et al.*, 1992).

Unlike FeSOD, and MnSOD, Cu/ZnSOD is almost absent in prokaryotic organisms. Its cloning, sequencing, mapping of the gene for this enzyme along with its periplasmic location have been described (Steinman and Ely, 1990). Cytosolic Cu/Zn SOD probably responds to cytosol localized reactions due to increased oxyradical formation. Recently, properties of Cu/Zn SOD regulation *in vivo*, apoenzyme-SOD levels in tissues, effect of Cu in acellular environment, role of cu-transport factors, ascorbic acid interference with binding of Cu by the enzyme protein etc. have been reviewed (Steinman and Ely, 1990; Harris, 1992; Jewell *et al.*, 1988). The salient features enlisted above have the following explanation. For its catalytic function, Cu/Zn SOD needs two ions of Cu. Yet only one Cu atom seems to regulate the functional activity. Deficiency of Cu in diet of animals quickly impairs the enzyme-catalytic function, indicating a constant depletion in the enzyme molecule-levels. The principal transport factors for Cu are ceruloplasmin and a specific albumin. Bovine erythrocytes-Cu/Zn SOD consists of two equal subunits with 151 residue each. Conformational studies show that the site of Cu is on the floor of a deep cleft of the protein, with Zn atom burried deep within. Cu and Zn communicate weakly through two-N-atoms on a common histidine residue. Cu is reduced with substrate O^{-}_2 radical to yield O_2 and then H_2O_2 through reoxidation. The function of Cu seems to be catalytic, while Zn is believed to have structural maintenance as also increasing redox potential. Mn deficiency leads to decreased Mn SOD but increases Cu/Zn SOD activities. Role of all the three SOD have been reviewed in detail in relation to various stress conditions (Bowler *et al.*, 1992).

3.3. Iron

Nutritional aspects of Fe require a separate treatise due to enormous data having vital results and implications. Hence, here very important biochemical observations in relation to Fe-acquisition and few other observations are narrated.

Roots are the first to respond to changes in availability of iron. Fe-deficiency induces formation of rhizodermal transfer cells which are the sites of both H^+. efflux pumps and release of phenolics into rhizosphere. In grasses, under Fe deficiency, non- proteinogenic amino acids like avenic acid, related compounds like nicotianamine are released which act as chelators for Fe (III) and Fe (II), respectively (Benes *et al.*, 1983). Brown and coworkers have proposed a series of events that occur as a result of iron-deficient states. They are : (a) chlorosis, (b) H^+ efflux causing reduction in pH of rhizosphere solution, (c) excretion of reductants that solubilize Fe^{3+} and then their reabsorption, (d) correction of chlorosis, (e) increase in pH of rhizospheric medium and cessation of reductant synthesis (Brown, 1978). Subsequent work has been summarised by Römheld and Marschner (1986). Kannan and co-workers from our laboratory have carried out extensive studies on this line and have identified dibutyl phthalate (DBP) as the Fe^{3+} reductant secreted by the plants for recovery from chlorosis in sorghum cultivars and jute plants and caffeic acid from tomato. Exogenous addition of these two compounds to nutrient medium reduced Fe-uptake. Tolerant cultivars of chickpea, and sesame reduced the pH of root medium with subsequent recovery from chlorosis (Kannan *et al.*, 1984; Kannan, 1986; Ramani *et al.*, 1984; Kannan and Ramani, 1987; Ramani and Kannan, 1987). They disagree with the view of Bienfait (1985) about lowering of pH of medium being specific and not a general phenomena. They interpret their findings to support chemical reductant theory of Olsen *et al.* (Olsen *et al.*, 1981).

Large accumulation of Fe and citrate in nicotianamine-less mutant of tomato indicated the role of this chemical in Fe-uptake and subsequently regulation of Fe-dependant at gene expression level (Pich *et al.*, 1991). Accumulation of nicotianamine in Fe- deficient barley root tips was shown to be essential for Fe- dependent mechanisms for response, Fe-enzyme regulation and phloem loading and unloading of Fe apart from being a component of strategy-II for Fe-acquisition (Scholz *et al.*, 1992). A direct correlation between reducing ability of some phenols to reduce ferrihydrite and their ability to release Fe from these compounds with pH optima as 5-6. They are caffeic acid, chlorogenic, ferulic and p-coumaric acids (Boyer *et al.*, 1989) Three isoforms of plasma-membrane associated, Fe-chelate reductase shown to reduce Fe-chelates from tomato having similar polypeptide profiles exhibited Michaellis - Menten kinetics with regard to Fe^{3+} citrate (Holden *et al.*, 1992). Thirty calcicolous and calcifugous grass species (Poaceae) were examined for Fe-solubilizing ability under different soils. Under Fe- deficiency the calcicolous and calcifugous species behaved differently in that the calcicoles released higher amount of Fe(III)-chelators, showing that the calcifuges could be excluded due to undercapacity of Fe-mobilization and acquisition. A wide variability was observed for Fe-demand and growth rate (Gries and Runge, 1992). This factor must be taken into account while fixing the Fe-efficiency of a species. While root apoplastic Fe-pool loaded by various types of siderophores in soil solutions can be an important Fe-source for graminaceous species, during limited Fe supply (Zhang *et al.*, 1991), presence of phytosiderophores released from Oat cv cocker 227 hindered the uptake of Fe by five soybean cultivars with varying Fe-efficiency (Hopkins *et al.*, 1992). This competitive phenomenan denoted a direct variation relationship between the rate of inhibition to the susceptibility for Fe-chlorosis as also their own capacity to reduce Fe^{3+} in rhizosphere. The synthesis and release of such compounds like mugineic acid and related substances have been shown to be under genetic control (Mori and Nishizawa, 1989).

Based on the observations on failure to respond to Fe-deficiency stress induced phytosiderophores in three out of five monocots (grasses) it is the considered opinion of the authors that "to state that under Fe-deficiency stress, mechanisms similar to those in dicots could account for iron-uptake in some graminaceous monocots and that strategy-II type response proposed for all, in this category of plants would just be an oversimplification" (Onyezilli and Ross, 1993. Phytosiderophores released from *Gloeophyllum trabeum* were shown to carry out single electron oxidation on lignocellulase (Chandhoke *et al.*, 1992).

An excellent review on Fe nutrition and interaction in plants covers: phytosiderophores, organic acids secretion, Fe-stress and genetic variation, plant-microbial interaction and related biochemical and physiological activities (Bienfait, 1988).

A review on leaf responses to Fe-deficiency (Abadia, 1992) describes changes and alterations in functions and structures of whole photosynthetic apparatus, thylakoid membrane, its composition, PS-I and PS-II stoichiomet.y, lipid composition, gaseous exchange, water status, chlorophyll fluorescence and new issues like misuse of membrane fractions etc.

Ferredoxin mediated reduction of NO^-_2 and of SO^{2-}_4 is essentially important for mineral nutrition, wherein Fe atoms bound to sulfur act as electron transmitter for reducing $NADP^+$ to NADPH. Evidences have been put forth for similar compound called flavodoxin, thought to be ferredoxin replacement e^- transport molecule produced in bacteria and algae. During Fe-stress, tobacco ferredoxin content reduction was equivalent to chlorophyll reduction with ferredoxin activity at status quo. This was suggested to be due to flavodoxin-like molecules. Such substances were found in roots functional in NR activity but not in $NADP^+$ reduction (Huang *et al.*, 1992). Fe-deficiency has been shown to create easily reversible photo-inhibition in pear (*Pyrus communis* L.) leaves due to sharp fall in PS-II activity (Morales *et al.*, 1992) and this may be attributed to altered structure of thylakoid membranes observed in Fe- deficient sugarbeet leaves (Quilez *et al.*, 1992).

3.4. Manganese

Mn has attained prominance due to its essentiality for Hill reaction - H_2O splitting and oxygen evolving system in photosynthetic processes. Manganoprotein - a member of PS-II has a requirement of 4 Mn atoms for the reaction centre PS-690. This whole complex and its functional potential in Young leaves is a sensitive marker for testing the characterization of nutritional status of Mn in plants (Nable *et al.*, 1984). Mn deficiency is known to lead to decline in rate of photosynthesis and a progressive dismemberment of lamellar system of the chloroplasts without any alteration in mitochondria (Possingham *et al.*, 1984). There has been considerable understanding of role of Mn in relation to organization, function and genetic control of PS-II. Complete removal of Mn from the reaction centre resulted into 40-50 fold decrease in photoreduction of $NADP^+$ which was restored back to 85-90% of the original by Mn. The authors have suggested an alternative pathway along with some problems concerning Z scheme (Allakhverdiev and Klimov, 1992). The ability of isolated PS-II to photoreduce 2,6, dichlorophenolindophenol (DCPIP) was strongly inhibited by Zn^{2+}, under protein depletion, which (inhibition) was significantly reduced by Mn^{2+} and H_2O_2 combination (Rashid *et al.*, 1991). Evidences are now available in relation to the pathway of electron transfer, by exploiting the thermoluminescent property of PS-II. The observed TL-glow at -55° C and again at -30°C (TL_{-55} & TL_{-30}) from PS-II disappeared as a result of Mn depletion via the action of tyrosine and histidine modifier respectively along with the disappearance of EPR signals. These data and the inhibition of e^- transport from the diphenylcarbazide to 2,6, DCPIP clearly demonstrate the participation of oxidized tyrosine and histidine residues in e^- transfer from Mn to the reaction centre Chl. P680 (Allakhverdiev *et al.*, 1992). A mutant of *Synechococcus* (a cyanobacterium) with inactivated PS680 gene did not contain any Mn-stabilizing protein. It could evolve photosynthetic oxygen on addition of $MnCl_2$ and $CaCl_2$ indicating that the origin of H_2O oxidizing enzyme may be in a flavoprotein with 1-amino dehydrogenase/oxidase activity (Bockholt *et al.*, 1991). Further, aspartic acid-170 of D1-polypeptide of PS-II reaction centre of *Synechocystis* sp. with 11 site directed mutation at asp 170, appeared to participate in the early steps of assembly of Mn-clusters in addition to tyr -161 (Nixon and Diner, 1992). This study is strengthened by observed relaxation kinetics of spinlattice of tyrosine radical in PS-II, wherein T_1-oscillations of tyrosine radical due to number of flashes (photo) was solely attributable to charge oscillations of Mn-cluster (Bosch *et al.*, 1991). Further evidences in support of Mn implication in electronic structure of O_2 - evolving complex include a new EPR signal due to a paramagnetic intermediate in resting state of PS-II (Dexheimer and Klein, 1992), origin of EPR signal at g~4 being either Mn(II) monomer or a coupled multimer, or a tetranuclear origin of Mn for multiline signals at g ~ 4 & g ~ 2, suggesting a different spin state configurations of single exchanged Mn cluster (Kim *et al.*, 1992). The role of Mn complex in the binding of extrinsic proteins of 17 and 23 KD polypeptide species has been shown from observed failure to rebind the PS-II membrane extrinsic protein consequent to destruction of isolated Mn-complex (Kavelaki and Ghanotakis, 1991).

Apart from the modulation of malic enzyme, malate dehydrogenase and isocitrate dehydrogenase, Mn similarly could modulate enolase also. Direct binding of Mn(II) to apoenolase denoted its competition with Mg(II). Two binding sites for Mn(II) were identified with site I influencing the catalytic function of the enzyme (Lee and Nowak, 1992). Mn also acts as enzyme activator. Initiation of DNA polymerization with a primer protein, Ø 29 DNA polymerase from *E. coli* exhibited a strong preference for Mn^{2+} as an activator, the molecular basis of activation being a significant upsurge in affinity of Mn^{2+} for dATP (Esteban *et al.*, 1992).

Apart from the properties of MnSOD as an oxy-radical scavanger, MnSOD has now been shown to afford protection against cold injury to rice seedlings, and imparting to them cold-tolerance. Increase in Mn^{2+} uptake was in consonance with increased SOD which was directly proportional to cold tolerance, which was specifically due to MnSOD and not due to Fe-or Cu/ZnSOD, as increased uptake of these three elements failed to elicit cold tolerance from the seedlings (Hahn *et al.*, 1991). Mn stress caused increased peroxidase activity which varied with different soybean cultivars. Mn-susceptible cv Bragg and Forrest registered higher peroxidase under Mn-toxicity, while Mn-tolerant cv Lee and T203 did so under Mn deficiency. Mn-Fe interaction caused an upsurge in total SOD activity, but MnSOD was modulated only by changes in Mn levels (Leidi *et al.*, 1987). Such enzymic disturbances at cellular and molecular levels due to Mn stress are bound to result into anomalous growth and development of tissues. Restricted supply of Mn (0.0055 mg/l) to maize plants resulted into poor tasseling, delayed anther- development, with fewer and smaller pollen grains containing decreased cytoplasmic content with poor germinability, but without affecting ovule fertility (Sharma *et al.*, 1991).

Mn ions can induce various fractions of DNA-Histone during cell-division through probably modification of conformational changes. It has been shown to affect the binding of ribosomes to endoplasmic reticulum. Other activities include changes in nucleic acid metabolism and modulation of genetic apparatus (Vlasyuk, 1971).

3.5. Molybdenum

Bergerson has proposed a scheme for the role of Mo in symbiotic N-fixation in legumes. It envisages the involvement of metabolic conversions in citric acid cycle contributing to amino acid pool into nodules containing bacteroids, releasing thereby electrons and also the generation of electrons received from bacteroidal cytochromes which finally are utilized by 180 KD Fe–Mo protein and 50 KD Fe–complex of nitrogenase which in turn fixes N_2 to NH_2 or NH_3. Though attractive, this hypothesis needs sound experimental evidences. There are other suggestions also worth thinking about the role of Mo in phospholipid synthesis for membrane, based on the observed inhibition of Pi influx as also the cell-surface acid phosphatase activity with its I_{50} (inhibition index) at 7.8 μm for enzyme and 216 μm for Pi influx in suspension cells of *Brassica nigra* (Lefebvre *et al.*, 1990). Mo also inhibits through selective photooxidation and irreversible photobleaching of B- carotene in presence of silicomolybdate in isolated PS-II reaction centre (Telfer *et al.*, 1991). On the other hand, stimulation by Mo was observed in N, P and K utilization by rice plants through seed dressing with 0.5% Mo (Sheudzhen and Logvina, 1991), increasing soluble sugars, chlorophylls and photosynthetic rate, number of fertile pods, fruitive branches and 1000 seeds weight in *Vicia faba* (Xia and Xiong, 1991). Amendment by 30 kg urea-N/ha increased the yield of winter wheat; crude protein and glutein by 0.9-1.1% and 1.6-2%, respectively, for the last two. Grain glassiness depression caused by urea was also abolished in presence of Mo, which enhanced Mn, Fe, Zn and Mo content in matured grains (Butorina *et al.*, 1991). Noteworthy breakthrough has been obtained to understand structure and function of genes involved in NO_3 assimilation, through the skillful use of herbicide chlorate - resistant mutants of *Arabidopsis thalliana* having low nitrate reductase activity and low levels of Mo–cofactor (Mo-cofactor-molybdenum - protein) synthesis-about 1/8th of wild type level. Experimental conclusions include the tungstate sensitivity, chlorate resistance, low NR activity, low levels of MoCO all being cosegregating, indicating that all are due to a single gene mutation that maps to chl-2 locus (Labrie *et al.,* 1992). Based on sequential composition and x-ray absorption data, fine structural studies and other relevant information, thiol groups of cystein-191 was seen to act as a ligand for MoCO binding with sulfite oxidase enzyme (Neame and Barber, 1989; Barber and Neame, 1990).

3.6. Zinc

A large number of Zn- metalloenzymes having diverse functional properties are present in almost all phylla and genera. Zn is present in atleast one enzyme of each of the six categories classified by the Commission on Enzyme - Nomenclature (Riordan, 1976). This is given as a summary in Table 1.

Table 1. Zinc Metalloenzymes

Class of enzyme	No. of enzymes containing Zn
E.C.1. oxidoreductases	Clarkson (1985)
E.C.2. Tansferases	Veen(1980)
E.C.3. Hydrolases	Sachs and Ho (1986)
E.C.4. Lyases	Huges (1972)
E.C.5. Isomerases	Woldendrop (1981)
E.C.6. Lipases	Woldendrop (1981)

Taking into account the association of Zn with more than 300 metalloenzymes and other proteins spread over entire living system, one may be inclined to think about Zn in terms of essentiality for these enzymes. There is some strangeness about this, e.g., aldolase in yeast contain Zn^{2+} as a cofactor, but not so in plant aldolases (Ochiai, 1977). There exists a relationship between Zn and other elements. There was a directly proportional relation observed between Zn and Fe in eelgrass leaves from Atlantic coast of France (Wasserman and Lavaax, 1991). Immunohistological investigations proved the presence of Zn and Cu in proliferative tissues of embryo and shoot apex of soybean, as metallothioneins, the role of which was metal transport, detoxification of heavy metals in polluted area (Chongpraditnun *et al.*, 1991). Traces of Zn along with Fe and Mn were found in globoid crystals of endosperm and embryo of *Datura stramonium*, Mn being confined to protodermic proteins only (Maldonado and Lott, 1991). Synergistic effects of externally added Zn to nitrohumic acid was observed in dry matter production of wheat (Cervellie *et al.*, 1991). Availability of free hydrated Fe^{3+}, Cu^{2+}, Mn^{2+} and Zn^{2+} was compared to their availability in bound form in chelator-bufferred solution, through a computerized programme for barley. Higher levels of metals absorbed in root and leaves in EDTA - bufferred solution conclusively indicated the role of total metals in external solution on their uptake in free state (Bell *et al.*, 1991).

Membrane preparation experiments revealed the presence of active Zn^{2+} inhibitory site on the donor side of PS-II. This site becomes more susceptible to Zn^{2+} when its concentration exceeds 1 mM in that the conformation of PS-II core complex gets perturbed (Rashid *et al.*, 1991). Environmental and other stresses as also the senescence cause a decrease in all the SOD enzymes, catalase and glutathione reductase, thus enhancing the processes leading to lipid breakdown. Functional and other aspects of SOD in general have been reviewed explaining role of Zn (Bannister *et al.*, 1987) in relation to other SOD (Bowler *et al.*, 1992). Hence, only mention is made of two recent references about this aspect. Glyoxysomal Cu/Zn SOD isolated from watermelon cotyledons with sp. activity 1211 units/mg protein at 400-fold purification (8 µg/g enzyme protein/g cotyledon) consisted of two subunits with 16,500 Daltons each. Unlike other Cu/Zn SOD, it had 1 g-atom Cu and Zn each per mol of dimer, but had UV and visible spectral properties similar to other SOD (Bueno *et al.*, 1992). A comparative competitive study on all three types of SOD for O^{-}_2 radicals at pH 7.8 revealed that the kinetics of enzyme - substrate reaction follows a function, $y = f[SOD^{0.5}]$. Further, the rate constant for Cu/Zn SOD was an order of magnitude higher than Mn or Fe SOD (Gray and Carmichael, 1992). Experiments on Cd = peptide ligands in tobacco suspension cells showed that excess of Zn can be accommodated by the cells through sequestration inside vacuoles as complexes with endogenous organic acids, irrespective of the presence or absence of Cd-peptides (Krotz *et al.*, 1989). The latest information on importance of Zn comes from the extensive use of Zn(Zinc) fingers transcription factor protein (TF III A) first investigated by Miller and co-workers (Miller *et al.*, 1985). A typical use of this method in molecular biology has helped in understanding and assessment of energy contribution to binding and the sites of interaction with 5S rRNA genes. At least nine zinc fingers TF-III-A showed a high degree of independence, the function of which are thermodynamically coupled in DNA- binding reactions, where Zn^{2+} association happened to be with coordinating histidine residues (Rio *et al.*, 1993).

It will only be appropriate to put on record certain interesting developments in the field of microbial molecular biology of zinc. Zinc-proteins binding to single stranded nucleic acids consists of Gene 32 protein from bacteriophage T_4, essential for replication, recombination and repair of phage DNA, wherein zinc ion is coordinated to four binding sites. It has been shown that the translation repression depends critically on the presence of zinc. Further, a number of workers have proposed that because nucleocapsid (NC) proteins of retroviruses, e.g. HIV retrovirus contained two repeats of an amino acid-sequence similar to Zn- binding sequence in gene 32 of phage T_4, the CCHC-containing sequences in NC proteins also bind zinc. It so,then zinc may have a significant role in subtending the reverse-transcriptase mediated replication. Zinc function has also been associated with proton transfer in reaction centres in photosynthetic bacteria, catalytic antibody proteins and metal cluster in the conformation of metallo thioneins involved in storage and transport of Zinc. (Hiraga, 1992; Benkovic, 1992; Arnheim and Erlich, 1992; Coleman, 1992).

4. CONCLUDING REMARKAS

The review reveals, through painstaking experimentals-par excellence, a deep insight in the functioning of only six of the total elements from periodic chart exerting a master control over numerous biochemical and biophysical events, some of which sometimes lasting for less than a microsecond or far less than that, during germination, vegetative growth, development and senescence. Their deficit or toxicity elicits a large number of reactions right in the rhizosphere, external to the plants, symbiotic organisms, as also at the cellular and molecular levels and many a time confined to organalles. They also are involved in the finer aspects of gene expression under normal and stress induced conditions created by environmental, biotic, abiotic and edific factors. However, compared to enormous amount of literature on this aspect available in relation to animal and human systems, the publications in the field of plant kingdom leaves much room for desire. But then, future awaits the plant scientists to step into this very interesting facet of life science. The author feels highly apologetic for not being able to accommodate a large number of references especially of agro- physiological importance due to many compulsions. An effort has been made to fill this lacuna by citing relevant reviews for the benefit of interested colleagues.

REFERENCES

Abadia, J. 1992. Leaf responses to Fe deficiency : A review. *J. Plant Nutr.*, **15 (10)** : 1699-1713.

Allakhverdiev, S.I. and Klimov, V.V. 1992, Photoreduction of $NADP^+$ in photosystem II of higher plants : requirement for manganese. *Z. Naturforsch. C., Biosci.*, **47** : 57-62.

Allakhverdiev, S.I., Klimov, V.V. and Demeter, S. 1992. Thermoluminescence evidence for light-induced oxidation of tyrosine and histidine residues in manganese depleted PS-II particles. *FEBS Lett.*, **297** : 51-54.

Amberger, A. 1973. Die Rolle des Mangans im stoffwechsel der pflanzen. *Agrochimica*, **17** : 165-183.

Arnheim, N. and Erlich, H. 1992. Polymerase chain reaction strategy. *Annu. Rev. Biochem.*, **61** : 131-156.

Asada, K., Kamematsu, S. and Uchida, K. 1977. Superoxide dismutases in photosynthetic organisms : Absence of the cuprozinc enzyme in eukaryotic algae. *Arch. Biochem. Biophysics.*, **179** : 243-256.

Bannister, J.V., Bannister, W.H. and Rotilio, G. 1987. Aspects of structure, function and applications of superoxide dismutase. *CRC Crit. Rev. Biochemistry*, **22** : 111-180.

Barber, M.J. and Neame, P.J., 1990. A conserved cysteine in molybdenum oxotransferases. *J. Biol. Chem.*, **265** : 20912- 20915.

Bell, P.F., Chaney, R.L. and Angle, J.S. 1991. Determination of the Copper $^{2+}$ activity required by maize using chelator-buffered nutrient solutions. *Dev. Plant Soil Sci.*, **43** : 69-80.

Benes, I., Schreiber, K., Ripperger, H. and Kircheiss, A. 1983. Metal complex formation by nicotianamine, a possible phytosiderophore. *Experientia*, **39** : 261-262.

Benkovic, S.J. 1992. Catalytic antibodies. *Annu. Rev. Biochem.*, **61** : 29-54.

Bienfait, H.F. 1988. Mechanisms in Fe-efficiency reactions of higher plants. *J. Plant Nutr.*, **11 (6-11)** : 605-632.

Bienfait, H.F., 1985. Regulated redox processes at the plasmalemma of plant root cells and their function in iron uptake. *J. Bioenerg, Biomember.*, **17** : 73-83.

Boardaman, N.K. 1975. Trace elements in photosynthesis. In *"Trace Elements in Soil-Plant-Animal Systems", ed. D.J.D. Nicholas, A.R. Egan. pp. 199-212 Academic Press. New York.*

Bockholt, R., Masepohl, B. and Pistorius, E.K. 1991, Insertional inactivation of the psbo gene encoding the manganese stabilizing protin of PS-II in the cyanobacterium *Synechococcus* PCC7942 . Effect on photosynthetic water oxidation and L - amino acid oxidase activity. *FEBS Lett.*, **294** : 59-63.

Bosch, M.K., Evelo, R.G., Styring, S., Rutherford, A.W. and Hoff A.J. 1991, ESE relaxation measurements in PS-II : The influence of the reaction center non-heme iron on the spin-lattice relaxation of Tyr D. *FEBS Lett.*, **292** : 279-283.

Bowden, F.L. 1975. The Biochemical function of molybdenum. In *"Techniques and Topics in Bioinorganic Chemistry,* ed. C.A. McAuliffe, pp. 207-267, MacMillan, New York.

Bowler, C., Montagu, M. Van and Inze, D. 1992. Superoxide dismutase and stress tolerance. *Ann. Rev. Plant Physiol. Plant Mol. Biol.,* **43** : 83-116.

Boyer, R.F., Clark, H.M. and Sanchez S. 1989. Solubilization of fernhydrate iron by plant phenolics : A model rhizosphere processes. *J. Plant Nutr.,* **12** (5) : 581-592.

Bray, R.C. and Swann, J.C. 1972. Molybdenum containing enzymes. *Struct. Bonding.,* **11** : 107-144.

Brown J.C. 1978. Mechanism of iron uptake by plants. *Plant, Cell, Environ.,* **1** : 249-257.

Buckingham, D.A. 1973. Structure and stereochemistry of coordination compounds. In *"Inorganic Biochemistry"* G.L. Eicchorn, ed. Vol. 2., pp 3-62. Elsevier, New York.

Bueno P., Del. R. and Luis, A. 1992. Purification and properties of glyoxysomal cuprozine superoxide dismutase from watermelon cotyledons. (*Citrullus vulgaris* Schrad). *Plant Physiol.,* **98**(1) : 331-336.

Butorina, E.P., Yagodin, B.A. and Feofanov, S.N. 1991. Effect of late foliar sprays of urea and molybdenum on winter wheat yield and quality. *Agrokhimiya,* **4** : 17-20.

Cervelli, S., Ragone, A.F., Di Giovanni F., Lubrano, L. and Petruzzelli, G., 1991. Nitrohumic acids as nitrogen fertilizers and their effect on zinc absorption by wheat seedlings. *Agrochimica,* **35**: 26-33.

Chandhoke, V., Goodell, B., Jellison, J. and Fekete, F.A. 1992. The isolation and immunolocalization of iron-binding compounds produced by *Gloeophyllum trabeum. FEMS Microbiol. Lett.,* **90** (3) : 263-266.

Chapin, F.S. III, 1980. The mineral nutrition of wild plants. *Ann. Rev. Ecol. Syst.,* **11** : 233-260.

Chongpraditnun, P., Suzuki, K., Kawaharada, U., Nakajima K. and Chino, M. 1991 Immunohistochemical localization of metallothionein in plant tissues. *Water, Air, Soil, Pollut.,* **57-58** : 545-553.

Clare, D.A., Rabinowitch, H.D. and Fridovich, I. 1984. Superoxide dismutase and chilling injury in *Chlorella ellipsoidea. Arch. Biochem. Biophys.,* **231** : 158-163.

Clarkson, D.T. 1985. Factors affecting mineral nutrient acquisition by plants. *Ann. Rev. Plant Physiol.,* **36** : 77- 115.

Clarkson, D.T. and Hanson, J.B. 1980. The mineral nutrition of higher plants. *Ann. Rev. Plant. Physiol.* **31** : 239-298.

Cohen, M.S. and Lepper, R.Jr. 1977. Effect of boron on cell elongation and division in squash roots. *Plant Physiol.,* **59** : 884-887.

Coleman, J.E. 1992. Zinc proteins : Enzymes, storage proteins, transcription factors, and replication proteins. *Annu. Rev. Biochem.,* **61**. 897-946.

Dave, I.C. 1987. Function & Metabolism of Trace Elements in Plants. *Proc. Natl. Symp. on Micronutrient Stresses in Crop Plants : Physiological and Genetical*

Approaches to Control them. Dec. 16-18. Rahuri, Maharashtra, India. pp 246-252.

Dave, I.C. and Kannan, S. 1980. Boron deficiency and its associated enhancement of RNase activity in bean plants. *Z. Pflanzenphysiol.*, **97** (3) : 261-263.

Dave, I.C. and Kannan, S. 1981. Influence of boron deficiency in micronutrients absorption by *Phaseolus vulgaris* and protein contents in cotyledons. *Acta Physiol. Planta,* **3** (1) : 27-32.

Dave, I.C. and Nair, C.K.K. 1993. Incorporation of H-thymidine into nuclear membrane complex (M-band) and DNA polymerase activity in nuclei of boron deficient bean seedings. *Proc. Intntl. Symp. on "Emerging Frontiers in Radiation Biology"*, Dec. 1-3. Abst. No. P-95. Bhabha Atomic Research Centre, Bombay, India.

De Vos, Ric, C.H., Vonk, M.J., Vooiys, R. and Schat, H. 1992. Glutathione depletion due to copper induced phytochelatin synthesis causes oxidative stress in *Silene cucubalus. Plant Physiol.*, **98** (3) : 853-858.

Dexheimer, S.L. and Klein, M.P. 1992. Detection of paramagnetic intermediate in the S1 state of the photosynthetic oxygen- evolving complex. *J. Am. Chem. Sco.,* **114** (8) : 2821-2826.

Dixon, R.K., Garrett, H.E. and Cox, G.S. 1989 *J. Plant Nutr.,* **12** (6) : 687-700.

Dugger, W.M. 1983. Boron in plant metabolism. *Encyclopedia of Plant Physiology,* New Series **15-B** : 626-650.

Esteban, J.A., Bernad, A., Salas, M. and Blanco, L. 1992. Metal activation of synthetic and degradative activites of 0 29-DNA- Polymerase, a model enzyme for protein primed DNA-replication. *Biochemistry,* **31 (2)** : 350-359.

F. da Silva, J.J.R. and Williams, R.J.P. 1976. The uptakes of elements by biological systems. *Struct. Bonding,* **29** : 67- 121.

Fox, R.L. and Hue, N.V. 1989. Boron requirements of macadamia seedlings *J.Plant Nutr.,* **12**(2) : 207-218.

Gerloff, G.C. 1963. Comparative mineral nutrition of plants. *Ann. Rev. Plant physiol.,* **14** : 107-124.

Glass, A.D.M. 1983. Regulation of ion transport. *Ann. Rev. Plant Physiol.,* **34** : 311-326.

Govindjee, Wyndrzynski, T., Marks, S.B. 1978. Manganese and Chloride : Their roles in Photosyntheses. In *"Photosynthetic Oxygen Evolution,* ed. H. Metzner. pp. 321-344. Academic Press, New York.

Gray, B. and Carmichael, A.J. 1992. Kinetics of superoxide scavanging by dismutase enzymes and manganese mimics determined by electron spin resonance. *Biochem. J.* **281**(3) : 795-802

Greppin, H., Panel, C. and Gaspar, T. (eds.) 1986. Molecular and Physiological Aspects of Plant Peroxidases. University of Geneva, Geneva.

Gries, D. and Runge M. 1992. The ecological significance of iron mobilization in wild grasses. *J. Plant Nutr.,* **15** (10) : 1727-1737.

Hahn, C.K., Kim, J.P. and Jung. J. 1991. Induction of antioxidant enzymes as defense systems in plant cells against low temperature stress. (II). Manganese (2+)-induced SOD activation and enhancement of cold tolerance in rice seedlings. *Hamguk Nonghwa Hakhoechi.*, **34** (2) : 168-173.

Harris, E.D. 1992. Copper as a cofactor and regulator of copper, zinc superoxide. *J. Nutri.*, **122** (35) : 636-640.

Hatanaka, H., Imaoka, H., Tajima, S. and Kasai, T. 1991. Purification and properties of an alpha-L-arabinofuranosidase from cotyledons of soybean seedlings. *Agri. Biol. Chem.*, **55** (10) : 2599-2605.

Heinen, M., De Jager, A. and Niers, H. 1991. Uptake of nutrients by lettuce on NFT with controlled composition of nutrient solution. *Neth. J. Agric. Sci.*, **39** (3) : 197-212.

Hiraga, S. 1992. Chromosome and plasmid partition in *Escherichia coli. Annu. Rev. Biochem.*, **61** : 283-306.

Hodgson, J.F., Lindsay, W.L. and Trierweiler, 1966. Micronutrient cation complexing in soil solution. II. Complexing of zinc and copper in displaced solution from calcareous soils. *Soil. Sci. Soc. Am. Proc.*, **30** : 723-726.

Holden, M.J., Luster, D.G., Chaney, R.L., and Buckhout, T.J. 1992. Enzymology of ferric chelate reduction at the root plasma membrane. *J. Plant Nutr.*, **15** (10) : 1667-1678.

Hopkins, B.G., Jolley, Von D. and Brown, J.C. 1992. Variable inhibition of iron uptake by oat phytosiderophore in five soybean cultivars. *J. Plant Nutr.*, **15** (1) : 125-135.

Huang, I-Jen, Welkie, G.W., and Miller, G.W. 1992. Ferredoxin and flavodoxin analysis in tobacco in responses to iron stress. *J.Plant Nutr.*, **15** (10) : 1765-1782.

Huges, M.N. 1972, The Inorganic Chemistry of Biological Processes. Wiley London.

Jewell, A.W., Murray, B.G. and Alloway, B.J. 1988. Light and electron microscopic studies on pollen development in barley (*Holdeum vulgare* L.) grown under copper-sufficient and deficient conditions. *Plant, Cell Environ.*, **11** (4) : 273-281.

Jin, Y. Qian, J. and Xiao, D. 1991, Studies on polyphenol oxidase in sweet potatoes *Shipin Kexue* (Beijing), **141** : 4-9.

Kannan, S. 1986. Foliar absorption and transport of inorganic nutrients. *CRC Crit. Rev. Plant Sci.*, **4** : 341-375

Kannan, S. and Ramani, S. 1987. Mechanisms of Fe-deficiency tolerance in crop cultivars : Effects of dibutyl phthlate and caffeic acid on Fe-chlorosis recovery. *J. Plant Nutr.*, **10** (9-16) : 1051-1058.

Kannan, S., 1986. Effects of dibutyl phthalate and phthalic acid on chlorosis recovery in iron-deficiency stressed sorghum cultivars. *J. Plant Nutr.*, **9** (12) : 1543-1551.

Kannan, S., Ramani, S. and Patankar, A.V. 1984. Excretion of dibutyl phthalate by sorghum roots under Fe-strees : evidence of its action on chlorosis recovery. *J. Plant Nutr.* **7** (12) : 1717-1729.

Kastori, R. and Petrovic, N. 1989. Effect of boron on nitrate reductase activity in young sunflower plants. *J. Plant Nutr.*, **12** (5) : 621-632.

Kavelaki, K. and Ghanotakis, D.F. 1991. Effect of Mn complex on the binding of the extrinsic proteins (17,23 & 33 KDa). *Photosynth. Res.,* **29** (3) : 149-155.

Kim. D.H., Britt, R.D., Klein, M.P. and Sauer, K. 1992. The Mn site of photosynthetic oxygen - evolving complex probed by EPR spectroscopy of oriented PS-II membranes : the g = 4 & g = 2 multiline signals. *Biochemistry,* **31** (2) : 541-547.

Krotz R.M., Evangelous, B.P. and Wagner, J.G. 1989. Relationship between cadmium Zinc, Cd-peptides, and organic acid in tobacco suspension cells. *Plant Physiol.,* **91** : 780-787.

Labrie, S.T., Wilkinson, J.Q., Tsay, Y.F., Feldman, K.A. and Crawford, N.M. Identification of two tungstate-sensitive molybdenum cofactor mutants chl2 and chl7 of *Arabidopsis thalliana.* 1992. *Mol. Gen. Genet.,* **233** (1-2) : 169-176.

Lee, M.E. and Nowak, T., 1992. Metal iron capacity of the catalytic site of yeast enolase. *Biochemistry,* **31** (7) : 2172-2180.

Lee, R.B. 1982. Selectivity and kinetics of ion uptake by barley plants following nutrient deficiency. *Ann. Bot.,* **50** : 429-449.

Lefebvre, D.D., Duff, S.M.G., Fife, C.A., Julien-Inalsingh, C. and Plaxton, W.C. 1990. Response to phosphate deprivation in *Brassica nigra* suspension Cells. *Plant Physiol.,* **93** : 504-511.

Leidi, E.O., Gomez, M. and Guardia, M.D. De La, 1987. Soybean genetic differences in response to Fe and Mn : Activity of metalloenzymes. *Plant Soil,* **99** (1) : 139-146.

Lewis, D.H. 1980. Boron, lignification and the origin of vascular plants - a unified hypothesis. *New Phytol.,* **84** : 209-229.

Loneragan, J.F., Snowball, K. and Robson, A.D. 1976. Re- mobilization and its significance in plant nutrition. In *"Transport and Transfer Processes in Plants".* I.F. Wardlaw, Passioura, J.B., eds. pp 463-469, Academic Press, New York.

Lovatt, J.C., Albert, L.S. and Trembley, G.C. 1981. Synthesis, salvage and catabolism of nucleoctides in boron deficient squash roots. *Plant Physiol.,* **68** : 1389-1394.

Lyashenko, T.E., Bozhkov, A.I. and Dogadina, T.V., 1991. The effect of copper ions on the content of nucleic acids and proteins in the cells of alga *Dunaliella. Biol. Nauki* (Moscow), **7** : 107-108.

Maldonado, S. and Lott., J.N.A. 1991. Protein bodies in *Datura stramonium* seeds : Structural and mineral nutrient composition. *Can. J. Bot.,* **69**(11) : 2545-2554.

Marschner, H. 1986. Mineral Nutrition of Higher Plants. Academic Press, London, New York.

Mattoo, A.K., Mehta, R.A., Baker, J.E., 1992. Copper induced ethylene biosynthesis in terrestrial (*Nicotiana tobacum*) and aquatic (*Spirodela oligorrhiza*) higher plants. *Phytochem.,* **31** (2): 403-409.

Mengel, K. and Kirkby, E.A. 1982. Principles of Plant Nutrition, Int. Potash Inst. Bern, Switzerland.

Miller, J., McLachlan, A.D. and Klug, A. 1985. Repetitive zinc- binding domains in the protein transcription factor IIIA from *Xenopus* oocytes. *EMBO* J., **4** : 1609-1614.

Moore, H.M. and Hirsch, A.M. 1983. Effects of boron deficiency on motosis and incorporation of tritiated thymidine into nuclei of sunflower root tips. *Am. J. Bot.*, **70** : 165-172.

Morales, F., Susin, S., Abadia, A., Carrera, M. and Abadia, J. 1992. Photosynthetic characteristics of iron chlorotic pear (*Pyrus communis* L.) *J. Plant Nutr.*, **15** (10) : 1783-1790.

Mori, S. and Nishizawa, N. 1989. Identification of barley chromosome No. 4, possible encoder of genes of mugineic acid synthesis from 2' deoxy-mugieneic acid using wheat-barley additional lines. *Plant & Cell Physiol.*, **30** (7) : 1057- 1061.

Nable, R.O., Bar-Akiva, A. and Loneragan, J.F. 1984. Functional manganese requirement and its use as a critical value for diagnosis of manganese deficiency in subterranean clover. *Ann. Bot.*, **54** : 39-49.

Neame, P.J. and Barber, M.J. 1989. 2 cysteines in N-terminal domain are probably involved in binding MoCO. *J. Biol. Chem.*, **264** : 20894-20901.

Nixon, P.J. and Diner, B.A. 1992. Aspartate-170 of PS-II reaction center polypeptide D1 is involved in the assembly of the oxygen-evolving Mn cluster. *Biochemistry.*, **31** (3) : 342- 348.

Ochiai, E.L. 1977. Bioinorganic Chemistry. Allyn and Bacon, Boston.

Olsen, R.H. Bennett, J.H., Blume, D. and Brown, J.C. 1981, Chemical aspects of the Fe stress response mechanism in tomatoes. *J. Plant Nutr.*, **3** : 905-921.

Onyezilli, F.N. and Ross J.D. 1993. Iron deficiency stress responses of five graminaceous monocots. *J. Plant Nutr.*, **16** (5) : 953-974.

Pich, A., Scholz, G. and Seifert, K. 1991. Effect of nicotianamine on iron uptake and citrate accumulation in two genotypes of tomato *Lycopersicon esculentum* Mill. *J.Plant Physiol.*, **137** (3) : 323-326.

Polster, J. Schwenk, M. and Bengsch, E. 1992. The role of boron, silicon and nucleic bases on pollen tube growth of *Lillium longiforum* L. *Z.Naturforsch. C : Bio. Sci.*, **47** : 102- 108.

Possingham J.V.,, Vesk, M. and Mercer, F.V. 1964. The fine structure of leaf cells of manganese deficient spinach. *J. Ultrastruct. Res.*, **11** : 68-83.

Price, C.A. 1968. Iron compounds and plant nutrition. *Ann. Rev. Plant Physiol.*, **19** : 239-248.

Przemeck, E. and Hasse, N.U. 1991. On the bonding of manganese, copper and cadmium to peptides of xylem sap of plant roots. *Water, Air, Soil. Pollut.*, **57-58** : 569-577.

Quilez, R., Abadia, A. and Abadia, J. 1992. Characteristics of thylakoids and photosystem II-membrane preparations from iron deficient and iron sufficient sugar beet (*Beta vulgaris* L.). *J. Plant Nutr.*, **15** (10) : 1809-1819.

Qunying, W., Changhao, H., Zhenlin, W., Mingjun, Y. and Naichui, Z. 1992. Effect of boron, manganese, molybdenum and copper deficiency on rate of photosynthesis and ultrastructure of chloroplasts in leaves. *Zhongguo Nongye Kexue* (Beijing)., **25** (1) : 32-36

Rains, D.W. 1972. Salt transport by plants in relation to salinity. *Ann. Rev. Plant Physiol.,* **23** : 367-388.

Rains, D.W. 1976. Mineral Metabolism. In "Plant Biochemistry" J. Bonner, J.E. Varner, eds. pp 561-597. Academic Press, New York.

Ramani, S. and Kannan, S. 1987. Enhancement of iron uptake by dibutyl phthalate in sorghum (*Sorghum bicolor* L. Moench) cultivars with different susceptibility to iron deficiency. *J. Plant Physiol.,* **127** : 379-383.

Ramani, S., Kannan, S. and Patankar, A.V. 1986. Identification of dibutyl phthalate in the root exudate of Fe-efficient (*Corchorus capsularis* L.) subjected to Fe-deficiency stress : Its action on Fe-chlorosis recovery. *J.Plant Nutr.,* **9** (12) 1533-1541.

Rashid, A., Bernier, M., Pazdernick, L. and Carpentier, R. 1991. Interaction of zinc (2+) with the donor side of photosystem II. *Photosynth. Res.,* **30** (2-3) : 123 - 130.

Rio, S. D., Monezes, S.R. and Setzer, D.R. 1993. The function of individual Zinc fingers in sequence-specific DNA recognition by transcription factor IIIA. Repetitive zinc-binding domains in the protein transcription factor IIIA from *Xenopus* oocytes. *J. Mol. Biol.,* **233** : 567-579.

Riordan, J.F. 1976. Biochemistry of Zinc. *Med. Clinics of N. America,* **60** : 661-674.

Robson, A.D., Hartley, R.D. and Jarvis, S.C. 1981. Effect of copper deficiency on phenolic and other constituents of Wheat cell-walls. *New Phytol.,* **89** : 361-373.

Romheld, V. and Marschner, H. 1986. Evidence for a specific uptake system for iron phytosiderophores in roots of grasses. *Plant Physiol.,* **80** : 175-180.

Sachs, M.M. and Ho, T.D. 1986. Alteration of gene expression during environmental stress in plants. *Ann. Rev. Plant Physiol.,* **37** : 363-376.

Sandmann, G. and Bogger, P. 1983. The enzymatological function of heavy metals and their role in electron transfer processes of plants. *Encyclopedia of Plant Physiology, New Series,* **15- A** : 563-596.

Schaecke, B., Makowski, N., Schroeder, G., Roehl, W. and Zajnoc, I. 1990. Micronutrient requirements of high-quality winter rape. *Wiss. Z. Univ. Rostock, Naturwiss. Reihe.,* **39 (2)** : 20-24.

Scholz, G., Becker, R., Pich, A. and Stephan, V.W. 1992. Nicotianamine - a common constituent of strategies I and II of iron acquisition by plants - a review. *J. Plant Nutr.,* **15 (10)** : 1647-1665.

Schon, M.K., Novacky, A. and Blevins, D.G. 1990. Boron induces hyperpolarization of sunflower root cell membranes and increases membrane permeability to K+. *Plant Physiol.,* **93** : 566- 571.

Sharma, S.P., Sharma, P.N., Chatterjee, C. and Agarwala, S.C., 1990. Manganese deficiency in maize affects pollen viability. *Plant Soil,* **138** (1) : 139-142.

Sharma, P.N. and Ramachandra, T. 1990. Water relations and photosynthesis in mustard plants subjected to boron deficiency. *Indian J. Plant Physiol.,* **33** (2) : 150-154.

Sheudzhen, A. Kh. and Logvina, T.B. 1991. Effect of trace elements on nitrogen P. and K utility coefficient in rice. *Khim. Selsk, Khoz.,* **12** : 44-46.

Shkolnik, M.Y. 1974. General conception of the physiological role of boron inn plants. *Sov. Plant Physiol.* (Eng.), **21** : 140-150.

Spence, J.T. 1976. Reactions of molybdenum coordination compounds as models for biological systems. *Encyclopedia of Plant Physiology,* **2A** : 297-324.

Steinitz, Y., Mazor, Z. and Shilo, M. 1979. A mutant of the cyanobacterium *Electonema boryanum* resistant to photo-oxidation. *Plant Sci. Lett.,* **16** : 327-335.

Steinman, H.M. and Ely, B. 1990. Copper-Zinc superoxide dismutase of *Caulobacter crescentus* : Cloning, sequencing, and mapping of the gene and periplasmic location of the enzyme. *J.Bacteriol.,* **172** : 2901-2910.

Stevenson, F.J. and Fitch, A. 1981. Reactions with organic matter. In *"Copper in Soils and Plants"*. J.F. Loneragan, A.D. Robson, R.D. Graham eds. pp 69-95, Academic Press, New York.

Stieffel, E.J., 1973. Proposed molecular mechanism for the action of molybdenum enzymes. *Proc. Natl. Acad. Sci. USA.,* **70** : 988-992.

Tanada, T. 1978. Boron - key element in the actions of phytochrome and gravity. *Planta,* **142** : 221-223.

Tanada, T. 1983. Localization of boron in membranes *J. Plant Nutr.,* **6** (9) : 743-749.

Teasdale, R.D., and Richards, D.K. 1990. Boron deficiency in cultured pine cells. *Plant Physiol.,* **93** : 1071-1077.

Telfer, A., De Las Rivas, J. and Barber, J. 1991. B-carotene within isolated photosystem II reaction centre : photo-oxidation and irreversible bleaching of this chromophore by oxidized P680. *Biochim. Biohys. Acta.,* **1060** (1) : 104-114.

Thellier, M., Duval, Y. and Demarty, M. 1979. Borate exchanges of *Lemna minor* L. as studied with the help of the enriched stable isotope and of a (n, a) nuclear reaction. *Plant Physiol.,* **63** : 283-288.

Tiffin, L.O. 1972. Translocation of micronutrients in plants. In *Micronutrients in Agriculture* J.J. Mortvedt, P.M. Giordano, W.L. Lindsey eds. pp 199-229. *Soil Sci. Soc. Am. Madison.*

Turvey, N.D., Carlyle, C. and Downes, G.M. 1992. Lignin distribution across tracheid cell walls of poorly lignified wood from deformed *Pinus radiata. Plant Soil,* **139** (1) : 59-65.

Veen, B.W. 1980. Energy cost of ion transport. In *"Genetic Engineering of Osmoregulation, Impact on Plant Productivity for Food,* Chemicals, and Energy, ed. D.W. Rains, R.C. Valentine, A. Hollander, pp 187-197, Plenum, New York.

Vlasyuk, P.A. 1971. *Fizioli. Biokhim. Kultur. Rasten.,* **3** : 276-286.

Wasserman, J.C. and Lavaax, G. 1991. Metal concentration of eelgrass (*Zostera noltii Hornem*). The importance of age structure in concentration of zinc, copper, iron and manganese. *Environ. Technol.,* **12**(8): 713-717.

Wild, A., Breeze, V. 1981. Nutrient uptake in relation to growth. In *"Physiological Processes Limiting Plant Productivity"*. ed. C.B.Johnson, pp 331-344. Butterworths, London.

Williams, R.J.P. 1973. Electron transfer and oxidative energy. *Biochem. Soc. Trans.*, **1** : 1-26.

Woldendrop. J.W. 1981. Nutrients in the rhizosphere. In 16th Colloquium, International Potash Inst. Agric. *"Yield Potentials in Continental Climates"*. pp 89-115 Warsaw : Int. Potash Institute.

Xia, M.Z. and Xiong, F.Q. 1991, Interaction of molybdenum, phosphorus and potassium on yield in *Vicia faba*. *J. Agri. Sci.*, **117** (1) : 85-89.

Zhang, F., Romheld, V. and Marschner, H. 1991. Role of the root apoplasm for iron acquisition by wheat-plants. *Plant Physiol.*, **97** (4) : 1302-05.

Zuberbukler, A.D. 1976. Interaction of Cu(I) complexes with dioxygen. In *"Metal ions in Biological Systems"* H. Siegal ed. Vol. **5**. 326-368. Dekker, New York.

Advancements In Micronutrient Research
Editor : A. Hemantaranjan
1996, Scientific Publishers, Jodhpur, pp. 203-365

9

PHYSIOLOGY AND BIOCHEMISTRY OF MICRONUTRIENT ELEMENTS

S.C. AGARWALA AND CHITRALEKHA CHATTERJEE
Botany Department, Lucknow University, Lucknow, INDIA

CONTENTS

1. INTRODUCTION

Micronutrients like macronutrients are essentail elements for plants. The former are required by the plants in very low concentration often in trace amounts. These elements are also referred to as oligo elements or microelements or spurene elements or trace elements. These alternative names for micronutrient elements except trace elements have been given up by most scientists. To avoid confusion and misuse of the term 'essential element' Arnon (1954) had stated that an element will be considered essential for plants if it fulfils the undermentioned criteria,

(a) the plant should not be able to grow or complete its life cycle in the absence of the element,

(b) the element could not be replaced by another element, and

(c) the element should play a direct role in the plant metabolism.

At present about 24 elements are considered as essential for all plants or for a limited number of plant species. These elements are the following : carbon, hydrogen, oxygen, nitrogen, calcium, phosphorus, potassium, magnesium, sulphur, iron, manganese, zinc, copper, boron, chlorine, molybdenum, sodium, cobalt, nickel, vanadium, silicon, iodine, selenium and aluminium. This list of essential elements can not be considered to be complete as with continuing improvement in technology of purification, cleaning of materials used in culture work, exclusion of dust and other contents of the atmosphere, analysis of nanogram quantities of elements and detection of the minute physiological and biochemical reactions which can be attributed to the presence of an element, many new elements may be added in future to the list of essential elements. But one may reasonably conclude that the requirement for new element will be lower than that for the known micronutrients and beneficial elements. To distinguish micronutrient elements from other essential elements Loomis and Shull (1937) stated that the elements which are required by plants in a concentration less than 1 part per million may be called micronutrient elements or simply micronutrients and the rest of the essential elements as macronutrient elements or simply macronutrients. In this classification the position of iron and chlorine seems to be anomalous. Iron and chlorine, although required by most higher plants in concentrations higher than 1 ppm are included in micronutrients. It is very difficult to classify nutrients according to principle laid down by Loomis and Shull (1937). The quantity of an element which is required by different groups of plants sometimes varies considerably and may even vary for the same plant species at different stages of growth, e.g., the requirement of iron for tomato, spinach and mangold was 0.1 ppm whereas for sorghum and millet it was 100 times more at early stages of growth. At the early stages of growth in many species, iron deficiency symptoms could be observed when the supply of iron was 5 ppm but after the plants are firmly established iron deficiency symptoms could not be observed even if the supply of iron was reduced to 0.5 ppm. Manganese, copper, zinc, molybdenum, boron, iron and chlorine are considered to be micronutrient elements for all higher plants. The essential elements, carbon, hydrogen, oxygen, calcium, magnesium, potassium, nitrogen and sulphur are called macronutrients. The essential elements other than macro– or micro– nutrients which are needed for a small group of plants or even for a single plant species, or for a particular essential activitity are called beneficial elements and these are cobalt, sodium, nickel, vanadium, silicon, iodine, selenium and aluminium.

Many physiologists do not agree with the classification of essential elements into macro, micro nutrient and beneficial elements, e.g., Mengel and Kirkby (1987) have proposed and entirely different classification of essential elements. They have suggested that plant nutrient elements may be classified on the basis of their biochemical behaviour and physiological functions rather than on their quantitative requirements. In this chapter, on physiology and biochemistry of micronutrients, the beneficial elements are also conisdered along with the micronutrients. It is very difficult to separate physiology from biochemistry of plants as the two disciplines deal with many similar problems, e.g., role of micro-nutrients in enzyme action, membrane proteins, action of light on enzyme synthesis, mechanism of uptake of micro-nutrients etc. For these reasons biochemistry and physiology of micronutrients will be dealt together in this chapter. Although good amount of literature from different parts of the World, particularly from Australia, U.K., U.S.A., European countries and the Botany Department, University of Lucknow, India, is available on the symptoms of micro-nutrients and their requirement in different crop plants of agricultural and horticultural importance yet these aspects have not been included in this chapter.

2. IRON

For the first time, Gris (1844) observed yellowing of grapevine leaves in absence of iron in the nutrient solution in which these plants were growing. Later, iron was reported to be essential for diverse group of plants, viz., higher plants (Sachs, 1860), fungi (Raulin, 1869), bacteria (Bortel, 1927) and blue green algae (Arnon, 1958). Detailed information on the differential iron requirement for different groups of plants has been described separately (Bowen, 1966). It has been observed that the requirement of iron for monocots was higher than that for dicots. Due to a great variation in iron requirement by different plant species, it is difficult to decide whether iron is a macro or micronutrient.

Non-availability of iron in the rooting medium results in the appearance of characteristic visible effects of the deficiency of the element as chlorosis of young emerging leaves (Hewitt, 1963; 1983). In many fruit trees under severe iron deficiency die back of young shoots have been reported by various workers (Wallace, 1961; Chapman, 1973).

Luo (1982) had reported that the deficiency of iron in wheat decreases the leaf area by significantly lowering the number of leaf cells. Deficiency of iron also causes abrupt cessation of cell division in the apical meristem of excised pea roots (Brown and Possingham, 1957). Iron deficiency also results in the reduction in the development of meristematic region, cortex, stele and pith and increases the number of xylem elements (Leela and Pathak, 1968).

2.1. Chloroplast Structure

In young leaves of iron deficient plants, the plastids broke down, their contents get mixed up and starch grains were lysed (Jacobson and Oertli, 1957).

Iron plays an important role in chloroplast development and maintenance of its integrity (Jacobson and Oertli, 1957; Spiller 1980; Platt - Aloia *et al.*, 1983; Bennett *et al.*, 1984; Terry and Abadia, 1986).

The concentration of chlorophyll in leaves is related to the level of iron supply to the plants (Jacobson, 1945; Terry and Low, 1982; Mehrotra *et al.*, 1992). It has been suggested that iron is not directly involved in the chlorophyll biosynthesis but it is required indirectly for its biosynthesis as it controls the formation of its precursor δ-amino levulinic acid (Jacobson and Oertli, 1956; Evans, 1959; Nason and McElroy, 1963; Marschner, 1986). On the other hand, iron is supposed to play a possible role in the synthesis of some specific RNA that regulated the chlorophyll synthesis (Noort and Wallace, 1966). Reports are also available on increased degradation of chlorophyll and decreased ratio of chlorophyll *a* and *b* due to increased activity of chlorophyllase in iron deficient maize plants (Almela *et al.*, 1983). Sugarcane is known to show restricted differentiation of mesophyll and bundle sheath chloroplasts of the cells of chlorotic leaves (Naik *et al.*,1985). Spiller (1980) found that iron deficiency did not decrease the number of chloroplasts per cell, however, reports are there on the retarded and abnormal development of chloroplast with decreased grana development, reduced number of stroma lamellae chloroplast and thylakoids/grana (Titu and Djendov, 1972; Spiller, 1980; Ji *et al.*, 1984; Kaliya *et al.*, 1989). As a result of low iron, the grana fret work is disorganised and thylakoids are dispersed along with disorganised membrane structure (Taylor *et al.*, 1982; Ji *et al.*, 1984; Platt - Aloia *et al.*, 1983). It is further suggested that iron deficiency affects chloroplast activity due to changes in components of chloroplast lamellae, decreased chlorophyll-protein complexes (Bennett *et al.*, 1984; Guikema and Sherman, 1983). Iron deficiency is known to cause qualitative as well as quantitative changes in the polypeptide composition in the chloroplasts of *Euglena gracilis* (Gaynor, 1982). Nishio *et al.* (1985) demonstrated that during chloroplast development proteins are inserted secondarily into galactolipid matrix of the thylakoid membrane and that iron deficiency diminished chlorophyll protein complex. Later Nishio (1986) reported that iron deficiency depressed the synthesis of thylakoid galactolipids as well as proteins. The chlorotic leaves of pea as a result of iron deficiency showed a decrease in chloroplast and cytoplasmic rRNAs and specific chloroplast and nucleus encoded mRNAs. These workers further suggest that iron supply regulates synthesis of nuclear mRNA encoding chlorophyll *a/b* binding protein and small subunit of *Rubisco* and chloroplast encoded mRNA for larger subunit of *Rubisco* (Spiller *et al.*, 1987). Iron deficiency is also known to decrease the synthesis of P700 molecules (Spiller, 1980), reaction centres (Nishio, 1986; Bordier and Lindberg, 1986), the primary electron acceptor complex of PSI (Malkin, 1987). Several workers (Terry and Abadia, 1986; Malkin, 1987; Hiyama, 1987) have reviewed the involvement of iron containing compounds in the photosynthetic activity in photosynthetic apparatus - the chloroplasts of higher plants. Bordier and Lindberg (1986) identified P700 chlorophyll *a*- protein I as the photosystem I (PS I) reaction centre and found that it carried at least 4 Fe atoms/P700. The Fe-S centres A and B of PS I have been suggested to be the (4 Fe-4S) ferredoxins and the X center as the (2Fe-2S) ferredoxin (McDermott *et al.*, 1987). Another Fe-S protein (8Fe-8S) has been described from the chloroplast membranes of spinach (Wynn and Malkin, 1988) whereas Hirasawa *et al.* (1988) have reported an Fe-S protein very much like ferredoxin from non-photosynthetic bean sprouts. Abadia *et al.* (1989) reported accumulation of xanthophylls in chloroplasts of iron deficient pea leaves. In addition to the presence of cytochromes-Cyt f557, b6, b553, b559 and Rieske Fe-S centre (Sandmann, 1985), evidences are obtained for two tightly bound Fe-Quinone complexes of PS II particles from pea chloroplast (Evans and Ford, 1986). Debus *et al.* (1986) could also substantially reconstitute the electron transfer from QA to QB due to iron deficiency by Fe^{2+} supply in photosynthetic bacteria, ***Rhodopseudomonas sphaeroides*** lost due to iron deficiency.

2.2. Uptake and Translocation

Iron is taken up by plants in ionic form as ferrous or ferric or as iron chelate but the most suitable form of iron supply to plants is usually believed to be in the chelated form because of its higher solubility (Wallace, 1962). Suggestions have been made that prior to its absorption by roots, iron is reduced from Fe^{3+} to Fe^{2+} (Brown *et al.*,1961; Chaney *et al.*,1972; Olsen and Brown, 1980).

Iron uptake follows a dual mechanism as the kinetics of iron absorption is different at low and high concentration of iron (Kannan, 1967, Nissen, 1974). The uptake of chelated iron by plants is a disputed topic and a large number of theories have been proposed. Jefferey and Wallace (1968) detected appreciable quantities of iron chelates in plants whereas others (Tiffin and Brown, 1959, 1961; Cheney *et al.*, 1972) reported that iron was separated from the chelates prior to its absorption. But it is now accepted that whether iron is supplied through roots or as foliar spray, it is absorbed by the plants after dissociation to metal and chelate (Schantz and Hunter, 1948; Steward and Leonard, 1952). After the chelate is broken at the root surface, iron and chelating agent may be taken up by the plant in equivalent (Wallace *et al.*, 1955) or non-equivalent (Wallace *et al.*, 1957; Brown and Tiffin, 1960) quantities. At least in some cases absorption of Fe^{3+} may not be ruled out (Olsen and Miller, 1986).

Iron is passively transported by mass flow in transpiration stream (Hylmo, 1958) or iron is translocated by diffusion (O'connor *et al.*, 1975). Several workers believe that iron uptake is an active process and is controlled metabolically (Epstein, 1960; Moore, 1972; Mandley, 1981). As the response of plants to the quantity of iron present in the rooting medium is not consistent which has given rise to lots of controversy regarding the mechanism of iron acquisition by plants. Brown (1978) classified the plants into iron-efficient and iron inefficient types. According to him, the efficient cultivars were less prone to iron deficiency effects due to the reduction of pH of the rooting medium, release of chemical reductants by the roots, to enable these cultivars to take up more iron.

Later two different strategies were proposed for iron acquisition in plants (Marschner *et al.*, 1986; Römheld and Marschner, 1986a, b; Römheld, 1987;

Chaney and Bell, 1987; Chaney, 1988). These were referred to as strategy I (in dicots) and strategy II (in monocots). Strategy I plants are characterised by higher rates of Fe^{3+} reduction at the plasmalemma, greater release of reducing agents such as phenolics and H^+ ions and increased ability of splitting chelates and reduction of Fe^{3+} to Fe^{2+} before uptake and changes in the morphology of apical root zone. The morphological changes in iron deficiency include increased root swelling (Stephen and Gruen, 1989), increased root hair formation (Rosenfield *et al.*, 1991) and appearance of rhizodermal transfer cells (Landsberg, 1989). The sites of Fe^{3+} reduction in these plants are located mainly in the young lateral roots (Ambler *et al.*, 1971; Brown and Ambler, 1973, 1974). The strategy II plants are characterized by the release of phytosiderophores (iron mobilizing non-proteinogenic amino acids), the Fe-phytosiderophore transport system regulation and conspicuous effect of pH (4 to 8) on iron uptake. Recently it has been realized that grasses show differences in their capacity to release phytosiderophores (Römheld and Marschner, 1990) and phytosiderophores also complex and mobilize other cations such as Zn, Mn and Cu (Römheld, 1991).

Scholz *et al.* (1988) recognized nicotianamine as iron sensitive sensor involved in a trigger mechanism that responds to the iron status in both strategies.

The site of iron reduction was identified as the plasmalemma of epidermal or cortical cells of roots (Bienfait *et al.*, 1982, 1983) and root haris (Bell *et al.*, 1988). Subsequent transport of iron is suggested to be through symplasm (Lauchli, 1972). Iron transport from cortex to endodermis is an active process, from where it is released into xylem and transported in the form of Fe^{3+} citrate (Tiffin, 1972). According to Scholz (1989) nicotiananime functions as a translocator of iron within symplast and phloem. Whether applied through roots or foliage, iron is fairly mobile from old to young tissues of sorghum, cotton and beans (Brown *et al.*, 1965). In another observation, iron was found to be not readily mobile from older to younger parts of soybean and tomatoes (Brown and Chaney, 1971). This was supported from the results when iron stressed plants developed chlorosis in the young developing leaves and at the same time older leaves remained green. This was concluded that iron was not redistributed between the old and young leaves. In chickpea, the transport of ^{59}Fe was restricted as major protion was accumulated in roots, viz., about 90% of it was bound to lipoprotein(Malewar *et al.*, 1982). The accumulation of iron in different plant parts was in the order roots > old leaves > young leaves > stem (Twyman, 1959, Abutalybov and Gummatov, 1966; Terry and Low, 1982). Iron distribution in leaf cells follows the order chloroplasts > mitochondria > nuclei (Manoryk and Hrodzynka, 1970). In plants, the different forms of iron identified are plastid ferritin, lamellar phospholipo-proteins and ribonuclease solubilizable fractions (Brown, 1961; Spiller and Terry, 1980; Terry, 1980).

The relationship between iron supply and tissue iron content within plants is a topic of debate. According to many workers iron content was decreased in plants showing iron chlorosis (Sommers and Shive, 1942; Weinstein and Robbins, 1955; Fujiwara and Tsutsumi, 1957; Twyman, 1959; DeKock *et al.*, 1960; Oertli and Jacobson, 1960; Patnaik and Bhadrachalam, 1965; Loop and Finck, 1984), whereas others found no relation between iron concentration in plants and iron supply (Jones and Hewitt, 1949; Twyman, 1951; Agarwala and Sharma, 1961b; Agarwala *et al.*, 1965; Knezek and Maier, 1966).

It is generally believed that total iron is not an indicator of iron status of plants because all iron present in the tissue is not in physiologically active form. Various extractants have been used by different workers to extract 'active' iron from plant tissues which include water, buffers, organic solvents, acids and chelating agents (Mehrotra *et al.*, 1970; Kaur *et al.*, 1984; Pierson *et al.*, 1986; Hamze *et al.*, 1987). Valenzuela and Romero (1988) characterized Biochemical indicator Index (BI) including catalase, peroxidase, chlorophyll, carotenoid, anthocyanin, active and total iron indices to evaluate the iron status of a plant species.

2.3. Role in Metabolism

2.3.1. Carbohydrate Metabolism

The role of iron in carbohydrate metabolism is not yet established. Sideris and Young (1949) reported that in iron deficiency total carbohydrates decreased, with a slight increase in reducing sugars in pine apple plants. In kidney bean (Ivanova, 1970), sunflower (Djendov, 1971) cabbage and cauliflower (Agarwala *et al.*, 1982) a decrease in total sugars was observed whereas in *Hibiscus esculentus* (Demetriades, 1956) and in cauliflower (Sharma and Sharma, 1987) reports are available on an increase in reducing, nonreducing and total sugars.

High iron supply has been shown to increase the synthesis of sugars and starch (Stocking, 1975; Chaturvedi and Singh, 1981).

2.3.2. Nitrogen Metabolism

Nitrogen metabolism of different plant species is affected differently by low iron supply marked decrease in total nitrogen content as obtained is capsicum (Carpena *et al.*, 1968) and contrary to this an increase in total N_2 concentraion was found in apples (Ivanova, 1964). Its concentration was not affected appreciably by iron deficiency in okra (Demetriades, 1955) and corn (Perur *et al.*, 1961). Iron being a constituent of nitrate reductase its activity is markedly reduced in the deficiency of iron (Nicholas, 1961; Nason and

McElroy, 1963; Zumft, 1972; Hageman and Reed, 1980, Hewitt and Notton, 1980). Some contrary reports on increased activity of nitrate reductase and glutamate synthase in iron deficient *Scenedesmus* are also available (Verstreate *et al.*, 1980; Omholt and Boyer, 1988). Iron is also an essential component of nitrogenase, leghaemoglobin, ferredoxin and flavodoxin and is therefore involved in dinitrogen fixation (Evans and Russel, 1971; Bothe *et al.*, 1983; Pardo *et al.*, 1990). Recently iron has been reported to be required for nodule initiation in *Lupinus* (Tang *et al.*, 1990), and as iron containing protein involved in nitrogen fixing symbiosis between rhizobia and host (Gurcinot, 1991). Some sporadic studies suggest the involvement of iron in protein synthesis as iron deficiency decreases protein nitrogen concentration (Bennett, 1945; Weinstein and Robbins, 1955; Agarwala *et al.*, 1965; 1982; Shetty and Miller, 1966; Carpena *et al.*, 1968). Iron deficiency reduced the concentration of chloroplastic proteins in corn (Perur *et al.*, 1961) whereas higher protein nitrogen was reported by Fujiwara and Tsutsumi (1957) in some low iron plants. Iron deficiency is known to decrease the concentration of some amino acids, viz., alanine, arginine, aspartic acid, glutamic acid, amino butyric acid, proline, lysine, threonine and valine (DeKock and Morrison, 1958; Steward *et al.*, 1959; Mehrotra and Bisht, 1990). The decreased number of ribosomes in iron deficient maize is indicative of the involvement of iron in protein synthesis (Lin and Stocking, 1978).

2.3.3. Organic Acids

Hsu and Miller (1968) reported accumulation of organic acids in iron deficient plants due to a decrease in aconitase activity, which catalyses isomerisation of citrate to isocitrate. In iron deficient plants, some organic acids like malic acid (Wadleigh and Brown, 1952), malonic acid (Hsu and Miller, 1961) and citric acid (McGeorge, 1949; Brown, 1966) are known to accumulate. DeKock and Morrison (1956) had observed that while normal and green plants accumulated dicarboxylic acids, chlorotic plants accumulated large amounts of tricarboxylic acids.

2.3.4. Nucleic Acid Metabolism

Iron is also known to be involved in the establishment of bonds between individual DNA helices and at the same time regulates denaturation of DNA (Glodstein and Gerasimova, 1963). The iron deficiency changes the concentration of nucleic acids and on spraying the field plants with adenine and guanine, the iron chlorosis was corrected showing indirect effect of iron on nucleic acid metabolism (Kessler, 1957). A marked decrease in the algal chloroplastic RNA and chloroplastic ribosomes (Price *et al.*, 1972) was also reported. In radish leaves iron deficiency decreased DNA content but had no significant effect on RNA content (Agrawala and Mehrotra, 1984).

2.3.5. Enzymes

Iron has a specific role as a constituent metal or co-factor and is known to be involved in a number of enzymes of metabolic processes in different plant species. Apart from this, the activities of many other enzymes, where iron is not directly involved affected by the deficiency of the metal. Iron, as a constituent, is usually associated with porphyrin, flain or as a simple metallic ion the role of the metal is well documented (McElroy and Nason, 1954; Hewitt, 1963, 1983; Sandmann and Böger, 1983). In catalase, iron is bound to porphyrin and protein probably by ionic linkages. Peroxidase is a heme protein with protoheme as prosthetic group. In cytochrome oxidase also iron is present in porphyrin undergoing continuous oxidation reduction into Fe^{3+} and Fe^{2+}. In aconitase, the bound iron is in ferrous form, whereas succinic dehydrogenase is an Fe-S protein with a tightly bound flavin component.

Decrease in the activity of catalase and peroxidase due to iron deficiency is well established (Brown and Steinberg 1953; Marsh *et al.,* 1963; Agarwala *et al.,* 1964, 1965, 1981; Nicholas and Goodman, 1958; BarAkiva and Lavon, 1965; Iyengar and Seshagiri Rao, 1971; Valeuzuela and Romero, 1988). On the other hand, the activity of peroxidase was not found to be related to the iron supply by many workers, e.g., Healy *et al.* (1955), Marsh *et al.* (1963) and Elmstrom and Howard (1969). Iron deficiency is also known to reduce the activities of cytochrome oxidase, succinic dehydrogenase, aconitase (Healy *et al.,* 1955; Weinstein and Robbins, 1955; Hsu and Miller, 1969; Suzuki *et al.,* 1976). Another iron containing enzyme superoxide dismutase which catalyses the conversion of superoxide radical (O_2) to H_2O_2, has been described by Salin and Bridges (1980). Certain ferrochelatases isolated from spinach chloroplasts catalyse incorporation of Fe^{2+} into protoporphyrin (Jones, 1968). Nitrate reductase also contains bound iron and its activity was also reported to decrease in iron deficiency (Nicholas, 1961; Agarwala *et al.,* 1978).

The deficiency of iron decrease the activity of many non-iron enzymes, viz. , aldolase (Bard and Gunsalus, 1950; Bar-Akiva *et al.,* 1971; Agarwala *et al.,* 1976), phosphorylase (Agarwala *et al.,* 1969; Suzuki *et al.,* 1976, 1965), ß-glycerophosphatase (Agarwala *et al.,* 1976). ATPase (Young and Wallace, 1976) and acid and alkaline phosphotases (Leela and Pathak, 1968).

In contrast to this, the activities of acid phosphatase ribonculease and transaminase were stimulated in radish and maize deficient in iron (Agarwala *et al.,* 1965).

2.3.6. Photosynthesis

Iron affects various aspects of photosynthetic activities directly or indirectly as it is a constituent of a number of heme and non-heme proteins in the chloroplast (Thomas, 1965; Arnon, 1965; Terry and Abadia, 1986; Hiyama, 1987; Pushnik and Miller, 1989). Suggestions have been made on decreased rate of photosynthesis in iron deficiency which is due to limited light harvesting and electron transport capacity of iron deficient plants (Terry, 1983). Iron deficiency is known to decrease the rate of net photosynthesis (Fujiwara and Tsutsumi, 1962; Naik *et al.*, 1985; Petrovic, 1987; Rueter, 1988). It appears that the decrease in the rate of carbon assimilation due to iron deficiency could be related to the decrease in chlorophyll concentration (Terry, 1983; Davis *et al.*, 1986; Rueter, 1988).

2.3.7. Respiration

In several plant species, the deficiency of iron is reported to decrease the rate of respiration (Sommers *et al.*, 1942; Hewitt, 1963). Iron is closely involved in the process of respiration evidently because of being a constituent of cytochromes, flavoproteins and enzymes like cytochrome oxidase, succinic dehydrogenase, aconitase and several other Fe-S centres, which carry out redox reactions in respiratory pathways (Hewitt, 1983).

In a review (Nason and McElroy, 1963) the role of iron in the functioning of cytochrome series in electron transfer system in respiration has been discussed. A specialised mitochondrial cytochrome c-560 reported from *Arum maculatum* (Bendall, 1958) is active as cytochrome c oxidase even in the presence of cyanide or carbon mono-oxide. In a separate study a Rieske Fe-S centre is reported from mung mitochordria (Prince *et al.*, 1981). In contrast to this, Terry (1983) observed that iron deficiency did not have any significant effect on respiratory CO_2 evolution and CO_2 compensation point in sugarbeet. Sandmann (1983) also did not observe any effect of iron deficiency on respiration in *Aphanocapsa*. But involvement of iron in respiration has been reviewed separately by Moore and Cottingham (1983) and Sandmann and Böger (1983).

2.4. Water Relations

The information on the involvement of iron in plant water relations is meagre and mostly pertains to the effect of iron deficiency or lime-induced chlorosis on leaf water deficit and stomatal opening. Hutchinson (1970) observed in iron deficiency an increase in transpirational water loss in the leaves of about 30 different plant species, whereas, Sharma *et al.* (1987) have observed that in iron deficient cauliflower, the effects on water relations are opposite to those of potassium deficiency and that combined deficiency of iron and potassium could alleviate the effects of potassium stress. In contrast to these reports the leaf permeability (Conductance) of groundnut was found to be inversely related to the degree of chlorosis. In sugarbeet, the stomatal conductance was found to decrease linearly with lowered chlorophyll concentration. Szlovak and Zoltanne (1981) recorded a positive linear correlation between iron content and the rate of transpiration in the leaves of hybrid corn throughout the growing season. But reports on no change in stomatal conductance in iron deficiency are also available (Kleinkopf *et al.*, 1976; Davis *et al.*, 1986).

2.5. Growth Substanes

Coconut milk iron, was found to cause stimulation of cell division in cultured plants of tomato, presumably by regulating protein syntheis (Steward *et al.*, 1968; Newman and Steward, 1969). Stimulated abscission of citrus fruits by formation of abscission layer by iron application has been suggested to be due to ethylene formation and destruction of IAA(Ben-Yehoshua and Biggs, 1970). Bozova and Stoeva (1981) have reported that iron deficiency in wheat decreased total auxin, gibberellin, and cytokinin content of leaves but increased the ABA content. Similar effects were reported in iron deficient maize having increased concentration of ABA type growth inhibitors and absence of gibberellin like substance (Stoyanov and Kha, 1981).

Iron nutrition of plants is suggested to affect the concentrations of vitamins, and co-factors in plants, e.g., decrease in ascorbic acid (Sharma, 1961), increase in riboflavin (Welkie and Miller, 1960; Nagarajah and Ulrich, 1966).

2.6. Reproductive Physiology

Effect of low iron on the development and physiology of pollen in plants has also been observed (Sharma *et al.*, 1984). Iron deficiency causes a marked decrease in the pollen producing capacity of anthers, pollen grain size and activities of catalase and peroxidase in maize anthers.

2.7. Interaction of Iron with Other Nutrients

Nitrogen

Irrespective of the Fe source (Fe-citrate, Fe-EDTA, Fe- EDDHA), iron requirement for normal growth of rice plants grown in refined sand and receiving N in the form of nitrate or amide (urea) was four times higher than that of plants receiving N as ammonium nitrate (Chatterjee *et al.*, 1983).

Application of iron as $FeSO_4$ to iron deficient soils significantly increased N, P and K contents of leaf and bulb of onion (Jawaharlal *et al.*, 1985).

Phosphorus

The interaction between Fe and P shows an antagonistic relationship. It has been suggested that excessive quantities of one may exhibit a depressive effect, both on yield and content of iron or phosphorus some times their uptake. Brown *et al.* (1959c) suggested that lime induced chlorosis may be partly connected with the combined effect of high Ca and P.

In several plants it was observed that Fe chlorosis induced by high levels of P did not usually depress iron concentration in plants but the ratio of P/Fe is larger in chlorotic plants than in plants receiving adequate P, perhaps the capacity of plants to absorb and hold Fe in a soluble, mobile form becomes less as the P concentration in the plants rises (Dekock, 1955; Brown *et al.,*1959). Excessive P appears to interfere with movement and metabolic functioning of Fe (Price, 1968).

Dekock, Hall and McDonald (1960) found evidence that P/Fe ratio were more indicative of Fe chlorosis than Fe concentration in mustard. There is some evidence which suggests that Fe is attached to phosphoproteins in both animals and plants. Phosphorus requirement of plants increases at high rates of Fe application. It seems that P : Fe ratio was not related to P or Fe content of the tissue or the soil (Chahel *et al.,*1979).

Potassium

In green house experiments, addition of K at low Fe concentration in soils increased grain and straw yield of barley (Singh and Singh, 1989). The interaction between Fe and K in different plant species is usually synergistic. In K deficient maize and sugarcane the iron deficiency is severer as iron is immobilised in the stem nodes (Hoffu, 1940; Hart, 1934). In potato also similar results were obtained by Hewitt and Bolle-Jones, 1953; Bolle - Jones, 1955. This interaction was also influenced by the magnitude of phosphorus supply. The effects of potassium deficiency in inducing iron deficiency can be explained by the accumulation of iron in roots and its restricted translocation to leaves.

Calcium

Application of lime reduced iron concentration in rice shoots and mature straw. Lime induced iron chlorosis has been reported for several plants (Wallace and Lunt, 1960; Brown, 1960; Hewitt, 1983).

Magnesium

In Mg deficient radish grown in refined sand, the dry matter was significantly and markedly decreased with an increase in Fe supply (Agarwala and Mehrotra, 1984). Excess Fe and Mg separately decreased dry matter significantly but when applied together each one counter-act in the action of the other element. This resulted in larger yield than in the excess supply of Fe and Mg. In cauliflower, the antagonstic effect of Fe and Mg was also reflected on the magnitude of changes in dry matter, leaf Fe, chlorophyll, starch, protein, RNA and the activities of catalase and peroxidase by iron - magnesium (Agarwala *et al.*,1988). Depression in aldolase activity in iron deficiency, became more pronounced by the combined deficiency of Fe-Mn-Mg and that of Mg-ATPase and RNase, which were enhanced by iron deficiency alone, were depressed by the combined deficiences of Fe-Mg and Fe-Mn-Mg. In these treatments and in Mg deficiency, the pyruvate kinase activity was also depressed.

Sulphur

Application of S increases availability and absorption of Fe due to synergism between iron and sulphur (Babaric and Patel, 1980). S application increased Fe concentration in all plant parts of soybean (Kumar and Sinha, 1981) and mustard (Agarwala *et al.*,1988). In mustard, it was also noted that increase in sulphur supply, increased the concentration of chlorophyll, and activities of catalase and peroxidase in iron deficient leaves by making iron available for different metabolic activities.

Manganese

In large number of plant species Mn is known to reduce tissue iron and its uptake (Kuo and Mikkelson, 1981; Baser and Somani, 1982) possibly through inhibition of the translocation of iron from root to shoot. In pot culture experiment in calcareous soils with peanut, Fe hampered Mn uptake and manganese inactivated the iron metabolic activity by decreasing Fe^{2+}concentration in plants (Zakharieva *et al.*, 1988) and Mn supply affected the physiological availability of iron in rice (Agarwala *et al.*,1986). Leidi *et al.* (1986) suggested that catalase and peroxidase as indicators of iron and manganese nutrition of soybean. In tobacco leaves, no indication of Mn-induced iron deficiency was observed by Nable *et al.* (1988) but in corn and clover there are reports of Mn induced iron deficiency (Amberger *et al.*, 1982). Increasing supply of iron alleviated Mn toxicity but iron concentration remained unaffected (Kohno and Foy, 1983).

Copper

In citrus, iron chlorosis was observed at high concentration of copper in nutrient solution (Chapman *et al.*, 1940; Hewitt, 1949) and also in soil and sand cultures (Smith, 1953; Spencer, 1966). Later Agarwala *et al.* (1961,1967) and Chaudhry and Wallace (1976) also found that iron deficiency is caused by excess Cu.

Iron chlorosis due to excess Cu was reduced by the addition of iron but all the adverse affects of excess Cu were not completely overcome by iron

(Spencer, 1966). In the presence of toxic levels of Cu, iron becomes inactive and is not available for metabolic functions thereby causing chlorosis of tobacco seedlings (Murthy *et al.*, 1988). Metal induced iron deficiency including that of Cu might be caused by formation of metallo-organic complexes (Hewitt, 1951; 1954). This hypothesis implies that metals including Cu competes with iron at some stage of chlorophyll formation and the form of iron is perhaps ferrous.

Zinc

It has been suggested that the metabolic functioning of Fe in plant is related in some manner with the supply of Zn to plants. In several studies it has been observed that application of zinc depresses the rate of Fe absorption as well as its translocation from the roots to the tops and *vice-versa*. This antagonism between Fe and Zn may arise from competition between the two ions at the absorption site. Many other results indicate a synergism between the two specially at low to optimum levels of zinc and iron (Tewari *et al.*, 1976; Mukh and Shukla, 1987).

Application of zinc to acid soils helped in alleviating the ill affects of high concentration of iron on paddy yield (Verma and Tripathi, 1985).

Molybdenum

Many earlier reports indicated an antagonistic interaction between Mo and Fe. In tomato, excess Mo accentuated iron chlorosis in tomato (Gerloff *et al.*, 1959) and red clover (Hanger, 1965). Induced iron deficiency was greater when N was supplied as ammonium nitrate or urea (amide) than as nitrate (Hewitt, 1955). In Mo deficient plants, iron uptake and its translocation to leaf margins away from central vein was reduced (Berry and Rai, 1967). The lower uptake of Fe was attributed to inadequate translocation of iron from roots to shoots in Mo - starved tomato. In tomato the deficiency of Fe accentuated the deficiency of Mo and *vice-versa* (Chatterjee and Agarwala, 1979). Using ^{59}Fe and ^{99}Mo they further observed that the deficiency of one element reduced the uptake and translocation of the other.

In mustard, the effects of Mo deficiency, viz., lowered activities of catalase, peroxidase, aldolase and nitrate reductase and increased activity of acid phosphatase, were enhanced when iron was also deficient (Chatterjee *et al.*, 1984). Similarly in *Aspergillus niger* (Nautiyal *et al.*, 1988) under iron deficiency the depression of a number of iron enzymes such as catalase, peroxidase, succinic dehydrogenase aconitase and nitrate reductase (which has both iron and Mo) were more depressed when both Fe and Mo were deficient than when iron was deficient alone. These observations confirm a synergistic rather than an antagonstic interaction between Fe and Mo.

3. MANGANESE

Bertrand (1897) initially reported that Mn was necessary for plant growth and suggested that Mn functions as a component of the oxidase system of plants. But it was Maze(1914) who observed that in absence of Mn, maize showed abnormal growth and chlorotic leaves. McHargue (1919) established the essentiality of Mn for higher plants. Bishop (1928) from his experiments in sand and soil cultures showed that low concentration of Mn added to culture stimulated the growth of radish, beans, corn and peas while high concentrations of Mn injured them.

3.1. Anatomical and Histological Effects

Manganese deficiency resulted in the hypertrophy of palisade cells and disintegration of plastids in subterrancan clover (Millikan, 1953). Brown *et al.* (1958) reported abnormal chloroplast division resulting in paired or tetrad clusters in *Lemna* and *Elodia canadensis.* Eltinge (1941) reported loss of starch grains associated with disorganisation of chloroplasts in manganese deficient tomato leaves. Manganese deficient plants had reduced number of xylem elements in stem which were often plugged with brown deposits thus restricting the transport and mobility of manganese (Eltinge, 1941; Millikan, 1953). Epidermal cells were smaller due to reduced volume with prominent cell walls in manganese deficiency (Bussler, 1958). In manganese deficient queen palm leaves the frequency of chloroplasts was lowered specially near the necrotic areas, fibres were thicker and more numerous (Meerow and Broschat, 1991). In soybean manganese deficiency reduced the number of chloroplasts per cell but had no effect on gross cell structure or organisation of leaves (Weiland *et al.,* 1975). The deficiency of Mn caused disorganisation of chloroplasts in spinach but nuclei, mitochondria and other organelles remained normal (Mercer *et al.,* 1962; Possingham *et al.,* 1964). Loss of granal frets and swellege of intergranal spaces in tomato chloroplasts is another characteristic of Mn deficiency (Vesk *et al.,* 1966). Mn deficiency resulted in reduced grana stacks or disorganised chloroplasts and as a consequence of this the concentration of chlorophyll was lowered in many plant species (Homann, 1967). On the other hand, Weiland *et al.* (1975) observed that in soybean, manganese deficiency had no effect on gross cell structure or organisation of leaves but only reduced the number of chloroplast per cell.

Chloroplasts contain Mn in two states of binding (i) loosely bound manganese in the membrane of stroma (4 atoms/centre) (Radmer and Cheniae, 1971) (ii) strongly bound manganese which may be of two types with separate function, viz., oxygen evolution at the weaker binding site and preservation of thylakoid membrane at the tightly bound site (Weiland *et al.,* 1975; Takahashi and Asada, 1977). The deficiency of tightly bound fraction is believed to be responsible for structural disorganisation of lamellar structure under manganese deficiency (Possingham *et al.*, 1964; Homann, 1967, Heath and Hinde, 1969; Teichler-Zallen, 1969; Weiland *et al.*, 1975). The disorganisation of chloroplast membrane (structure) in manganese deficient plants possibly resulted in inhibition of photosystem II activity (Possingham *et al.*, 1964, Teichler - Zallen, 1969). The ultrastructure of chloroplast in mild manganese deficiency was not affected but photosynthetic oxygen evolution was reduced (Cheniae and Martin, 1968; Simpson and Robinson, 1984; Nable *et al.*, 1984). The Mn deficiency effect on membrane structure seems to be secondary and its effect on water oxidation site seems to be the primary effect.

In Mn deficient maize an increase in osmophilic globules was observed which indicated the destructive changes causing decrease in ATP synthesis by the non-cyclic photophosphorylation chain (Milivojevic and Restovic, 1983).

3.2. Uptake and Distribution

The uptake of Mn in decapitated root tips of *Phaseolus multiflorus* took place against concentration gradient (Laine, 1934). Two phases of manganese uptake were observed in corrot tissues (Stiles and Skelding, 1940); one corresponding to rapid uptake while another to slower. The rate of Mn uptake differs considerably in different plants species (Collander, 1941). Mn is absorbed mainly as Mn^{2+} from the medium and is translocated predominantly as free divalent cation in xylem from roots to the shoots (Graham, 1979; Clarkson and Hanson, 1980). Manganese is often stored in stem and later distributed to other tissues/organs (Hannam *et al.*, 1985). Manganese is also suggested to move passively in different plant organs (Ramani and Kannan, 1974). Manganese is relatively immobile in plants (Wittwer and Teubner, 1959) and is preferentially translocated to meristematic tissues as a consequence of which the young plant organs are rich in manganese (Amberger, 1973).

In soybean leaves a linear relationship was found (Kluthcouski and Nelson, 1979) between supply and concentration of Mn. In Mn deficient oats, subterranean clover and lupin (Munns *et al.*, 1963; Nable and Loneragan, 1984; Hannam *et al.*, 1985), the Mn content of roots is redistributed but large amounts of Mn may accumulate in roots, stem and leaves of plants grown initially under high Mn supply, which can not be remobilised (Loneragan, 1988).

Mn uptake is dependent on various factors–pH (Maas *et al.*, 1968), temperature (Heenan and Campbell, 1990), manganese supply (Reid and Rack, 1985), soil acidity and soil types (Hynes and Swift, 1985).

The effect of pH on Mn uptake was datermined and pH 6 was found to be optimum for Mn uptake (Maas *et al.*, 1968). In another view, the uptake of

Mn is metabolically mediated (Moore, 1972). Mg a divalent cation is also known to depress the uptake of Mn in several crop plants (Lohnis, 1940; Maas *et al.*, 1969). Liming also reduces Mn uptake as a result of pH increase as well as by the direct presence of Ca^{2+} ions in the soil (Mengel and Kirkby, 1982). At adequate and high levels of manganese supply, manganese concentration is generally higher in mature (old) leaves than in young leaves (Nable, 1983; Heenan and Campbell, 1990). In barley leaves, manganese tended to be accumulated at the tip (Nakanishi and Matsumoto, 1991). In barley (Williams and Vlamis, 1957) and sugarcane (Samuels and Alexander, 1969), addition of silicon to the culture solution improved the distribution of Mn.

3.3. Physiological and Biochemical Effects

3.3.1. Photosynthesis

Mn deficiency reduced the photosynthetic efficiency and chlorophyll content in tree seedlings (Angadi and Kamala, 1988) and many other plants (Weiland *et al.*, 1975; Ohki, 1985; Sharma *et al.*, 1986; Leidi *et al.*, 1987).

Photosynthetic oxygen evolution (POE) declined markedly when subterranean clover was transferred to a solution without manganese (Nable *et al.*, 1984), similar observations were made in some other plants also (Lerer and Bar-Akiva, 1979; Burnell, 1988; Sharma *et al.*, 1988).

In *Picea abies* plants, Mn application resulted in a greater increase in the concentration of chlorophyll and carotene than in xanthophyll but the ratio between chlorophyll *a* and *b* and between total chlorophyll and carotenes did not change (Kreutzer, 1970). In sugarcane, total (Kumar and Alexander, 1972) chlorophyll as well as chlorophyll *a* and *b* concentrations (Sharma *et al.*, 1986) were depressed in Mn deficiency but in maize, Mn deficiency inhibited the synthesis of both chlorophyll *a* and *b* (Popovic *et al.*, 1977). The reduced rate of photosynthesis and low chlorophyll concentration in Mn deficient tissues has been attributed to the involvement of Mn in photooxidation of water (Gerretsen, 1950). Arnon (1958) suggested that Mn plays a catalytic role in the primary photochemical reaction of photosynthesis. Mn might be involved in oxidation-reduction reaction in photosynthesis in light isolated chloroplasts as the divalent manganese is oxidised to trivalent state (Kenten and Mann, 1955).

Before affecting the chlorophyll concentration the photosynthetic oxygen evolution is inhibited in Mn deficiency (Cooper and Girton, 1963; Cheniae and Martin, 1968; Weiland and others, 1975; Nable *et al.*, 1984). In this case, the decrease in chlorophyll concentration was the result of photooxidation and photodestruction of chlorophyll due to increased activity of polyphenol oxidase in low Mn conditions. Photosynthetic oxygen evolution required a manganese complex which was activated by sunlight (Cheniae and Martin, 1971). A manganese containing protein, as a component of oxygen evolving enzyme was obtained from isolated chloroplasts which catalysed dehydration of water (Okada and Asada, 1983). This complex was later studied by Yachandra *et al.* (1986). In Mn deficiency not only the electron transport activity was reduced but there was reduction in the concentration of both chlorophyll *a* and *b* (Simpson and Robinson, 1984). A chlorophyll-protein complex containing 3-4 Mn atoms per molecule has been isolated from spinach thylakoids (Yamada *et al.*, 1987). The protein which catalyses the water splitting reaction has four Mn atoms at its active centre and these are thought to act as charge accumulators so that the four - electron oxidation of water can occur as one event (Prince, 1986; Allakhverdiev *et al.*, 1986). The involvement of Mn in water splitting process of photosystem II has been shown in figure (1) (Edwards and Walker, 1983)

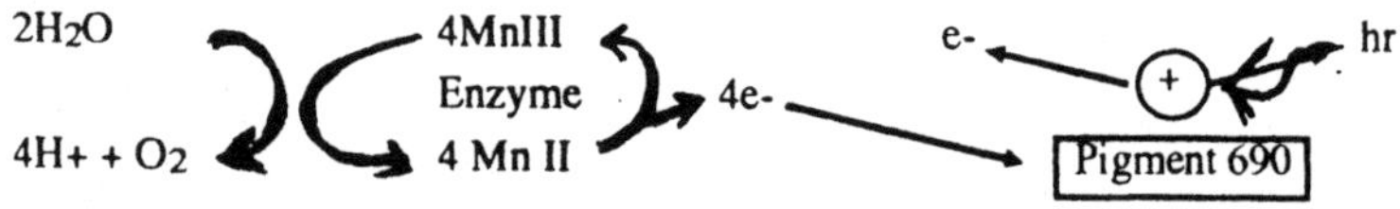

Activity of photosystem II was also decreased in corn leaves but size and number of grana were not affected (Milivojevic and Rastovic, 1983).

3.3.2. Respiration

In manganese deficient wheat roots no stimulation in oxygen uptake was observed but on addition of Mn to the deficient roots, oxygen uptake was increased (Lundegardh, 1939). No effect of Mn deficiency on respiration has been suggested by many workers (Ruck and Bolas, 1954; Terry and Ulrich, 1974; Ohki, 1981) but under severe or long term Mn deficiency, a decrease in respiration was observed which was attributed to decreased availability of substrates, through reduced photosynthesis (Ohki, 1984). Addition of Mn to meadow brown soils increased the intensity of plant respiration and crop yields (Kobakhidzle *et al.*, 1971). In contrast to this, Mn deficiency reduced the photorespiratory CO_2 exchange but had no effect on respiration in dark in sugarbeet (Terry and Ulrich, 1974).

3.3.3. Carbohydrate Metabolism

Carbohydrate metabolism is disturbed in Mn–deficient tissues by reduction in the rate of photosynthesis in algal culture and that a supply of manganese to manganese deficient cultures restored the normal rate of photosynthesis (Miller, 1933; Pierson, 1937). In Mn deficiency conditions, the sensitivity of chlorophyll to light destruction increased which ultimately lead to chlorosis in *Chlorella pyrenoidosa* (Eyster *et al.*, 1958), whereas it was also observed that on withdrawing Mn from culture solution, the rate of photosynthetic CO_2 assimilation was decreased in sugarbeet (Terry and Ulrich, 1974). Almost similar results were obtained in maize seedlings grown in water culture (Popovic *et al.*, 1977).

Application of Mn either through roots or as foliar spray to Mn deficient sugarbeet increased the sugar yield (Antonov, 1970; Emel'yanova *et al.*, 1971; Draycott and Farley, 1973). The deficiency of the metal resulted in an accumulation of sugars in the leaves of grape vines as it inhibited translocation of photosynthates (Milosavljevic and Popovic, 1970). An increase in amylase activity under Mn deficient conditions was paralleled by low sucrose values in sugarcane leaves (Kumar and Alexander, 1972).

Reduced carboxylation in Mn deficient tissues were mainly explained on the basis of various pools of Mn (Robinson *et al.*, 1980; Fridovich, 1983; Burnell, 1988). In C4 plants, out of three carboxylation, two of them, NAD - malic enzymes and pyurvate carboxykinase (PCK) are entirely dependent on Mn supply (Hatch and Mau, 1977; Burnell, 1986). The manganese containing complex may be one of the active site component of photosystem II (Glikman and Zavgorodnyaya, 1973). On the other hand, excess supply of Mn altered the structure of the electron transport chain of photosynthesis in chloroplasts of oat plants (Chernavina *et al.*, 1973).

3.3.4. Nitrogen Metabolism

Low manganese affected the soluble nitrogen levels of plants and resulted in excessive accumulation of nitrates (Burstrom, 1939; Leeper, 1941; Hewitt, 1949; Lerer and Bar-Akiva, 1976), nitrites (Jones *et al.*, 1949; Amberger, 1974), amides (Steward and Margolis, 1962), free aminoacids and water soluble proteins (Nicholas, 1961; Steward and Margolis, 1962; Navarro *et al.*, 1972). The role of Mn in nitrate reduction has also been emphasized due to suppressed nitrate reductase activity (Mulder and Gerretsen, 1952). Manganese toxicity is also known to result in accumulation of nitrates (Elamin and Wilcox, 1986 a, b).

In cauliflower, low Mn caused an accumulation of arginine, aspartic acid, glutamic acid and asparagine but the concentration of α-aminobutyric acid was decreased (Hewitt *et al.*, 1949). Similar observations have been made in tobacco (Steinberg, 1956), tomato (Possingham, 1956) and sugarcane (Rao and Gupta, 1979).

Application of Mn is known to increase nitrogen content and crude protein in corn (Nikolov, 1968). Mn is also reported to affect the structure of proteins of photosynthetic membranes and non-histone proteins (Vlasyuk,

1980). There was considerable reduction in the amount of protein produced by chloroplasts in pea leaves where Mn deficiency had damaged the photosynthetic apparatus (Vlasyuk and Zorya, 1970).

3.3.5. Nucleic Acids

The concentration of both deoxyribonucleic acid (DNA) and ribonucleic acid (RNA) decreased in Mn deficiency in sugarcane (Sharma *et al.*, 1986) possibly because Mn is required both for the synthesis and maintenance of the integrity of the nucleic acids (Lyttleton, 1960; Vlasyuk *et al.*, 1973; Klimovitskaya and Lobanova, 1975). The increase in Mn supply increased the concentration of all RNA types in pea leaves (Klimovitskaya *et al.*, 1969). These changes could be explained in terms of the role of manganese in activation of RNA polymerase of *Phaseolus* chloroplasts (Ness and Woolhouse, 1980).

3.3.6. Lipids, Phenols and Lignin

Manganese has favoured the accumulation of protein and lipids in the kernels of maize and increased RNA polymerase activity and ascorbic acid content (Kobakhidzie *et al.*, 1971). A marked decrease in the lignin concentration in roots and shoots and that of phenols in shoots were obtained in Mn deficient wheat (Brown *et al.*, 1984). Manganese deficiency is known to decrease the polyunsaturated fatty acid, glycolipid of chloroplasts (Cooper and Girton, 1963) and oil content in seeds (Heenan and Campbell, 1980; Wilson *et al.*, 1982). Manganese is known to act as a co-factor for a number of key reactions of shikimic acid pathway (Gross, 1980) which leads to the synthesis of secondary metabolites (Burnell, 1988). In shikimic acid pathway, the first product is deoxy - D - arabinoheptulosonate -7- phosphate (DAHP) which was further converted into shikimic acid. A number of intermediates are formed in this pathway, further synthesizes different phenolics like, ferulic, caffeic, coumaric, chlorogenic, protocatachuic and quinic acid which are widely distributed in plants and have been known to act as disease resistant compounds. The synthesis of primary product DAHP catalyzed by enzyme DAHP-synthase is stimulated by Mn^{2+}.

3.3.7. Growth Hormones

Manganese deficient tissues of cotton demonstrated abnormally high indole acetic acid (IAA) oxidase activity without any detectable inhibitor activity (Taylor *et al.*, 1968). It was further observed that presence of manganese in low or high amounts decreased the level of auxin in cotton plants by increasing the IAA oxidase activity with concomitant decrease in the concentration of IAA - oxidase inhibitor (Morgan *et al.*, 1976; Horst and

Marschner, 1978). Peroxidase might oxidize Mn^{2+} to Mn^{3+} which inturn oxidizes IAA (Kenten, 1955; Meclachlan and Waygood, 1956a and b). In pea, a system where manganese ion directly oxidizes IAA to produce CO_2 has been proposed (Waygood *et al.*, 1956).

3.3.8. Enzymes

Manganese acts as an activator of a large group of enzymes. Manganese activated enzymes are widely distributed in various groups of plants and are concerned with major pathways of plant metabolism - photosynthesis, respiration and protein, lipid, nitrogen and nucleic acid metabolism. Manganese can be substituted by magnesium in many enzymes related to phosphorylation, group transfer reactions and kinases (Nason and McElroy, 1963). So far only four manganese containing enzymes are known, purple acid phosphatase (Amberger, 1973; Fujimoto *et al.*, 1977), superoxide dismutase (Michelson *et al.*, 1977; Sevilla *et al.*, 1980), malic dehydrogenase (Hewitt, 1983) and the photosynthetic manganese containing complex (Yamada *et al.*, 1987). The enzymes which have been reported to be activated by manganese are given in the table below.

Enzyme	Reference
Hydroxylamine reductase, pyruvate oxidase, glucose – 6 - phosphate dehydrogenase, NAD - kinase, Adenosine kinase, arginine kinase, lecithinase, Hexokinase, DNase I, Nitrite reductase	Nason and McElroy (1963)
Pyruvate kinase, phospho-glutarate - dehydrogenase, alkaline and acid phosphatases, enolase, phosphoglucomutase	Amberger (1973)
NADP - malic enzyme	Johnson and Hatch (1970)
Mn - superoxide dismutase	Michelson *et al.* (1977)
Mn - peroxidase, IAA oxidase	Maclachlan and Waygood (1956)
NADP - isocitrate dehydrogenase	Maloney and Dennis (1978)
NAD - isocitrate dehydrogenase	Nason (1952), Coultate and Dennis (1969)
PEP - carboxylase	Dittrich *et al.* (1973) O'Leary (1982)
RuBP - carboxylase	Ryan and Tolbert (1975)
PEP - carboxykinase	Burnell (1986)

Application of 1 ppm manganese to Mn deficient alfa-alfa depressed the activity of catalase, invertase obtained in the deficiency of the metal (Bailey and McHargue, 1944) whereas in barley, coffee, cotton and other plants both deficiency and excess of Mn were known to decrease the activity of catalase (Agarwala *et al.*, 1964; Bar-Akiva, 1964; Iyenger and Seshagiri Rao, 1971b; Sirkar and Amin, 1974) or the activity remained unchanged (Vielemeyer *et al.*, 1966; Bar - Akiva and Lavon, 1967). On addition of 10 ppm Mn to culture solution the activity of catalase was increased in rice to a maximum (Pattanaik, 1950) whereas Sharma *et al.* (1986) reported stimulated catalase activity in manganese deficient sugarcane grown in refined sand.

In several crop plants, manganese deficiency is known to stimulate peroxidase activity (Bar-Akiva, 1961, 1964; Agarwala *et al.*, 1964; Vielemeyer *et al.*,1966; Bar-Akiva and Lavon, 1967; Sharma *et al.*, 1986). Bar-Akiva (1961) considered increased peroxidase activity as an indicator of Mn deficiency. Excess supply of manganese increased the peroxidase activity in cotton and other crops (Sirkar and Amin, 1974; Bisht *et al.*, 1976; Leidi *et al.*, 1986; Sharma *et al.*, 1988). Mn deficiency reduced acid phosphatase activity and increased amylase activity in sugarcane (Kumar and Alexander, 1972). Leela and Pathak (1968) found decreased succinate degydrogenase activity and low alkaline phosphatase activity in Mn deficient leaves of *Trigonella foenum graecum*. The application of Mn++ and Mg++ stimulated the sucrose synthetase activity in sugarcane and sugarbeet (Patil and Joshi, 1972; Prasolova *et al.*, 1976). In Manganese deficient radish leaves, Agarwala *et al.* (1976) observed a decrease in the activity of FDP-aldolase and this depression could not be accounted for by depression in protein content.

Manganese is known to act as a metal component of superoxide dismutase (Mn-SOD)(Sandmann and Böger, 1983) which protects the photosynthetic apparatus from the deleterious effects of Oxygen free radical $O_2^{-\cdot}$ (superoxide) formed in various enzyme reactions in which a single electron is transmitted to O2 (Fridovich,1983):

$$O_2^{-\cdot} + e- \text{---------------->} O2^{-}.\ \text{(superoxide)}$$

$$O_2^{-\cdot} + O_2^{-\cdot} + 2H+ \xrightarrow[\text{dismutase (SOD)}]{\text{superoxide}} H_2O_2\ \text{(Hydrogen peroxide)} + O_2$$

$$2H_2O_2 \xrightarrow{\text{Catalase}} 2H_2O + O_2$$

In several plant species, both deficiency and excess supply of manganese were known to increase the activity of IAA oxidase (Taylor *et al.*, 1968; Burnell, 1988). In cotton, ethylene production and auxin destruction were found to be associated with the presence of manganese and activity of peroxidase (Fowler and Morgan, 1972, Morgan *et al.*, 1976). Further suggestions have been made on the oxidation of Mn^{2+} to Mn^{3+} as a result of the activity of peroxidase, which in turn oxidizes IAA (Kenten and Mann, 1949; Maclachlan and Waygood, 1956) and high oxidase activity caused auxin deficiency and this could be attributed to manganese toxicity effect also (Morgan and Hall, 1963; Morgan *et al.,* 1966). Manganese is known to be an essential cofactor of malate dehydrogenase (Anderson and Evans, 1956). Mitochondrial NAD-malic enzyme the predominant C_4 acid decarboxylating enzyme in NAD-ME type C_4 plants has a specific requirement for Mn^{2+}(Hatch and Mau, 1977; Hatch *et al.,* 1982). In addition to NAD-ME type, in PCK (Phosphoenolpyruvate carboxykinase) type, C_4 plants also have an absolute requirement for Mn^{2+} which can not be replaced by Mg^{2+} or any other divalent metal cation (Burnell, 1987). NADP malic enzyme is known to have a higher affinity for Mn^{2+} than Mg^{2+} but exhibited a higher activity with Mg^{2+} (Johnson and Hatch, 1970).

Manganese appears to be structural constituent of ribosome (Lyttleton, 1960) and also activates RNA polymerase in sugarbeet root tissue (Duda and Cherry, 1971). Manganese deficiency caused an increase in RNAse activity and strong inhibition of DNAse activity in pea (Sharma *et al.,* 1986; Vlasyuk and Kovalchuk, 1971).

Manganese as chloride is an effective co-factor of ribulose- diphosphate oxygenase (Ryan and Tolbert, 1975). Manganese along with Mg^{2+} and sufficient ATP is required for the maximum activity of phosphoenol pyruvate carboxykinase (Dittrich *et al.,* 1973). In excess Mn supply, the activities of catalase, cytochrome c oxidase, ascorbate oxidase and glutathione oxidase were increased but those of peroxidase and polyphenol oxidase were decreased (Sirkar and Amin, 1974; Nable *et al.,* 1988).

3.4. Reproductive Physiology

Manganese deficiency in wheat resulted in reduction in grain yield and poor viability of pollen when the plants were grown in soil pot culture with low manganese supply (Kaur *et al.,* 1991). They have further indicated that the reduction in grain yield in manganese deficiency is due to presence of higher oxidative enzymes and reduced translocation of photosynthates from flag leaf to reproductive organs or due to lower sink activity. Whereas wheat when grown is refined sand (Sharma, 1992) at low Mn levels, produced smaller anthers with fewer, smaller and unfertile pollen grains. In maize, Mn deficiency resulted in poor tasseling delayed anther development, reduced seed setting and seed development (Sharma *et al.,* 1991). Mn deficiency reduced the *in vitro* germination of pollen. These observations indicate the role of Mn more in the reproductive phase of plants than in the vegetative phase.

Hannam *et al.* (1984) had reported that the split seed symptoms in mature grains of Mn deficient *lupinus* which can be used as a sensitive indicator of Mn deficiency.

3.5. Interaction of Manganese with other Elements

Nitrogen

Mn uptake by plants is partially determined by the forms of nitrogen supply. In sand culture experiments it was observed that watermelon has better growth when supplied with nitrate than with ammonium or mixture of both nitrogen sources (Elamin ane Wilcox, 1986). However, symptoms of excess Mn were observed initially under nitrate nutrition and were absent when ammonium alone was supplied. In marigold, Mn uptake was enhanced by nitrate nitrogen whereas ammonium nitrate reduced it (Reddy and Mills, 1991). In contrast to this, in green house experiments using silt loam soil, nitrate nitrogen resulted in less vigorous plant growth with lower concentration of Mn in snapbean (Palaniyandi and Smith, 1979).

Presence of Mn hampered absorption of nitrate and promoted intensive accumulation of sulphate in grapevines (Khatiashvili *et al.*, 1985). In tomato, spinach and cabbage, the effects of Mn toxicity were less severe in presence of N as NH^{4+} and NO_3 than the control Oswa and Ikeda, 1980). Reports are also available on increased Mn content in vegetable as a consequence to nitrogen application (Taylor *et al.*, 1983).

Phosphorus

In sorghum, high Mn increased P in shoots but depressed its uptake. Addition of P can reduce Mn toxicity by rendering it inactive within plant (Foy, 1984). In soybean, manganese fertilizers with phosphorus fertilizers enhanced phosphorus absorption and increased DNA and RNA formation, P promoted manganese absorption (Qiu *et al.*, 1990).

Phosphorus decreased the Mn content in grain and straw (Gupta and Bhadra, 1980) and Mn uptake by grapevines (Khatiashvili and Abuleshvili, 1984).

In zinc deficient soils, P fertilizers induced zinc deficiency and intensified Mn uptake suggesting a direct relationship between Mn content of leaves and P/Zn ratio (Bajescu *et al.*, 1983).

Potassium

Manganese toxicity symptoms were decreased by the reduced rate of Mn absorption at high K supply (Heenan and Campbell, 1981) but there was no effect of K supply when Mn was absent. In wheat both deficiency and toxicity of Mn reduced K concentration (Okhi, 1984) but by increasing the Mn concentration the uptake of K was reduced (Heenan and Campbell, 1981). In tomato, the uptake and distribution of K concentration was independent of Mn concentration specially at non-toxic levels (<5 ppm) of Mn (Alverz-Tinaut *et al.*, 1979).

Calcium

Bowen (1981) reported antagonistic interaction between manganese-calcium in sugarcane grown in sand culture. Crinckle leaf is one of the well known symptoms of calcium deficiency induced by Mn toxicity in dicotyledonous species (Foy *et al.*, 1981; Horst and Marschner, 1978; Horst, 1988).

An increase in calcium supply severely inhibited manganese absorption in *Medicago* (Robson and Loneragan, 1970), inhibited absorption and translocation in excised barley roots (Kawaski and Moritsugu, 1987), Legumes (Bell *et al.*, 1989), uptake in rice (Nelson, 1983) and many plant species (Robinson, 1986) and reduced the toxicity of Mn in *Epilobium hirsutum* (Nazrul-Islam, 1986).

In Ryegrass, Kazak *et al.* (1984) noticed that application of $CaCO_3$ prevented the soil acidification and decreased manganese content of the plants below toxicity level.

Application of Mn stimulated calcium dependent ATPase activity and increased root and leaf calcium in cotton (Shchiteava *et al.*, 1984). In spinach, Qian *et al.* (1986) reported that the activities of oxygen evolution and H+ release from water in photosystem II particles were stimulated by adding Mn^{2+} or Ca^{2+} but their action was not additive. Alexander and Samuels (1967) reported that high Mn depressed the activity of amylase at low Ca and was not affected at excess calcium.

Magnesium

Mn induces Mg deficiency in roots by competing at the binding site for Mg (Heenan and Campbell, 1981), Mn toxicity can be counteracted only by a large Mg supply (Lohnis, 1960). Deficiency as well as toxicity of Mn reduced Mg concentration (Ohki, 1984). Increase in Mn supply, decreased the leaf concentration of Mg in soybean (Heenan and Campbell, 1981) and depressed the uptake of Mg (Turner and Barkus, 1983) in banana. Excess manganese induced 'Golden Tip' chlorosis which is a Mg deficiency in Norway Spruce (Hecht-Buchholz *et al.*, 1987).

Sulphur

Bowen (1981) has reported synergistic interaction between Mn and S in sugarcane but Ghosh *et al.* (1990) showed antagonism between Mn and S. Increasing rates of S applied to calcareous soils decreased Mn uptake by corn, sorghum and soybean (Mahmoud *et al.*, 1988).

Sodium

Reports are available on increased manganese levels in *Phaseolus vulgaris* leaves on increasing NaCl levels in nutrient solution (D'Arrgo *et al.*, 1983).

Zinc concentration in leaves of bushbean and soybean increased with increasing levels of manganese supply (Kohno *et al.*, 1984; Abd-Elgawad and Knezek, 1983). On the other hand, presence of zinc inhibited manganese uptake in bushbeans (Wallace and Wallace, 1983) and corn (Peneva and Velinova, 1981).

Copper

In excised *Hordeum vulgare* root, copper concentration was reduced in the presence of Mn (Harrison *et al.*, 1983) whereas application of copper decreased the uptake of Mn (Hulagur and Danagarwala, 1982) in maize. In a separate report on onion it was observed that concentration of Mn increased following copper treatment but opposite was not true for onoin bulbs (Frigany *et al.*, 1980).

Boron

Alvarez-Tinaut *et al.* (1981) reported that the symptoms of boron deficiency were accentuated in the presence of normal levels of Mn and symptoms of Mn toxicity were accentuated in presence of even moderate excess boron in tomato, spanish groundnut leaves showing boron toxicity symptoms contained high concentration of Mn (Blamey and Chapman, 1979). In sunflower, a significant inverse correlation between boron concentration in the solution and manganese content of leaves was obtained (Gomez *et al.*, 1981).

Aluminium

In peach seedlings, high Al concentration were associated with low Mn concentration (Kirkpatrik *et al.*, 1974). Al is known to decrease the Mn concentration both in roots and shoots of maize (Mathan, 1980) and also total Mn uptake is decreased in ryegrasses (Rengel and Robinson, 1959) and rice (Alam, 1983, Tang *et al.*, 1989). It has been suggested by several authors that Mn competes effectively with Al for root absorption sites (Alam, 1983; Mukhopadhyay and Sharma, 1991). High Mn levels replace Al toxicity in acid oxisols of Pureto Rico (Abruna, 1980). Conversely, Al reduces Mn toxicity (Rorison, 1986; Thung *et al.*, 1987).

Nickel

Plants given excess supply of cobalt and nickel showed possible Mn deficiency symptoms (Hewitt, 1954). In mulberry seedlings, application of Mn increased the concentration of Ni in each organ and its uptake with a slight decrease in the rate of transport (Low *et al.,* 1991). In wheat an antagonistic interaction between Mn and Ni has been reported (Taylor and Stodt, 1990).

4. COPPER

The essentiality of copper for higher plants was established separately by Sommer (1931), Lipman and Mackinney (1931). The yield of three different plants, viz., tomato, sunflower and barley was markedly reduced when copper was omitted from nutrient medium (Sommer, 1931). On the other hand, Lipman and Mackinney (1931) observed that when barley was grown in a nutrient solution without copper, the plants failed to produce proper fruits but when grown in complete nutrient solution the growth of barley plants was normal.

4.1. Uptake, Requirement and Distribution

It appears that uptake of copper is a metabolically mediated process and has been shown to be largely independent of competitive effects and depends primarily on the level of available copper in the soil (Mengel and Kirkby, 1982). Copper is usually rendered unavailable in soils rich in humus (Mercer and Richmond, 1970) which results in copper deficiency in plants grown on such soils. Mostly the availability of copper in organic soils depends not only on copper concentration in soil solution, but also on the form in which it occurred.

The relative requirement of copper by different plant species has been reported by several workers. According to Hewitt (1952), most plants grow normally when supplied 0.01 to 0.05 ppm Cu in the nutrient solution. At the tenth to one hundredth of the normal copper supply, most plants develop deficiency effects.

Usually the copper concentration for normal growth of most plant species, is 5 - 20 μg per g dry weight. Plants having less than 4 ppm Cu in dry matter develop copper deficiency effects and those with more than 20 ppm Cu show effects of copper toxicity.

Apart from this 57 ppm Cu in leaves from filed grown cherry (***Prunus cerasus***) has been reported by Kentworthy (1961).

Rahimi and Bussler (1973) observed an increase in copper concentration in different plant parts as a result of increase in copper supply, the highest was in roots and lowest in the stem. Similar observations were made in roots of

Agrostis stolonifera (Wu *et al.*, 1975). But in copper sufficient plants, the highest conentration of copper was in shoot tips and young leaves and the lowerst concentration was in the old leaves, stem and petioles (Wlaker and Loneragan, 1981). Kotze and Villiers (1989) studied the uptake and distirbution of copper in vines and observed that copper was not ranslocated from leaves to fruits before harvest, indicating lower mobility of element in the plants and the amount of Cu was maximum in roots.

4.2. Physiological and Biochemical Effects

Several reports have been made on chanages in a wide range of physiological processes due to copper deficiency such as photosynthesis, metabolism of carbohydrates, nitrogen, cell wall constituents and lipids etc. (Bussler, 1981).

4.2.1. Photosynthesis

Copper deficieny is also known to affect photosynthetic activity in wheat (*Triticum aestivum* cv. Nazareno stramppeli.) The flag leaf of Cu-deficient plants showed 50% reduction in photosynthetic rate. The activites of photosystems I and II determined for isolated chloroplasts as well as flourescence measurement on intact leaves of copper deficient plants, indicated a low activity of photosynthetic electron transport. Ribulose-bisphosphate carboxylase/oxygenase (Rubisco) acitivity was not affected by Cu deficiency but Cu deficiency affected the chloroplast ultrastructure, especially at the level of grana where a disorganisation of thylakoids is also evident (Casimiro *et al.*, 1990). Ribulose bisphosphate carboxylase which has been isolated from spinach has been reported to contain copper (1 atom mol^{-1}) (Wishnick *et al.*, 1969, 1970). But the role of copper in this enzyme has been attributed to oxidation rather than to CO_2 fixation. The enzyme which is activated by copper may be functional in CO_2 fixation as suggested by CO_2 / O_2 competition. The oxygenase activity of Ribulose-biphosphate carboxylase is now attributed to a separate copper containing protein (Branden, 1978).

In sugarbeet (Henriques, 1989), copper deificency decreased the concentration of chlorophyll but the ratio of chlorophyll *a* : *b* was increased. The chloroplasts exhibited significant changes in their ultrastructure and photosystsem I was affected more than photosystem II. The internal membrane system of Cu-deficient chloroplasts was greatly reudced. It was further suggested by the authors that these changes in the internal organization was accounted for some of the observed reduction in pigment and photochemical capability shown by chloroplasts from Cu deficient plants.

In copper deficient peas, the chloroplastic membranes were disturbed affecting the PSII dependent photosynthetic electron transport rate (Baron *et al.*, 1990).

In another report on tomato, copper deficiency resulted in discontinuous distribution of chlorophyll in the plastids which was later collected as droplets (Reeds, 1929). A marked change in the fine structure of the leaf cells of copper deficient oats was also reported (Dekock *et al.,* 1971).

In copper deficient rubber plants, the chloroplasts were seen to contain large and numerous starch vesicles (Hamzah and Gomej, 1980). Yakout *et al.* (1981) and Tasiewicz (1981) observed an increase in the chloroplastic activity and the photosynthetic pigment content as a result of foliar application of copper. With the onset of copper deficiency in sugarbeet, extra chloroplastic copper decreased more rapidly than the chloroplastic copper (Magdolna *et al.,* 1984). According to these workers, during copper deficiency and recovery from the deficiency, the chlorophyll *a, b,* carotenoids and plastoquinone contents were increased concomitantly. Spraying of wheat with copper increased the concentration of chl *a* and chl *b* (El-Hadidi *et al.,* 1980).

In cotton leaves the deficiency as well as copper excess decreased the rate of photosynthesis (Isaev and Rustamov, 1975). In a separate study, the involvement of the copper compound plastocyanin in the photosynthetic electron transport system has been reported (Plesnicar and Bendall, 1970). Several workers have reported evidences suggestive of involvement of copper in the photosynthetic reactions (Arnon, 1950; Trebst and Eck, 1963 ; Bishop, 1964, 1966; Neilson *et al.,* 1969; Gurevich and Yakushkina, 1989). It has been observed that many chelating agents when complexed with copper increase both photosynthesis and Hill reaction activity. Reports are also available on decreased Hill reaction activity in isolated chloroplasts of copper deficient tomato (Spencer and Possingham, 1960). Bishop (1964, 1966) observed that copper deficiency had a marked effect on photoreduction but only a slight influence on the Hill reaction activity. He further suggested that copper is an essential part of the electron transport mechanism of Pigment system I.

Plastocyanin, a protein of the flavin type required for the light induced production of $NADPH_2$, contains 1 atom of Cu per molecule of the protein (Katoh and Takamiya, 1961; Blumberg *et al.,* 1967; Ramshaw *et al.,* 1974) and is an essential component of the electron transport chain of pigment system I (Fork and Urbach, 1965; Akulova and Mukhin, 1968; Kok and Rurainski, 1965; Elstner *et al.,* 1968; Malkin and Bearden, 1973). In two separate reports it was shown that plastocyanin is not only required for NADP reduction but has been suggested to function as an obligate electron carrier in photosynthesis (Trebst and Elstner, 1965; Boardman, 1975). It has been reported that copper deficiency reduces the plastocyanin content and the total copper content of chloroplasts to less than half of that found in copper sufficient plants (Baszynski *et al.,* 1978). Copper deficiency inhibited photosystem 1 and cyclic phosphorylation causing a decrease in the rate of photosynthesis. In spinach, the effects of copper chloride on photosynthetic electron transport and chlorophyll protein complexes was observed and it was concluded that copper causes a stimulation in the rate of overall electron transport from water to NADP (Samuelson and Ocquist, 1980).

Copper deficiency is also known to inhibit the whole chain of electron transport in PS I and PS II in sugarbeet (Horvath *et al.*, 1983). The minimum copper requirement to maintain the normal contents of chlorophyll, carotenoids and plastoquinones was 0.5n atom Cu/cm^2. An inhibitory effect of copper was shown on photosynthesis in *Chlorella* cells where PS II was more susceptible to copper than PSI (Wu and Lorenzen, 1984).

Sandmann and Böger (1980) reported that the lipid peroxidation processes in the photosynthetic membranes in spinach chloroplasts is mediated by copper.

4.2.2. Carbohydrate Metabolism

Application of copper (alone or with boron) has increased the carbohydrate content of onion leaves significantly (Rao, 1971). The effect of some copper containing fungicides on the carbohydrate metabolism in leaves and tubers of potato have been studied and it was observed that application of bordeaux mixture or copper oxychloride raised the level of soluble sugars in the leaves and tubers of potato (Gudkova, 1969). A positive effect of copper (boron and molybdenum) on the carbohydrate content of some plants have been observed (Valentina *et al.*, 1969; Graham, 1980). In the culms and leaves of copper deficient wheat a marked decrease in the concentration of soluble carbohydrates was observed at vegetative stage (Mizuno *et al.*, 1982) which would explain growth hindrance probably because of reduced supply of sugars to the growing points. On the other hand, after seed setting Cu deficient wheat leaves exude honey like substance (Graham, 1980). This exudation may be explained by reduced demand for sugars as the number of grains formed in Cu deficient wheat is reduced and therefore leaves accumulate excess amount of sugars which are exuded in the form of honey. From these results it appears that the plants growing in the copper deficient conditions developed visible symptoms of copper deficiency and their vegetative growth was hindered because of insufficient supply of carbohydrates to the growing point.

4.2.3. Lipids

Copper is also known to mediate the lipid peroxidation process in photosynthetic membranes of spinach chloroplasts (Sandmann and Böger, 1980). Reports are available on pre-sowing treatment of copper given to wilt resistant varieties of cotton which accelerated the synthesis of lipids and caused an increase in the oil content of cotton seeds (Sharafutdinova *et al.*, 1974). Copper deficiency is known to decrease the total lipid content in the cells of *Candida utilis* (Khovrychev *et al.*, 1974). Application of copper increased the accumulation of starch and proteins in maturing barley seeds

(Livdane, 1978). He further observed that on addition of Cu to copper deficient barley plants, the rate of starch accumulation and its concentration in maturing ripened grains was increased.

4.2.4. Nitrogen Metabolism

Low copper results in accumulation of soluble amino compounds (Possingham, 1956). Some evidences suggest that copper forms co-ordinate compounds with NH_2 groups in amino acids (Reilly, 1969). A direct correlation was observed between the tissue concentration of copper and the protein content (Echevarria *et al.*, 1970). In another experiment it was reported that in a presowing treatment of soybean seed with 0.04% copper solution, the concentration of nitrogenous compounds in leaves was increased (Kvariani, 1971). Apart from this, reports on highly increased alanine content in leaves of severely copper deficient *Setaria sphacelate* has also been made (Court *et al.*, 1972). As a result of copper application, concentration of some amino acids, viz., arginine, valine, leucine, phenylalanine, lysine, histidine, methionine and threonine, increased in lucerne (Pillar, 1977). Deficiency of copper is also known to bring about changes in the metabolism of proteins by increasing the accumulation of nitrogen and proteins in barley seeds (Livdane, 1978). Similar reports were also made by Avakyan (1971) on *Candida utilis* and by Baranov (1980) for some plants other than *C. utilis*. In copper deficiency, some plants show low nitrogen fixing activity (Cartwright and Hallsworth, 1970). Copper deficiency results in an increased activity of nitrate reductase (Medina and Nicholas, 1957 ; Peive *et al.*, 1964, 1965). In both maize and flax the level and type of nitrogen nutrition and copper requirement are inter-dependent (Ozolinya and Lepinya, 1965). Later, the importance of copper in controlling protein biosynthesis in maturing barley seeds was also emphasized (Ozolinya and Livdane, 1970). In another study, it was observed that copper along with iron and molybdenum affects the concentration of amino acid pool in root nodules of leguminous plants (Zhiznevskaya, 1972).

Beneficial effects of copper was also notcied on fixation of molecular N_2 (Gribanov, 1954). In oat leaves the requirement of copper as an essential factor for the assimilation of molecular N_2 and protein sysnthesis was studied using ^{15}N (Ostrovaskaya and Geller, 1955). They observed that amides accumulate in low Cu, thus, disturbing protein synthesis.

4.2.5. Nucleic Acids

Copper plays an important role in the synthesis of DNA and RNA. Low concentration of DNA has been reported in young growing organs of copper deficient maize plants (Ozolinya and Lapinya, 1965). DNA preparation isolated from microorganisms have been shown to produce complexes with copper and also with Mn and other metals (Dove and Davison, 1962). In some cases highly purified preparations of nucleic acids were found to contain traces of copper besides iron, manganese, zinc, cobalt, vanadium, etc., bound to the DNA molecule suggesting a role of the metals in maintaining the conformation of DNA in which purine and pyrimidine bases formed covalent bonds with metals (Wacker and Vallee, 1959a, 1959b; Belokopylosky *et al.*, 1968). Purine derivatives may form complexes with copper (Wiess, 1961). Some reports have been made on the ability of copper ions to form complexes with DNA (Zakharenko and Moshkovsky, 1966)

4.2.6. Lignin

In Cu deficiency, the lignification of the cell is reduced. This leads to lodging of cereals specially when nitrogen supply is in excess (Vetter and Teijchawne, 1968). In Cu deficiency the leaves are distorted and stem twisted, and give rise to pendule forms in trees. Lignin content of Cu deficient plants decreases and is often less than half of that found in corresponding plants with adequate copper supply. This effect can be made out in sclerenchyma and xylem vessels of Cu deficient plants. Some workers (Rahimi and Bussler, 1974) recorded decrease in lignification in low Cu as a diagnostic feature of Cu deficiency in plants. The decrease in lignification in Cu deficiency is attributed to lowered activity of Cu containing enzyme phenolase. This enzyme is responsible for the oxidation of phenols which are prescursors of lignin. In Cu deficient plants phenols accumulate.

4.2.7. Enzymes

Copper is a consitituent of polyphenol oxidase (Keilen and Mann, 1938 Kubowitz, 1938) and ascorbic acid oxidase (Powers *et al.*, 1944). Cu^{2+} is also a constituent of mono-amino oxidase, uricase, cytochrome oxidase and galactose oxidase as well as that of plastocyanin (Evans and Sorger, 1966). Copper is also known to be tightly bound to ribulose-diphosphate carboxylase (Wishnick and Mildvan, 1969). A shortage of copper causes a decrease in the activities of polyphenol oxidase, ascorbate oxidase and other copper enzymes (Okanenko and Ostrovskaya, 1950; Rubin and Lognova, 1968; Soboleva and Bokuchava, 1969; Peive, 1970).

The catalysis of ascrobic acid oxidation by Cu^{2+} is increased by the copper enzyme ascorbic acid oxidase (Dawson, 1950; Joselow and Dawson, 1951; Dawson and Tarpley, 1951) in which copper is in the divalent state and shuttles reversibly between the divalent and monovalent forms during the enzyme catalysis.

Polyphenol oxidase, also called catecholase or tyrosinase, catalyses the oxidation of O-diphenols by molecular oxygen to form the corresponding quinones and H_2O (Raper, 1928; Gunsalus *et al.*, 1975). Copper is also involved in the mono-phenolase activity (Dawson and Tarpley 1951; Mason, 1957). Dressler and Dawson (1960) made a study of the exchange between radioactive cupric ions (Cu^{64}) and the copper containing protein tyrosinase. They further observed that copper at the functional site of catecholase is exchangeable but that at the *cresolase* site is non-exchangeable.

In copper deficient plants, observations have been made on decreased activities of the copper enzyme polyphenol oxidase (Nason *et al.,* 1952). In another report decreased activity of phenolase in the leaves of copper deficient *Chrysanthemum morifolium* has been reported (Graves *et al.,* 1979). Copper has a role in the activities of oxalosuccinic carboxylase and isocitric dehydrogenase (Grafflin and Ochoa, 1950). In cytochrome oxidase enzyme, Williams *et al.* (1968) have shown that both copper and iron are found at a single active site. Purified cytochrome oxidase preparations contain one atom each of copper and iron per mole of cytochrome C (Sands and Beinert, 1959; Takemori, 1960; Griffiths and Wharton, 1961). Some plant amino oxidases are copper enzymes (Pirie, 1962) and these enzymes carry out the oxidative deamination of mono- and diamines. Another enzyme, galactose oxidase from *Polyporus cincinatus* is also known to be a copper enzyme. This enzyme is concerned with the oxidation of substrates by atmospheric oxygen. Tryptophan pyrrolase from *Pseudomonas* is yet another copper enzyme. Along with copper, this enzyme contains a dissociating haemin co-enzyme (Maeno and Feigelson, 1965). Pyridoxal phosphate is a constituent of benzylamine oxidase. Copper, in this enzyme does not undergo any valency changes during catalysis (Kovalsky and Rish, 1970).

Depression of the ascorbic acid oxidase activity is regarded as one of the physiological indicators of copper deficiency (Ruszkowska and Lyszez, 1975; Delhaize *et al.*, 1982).

In some separate reports copper deficiency has been shown to cause not only a decrease in the avtivity of ascorbate oxidase but also in the activities of diamine oxidase and O-diphenol oxidase (Hill, 1973; Zarin *et al.*, 1978; Walker and Loneragan, 1981). According to Ruszkowska *et al.* (1975), the activity of catechol oxidase decreased in leaves of sunflower with a decrease in copper supply. On the other hand, Isaev and Rustamov (1975) showed an increase in the peroxidase activity in copper deficient plants. In radish, deficiency of copper (so also that of iron, zinc and Manganese) decreased the activity of FDP aldolase (Agarwala *et al.* 1976). This decrease in the avtivity of enzyme was overcome when deficient plants were supplied with adequate copper. Agarwala *et al.* (1980) have further reported a decrease in the specific activities of catalase, peroxidase and cytochrome C oxidase and a marked increase in that of ribonuclease and aldolase in anthers of copper deficient wheat plants.

In copper deficient sugarbeet (Agarwala *et al.*, 1985) the activities of Cu containing enzymes, cytochrome C oxidase and polyphenol oxidase were decreased. The activity of the former was restored when additional Cu was infiltrated to the leaf discs of Cu deficient plants even in the presence of cycloheximide and of the latter only in its absence.

In recent years another copper containing enzyme, superoxide dismutase has been a topic of great interest. In higher plants, this enzyme is found in two different forms, one contains copper and zinc (Mc Cord *et al.*, 1971 ; Fridovich, 1975; Kanematsu and Asada, 1989) and the other containing copper and manganese (Sevilla *et al.*, 1980). The former contains 2 atoms of Cu and 2 atoms of Zn and has a molecular weight of 32,000 daltons. The acitivity of this enzyme was initially detected in leaves of wheat and barley (Ozolinya *et al.*, 1979). The enzyme activity was the highest in the leaves of young plants and was a sensitive indicator of copper deficiency. In barley also, Ozolinya *et al.*, 1979) found superoxide dismutase and ascorbate oxidase to be very sensitive to decreasing levels of copper. Ozolinya and Klavina (1981) found that in several plants the activity of ascorbate oxidase, O-diphenol oxidase, diamine oxidase and superoxide dismutase to be related to the level of copper supply.

Copper is also reported to be involved in the enzyme concerned with the desaturation and hydroxylation reactions of fatty acids (Wahle and Davies, 1977). The desaturation of stearic acid by 9-desaturase in presence of Cu and O_2 which results in the production of oleic acid.

Addition of 3 μM $CuSO_4$ to the medium in which callus was induced from the peel tissue of cucumber fruit (*Cucumis sativus* cv. Suyo), enhanced the activity of ascorbate oxidase without inhibiting the callus growth (Sekiya *et al.*, 1990), in another sudy it was observed that Cu affects the stability and not the protein conformation of the enzyme ascorbate oxidase (Sevini *et al.*, 1990).

4.2.8. Effect on Water Relations

When soybean and corn seedlings were soaked in 0.005% $CuSO_4$ solution, the rate of transpiration decreased considerably (Ninh, 1971), whereas it was also observed that a spray of 0.1% solution of $CuSO_4$ to ash seedlings increased the rate of transpiration in morning and evening but decreased it during the day time provided the plants are under water stress (Pudova, 1970). In another report it was observed that low (0.005%) concentration of copper increased the amount of bound water and decreased the rate of transpiration (Gaina and Silli, 1973). A decrease in the lignification of the tracheids of copper deficient plants was reported by Pissarek (1974), because of which these trachied cells could not withstand transpiration suction and collapse, restricting water supply to the upper parts of the shoot, causing wilting of the apical leaves.

The effect of copper deficiency on water stress of wheat plants was examined by Graham (1976) who observed that in severe deficiency of copper, there is total loss in grain yield, along with induced wilting of the plants, increased leaf diffusive resistance and at the same time increased leaf water potentials. This wilting in wheat plants due to copper deficiency was attributed by Graham to the structural weakness resulting from decreased lignification and not to the water status of the plants. In *Pinus radiata*, copper deficiency resulted in stem deformation associated with poor lignification and low concentration of copper but not with water stress (Downes and Turoey, 1990). The increase in leaf diffusive resistance in copper deficient plants was explained in terms of the effect of copper deficiency on guard cells. In another report on wheat (Mizuno and Dobashi, 1982; Mizuno *et al.*, 1983), it was observed that leaves in copper deficiency show curling due to decrease in osmotic pressure. A sequential loss of stomatal functioning in leaves of sunflower was reported in copper deficiency (Habermann and Shoemaker, 1983). Copper deficiency resulted in a decrease in the total leaf diffusive resistance in some plants and which suggested that this accounts for the apparent decrease in photosynthesis (Assche and Clijsters, 1983). Copper deficiency is also know to decrease the rate of transpiration, respiration and photosynthesis (Koula and Cingrosova, 1975).

In wheat, copper deficiency decreased the growth, root: shoot ratio, nutrient uptake, stomatal conductance and water potential (Casimiro, 1987). He also reported reduced lignification of xylem vessels in the leaves of copper deficient plants.

4.2.9. Reproductive Physiology

Copper plays an important role in the reproduction of higher plants. In cereals particularly copper deficiency affects the reproductive phase. In severe Cu deficiency the number of tillers increased and so also straw yield but no grains were formed. And with increase in cooper supply, the grain yield rises sharply (Marschner, 1986). In wheat, copper deficiency was found to inhibit anther formation and produced smaller number of pollen grains (Graham, 1975). Due to non-viability of pollens in low copper, grain setting was disturbed. According to Graham (1975), under such conditions the synchronous meiotic division of a large number of pollen mother cells may create a high localised demand for Cu which is not available in sufficient amounts in deficient plants. This ultimately, causes retardation in the onset of pollen formation. The inhibition in lignification of anther cell walls in Cu deficiency might also be involved in the lack of opening of anthers. Further reports on wheat by Agarwala *et al.* (1980) suggest that Cu deficiency retarded the development of ears, anthers and pollen grains. In most cases, in Cu deficiency anthers were small, empty and failed to open. The copper deficient pollen grains of wheat showed decreased activities of catalase, peroxidase and cytochrome C oxidase and increased the activities of ribonuclease and aldolase. The germination capactiy of these deficient pollen grains was reduced and therefore they could not fertilize the ovules. In a separate report on wheat, Galaro (1988) observed that application of copper (2 kg/ha) to soil increased that biomase and reduced the male streility in wheat. In rice (Aleshin *et al.*, 1989), it was found that copper treatment per panicle, 1000 grain weight and biomass but also decreased spikelet sterility.

4.3. Interaction of Copper with Other Nutrients

Nitrogen

Copper deficiency symptoms in citrus were increased when N was applied to copper deficient soils (Camp and Fudge, 1939). On the other hand, Cu accumulation in citrus roots was directly related to the proteins in roots (Smith, 1953), perhaps copper forms a stable protein complex in roots. But in raya (*Brassica juncea*) an antagonistic relation was observed when both Cu and N were in excess supply (Antil *et al.*, 1988). In wheat, Hill *et al.* (1979) had reported a synergistic interaction between the two elemenets.

Phosphorus

In one of the initial reports it has been observed that application of P to soils depressed Cu concentration in plants. The concentration of Cu in sour orange seedlings decreased on increasing the levels of P in refined sand (Bingham, 1963) while Cu inhibited the uptake of P in excised sugar maple roots (Gysi *et al.*, 1975), a two-told increase in P content of Cu deficient wheat plants in the flag leaf at flowering stage was reported by Mizuno and Dobashi (1982). Phosphorus application reduced copper toxicity (Spencer, 1966) in citrus whereas in sand culture high levels of P were inhibitory for the growth of clover at high Cu levels but not at low copper levels (Greenwood and Hallsworth, 1960). An antagonistic effect of Cu and P in rice was observed when one of the nutrients was applied in large quantity (Alam, 1983). In a separate experiment with rice it was observed that application of copper increased the dry matter, total leaf chlorophyll and decreased the P concentration in rice (Eun, 1981). In acid soils, P levels did not influence the copper concentration in maize (Misra and Dwivedi, 1976).

Sulphur

Application of sufficient S alleviates induced Cu deficiency in mustard grown in refined sand, Cu deficiency was partially alleviated at high levels of S. This effect was more pronounced in the nitrate than in ammounium nitrate treatment (Chatterjee *et al.*, 1992).

Calcium

In a number of plants the effect of Cu toxicity has been reported to be alleviated by increasing the Ca concentration (Wallace *et al.*, 1966; Osawa and Tazuku, 1990). Increase in Ca supply reduced Cu concentration in roots of bean, spinach, lettuce and radish (Osawa and Tazuk, 1990).

In tomato, Cu concentration increased an increasing Ca concentration between 100 to 300 ppm but decreased at 400 ppm Ca (Dechen *et al.*, 1980).

Magnesium

An antagonistic interaction between Cu and Mg was reported in sugarcane on the basis of leaf Sheath analysis under field conditions (Bowen, 1981). In apple and spruce, excess Cu reduced the Mg concentration in the tissue (Buston *et al.*, 1955; Watkins and Ferguson, 1982). In *Anabaena variabilis* copper was found to substitute for Mg in the chlorophyll porphyrin ring when grown in presence of sufficient Cu under laboratory conditions (Kowaleska *et al.*, 1987).

Boron

Interaction between Cu and B in plants is rather controvercial. Copper concentration in sunflower remained unaffected by variable S supply but translocation of Cu was stimulated by excess B (Alvarez *et al.*, 1979; Gomez *et al.*, 1981). In contrast to this, excess boron decreased Cu concentration in both leaves and roots of dwarf bean due to harmful effects of B in excess amounts (Avila *ét al.*, 1981).

On the other hand, in cotton, concentration of Cu was increased with an increase in boron supply (El-Gharably and Bussler, 1986).

In boron deficient sugarcane the concentration of copper remained unaffected but the activity of Cu containing enzymes, cytochrome and polyphenol oxidases were decreased (Agarwala *et al.*,1986). It was further observed that boron and Cu acted synergistically on various metabolic parameters of the plants when grown in refined sand. Apart from this, boron deficiency alleviated the effects of copper excess.

Boron deficient tomato plants showed higher activity of ascorbic acid oxidase a copper containing enzyme as compared to boron sufficient plants (Brown, 1979). This was more pronounced in boron inefficient plants than in boron efficient plants.

Manganese

Interaction between Cu and Mn is reported to be antagonistic (Gupta, 1972). Reports are available on excised roots that Cu and Mn compete for a common site for which copper showed 307 times greater affinity than Mn (Harrison *et al.*, 1983). Other workers have reported that Cu inhibits uptake of Mn^{2+} in a non-competitive fashion but no apparent effect of Mn^{2+} on Cu^{2+} uptake was observed (Landi and Fagioli, 1983). In canola (*Brassica napus* Wister) Mn and Cu showed antagonistic effects specially at low S levels (Karamonos *et al.*, 1989).

Zinc

In different plant species copper acts antagonisticaly with zinc (Adriano *et al.*, 1971; Levesque and Mathur, 1983). Kinetic studies on copper uptake in intact barley roots for 2h and 24h with or without zinc in the uptake medium showed that Cu uptake was mediated by a mechanism which was not influenced by Zn (Vettrup, 1977). This report is contrary to the suggestion of Chowdhry and Loneragan (1972) that Cu and Zn act competitively at the site of their uptake. The absorption of zinc was strongly inhibited by Cu; similar results were obtained in beans but was shown to increase in rice (Sedberry *et al.*, 1980). Some other reports are also available on decreased uptake of Zn when high levels of Cu was applied (Chandel *et al.*, 1978). In a separate report on rice suggests a synergistic interaction between Zn and Cu in boosting the yield of rice.

Molybdenum

The direct interaction between Cu and Mo mostly reported to be antagonistic (Gilbert, 1952; Mackay *et al.*, 1966; Gupta and Ram, 1980). In field trials, an increasing dose of Mo, the concentration of Cu in leaves was reduced in alfalfa and timothy (Butorac and Uscumlic, 1977; Rayment and Gupta, 1984; Urvas, 1989). In maize, application of Cu decreased that uptake of Mo and Fe (Hulagur and Danagarwala, 1982). Application of high doses of Cu decreased Mo concentration in lucerne (Butorac, 1981). In rice, increase in Mo application resulted in a decrease in Cu, thus showing an antagonistic relation between Mo and Cu (Gupta and Singh, 1981). Agarwala *et al.* (1989) had reported in maize that both elements behaved synergistically at high levels as the toxicity effects of one element were accentuated by the excess of the other. On the other hand, they further reported that excess Cu decreased Mo concentration in maize leaves at low Mo and also inhibited its metabolic functioning by further decreasing the concentration of protein nitrogen and nitrate reductase activity and increasing the peroxidase activity.

5. ZINC

For the first time Maze (1914) reported the essentiality of zinc for the growth of maize. Later the essentiality of zinc was established for a wide variety of crop plants by several workers (Sommer and Lipman, 1926; Sommer, 1928; Van Schrevan, 1937) and fruit trees (Chandler *et al.*, 1931; Hoagland *et al.*, 1936; Chapman, 1966). Hoagland *et al.* (1936) showed that zinc is essential for the normal growth of tomato, squash, corn, mustard, sunflower, cotton, alfa-lfa and a number of fruit trees.

5.1. Uptake, Distribution and Requirement

The uptake of zinc in barley roots (Schmid *et al.*, 1965) was found to have the characteristics of an active metabolic uptake. Additional evidence also supported that zinc uptake is metabolically controlled (Moore, 1974; Lindsay, 1972). Giordano (1974) showed severe retardation of zinc absorption by metabolic inhibitors.

The optimum yield of beans was found to be associated with 20 ppm or more zinc in leaves or top parts (Boawn *et al.*, 1950) whereas in tomato, 11 ppm zinc has been reported as the critical level for zinc deficiency (Mohapatra and Kibe, 1971). Berry (1971) observed poor growth of lettuce when leaf petioles contained less than 8 ppm zinc. In cotton, 3 ppm zinc represented the threshold of zinc deficiency and 200 ppm zinc represented the threshold of zinc toxicity (Ohki, 1975). Millet and soybean exhibited zinc deficiency symptoms when the young leaves of these plants had less than 8-5 ppm zinc but in flax, the visible effects were observed when the leaves contained less than 21.5 ppm zinc (Rahimi and Bussler, 1979).

5.2. Carbohydrate Metabolism

A requirement of zinc for the synthesis and utilization of carbohydrates in plants has been suggested (Chesters and Rolison, 1951). Plants deficient in zinc show increase in the concentration of reducing sugars and decrease in the concentration of sucrose and starch. Similar reports were made on increased concentration of reducing sugars and decreased concentrations of non-reducing sugars in zinc deficient leaves of *Citrullus* (Sharma *et al.*, 1981). In bean plants subjected to zinc deficiency a decrease in the activity of starch synthase accompanied by a reduction in the number of starch grains and starch was noted (Jyung *et al.*, 1975). Shrotri *et al.* (1980) studied the effect of zinc nutrition on sucrose biosynthesis in maize. They found that in zinc deficiency there was an increase in ^{14}C labelling in malic acid, sugar phosphates, sugar nucleotides, glucose and fructose and a decrease in the ^{14}C labelling in sucrose. Severe deficiency of zinc reduced the biosynthesis of many proteins including that of an enzyme protein sucrose synthase.

Zinc deficiency reduced the synthesis and hydrolysis of carbohydrates and which was also influenced by the form of nitrogen supply (Shkolnik and Greshischeva, 1968). Increase in zinc supply to wheat and buckwheat increased the amount of sucrose when nitrogen was supplied as ammonium but the effect was reversed when nitrogen was supplied as nitrate. Spraying of grapevine with zinc enhanced the sugar content and improved the ratio between sugars and acid in grapes (Tanukov, 1981). Takahashi and Ito (1976) reported that maximum carbohydrate content in leaves and grain were formed in maize when the supply was 0.05 ppm zinc.

5.3. Nitrogen Metabolism

Zinc deficiency is known to increase the concentration of amides and total amino nitrogen in tomato (Possingham, 1956). Increased accumulation of glutamine and asparagin, followed by aspartic acid, lysine, histidine, arginine, serine, threonine, alanine, phenylalanine, valine and leucine was observed in zinc deficient plants (Steinberg, 1956). Similarly, concentration of several amino acids, viz., glycine, alanine, valine, leucine, isoleucine, serine, threonine, glutamic acid, lysine, cystine, methionine, proline, phenylalanine, tyrosine and tryptophan were increased and that of total aspartic acid and arginine decreased in zinc deficient cotton plants (Ohki and Young, 1978). These changes in amino acids and amides can be related to overall effects of impaired protein synthesis which cause large accumulation of nitrogen storage forms, the amides and ureides (Hewitt, 1983). In maize, zinc deficiency resulted in a higher accumulation of most of the free amino acids and a decrease in the total protein concentration along with reduced activity of glutamate dehydrogenase (Polikarpochkina and Khavbin, 1972).

On the other hand, Kositsyn and Igoshina (1971) have reported low concentration of soluble proteins in the chloroplasts of tomato and also in some other plants deficient in zinc (Schneider and Price, 1962; Shkolnik and Davydova, 1965). Soaking of seeds in zinc sulphate before sowing resulted in an increase in the protein content in grains of barley (Gafarova, 1973). Reduced weight and leghaemoglobin concentration in zinc deficient root nodules was noted which ultimately caused a decrease in N_2 fixation (Demeterio *et al.*, 1972). In other observations, it was reported that zinc deficiency results in decreased concentration of total and soluble proteins in many plants (Lee, 1978). Reddy and Rao (1979) and Sharma *et al.* (1981) have reported accumulation of non-protein nitrogen in zinc deficient *Cicer* and *Citrullus* respectively. In garden pea plants, zinc deficiency caused a decrease in the total nitrogen concentration with an increase in zinc concentration (Paivoke, 1983). According to Praske and Plocke (1971), in absence of zinc there was complete degradation of ribosomes in *Euglena gracilis*.

It has been reported that in zinc deficient beans (Boawn *et al.*, 1954), *Phaseolus vulgaris* (Lessman and Ellis, 1971; Adriano, 1969), paddy (Singh and Singh, 1983), corn (Rudgers *et al.*, 1970; Ganiron *et al.*, 1969; Giordano and Mortvedt, 1970; Stukenholtz *et al.*, 1966) phosphorus accumulates in high concentration. In tomato and pea plants zinc deficiency disturbed normal utilization of phosphorus (Paribok and Alkseyevapopova, 1965).

According to Paribok and Alkseyevapopova (1965) there is several fold increase in the concentration of inorganic phosphorus and decrease in the concentration of lipid and nucleic acid P-fractions in the zinc deficient plants.

Stashauskaite *et al.* (1972) observed that in maize roots, zinc supply suppressed the activity of ATPase and acid β-glycerophosphatase. Zakharchisina and Klyuchko (1970) have shown that inadequate supply of zinc causes uncoupling of oxidation and phosphorylation in zinc deficient barley plants. Apart from this, observations have been made on the increase in organic phosphorus content in maize leaves by the addition of zinc to zinc deficient plants (Yagodin *et al.*, 1979). According to Johnson and Simons (1979), zinc deficient legumes contained low concentration of organic phosphorus fraction and high concentration of inorganic and total phosphorus.

5.4. Nucleic Acids

Zinc is involved in the nucleic acid metabolism. The essential nature of zinc for the synthesis of nucleic acids was first reported for *Myobacterium smeqmatis* (Winder and Denney, 1959). It has been also realised that one of the earliest effects of zinc deficiency is a sharp decrease in the RNA and ribosome content of the cells (Price *et al.*, 1972). Low levels of RNA usually reduces the protein synthesis. A suppression of ribosomal RNA synthesis, consistent with the slowing down of protein synthesis has been reported in zinc deficient *Rhizopus nigricans* (Wegenur and Romano, 1963). On addition of zinc thermal stability of ribosomes was enhanced significantly (Tal, 1969). The cytoplasmic ribosomes of *Euglena gracilis* which usually contain substantial amounts of zinc becomes extremely unstable in zinc deficiency (Praske and Plocke, 1971). Low zinc resulted in a decreased concentration of RNA in citrus leaves (Kessler and Monselise, 1959). The requirement of zinc for the synthesis of ribose and deoxyribose in *Aspergillus niger* is well documented (Bertrand and Wolf, 1961), whereas Wacker (1962) has reported a decrease in the RNA content but not in DNA content in zinc deficient *Euglena gracilis*. In fact, the latter was found to be increased by zinc deficiency (Price, 1962). In barley leaves, severe symptoms of zinc deficiency were accompanied by decrease in the RNA concentration in leaves (Tsutsumi and Fujiwara, 1964). Simultaneous decrease in both RNA and protein contents has been observed in zinc deficient tomato (Davydova, 1967) and maize (Vlasyuk *et al.*, 1969).

In *Chlorella pyrenoidosa*, a strong correlation between variation in zinc concentration and the rate of guanine incorporation in mRNA was obtained through the mitotic cycle (Altman *et al.*, 1968) and which suggested a role of zinc ions in the synthesis of RNA. Using Heidelberg scanning proton microprobe, Ender *et al.* (1983) have shown that zinc is associated with a special zone of the smooth endoplasmic reticulum. The effect of zinc nutrition on protein and ribosome metabolism in *Phaseolus vulgaris* leaves was studied by Lee (1978). He found that zinc deficiency reduced the level of RNA which accounted for the decrease in the rate of protein synthesis. It was further observed that under zinc deficiency the rate of synthesis of ribosome was slightly lower and the rate of degradation or ribosomes was higher than in plants supplied adequate amount of zinc. Increase in the concentration of free nucleotides in culture of *Rhizopus nigricans* following addition of zinc is suggestive of a role of zinc in the synthesis of nucleotides (Wegenur and Romano, 1963). An increase in the nucleic acid content in spinach was obtained when the seeds were treated with 0. 05% $ZnSO_4$ solution for 24 hours (Vekirchik, 1981).

5.5. Enzymes

Day and Franklin (1946) established a requirement of zinc for carbonic anhydrase activity in plants. They found that zinc bound to enzyme protein can be removed by dialysis. Lindskog and Malmstrom (1960) have reported that addition of zinc fully restored the activity of the enzyme carbonic anhydrase. Later several reports on decreased carbonic anhydrase activity in zinc deficiency in various plants have been made (Bradfield, 1947; Wood and Silby, 1952; Bar-Akiva and Lavon, 1969; Edward and Mohamad, 1973; Randall and Bouma, 1973; Ohki, 1976, 1978; Sharma *et al.*, 1981; Gibson and Leece, 1981; Snir, 1983). Many other enzymes are also known to have a zinc requirement for their activity, e.g. , several dehydrogenases-glutamate dehydrogenase, lactic acid dehydrogenase, alcohol dehydrogenase and proteinases, peptidases (Vallee and Wacker, 1976). Presence of zinc in significant amounts in glyceraldehyde 3-phosphate dehydrogenase has been noted (Hoch and Vallee, 1958). Bertrand and Wolf (1957) have reported that glucose-6-phosphate dehydrogenase and 6-phosphogluconate dehydrogenase in *Aspergillus niger* also contain Zn. Komai and Neilands (1968) have shown δ-aminolevulinate dehydrogenase to be a zinc enzyme. The enzyme L-mannosidase of Jackbean is a zinc metalloenzyme and is inhibited by chelating agents and the enzyme activity was resorted on addition of zinc (Snaith, 1975).

According to Parisi and Vallee (1969) glycylglycinedipeptidase, arginase, dehydropeptidase, lecithinase, C, 1-mannosidase, oxalacetate carboxylase and enolase are all zinc activated metallo-enzymes. According to Vallee (1960) the NAD dependent malate dehydrogenase requires zinc for the maintenance of its quaternary structure.

Zinc deficiency has been reported to cause a decrease in aldolase activity in several plants (Warburg and Christian, 1943; Quinland-Watson, 1953;

Hewitt and Tatham, 1960; Shaltenene, 1965) and an increase in the acid phosphatase activity in tomato (Reed, 1946). According to O'Sullivan (1970) zinc deficiency drastically and specifically reduces the aldolase activity in several plant species and which can be used as an indicator of zinc status of plants. Jyung and Camp (1976) found a decrease in RuBP carboxylase activity, with an increase in severity of zinc deficiency.

In several plants, zinc deficiency resulted in a decreased concentration of tryptophan (Hoagland, 1944; Nason, 1950). Since tryptophan is known to be a precursor of auxin biosynthesis, decrease in tryptophan concentration due to zinc deficiency was advanced as a possible explanation for decrease in auxin concentration in zinc deficient plants. Karakis and Rudakova (1971) and Vlasyuke *et al.* (1972) showed that zinc deficiency limited the biosynthesis of auxin from tryptophan. Zinc has been found to inhibit ribonuclease activity (Holden and Pirie, 1955; Kessler and Monselisc, 1959; Wojnar and Roth, 1964). Breslow and Girotti (1966) discussed a possible mechanism for the inhibition of ribonuclease by zinc (and copper) ions. Ability of zinc to form complexes with RNA that are resistant to ribonuclease effect, modification of the enzyme conformation and alterations involving the active site of the enzyme have been advanced as possible explanations for inhibition of ribonuclease activity by zinc.

Apart from this, zinc deficiency reduced significantly the activities of catalase and peroxidase in wheat anthers (Sharma *et al.*, 1979). Patner *et al.* (1983) found aldolase, carbonic anhydrase and alcohol dehydrogenase to be significantly related to zinc supply. Reports are also available on reduced activities of fructose diphosphatase and carbonic anhydrase in zinc deficient plants (Shrotri *et al.*, 1983). According to Michalski and Kaninga (1982) a decrease in zinc and copper content of chloroplastic preparations of tomato was associated with inactivation of the cyanide sensitive Cu and Zn containing superoxide dismutase activity in the chloroplasts of the detached or growing leaves of tomato plants. Duke and Salin (1983) isolated cyanide sensitive and organic solvent inactivated superoxide dismutase isoenzymes from pea leaves. These contained 2 atoms of Cu and 2 atoms of Zn per dimer. In rice seedlings, the activity of superoxide dismutase was reduced markedly in Zn and zinc + copper deficient plants (Wu and Wu, 1990).

5.6. Photosynthesis

A zinc protoporphyrin in yeast cells and maize leaves was detected (Shuvalov and Krasnovsky, 1971) and which was essential for the production and function of δ-amino levulinic acid dehydratase, which catalyses the condensation of two molecules of δ-amino levulinic acid to yield a molecule of ix porphobilinogen, a precursor of protoporphyrin which is an intermediary in chlorophyll synthesis (Komai and Neilands, 1968). On the other hand, zinc is known to cause a selective activation of δ-amino levulinic acid dehydratase

(Zaitseva *et al.*, 1970). Ohki (1976) reported that for photosynthesis and for formation of chlorophyll a minimum of 13 and 14 μg Zn/g respectively was required for maximum activity. Deficiency of zinc is known to cause a marked decrease in the concentration of chlorophyll and the rate of photosynthesis in pepper (Pandev *et al.*, 1981) and in rice (Wu and Wu, 1990). In garden pea seedlings, zinc deficiency caused a significant decrease in the concentrations of chlorophyll *a* and *b* (Paivoke, 1983).

The role of zinc in photosynthesis is well established as in the deficiency of the element, the rate of photosynthesis decreased (Fujiwara and Tsutsumi, 1962) but direct involvement of zinc in photosynthetic reactions has not yet been confirmed. It has, however, been suggested that low concentration of zinc in chloroplasts may adversely affect the chloroplast structure (Vesk *et al.*, 1966). The number of grana and stroma in thylakcid lamellae were considerably decreased in Zn deficient seedlings (Wu and Wu, 1990). They have also reported indirect effect of Zn deficiency on thylakoid membranes by altering SOD activity, thus leading to peroxidation of membrane lipids. The ability of zinc to complex with the lipid compounds of the chloroplasts (Igoshina, 1970) is suggestive of zinc involvement in the photochemical reactions occurring in chloroplasts. According to Spencer and Possingham (1960) zinc deficiency reduces the Hill reaction activity of tomato. Whereas Hatch and Slack (1970) suggested that zinc dependent carbonic anhydrase facilitates diffusion of the CO_2 assimilated in photosynthesis on its path through the liquid phase of chloroplasts. A decrease in photosynthesis was observed when carbonic anhydrase was severely reduced due to zinc deficiency (Randall and Bouma, 1973). In another report on *Phaseolus vulgaris* it was observed that supra optimal levels of zinc in primary leaves increased the CO_2 compensation point and inhibited net photosynthesis (Assche *et al.*, 1980). Application of zinc alongwith Mn and Fe caused an increase in photosynthesis in grapevines (Tesu *et al.*, 1980).

Photosynthesis is severely depressed in Zn deficiency (Lonstalot *et al.*, 1945) which may be partly explained by the depression in the activity of carbonic anhydrase which is a Zn dependent enzyme (Waygood, 1955; Tobin, 1970; Ohki, 1976). An additional reason for the reduction in photosynthesis in Zn deficient plants may be the depression in the activity of RNA polymerase which in wheat germ contains 7 atoms of Zn mol^{-1} protein (Petrany *et al.*, 1977) and which controls the synthesis of many enzymes including ribulose-bis - phosphate carboxylase and aldolase (Hewitt, 1983). The latter two enzymes form the bulk of chloroplastic proteins and are key enzymes in photosynthesis. Depression in these enzymes will reduce the rate of photosynthesis in Zn deficiency. Further Shrotri *et al.* (1989) have reported that in bajra (*Pennisetum typhoides*), Zn deficiency lowered the chlorophyll/carotenoid ratio, the activity of carbonic anhydrase which further declined with age, the activities of phosphoenol-pyruvate carboxylase and ribulose-bis-phosphate carboxylase which were considerably depressed with increase in age. The depression in these enzymes in Zn deficiency would depress the rate of photosynethsis.

Zinc at a concentration of 4 mM inhibited electron flow at the oxidising site of PS II and this inhibition could be removed upon removal of zinc from chloroplasts by washing (Tripathy and Mohanty, 1980). In another report (Baker *et al.*, 1982), the effect of 400 μM zinc concentration was studied on 3 photosynthetic electron transport reactions of thylakoid isolated from *Pisum sativum*. They suggested that zinc acts at 2 sites, site I is on the oxidising side of photosystem II and site II is between photosystem I and II and this electron flow can be positively or negatively affected by zinc depending on the light intensity.

Some experimental evidences indicate that when zinc is in excess supply, it impairs photophosphorylation capacity and NADPH production (Assche and Clijsters, 1983).

In a separate observation it was stated that supraoptimal level of zinc decreased the growth, dry matter production and total leaf chlorophyll and increase the total leaf diffusive resistance which could account for the reduced photosynthesis (Assche *et al.*, 1979). They further observed that non-cyclic electron transport in chloroplast fragments was reduced by supra optimal zinc concentration.

Sharma *et al.* (1982) found that zinc deficiency in cabbage caused a marked increase in water saturation deficit and a decrease in water potential and the rate of water loss by leaves. Most of these changes could be reversed when zinc was supplied to zinc deficient plants.

5.7. Reproductive Physiology

Recently the role of zinc has been emphasised in the reproduction of higher plants. Sharma *et al.* (1979, 1987) reported that the development of anthers and pollen producing capacity was adversely affected by zinc deficiency in wheat and maize. They observed that in severe zinc deficiency, florets failed to open and anthers lacked sporogenous tissue in wheat (Sharma *et al.*, 1979) whereas in maize vessels were formed in place of the sporogenous tissue (Sharma *et al.*, 1987). The pollen grains formed in moderate zinc deficiency had less dense cytoplasmic contents and poor viability and anthers failed to develop beyond the freshly liberated young pollen grains stage.

5.8. Interaction of Zinc with other Elements

Nitrogen

With an increase in N supply, Zn deficiency became more acute in rainfed low land rice (Singh and Singh, 1985) and citrus (Haas, 1936; Chapman *et al.*, 1937; Camp and Fudge, 1945). In several reports the interaction between Zn and N was found to be synergistic (Gupta and Potalic, 1991; Sakal *et al.*, 1988). Nitrogen application has been reported to influence Zn absorption by plants and vice versa (Dev and Shukla, 1980). They showed that for maximum utilization of applied Zn, the presence of adequate amount of N is essential. They further observed that the increase in Zn concentration in the plant was the highest when the source of N was NH_4NO_3 followed by urea and $(NH_4)_2SO_4$. Ozanne (1955) observed an increased severity of Zn deficiency in subterranean clover as the N supply increased regardless of N source, and the effect was not due to an increase in growth rate. He further noticed that Zn concentration in the roots was correlated with percentage of protein N. In later stages with $(NH_4)_2SO_4$, NH_4NO_3 and $Ca(NO_3)_2$ it was observed that the Zn concentration and uptake increased with increasing rate of N independent of soil pH (Boawn *et al.*, 1960). Soltanpour (1969) observed that N accentuated uptake of P and Zn by potatoes.

The differences among N carrier with regard to efficiency of Zn utilization may be attributed to the effect of the accompanying anion on the mobility of Zn as well as to the reduction in soil pH which enhances Zn availability. Zinc application has also been reported to increase the efficiency of legumes to fix atmospheric N_2.

A significant increase in total N and true protein besides a marked increase in the yield of sorghum with Zn application was observed (Kene and Despande, 1980). Zinc application also increased the content of amino acids, lysine, and histidine in chickpea and wheat. The content of arginine increased in chickpea but decreased in wheat with Zn application (Nayyar *et al.*, 1990).

Phosphorus

This interaction has been studied by various workers since 1936 (Barnett *et al.*, 1936; West, 1938; Stuckenholtz *et al.*, 1966), usually the interaction is turned as P-induced Zn deficiency. This interaction is commonly associated with high levels of available P or with the application of high dozes of P (Reddy *et al.*, 1973; Olsen, 1972; Takkar *et al.*, 1976; Verma and Tripathi, 1986). The symptoms can be prevented or corrected by Zn fertilization. The interaction between P and Zn might be due to (1) a slower rate of translocation of Zn from the roots to the tops, (2) a P-Zn interaction in the soil, (3) a simple dilution effect on Zn concentration in the tops owing to the growth response of P and (4) a metabolic disorder within plant cells related to an imbalance between P and Zn or an excessive concentration of P which interferes with the matobolic functions of Zn at certain sites in the cell.

It has been claimed that P affects the absorption of Zn by roots in soil in some other way than by precipitation of an insoluble zinc phosphate (Bingham and Garben, 1969; Bingham, 1963). Reports are also available on P application inducing or accentuating Zn deficiency symptoms in plants (Loneragan, 1951; Brown *et al.*, 1970). In some separate reports it has been observed that applied P decreased Zn concentration but the total uptake of Zn was the same or increased (Millikan, 1963). On the other hand the cause of the P induced Zn deficiency has been suggested to be due to an interference by P with the uptake, translocation or utilization of Zn (Thorne, 1957; Adriano *et al.*, 1971).

In several observations it was found that Zn fertilization depressed P concentration in plants but the interactive effect of P on Zn was generally more pronounced than that of Zn on P (Prasad *et al.*, 1971; Kashyap *et al.*, 1986; Gupta and Raj, 1980). High P supply interfered with some metabolic functions of Zn in plant cells (Dwivedi *et al.*, 1975). They further indicated that in the presence of excess P some organic substances are synthesized in plant cells which hinder the translocation of Zn (Dwivedi and Randhawa, 1974).

Potassium

The literature on this aspect is meagre. In one of the experiments using sandy soil, Gupta and Raj (1983) observed a positive and significant interaction between K and Zn on the grain yield of wheat. In the absence of K, Zn application increased the yield which signifies that Zn supply was more limiting than the K supply for crop growth in the soil. Similarly in potatoes also a synergism was observed between the two elements. K application promoted Zn absorption and its translocation in plants (Biswas *et al.*, 1977).

Sulphur

Application of S enhanced Zn concentration in potato (Aulakh *et al.*, 1977) and sunflower (Gangadhara *et al.*,1990). Likewise, Zn addition to soils has been reported to increase sulphur content in rapeseed (Shukla *et al.*, 1983) and its uptake in maize (Hulagur and Dangarwala, 1983). Mostly the interaction between the two is synergistic. A significant synergistic effect of S and Zn on yield as well as on their concentration and uptake has been reported for wheat (Gupta *et al.*, 1984) and groundnut (Behl *et al.*, 1986).

Calcium

Liming of acid soils increased soil pH and reduced Zn availability to plants. A marked and significant decrease in ^{65}Zn absorption by sweet orange leaves occurred when increasing rates of lime were added to the zinc sulphate spray in solution (Mann and Takker,1982).

Zinc-Ca relationship is somewhat complicated in sodic soils due to prepondrance of sodium which adversely affects plant growth.

Magnesium

Seatz (1960) had reported that the interaction of Mg with Zn occurred to a greater degree within the plant than in the soil.

Mostly the studies are limited and conflicting. Zn concentration in plant tissues, decreased with Mg application in some studies (Kumar *et al.,* 1981; Gupta and Gupta, 1984) whereas it increased with Mg addition either in the absence of Zn or at low levels of applied Zn (Gupta and Singh, 1985).

A hypothesis suggests that since Mg and Zn have nearly the same ionic radii, the Mg ion may interact with a relatively insoluble Zn compound in the soil to release Zn in a plant available form (Olsen, 1972).

Manganese

In several crops, Mn concentration is decreased with zinc application (Gangwar and Mann, 1972; Sidhu, 1989) as well as decrease in zinc concentration in wheat by high rates of Mn application further indicates an antagonistic relationship. On the contrary, increasing Mn concentration in the nutrient solution decreased translocation of Zn but not its absorption by pea (Chandel and Saxena, 1985).

6. MOLYBDENUM

Bortels (1930) had indicated the essentiality of molybdenum (Mo) for gaseous N_2 fixation by the aerobic bacterium, *Azotobacter chrococcum.* In water culture experiments with tomato, Arnon and Stout (1939) demonstrated the essentiality of traces of Mo for higher plants which showed mottling of leaves in absence of Mo which was alleviated by the addition of Mo. These results were confirmed by several other workers for many other plant species (Nason,1958; Evans,1956). Molybdenum acts as a metal component of enzymes, nitrate reductase (Nicholas and Nason, 1954, 1955; Evans and Hall, 1955) and nitrogenase (Bulen and LeCompte,1966; Klucas *et al.,* 1968; Burris, 1971) and thus, it is closely associated with nitrogen metabolism (Mulder,1948). When the source of nitrogen is nitrate, in Mo deficiency in brassicas, particularly in cauliflower, two distinct types of symptoms are produced, one in acute deficiency (0.000005ppm) and the other at intermediate level (at 5 or 10 times of the Mo supplied in acute deficiency). In acute deficiency of Mo in cauliflower the second leaf may show bright yellow green interveinal mottling which is distributed fairly evenly over the leaf. Mottling seems to be related to uneven distribution of Mo as shown by experiments using Mo (Stout and Meagher, 1948), the mottled area contained more nitrate and less Mo than the greener regions in the leaf. This is followed by marginal wilting and cupping and ultimately leaf becomes severely necrotic and gets withered. These symptoms progress to the apex and ultimately plant collapses. The second type of symptoms are produced at intermediate level of Mo when the source of nitrogen is nitrate. These symptoms commence as one or several transluscent areas usually situated in a row on a young leaf. These areas become chlorotic, necrotic and papery. The papery areas perforate and the holes produced enlarge (Agarwala and Hewitt, 1954a, 1955b; Hewitt and Bolle-Jones, 1952). Later these areas appear near the margins of the leaf and margins become torn and ragged. Very little portion of the lamina is left on the leaf but midrib of the leaf grows. In still younger leaves the symptoms appear at a younger stage, leaves become brown, necrotic and deliquescent. But these leaves elongate although there is little lamina left. Finally growing point is killed. These symptoms have been described as 'whip tail' disease. These symptoms could be produced at 0.00005 ppm molybdenum when the source of nitrogen was nitrate (Agarwala and Hewitt, 1954a). But when source of nitrogen was ammonium or urea or glutamic acid, in acute deficiency (0.000005ppm) of Mo, no mottling was observed but instead whip tail type of effects appeared (Agarwala, 1952; Agarwala and Hewitt, 1955b). This pointed out that molybdenum was essential with source of nitrogen other than nitrate but its requirement was less than with nitrate. Some aspects of metabolism were also affected in these plants which will be discussed later. By their experiments with sand culture on tomato, Hewitt and McCready (1956), confirmed the results on cauliflower that molybdenum is required even with reduced form of nitrogen. The metabolic effects in deficiency of Mo were of similar in two plants, cauliflower and tomato (Hewitt,1963). Hewitt (1963) pointed out that the role of molybdenum was not only in the activity of nitrate reductase of which Mo and iron in the form of cytochrome C_{659} were constituent part, but also in many other aspects of plant metabolism,some of which were independent of nitrogen source. In 1970, Hewitt and Gundry on the basis of their sand culture experiments carried out in sterile conditions stated that molybdenum was not required for the growth of cauliflower when the source of nitrogen was ammonium. Hewitt (1983) attributed the symptoms that were observed in acute deficiency of molybdenum with ammonia, urea or glutamic acid, to the injury caused, principally, by dehydrogenase activity of nitrate reductase in the absence of Mo. It is rather difficult to accept this view of Hewitt as no concercete evidence exists which will show that dehydrogenase activity of nitrate reductase could cause the reported changes in acute molybdenum deficiency in cauliflower plants supplied ammonium, urea or glutamic acid. Even Hewitt (1983) has stated that when ammonium nitrogen is provided in sterile culture, molybdenum requirement is not normally observed. Perhaps,

Hewitt (1983) by this statement means that sometimes molybdenum requirement can be observed in cauliflower plants even with the reduced form of nitrogen. Rains (1971) considers that the molybdenum requirement with reduced sources of nitrogen is much less than that with NO_3 (Mulder, 1948; Ichioka and Arnon, 1955; Sheat *et al,* 1959). Marschner (1986) agrees with this view. Therefore, it seems that in regard to the molybdenum requirements with sources of nitrogen other than nitrate several prominent workers in the field hold conflicting views and the matter seems to be controversial. Besides changes in metabolism which have been observed in molybdenum deficiency irrespective of nitrogen supply, there are many changes other than the reduction of NO_3 due to Mo deficiency which have been worked out only with NO_3 supply. But these changes do not seem to be the result of low nitrate reductase activity.

Molybdenum deficiency is known to result in disorganisation of chloroplast (Fido *et al.,* 1977; Notton, 1972) so that the chloroplast material especially protein and RNA become scattered throughout the cytoplasm (Hewitt,1983). These changes were accompanied by copious formation of lipid droplets in the cytoplasm. In molybdenum deficient cauliflower, the chloroplasts from the chlorotic leaves became bulbous and enlarged (Hucklesby,1961). The starch grains of these chloroplasts shrank and gradually disappeared (Hewitt,1983).

Molybdenum is usually taken up by plants in the form of molybdate (Mengel and Kirkby,1987). Moore (1972) suggested that Mo uptake is metabolically controlled process which was reduced by competitive effects of SO_4 (Stout *et al.,* 1951; Reisenauer, 1963) and enhanced by phosphate ions (Stout *et al.,* 1951). In the xylem elements, Mo is usually translocated as molybdate, Mo-S amino acid complex or as a molybdate complex with sugars and other polyhydroxy compounds (Tiffin,1972). Molybdenum is located in the phloem and vascular parenchyma and is only moderately mobile in the plant (Hewitt and Agarwala,1952). Giordano *et al.* (1966) reported that the uptake of Mo is greater in the presence of nitrate nitrogen than ammonium nitrogen. In contrast to other micronutrients Mo can be taken up in much higher amounts by plants without normally resulting in toxic effects. The physiological Mo requirement by several plant species is very low and is less than 1 ppm in dry matter (Stout and Meagher, 1948).

6.1. Enzymes

Molybdenum is known to act as a metal component of two major enzymes in plants-nitrogenase and nitrate reductase (Mengel and Kirkby,1987). The nitrate reductase enzyme contains FAD, cytochrome b557 and molybdenum (Notton and Hewitt, 1979; Hewitt and Notton,1980). It catalyzes the reduction of nitrate to nitrite. The electron flow in the reduction of nitrate is as follows:

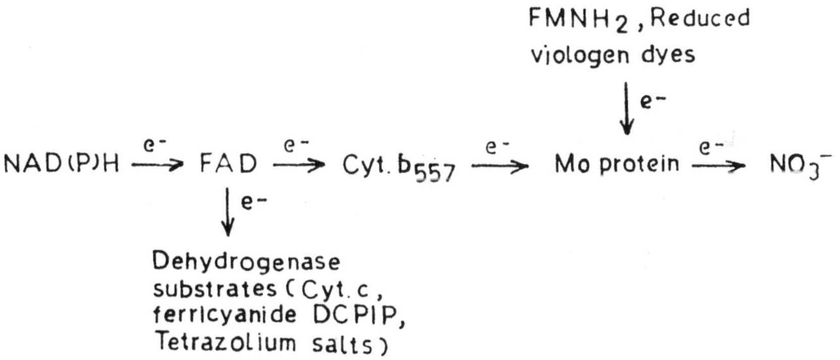

Fig. 2. Flow of electrons in reduction of nitrate (after Hewitt et al., 1976)

Nitrogenase is a iron molybdenum containing N_2 fixing protein present in numerous free living and symbiotic nitrogen fixing bacteria,cyanophyceae and some photosynthetic bacteria,in legumes and this protein is always associated with another type of protein which has only iron as a prosthetic group (Hewitt,1983). Atmospheric nitrogen is reduced by the activity of nitrogenase enzyme. A number of reactions can be catalysed by it. H_2 can act as a competitive inhibitor of nitrogen in the action of the enzyme. Its activity can be conveniently measured by the reduction of acetylene to ethylene in a gas chromatograph with a flame ionising detector. ATP forms a salt with Mg. MgATP can be bound to Fe-protein which is one of the proteins present in nitrogenase enzyme. This complex acquires a low potential and is capable of reducing Mo-Fe protein which is the other protein in nitrogenase enzyme (fig. 3). Vanadium is known to replace molybdenum in nitrogenase and be functional with reduced efficiency. Now besides nitrogenases enzymes, two other alternative nitrogenases have been identified (Yakunin *et al.*, 1990). These are known as nitrogenase-2 and nitrogenase-E. Nitrogenase-2 contains V instead of Mo (Garner *et al.*, 1989) and can reduce acetylene to ethylene and further to ethane whereas nitrogenase enzyme can reduce acetylene to ethylene only.

Nitrogenase-E does not contain Mo or V. The physiological significance of alternative nitrogenase is not yet known. The mode of action of nitrogenase enzyme with atmospheric nitrogen is desribed below; nitrogenase enzyme is active only under reduced pressure of O_2 which is maintained by the action of leghaemoglobin which is synthesized in nodule of leguminous plants.

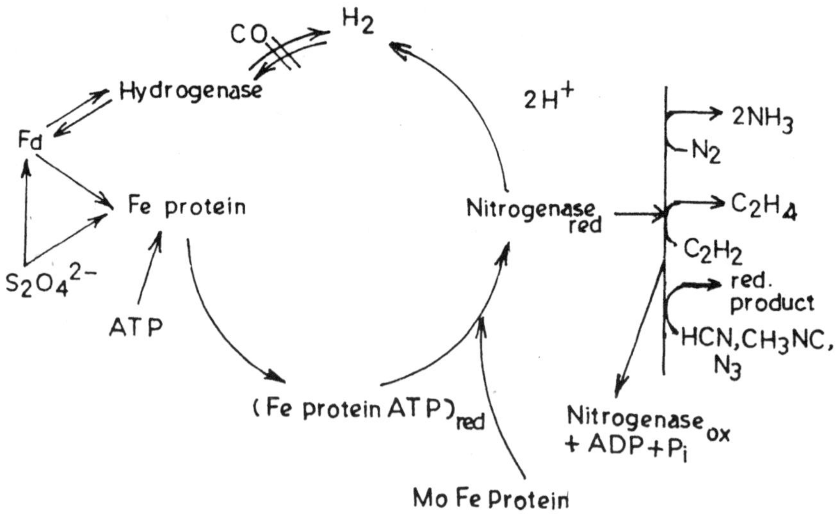

Fig. 3. Diagrammatic representation of reactions catalysed by nitrogenase enzyme (after Burris, 1976)

Another nitrogenase enzyme which has Mo-Fe as protein contains acid-labile sulphur (Shah and Brill, 1977; Pienkos *et al*, 1980). Besides these two enzymes, there are several other molybdenum proteins, viz., xanthine oxidase or dehydrogenase of some fungi and in few higher plants (Mendel and Muller,1976; Nguyen and Feierabend,1978), sulphite oxidase (Hewitt, 1983), aldolase and formate dehydrogenase. Xanthine dehydrogenase is present in animal system and also in pea and tobacco plants. But as its function is mainly catalytic it may not explain the results obtained with reduced form of nitrogen in molybdenum deficiency in sterile conditions. In sulphite oxidase, the Mo component is reported to be an unidentified pteridin complex (Johnson *et al.*,

1980). The basic pteridyl structure is formed by several steps from guanosine monophosphate and is related to the folic acid co-enzyme which are involved in diverse reactions including methyl (one carbon) transfer in reactions which are independent of the cobalt B_{12} factors and in tyrosine formation by hydroxylation of phenyl alanine. Sulphite oxidase plays an important role in metabolism of animals and is abundant in liver of many animals (Johnson,1980; Rajagopalan, 1980). Its presence has been observed in many plants, e.g., wheat germ. If present in *Brassicas*, it may play an important part in the metabolism of these plants where S metabolism is very active (Hewitt,1983). In many higher plants, Mo deficiency is known to create variable changes both in organic constituents and in the activities of certain enzymes which are not activated by Mo.

Molybdenum deficiency leads to accumulation of nitrate in many plants grown in culture (Hewitt and Jones,1947; Mulder,1948; Agarwala and Hewitt, 1955a) or in the field conditions (Wilson, 1948; Wilson and Waring,1948). In molybdenum deficiency with NO_3 as source of nitrogen, lower concentration of protein, soluble organic nitrogen compounds and ammonium nitrogen has been reported by Agarwala and Hewitt (1955a) and Hewitt *et al.* (1957, 1949). When NO_3 is being supplied in molybdenum deficiency it is known to decrease some free amino acids, viz., glutamic acid, glutamine, aspartic acid, asparagine, glycine, ß-alanine and lysine in cauliflower but the concentration of arginine usually increases (Hewitt *et al.,* 1957) and even in Mo deficient plants such as tomato some amino acids were increased (Possingham, 1956), viz., r-aminobutyric acid, β-alanine and arginine. When the source of N was ammonium or urea or glutamic acid in molybdenum deficient cauliflower, the concentration of many amino acids specially arginine and serine were found to be high.

Protein nitrogen content in mustard (Chatterjee *et al.*, 1985), sorghum (Agarwala, *et al.,* 1988), sugarcane (Agarwala, *et al.,* 1985), and papaya (Agarwala *et al.,*1986) has been found to be depressed in low molybdenum supply. Mulder (1954) reported decreased activity of malic and succinic dehydrogenases in molybdenum deficient tobacco. In molybdenum deficient tomato leaf extracts, Nason *et al.* (1951) have reported increased activity of enzymes, polyphenol oxidase and peroxidase.

In molybdenum deficient maize, the low activity of alanine aminotransferase (Agarwala *et al.,* 1978) may indicate the involvement of molybdenum in protein synthesis at steps other than nitrate reduction. Apart from this in different species of *Aspergillus,* Mo deficiency caused a significant decrease both in dry weight of mycelia and protein content of felts (Agarwala *et al.,* 1986). This indicates a role of Mo in protein synthesis which was also supported by Hewitt *et al.* (1979). The depression in the aldolase activity in Mo deficiency in maize (Agarwala *et al.,* 1978) in which dry matter yield is also depressed might be related to its role in photosynthesis.

TABLE 1. Enzyme activities expressed as per mg protein in leaves of plants supplied adequate molybdenum and Mo deficient plants without and with additional molybdenum.

Enzyme	Plant species	Plants supplied adequate Mo	Mo deficient plants	Mo deficient plants + Mo
		n moles NO_2 formed/h		
Nitrate reductase	Maize[1]	21	12	21
	Mustard[2]	1560	362	1647
	Sorghum[3]	162	32	114
Peroxidase	Mustard	3.75	4.21	4.69
	Sorghum	3.81	1.03	2.21
		μ moles H_2O_2 decomposed		
Catalase	Maize	354	277	338
	Mustard	1827	1621	1956
	Sorghum	167	130	160
		μg triose P formed		
Aldolase	Maize	144	89	89
	Mustard	386	421	469
	Sorghum	296	216	247
		μg P solubilised		
Ribonuclease	Maize	19.2	33.8	40.8
		μg Pi liberated		
Acid phosphatase	Maize	26	54	51
	Mustard	84	224	197
	Sorghum	32	53	37
		μ mol pyruvate formed		
Alanine aminotransferase	Maize	1.92	0.85	1.31

1. Agarwala *et al.* (1978), 2 Chatterjee *et al.* (1985), 3. Agarwala *et al.* (1988)

Several iron enzymes, catalase, peroxidase, succinic dehydrogenase, aconitase and cytochrome oxidase were found to be depressed in a number of plant species (Table 1). It appears that molybdenum deficiency depresses the physiological availability of iron. As in *Aspergillus niger* on addition of Mo aseptically to cultures of Mo deficient *A. niger,* the activity of catalase, peroxidase, aconitase and succinic dehydrogenase were significantly increased in absence of cycloheximide but not in its presence indicating that the Mo may be involved or Mo is acting as an inducer in the synthesis of the protein moiety of these enzymes (Agarwala *et al.,* 1978,1986,1988). On addition of adequate Mo to Mo deficient maize, sorghum and *Aspergillus,* significant increase in the activity of nitrate reductase,peroxidase, catalse and aldolase was observed which may be due to the role of Mo in increasing the protein moiety of the enzymes (Table. 2).

TABLE 2. Recovery of the specific activity of certain enzymes on addition of molybdenum, with or without cycloheximide, to molybdenum-deficient cultures of *A. niger*

Control	Additions to Mo-deficient cultures			
	None	+ Mo (to extracts	+ Mo (to medium)	+ Mo + cycloheximide (to medium)
Nitrate reductase, n mol nitrite formed. h^{-1}				
164 ± 4	70 ± 2	90 ± 4	216 ± 6	192 ± 7
Catalase, µmol H_2O_2 decomposed				
248 ± 4	42 ± 2	38 ± 2	328 ± 3	42 ± 3
Peroxidase, mg purpuogallin formed				
0.10 ± 0.01	0.03 ± 0.01	0.02 ± 0	0.10 ± 0.01	0.03 ± 0.01
Succinic dehydrogenase, unit				
20.8 ± 0.3	6.0 ± 0.3	5.9 ± 0.3	12.3 ± 0.3	4.0 ± 0.3
Aconitase, unit				
123 ± 6	107 ± 5	105 ± 3	107 ± 6	82 ± 2
Acid phosphatase, µg Pi liberated				
57 ± 2	18 ± 1	16 ± 1	18 ± 1	15 ± 1
Protein, mg protein .100 fresh weight $^{-1}$				
1.94 ± 0.06	1.71 ± 0.06		1.04 ± 0.01	1.45 ± 0.01

NOTE : Growth media for the control contained 1.0 ppb Mo. Molybdenum (5 ppb) and cycloheximide (50 ppm) were added to cultures after growth for 72h in media containing 0.02 ppb Mo (Mo-deficient cultures). Enzyme activity was determined after a further 18 h of incubaton. Each value is the mean of three determinations ± SE.

The stimulation in acid phosphatase in several higher and lower plants deficient in Mo (Agarwala *et al.*, 1978, 1986, 1988) has been attributed to either accumulation of inorganic phosphorus in low Mo or Mo acting as a competitive inhibitor for the enzyme activity as in presence of sufficient Mo the phosphatase activity is reduced (Hewitt and Tatham, 1960; Spencer, 1954; Nicholas and Commissiong, 1957).

In molybdenum deficient cauliflower when the source of nitrogen was nitrate, ascorbic acid content is severely depressed. Even when the source of nitrogen was ammonia or glutamic acid or urea in Mo deficient cauliflower the ascorbic acid content was depressed but the extent of depression was much less than with nitrate (Agarwala and Hewitt, 1954 b).

DNA and RNA contents were reduced in Mo deficient mustard leaves (Chatterjee *et al.*, 1985). It seems that the reduction in DNA content was due to the reduction in number of cells in interphase whereas reduction in RNA content could be the result of high ribonuclease activity indirectly resulting low protein synthesis.

In molybdenum deficient sugarcane not only the yield of canes was reduced but the sucrose concentration in cane juice was significantly reduced (Agarwala *et al.,* 1985). This could be attributed to lowered rate of photosynthesis in Mo deficient conditions.

The yield of melons and watermelons increased to highest when NPK fertilizers were added along with Mn, Cu and Mo to soils in Russia (Ragimova,1987). Apart from this, in several field trials it has been further observed that application of Mo as sodium molybdate along with inorganic fertilizers has increased not only the total and seed yield of different plants but also increased the protein and total nitrogen content of seeds or leaves or both (Domska *et al.,* 1989; Fomasieri *et al.,* 1988; Kalia and Sharma, 1988; Chandel *et al.,*1989;Raju and Reddy, 1989; Mondy and Munshi,1988). Agarwala *et al.* (1988) have reported that in sorghum (*Sorghum bicolor*) Mo deficiency reduced the seed yield more than the biomass and affected the metabolism also. Molybdenum requirement for maximum growth and seed yield of mustard is ten times higher (0. 2ppm) than for other plant species (Chatterjee *et al.,* 1985). It appears from these results that nontoxic (high)levels of Mo in plants favour seed production in legumes, brassica and cereals and not in forage yielding plants (Marschner,1986). In addition to this, considerable deficiency in the molybdenum requirement was observed in species of *Aspergillus.* It was 1 ppb for *A. niger* but for *A. flavus, A. sulphureus* and *A. terreus* it was 0.2 ppb (Agarwala *et al.,* 1986). In tobacco, application of Mo has been observed to increase the growth of plant, its foliar development and fresh and dry weights. This positive effect was related directly to the amount of Mo which has also helped in improving the localisation of nicotine in tobacco leaves (Martin *et al.,* 1988). Whereas application of Mo reduced the content of NO_3-N and total glycoalkaloids in potato tubers both at harvest and after storage for sometime (Munshi and Mondy, 1988).

The response of Mo application to different plant species appears to be variable. Sometimes high application of Mo might also result in toxic Mo levels (Graupe, 1966) in plants.

6.2. Reproductive Physiology

The results obtained by Agarwala *et al.* (1979) on maize suggest the important role of Mo in male reproductive phase of plants. In maize deficient in Mo, tasselling was poor and delayed, most of the flowers failed to open and pollen producing capacity of the anthers was reduced. The pollen grains were small, free of starch had lower activity of invertase and starch phosphorylase and showed poor germination.

6.3. Interaction of Molybdenum with other Elements

Nitrogen

Mo and N are known to interact within a certain range of concentration. In legumes there is a resemblance between Mo deficiency and N deficiency (Magstrom and Berger, 1965). This has been attributed to restricted N assimilation in low Mo conditions by affecting both nitrate reduction and N_2 fixation as Mo is a constituent of enzymes - NO_3 reductase and nitrogenase. In tomato, Mo uptake was found to be greater in the presence of NO_3- N than ammonium N (Giordano *et al.,* 1960). Concentration of Mo in maize, sunflower and soybean grown in different soils decreased with increased rate of N application suggesting an antagonism between the two (Mircher and Gezencheva, 1986). The effects of excess Mo differed with different sources of N, slight with nitrate, severe with ammonium sulphate or urea and intermediate with ammonium nitrate in cauliflower, grown in refined sand (Agarwala and Hewitt, 1955). When molybdenum deficient cauliflower were grown with nitrate N, the leaves had lower concentration of protein, soluble organic N compounds and ammonium nitrate than plants grown with adequate Mo.

Phosphorus

In tomato, uptake and translocation of Mo was enhanced by P application (Stout *et al.,* 1951). This was attributed to the formation of phosphomolybdate complex which is taken up more readily by the plants than molybdate (Barshed, 1951). Application of P and Mo increased the yield of groundnut besides affecting a number of growth parameters (Muralidhaman and George, 1971). Application of Mo to soils resulted in small increase in Mo concentration of soybean but a remarkable increase in P concentration (Karbi *et al.,* 1974).

Potassium

Application of Mo to soils has been shown to increase K content in soybean (Karbi *et al.,* 1974) and in pea (Bilteanu *et al.,* 1973). In field trials potassium fertilizer when applied to alfalfa growing in soil which had high available Mo, decreased Mo in the herbage but even then this decreased concentration of Mo in herbage was found toxic to ruminants (Ghalaic, 1985) which were fed on this herbage. The ruminants showed symptoms of Cu deficiency induced by the excess Mo in the herbage.

Sulphur

Several reports indicate a negetive correlation between S and Mo concentration in a number of plant species (Gupta and Munro, 1969; Shukla and Pathak, 1973). In several glass house experiments it was observed that S application drastically reduced Mo concentration in plants (Gupta and Macleod, 1975; Sisodia *et al.*, 1975). A significant correlation between Mo and S was observed in burley tobacco (Sim *et al.*, 1979) which suggests that addition of SO_4 fertilizer to low Mo soil had detrimental effects but to soils rich in Mo it has positive effects as Mo toxicity is decreased in herbage. These studies show an antagonistic interaction between Mo and S.

In mustard, excess sulphur at adequate molybdenum level partially alleviated copper deficiency symptoms and decreased the dry weight, seed yield and activity of cytochrome oxidase. The alleviation of copper deficiency effects by excess sulphur was also discernible at excess molybdenum level. Excess sulphur interacted with excess Mo and decreased tissue molybdenum content at both NO_3 and NH^{-}_4 nitrogen sources (Chatterjee *et al.*, 1991).

Calcium

Reports on calcium-molybdenum interaction show an indirect effect of lime application, soil pH increases which in turn increases the availability and uptake of Mo (Datir *et al.*, 1976; Falke *et al.*, 1988).

Manganese

Some studies on cauliflower showed that the effect of Mo deficiency was accentuated by excess Mn (Hewitt *et al.*, 1949). In Mo-deficient plants grown with excess Mn, the concentration of arginine, alanine, aspartic acid, glutamic acid, proline, aspargine and glutamine were further decreased in comparison to their concentration in Mo deficient plants with normal Mn.

In soil pot culture with barley, Mo was found to act antagonistically with Mn, Mn application decreased the concentration of Mo (Singh and Steenberg, 1975).

Zinc

Increased amounts of Mo associated with NPK decreased Zn absorption in pea (Bilteaner *et al.*, 1973). Mo application to rice increased Zn uptake and its concentration in leaves but decreaed it in stem and roots (Gupta and Singh, 1981).

Mo application to soils increased Zn uptake in peanut whereas Zn application above 2.5 ppm supprssed Mo concentration and uptake by shoot (Gupta and Singh, 1984).

7. Boron

The essentiality of boron for the growth of higher plants was indicated by Aghulon (1910); Maze (1915) could not get normal growth of maize when grown in solution culture omitting boron. Warington (1923) was credited with establishing the essentiality of boron for higher plants. Later boron was shown to be essential for a wide variety of plants (Hewitt, 1966).

Pilbeam and Kirkby (1983) studied the physiological role of boron in plants and found that the presence of boron was essential to maintain the structural integrity of plant membranes and that many symptoms of boron deficiency in plants were secondary effects caused by changes in membrane permeability. The deficiency of boron retards cell division and cell enlargement in *Vicia faba* (Hudak, 1973) and tomato (Kouchi and Kumazawa, 1975b) which might be due to suppression of mitosis (Whittington, 1957). On the other hand, Skok (1958) proposed that boron functions primarily in differentiation and maturation of plant cells rather than in cell division. In some plants, boron deficiency suppreses apical meristem, in others, premature differentiation of immature tissues, takes place (Lohnis, 1940; Whittington, 1959).

7.1. Uptake and Distribution

The uptake of boron was studied (Bowen,1968, 1969, 1972) in leaf, meristematic tissue and excised roots of sugarcane and it was observed that only a fraction of boron uptake was regulated metabolically and had the characteristic of carrier mediated absorption process. In contrast to this, a high proportion of boron was present in the water free space bound to cell wall as borate polysaccharide complex and its active uptake is nil at 2 degree C (Bowen and Nissen, 1976).

In maize, the uptake of boron was not found to be affected by change in pH (Lutz *et al.*, 1972). But Scheffer and Welte (1955) have reported that the concentration of boron in barley roots was decreased with an increase in pH and soil calcium carbonate levels and maximum boron uptake was at pH 6. 0 (Oertli and Grgurevic, 1975).

Boron is relatively 'immobile' in plants but frequently the boron content increases from the older to younger plant parts (Cripps, 1956; Wilkinson, 1957; Al-Malla, 1986). Later evidences show that in boron deficiency the element is remobilised to younger leaves but otherwise it is relatively phloem immobile (Shelp *et al.*, 1987). In boron deficiency inhibition of root elongation in sunflower due to cessation of mitosis is a secondary effect whereas the primary effect being the cell membrane alteration (Moore and Hirsch, 1983). In a second view the inhibition of root elongation was also attributed to the change in the direction of cell expansion from longitudinal to radial (Hirsch and Torrey, 1980). The deficiency of boron results in the formation of lignified cell walls of potato tubers which developed a brown colour (Combrink and Hammes, 1972). The cells surrounding the thick walled cells showed abnormal cell division and often showed damage and discolouration leading to enlargement of the brown flecks. Boron is involved in structural organisation of heterochromatin in interphase nucleus in sunflower (Shkol'nik *et al.*,1972).

Boron is apparently fixed in the tissue and constant supply of the element is necessary for normal growth (Johnson and Fischer, 1930). In another view it was suggested that boron was highly mobile in redistribution (Goore and Lune, 1980). Boron concentration of the green tissue in *Dianthus caryophyllus,* grown in sand culture increased with an increase in boron supply but its concentration in reproductive parts was not appreciably affected by the supply (Heuterman *et al.*, 1973). Boron is believed to be distributed in acropetal manner in boron deficient conditions but when its concentration is optimum it is distributed in the order, leaves > pods > roots > stem (Golov and Kazachkov, 1979). Floral parts, i.e., styles, sepals, patals, anthers, filaments and receptacles in pea, apple and cherry were reported to have higher boron concentration than petioles and leaves (Woodbridge *et al.*, 1971; Lotti *et al.*, 1989). However, reports have also been made on higher boron content in seeds than other plant parts but highest content was in leaves (Touchton *et al.*,1980). Among the cell organelles, higher boron concentration in nucleie, plastids, and dialysable fraction of the supernatant was reported than in mitochondria and microsomes in mung bean and sunflower (Skok and McIlrath, 1958).

7.2. Effect on Water Relations

Some evidences show that boron regulates the water relations in plants. By addition of small amounts of boron to the medium in which pollen grains were kept, bursting of the pollen grains due to excessive uptake of water could be prevented. The leaves of soybean (Minarik and Shive, 1939) and tobacoo (Leaf, 1953) had a higher moisture content in boron deficiency than in normal plants. The decrease in water loss in boron deficiency has been attributed to the presence of higher concertration of hydrophilic colloids in boron deficiency (Baker *et al.*, 1956).

In boron deficient leaves of cabbage (Sharma *et al.*, 1984) and cauliflower (Sharma and Sharma, 1987), a decrease in water potential, stomatal pore opening and transpiration rate and an increase in tissue hydration and lowering of water saturation deficit was obseved. Apart from this in safflower and mustard leaves a report (Sharma and Ram Chandra, 1989, 1990) on an increased diffusive resistance and decreased rate of transpiration, water potential and water saturation deficit was made. Application of boron increased the percentage of bound water (Dvornic *et al.*, 1972) and total water content and improved water utilization but decreased water deficit in alfalfa (Radeve *et al.*, 1980).

7.3. Physiological and Biochemical effects

7.3.1. Respiration

Boron deficiency resulted in a higher rate of O_2 uptake in leaves or leaf extracts of tomato (MacVicar and Burris, 1948, Nason *et al.,* 1952). At the same time, a decrease in respiratory phosphorylation in boron deficient sunflower tissue was also reported (Timashove, 1970). In boron deficient terminal buds of sunflower the consumption of O_2 increased even before the appearance of deficiency symptoms on leaves and just after the appearance of visible effects a decrease in O_2 uptake was found (Augsten and Hundt, 1970). It was further established that a phenol oxidase system was operative and the capacity of phosphorylation is diminished in boron deficient conditions. On the other hand, when boron is in excess supply, it suppressed the oxygen evolution and respiratory oxygen uptake in *Dunaliella tertiolecta* (Ahmed *et al.*, 1988).

7.3.2. Carbohydrate Metabolism

Boron plays an important role in the carbohydrate metabolism. In several crops in boron deficiency total and reducing sugars are known to accumulate either in leaves or in roots (Haas and Klotz, 1931; Scripture and McHargue, 1943, 1945; Milosavljevic and Popovic, 1970; Agarwala *et al.,* 1978; Chatterjee *et al.,* 1990; Neales, 1959; Esteban *et al.,* 1985). The accumulation of starch in boron deficiency in several plant species have also been reported, viz., in beans (Dvornic *et al.,* 1972), *Lemna minor* (Dugger, 1983) and mustard (Sharma and Ramchandra, 1990). In contrast to this, a reduction in starch concentration in boron deficient cowpea leaves was observed (Chatterjee *et al.,* 1990). A direct relationship has been noticed in the boron supply and sugar content in the storage roots of sugarbeat (Bonilla *et al.,* 1980; Agarwala *et al.,* 1991), the concentration of sugar dropped both in deficiency and toxicity of boron. Boron supply improved the seed quality in respect to protein, fat and carbohydrate concentration in corn (Gaina and Silli, 1973). An observation was made on a positive correlation between boron supply and sucrose concentration in apple and tomato fruits (Lopez-Andreu, 1988). But almost opposite effect was observed in legumes where the sugar content decreased with increased rates of boron application (Saeed and Woodbridge, 1981).

A hypothesis has been deduced by Gauch and Dugger (1954) which suggests that boron plays a key role in sugar translocation. Boron forms a polyhydroxy complex with sugars which is more readily translocated through cellular membranes than the non-borated sugar molecules (Dugger, 1983). Cessation of growth and breakdown of meristematic activity in boron deficient plants was accounted for by the deficiency of carbohydrates in the meristematic region. This hypothesis was supported by some later work (Sisler *et al.*, 1956; Shkol'nik and Abdarashitov, 1958; Whittington, 1957; Starck, 1963). An alternate theory was proposed by Skok (1956, 1958) which states that in boron deficient plants sugars are not utilized and the gradient favouring sugar translocation no longer exists, therefore, the cellular activities are reduced in growing points of both tops and roots. The accumulation of carbohydrates in leaves is most probably due to destruction of phloem tissue (McIlrath and Palser, 1956). Boron might play a role in translocation of growth regulators which facilitate sugar transport (Mitchell *et al.*, 1953). Boron deficiency most probably did not interfere with transport of photosynthetes or precursors required for citrate synthesis (Brown and Ambler, 1969). Dugger (1983) suggested that boron influences carbohydrate metabolism rather than its translocation. Water soluble sugars increased with an increase in boron supply in lucerne (*M. Sativa* L.) flowers (Misra and Patil, 1987). It was further observed that a foliar spray of boron at reproductive stage in boron deficient sorghum increased the water soluble sugars of seed coatings (Misra *et al.*, 1991).

7.3.3. Phenolic Compounds and Lignin Biosynthesis

Several reports have been made on the accumulation of oxidised phenolic compounds in boron deficient tissues (Reed, 1947). Spurr (1952) observed in boron deficiency a characteristic flourescences in the leaves of celery, due to the presence of phenolic substances identified as chlorogenic and caffeic acids, which accumulated around the necrotic areas in boron deficient tomato, lettuce, radish and sunflower (Perkins and Aronoff, 1956). Later two major blue coloured flourescing compounds as scopolin and glucose derivatives of gentisic acid in boron deficient sunflower leaves was also identified (Watanabe *et al.*, 1964). Some other compounds that have been identified are, e.g., 2,5-dihydroxybenzoic acid and a glycoside of 2, 3, dihydroxybenzoic acid (Troitskaya *et al.*, 1971). Increased levels of flavonals, catechins leucoanthocyanins in buck wheat (Shkol'nik and Abysheva, 1971) flavonoids in dicotyledonous plants (Carpena *et al.*, 1984; Zornaza and Esteban, 1984) and in *Amelandria melanocarpa* (Martynov, 1980) were obtained in boron deficiency. Some other phenolic compounds are also reported to accumulate in low boron conditions are nicotine, alkaloids, aromatic amino acids, and total phenols (Steinberg, 1955; Grinkevich *et al.*, 1970; Susarezyk, 1975; Krupnikova and Smirnov, 1981). A comparative picture of differential susceptibility of several boron deficient plants was drawn specially on the basis of the presence of increased total phenolic concentrations (Krupnikova and Smirnov, 1981).

Suggestions on the role of boron as borate in partitioning and metabolism between glycolytic and pentose shunt pathway have also been made (Lee and

Aronoff, 1967). It was further observed that in association with 6-phosphogluconic acids a substrate is formed which inhibits the action of 6-phosphoglucorate dehydrogenase. In the absence of borate the inhibition of enzyme is removed and excess phenolic acid is formed resulting in necrosis and death of the tissue. The inhibition of growth of plants in boron deficiency has been attributed to the accumulation of phenolic compounds (Shkol'nik, 1974). Boron protects the plants against accumulation of quinonic products of phenolic oxidation as the homogenates of boron deficient plants form quinones from exogenously supplied polypherols (Shkol'nik and Kruphikova, 1981). In boron deficient tomato, an increase in total flavonoid in fruits and leaves is an indicative of the possible interaction of boron with phenylalanine and phenylalanine ammonia lyase responsible for flavonoid biosynthesis (Carpena *et al.*, 1979, 1982; Lopez-Andreu *et al.*, 1988). In another view, it has been suggested that borate plays a potential role in regulating the hydroxylase and oxidase activities of phenolases involved in the biosynthesis of caffeic and hydroxyferulic acids (Lewis, 1980a).

The absence of caffeic acid and esculenatin in tomato could be used as an indicator of boron toxicity (Revilla, 1985). Boron plays an important role in xylem formation and lignification (Skok, 1958, 1968). Lignin metabolism is affected indirectly as boron is involved in the synthesis of aromatic structure of phenols from glucose which is transformed into lignin (Acerbo *et al.*, 1973). Boron has been suggested to play a primary role in the biosynthesis of lignin in conjuction with auxin in the differentiation of xylem (Lewis, 1980b).

7.3.4. Nitrogen Metabolism

In a number of plant species boron deficiency is known to decrease protein nitrogen and increase non-protein nitrogen (Wadleigh and Shive, 1939; Baumeister, 1941; Brigg's, 1943; Scripture and McHargue, 1943, 1944; Kibalenko *et al.*, 1973; Agarwala *et al.*, 1978, 1991; Saeed and Woodbridge, 1981).

In pea roots the increase in non-protein nitrogen was attributed to the decrease in the rate of protein synthesis in boron deficiency (Rapota, 1970; Saeed and Woodbridge, 1981) and increased concentration of free amino acids (Steinberg, 1956; Tso and McMurtrey, 1960; Tso and Engelhaupt, 1963; Pillar, 1969; Shiralipour *et al.*, 1969; Carpena *et al.*, 1979; Agarwala *et al.*, 1978; Pedash and Sereda, 1979, 1983). Total nitrogen content was lowered in stalks and increased in roots of oats in boron deficiency (Voica and Mihaila, 1982). On the other hand, it has been reported that boron supply to boron deficient soils increased protein N, absorption and transformation of N, nodulation and N_2-fixation in red clover roots (Ma *et al.*, 1990). In wheat (Dami *et al.*, 1970) and rice (Swamy, 1980) it has been observed that protein content increased with an increase in boron supply. A negative linear relationship was observed between boron and non essential amino acids except histidine and tyrosine (Iqtidar and Rehman, 1984). In boron sufficient sugarbeet the quality of protein fractions was improved and there was inhibition in the synthesis of individual protein groups in boron deficiency and toxicity (Kibalenko *et al.*, 1973). Some amino acids, viz., aspartic acid, glutamic acid, arginine, glycine, α-alanine, serine, leucine and methionine are known to accumulate in boron deficiency (Gopal, 1971; Agarwala *et al.*, 1978). Highest increase was obtained in the concentration of glycine and aspartic acid in boron deficiency and threonine and glycine in boron toxicity in tomato (Carpena *et al.*, 1979).

In deficiency and toxicity of boron, in sugarbeet and sunflower roots and shoots, a marked accumulation in nitrate and a decrease in nitrate reductase activity was noticed (Bonilla *et al.*, 1980). Boron application to *Rhizobium* treated clover plants resulted in an increase in total and protein nitrogen (Roden and Allen, 1982) whereas boron application caused quantitative changes in leaf amino acids (Karzhasova and Kalekenov, 1982). A foliar spray of boron to boron deficient sorghum at reproductive stage increased the nitrogen content and nitrate reductase activity (Misra *et al.*, 1991). It has been claimed that effect of boron fertilizers depend upon the nature of nitrogen fertilizers (Starovoitova, 1983). Boron plays an indispensable role in nodule formation, nitrogen fixation, nitrogen assimilation in grain legumes of soybean (Srivastava *et al.*, 1982). Reports are also available on an increased N concentration in vascular tissues of tomato in boron deficiency (Penalosa *et al.*, 1987).

7.3.5. Nucleic Acids

Boron in many ways influences the nucleic acid metabolism in plants (Dugger, 1983). The deficiency of boron causes a decrease in RNA content in roots (Whittington, 1957; Agarwala *et al.*, 1991) and root tips (Albert, 1965). Complete cessation of RNA synthesis was reported in the root tip cells of pea grown in boron deficiency (Rapota, 1970). Application of boron is also known to increase the RNA content in barley (Dami *et al.*, 1970).

High temperature along with boron deficiency resulted in an increased ratio of purine and pyrimidine nucleotides in total and ribosomal RNA in sunflower roots and stem tips (Bozhenko *et al.*, 1973).

Addition of RNA to boron deficient medium alleviated the effects of boron deficiency in sunflower and beans (Shkol'nik and Soloviyova Troitzkaya, 1962; Shkol'nik and Mayevskaya, 1962). Boron deficiency is known to increase the incorporation of labelled precursors into RNA in the root tips of *Phaseolus* (Chapman and Jakcson, 1974). An increase in the rate of labelling within 6-10 h after transferring the roots to boron deficient medium was further noticed. The decrease in the RNA level was observed concurrent with the appearance of boron deficiency symptoms and cessation of the root growth. Addition of nucleoside base analogues to the cultures of cotton ovules led us to believe that boron deficiency symptoms were not related to a reduction in the synthesis of nucleic acids but caused by a reduction in UDP glucose synthesis (Birnbaum *et al.*, 1977). On the other hand, no appreciable change was observed in the DNA content of boron deficient and sufficient roots (Whittington, 1957) but involvement of boron in DNA synthesis can also not be ruled out (Parmar and Hammond, 1971). Boron deficiency inhibited DNA synthesis, mitosis and eventually elongation of the roots of sunflower (Moore and Hirsch, 1983) and in squash (Krueger *et al.*, 1987) was observed to stop. Absence of adequate boron increases the rate of incorporation of 32p into nucleotides in the seedlings of broad bean (Cory *et al.*, 1966). On the other hand boron sufficient roots of *Vicia faba* showed higher incorporation of 32p phosphate into nucleotides (Robertson and Longhman, 1974). Treatment of intact roots of squash to boron free solutions for 20 h reduced the rate of incorporation of tritiated thymidine into DNA and root elongation was ceased within 6.5 h (Cohen and Albert, 1974). When boron was resupplied, the cell of squash roots resumed their ability to synthesize DNA, whereas cell division and root elongation were restored later (Lovatt *et al.*, 1981). Krueger *et al.* (1987) suggested that boron is required for continuous DNA synthesis, cell division and root elongation. It has been further established that there is requirement of boron in the first hour of germination of pollen grains of *Petunia* for the formation and supply of precursors of DNA before the onset of pollen tube growth.

7.3.6. Growth Hormones

Evidences indicate that in the deficiency of boron there is accumulation of auxins. It has been suggested that boron plays a significant role in auxin metabolism (Dyar and Webb, 1961). When auxin was applied to the necrotic growing points in boron deficient plants, it ameliorated the effects of boron deficiency in sunflower roots, the effects of boron deficiency, viz., increase in the thickness of cell wall and loss of membrane integrity (Hirsch and Torrey, 1980). Further observations were made on a correlation between the elevated levels of endogenous IAA and boron deficiency symptoms in sunflower (Hirsch *et al.*, 1982). In contrast to this, no interaction was obtained between boron and the endogenous auxin level in tomato pollen germination (McLeod, 1975). But in another observation it was reported that the growth retardation in sugarbeet could be corrected by foliar application of IAA and GA (Agarwala *et al.*, 1978).

The internal browning in boron deficient *Barbera berries* has been attributed to higher levels of auxins (Scienza *et al.*, 1981). They reported higher than normal auxin and ABA content immediately after berry set. The contents of IAA and ABA were reduced in the shoot tips just after transferring them to boron deficient medium (Fackler *et al.*, 1985). Boron also affects cytokinins as its synthesis was found to be decreased in sunflower roots supplied with insufficient boron (Wagner and Michael, 1971). Inhibition in ethylene production and delayed senescence were observed when cut flowers of *Dianthus* were kept in a solution of excess (500 ppm) sodium borate solution (Camprubi *et al.*, 1981). High doses of boron also resulted in accumulation of growth inhibitiors such as phenolics and disturbances in hormonal regulation (Shkol'nik and Abysheva, 1982) which were caused by accumulation of auxin and oxidised phenols in that region (Coke and Whittington, 1968; Shkol'nik, 1974; Rajaratnam and Lowry, 1974; Robertson and Loughman, 1974; Crisp *et al.*, 1976). The reaction of plants to boron deficiency has further been relatea to the total level of IAA in the cells (Pilipenko, 1983). Apart from this, a decrease in free auxins and an increase in bound auxins was found in boron deficient sunflower (Krupnikova and Davydova, 1972). It has been suggested that boron protects the IAA oxidase system by complexing the inhibitors of the enzymes (Coke and Whittington, 1968). The activity of IAA oxidase was affected differently in plants differing in boron requirement (Krupnikova *et al.*, 1975). It was further deduced that boron deficiency affected the accumulation of inhibitors of IAA oxidase.

Reports have been made on the enhancement of appearance of visible boron deficiency symptoms and changes in growth by the application of GA(Skok, 1961). Parmar and Hammond (1971) studied the effect of boron and GA on the senescence of leaf discs of tomato and found that the application of boron and GA separately or together to plants deficient in boron delayed the degradation of chlorophyll a and b, protein, DNA and RNA and decrease in fresh weight and disc diameter. In addition, some reports on the effect of foliar application of boron overcoming the negative effect of GA on photosynthetic apparatus of plants by enhancing the decreased chlorophyll *a* and *b* and protein content are also available (Ashour and Thalooth, 1974).

Boric acid when applied to *Pinus contorta* cuttings along with NAA and ABA increased the rooting but remained ineffective when applied alone (Bowen *et al.*, 1975).

7.3.7. Enzymes

So far boron is not reported to be an integral part of any enzyme protein but there are numerous reports that it influences the activity of many enzymes. In the deficiency of boron a decrease was observed in the activities of peroxidase in sunflower (Dutta and McIlrath, 1964), amylase and invertase in potato (Buzover, 1951), acid phosphatase, invertase, and starch phosphorylase in maize pollen (Agarwala *et al.*, 1981), nitrate reductase in sugarbeet and tomato (Carpena *et al.*, 1978; Bonilla *et al.* 1980; Ramon *et al.*, 1989) and starch phosphorylase in cowpea (Chatterjee *et al.* 1990). In contrast to this, from histochemical studies it was inferred that boron deficiency resulted in a stimulation in the activities of phosphatase, polyphenol oxidase (Reed, 1947) and catalase in squash (Alexander, 1942). Nason *et al.* (1952) reported in tomato an increased activity of polyphenol oxidase and peroxidase and so also in beans (Odhnoff, 1957).

The behaviour of activity of phenyl alanine ammonia lyase in different plant species was variable thus indicating differential susceptibility of the test plants subjected to boron deficiency (Shkol'nik *et al.*, 1980). The enzyme activity was enhanced in boron deficiency in susceptible plants such as beans, cucumber and corn and was not affected in less sensitive plants such as wheat and oat. Similarly, Shkol'nik *et al.* (1981) further reported that the activity of polyphenol oxidase was increased in all dicotyledonous plants and corn and was not affected in boron deficient wheat and sorghum. In *Zea mays* seedlings, enhancement of quinone level in the deficient and excess supply of boron may be due to the increase in ATPase activity (Besong and Lawanson, 1991). Whereas a direct relationship was observed between a supply of 0. 1 to 10 ppm B and increase in the activity of invertase (Bailey and McHargue, 1944). Presence of boron in the germinating medium of *Themeda triandra* seeds not only increased the RNA level but also stimulated the activity of α-amylase (Cresswell and Nelson, 1973). Similarly, treatment of corn seeds with 0. 001% H_3BO_3 enhanced the activity of nitrate reductase enzyme.

In vitro boron addition inhibits aldehyde oxidase and tyrosinase (Winfield, 1945), starch phosphorylase (Dugger *et al.*, 1957; Scott, 1960), reaction between UDPG and fructose in sucrose synthesis (Dugger and Humphreys, 1960) and the activity of phosphoglucomutase (Loughman, 1963). The activity of polyphenol oxidase increased in the leaves of plants grown in nutrient solution enriched with H_3BO_3 (Smirnov, 1978). Excess supply of boron (10 mM) favours the conversion of glucose-1-phosphate to uridine biphosphate glucose by UDPG pyrophosphorylase (Dugger and Humphreys, 1960). In *Crotolaria juncea* treated with boric acid, the activities of several enzymes increased to a variable extent. The enzymes that were affected are; succinic dehydrogenase, malic dehydrogenase, glucose-6-phosphate dehydrogenase, acid phosphatase, B-glycosidase and ß-galactosidase. Boron is also known to inhibit the activity of lipoxygenase in 6 day old sunflower cotyledons (Belver and Donaire, 1983). In mature curd of cauliflower, pre transplant incorporation of borax increased α-amylase, and decreased catalase, peroxidase, polyphenol oxidase activity (Singh *et al.*, 1988).

7.4. Reproductive Physiology

Boron plays an important role in the development of reproductive parts in plants. Generally the concentration of boron in plants is comparatively higher in flowers, anthers, ovary and stigma (Gauch and Dugger, 1954; Syworotkin, 1958). The deficiency of boron is known to cause reduction in flower production, bloom diameter and increased incidence of calyx splitting (Adams *et al.,* 1980). The flowers formed in boron deficiency may either develop into fruits parthenocarpically (Scott, 1944) or become malformed and show a variety of disorders (Hewitt, 1963, 1983). In *Cola nitida* boron deficiency resulted in abnormalities in floral parts, preponderance of female flowers, reduced fruit setting and parthenocarpy (Jacob *et al.,* 1972). Premature dropping of fruits of peach (Kim *et al.,* 1970) and other trees/plants is a common feature of boron deficiency. Pod and seed development is more sensitive to deficiency of the element than the development of flowers (Bell *et al.,* 1990). Boron deficient pods caused withering and abscission and decreased the number of fully developed seeds in green gram grown in solution culture. Lopez-Andreu *et al.* (1988) reported that boron deficiency not only reduced the number of fruits per plant but also reduced the weight, diameter, reducing sugar content and acidity in fruits of tomato when grown in solution culture. On the ohter hand, application of boron increased the fruit weight, sugar accumulation in fruits, seeds setting and vitamin content of fruits (Badawi *et al.,* 1981, Zaki *et al.,* 1981). Foliar spray of boric acid at bud formation stage increased the seed yield in lucerne (Krivoshei and Kononovich, 1978), red clover (Eriksson, 1979), rape (Teutebeng, 1979) and alfalfa (Radeva and Kudrev, 1984). Similarly, in *Medicago sativa,* foliar spray of boron at the end of flowering increased the 100 seed weight per plant in *Onobrychis arenaria* and soybean (Radeve *et al.,* 1991; Buzetti *et al.,* 1990).

Boron deficiency affects the male reproductive parts more than any other plant part. Lohnis (1937) reported atrophied anthers in cereals deficient in boron. In another report (Agarwala *et al.,*1981) it has been observed that in boron deficient maize, the emergence of tassels and anthers was delayed and suppressed, the stamens lacked sporogenous tissue and appeared as staminodes or floral appendages. Stamens of boron deficient maize plants showed a marked decrease in pollen producing capacity, pollen size and pollen germination. Pollen grains of plants marginally deficient in boron also showed poor germination.

Various compounds of boron were found to be beneficial for the germination and growth of pollen tube (Stanley and Lichtenberg, 1963; Vasil, 1964). Among the various concentrations of boron, 5 ppm boron was found to be most suitable for stimulating pollen germination in *Amaryllis hybride* (Stanley, 1961), 0.1 ppm boron was found optimum for stimulation of pollen germination in *Ginkgo biloba* (Roy *et al.,* 1966).

Reports are also available on reduced number of flowers, buds, pollen viability and fruit set in B excess (Arora *et al.*, 1982). On increasing boron supply, pollen germination, number of seeds/pod and seeds/inflorescence increased but pollen fertility decreased. Pollen size and 1000 seed weight were not affected by boron concentration (Misra and Patil, 1987).

7.5. Interaction of Boron with other Elements

Nitrogen

A highly significant interaction was observed between boron and nitrogen in *Carica papaya* (Perez-Lopez and Reyes-Juredo, 1983).

Sugarbeet plants grown in nutrient solution deficient in boron accumulated nitrogen in the form of soluble nitrogen compounds and nitrate but not the protein content which showed a decline (Kibalenko *et al.*, 1973). Radish plants supplied with ammonium nitrate were able to absorb and accumulate more boron than when supplied with nitrate (Agarwala *et al.*, 1977). Fertilization of boron decreased the nitrate content of forage plants (Starovoitova, 1983) and total nitrogen in all parts of wheat plants (except roots) when applied along with phosphorus (Dami *et al.*, 1971).

Nitrogen deficiency in sugarbeet decreased the tissue boron content (Reeve and Shive, 1944). Nitrogen application increased the boron concentration in cauliflower leaves (Cutcliffe and Gupta, 1980). On the other hand, Willett *et al.* (1985) reported a decrease in boron concentration and uptake by the application of nitrogen to lucerne plants grown in boron deficient soils in a glass house. In boron deficient soils, application of boron fertilizer and nitrogen fertilizer increased the yield and lipid content of rape seed (Zhu and Zhao, 1990).

Phosphorus

Phosphorus improves the efficiency of boron use when applied along with boron (Nelyubova and Sychov, 1969). Grant and Gupta (1972) failed to observe any interaction between the two elements in corn.

Reports reveal that boron deficiency affects phosphorus metabolism. Reduced phosphate absorption was observed in *Vicia faba* in boron deficiency which was counteracted by the addition of 10 µg B for 1-2 h (Robertson and Loughman, 1973). Teare (1974) also reported a decrease in phosphorus content specially that of inorganic phosphorus in the roots of boron deficient *Phaseolus vulgaris*. Boron concentration was found to decrease with an increase in phosphorus supply (Yamanouchi, 1980) or when lime was applied with high phosphate (Adams and Winsor, 1974). Leaves of spanish peanut plants showing boron toxicity symptoms had low phosphorus content (Blamey and Chapman, 1979). In maize plant, effect of phosphorus deficiency (i.e., reduction in dry matter, soluble protein, DNA, activity of ribonuclease and increase in the activities of peroxidase, acid phosphatase and polyphenol oxidase) were intensified by combined deficiency of boron and phosphorus. The effect of boron deficiency (i.e., reduction in dry weight, leaf boron and DNA and increase in starch content and in the activities of starch phosphorylase, peroxidase and polyphenol oxidase) become more intense in deficient B-excess P. The decrease caused by excess phosphorus (i.e., dry weight, starch and sugar content, DNA, RNA and activity of ribonuclease) were aggravated by combined excess boron and phosphorus (Chatterjee *et al.*, 1990).

Potassium

Alvarez-Tinaut *et al.* (1979) reported that boron deficiency influences the absorption and distribution of potassium in tomato plants. Total potassium in cotton plants was found to be decreased in both low and high boron (Liu *et al.*, 1987). Pinero *et al.* (1977) suggested that boron supply alters the balance between potassium, boron and calcium. No significant interaction was observed between boron and potassium in spanish peanut (Hill and Morrill, 1975).

Reeve and Shive (1944) found a decrease in boron concentration in potassium deficiency. Potassium application at higher rate depressed the concentration of boron (Salinas *et al.*, 1986; Liu *et al.*, 1987). Xie *et al.* (1989) reported that boron and high potassium application decreased tha rate of dry matter and starch accumulation but had no effect on protein and amino acid content in sweet potato.

Calcium

Boron-calcium interaction is one of the most widely worked out interaction of boron than with other elements. Brenchley and Warington (1927) reported that presence of boron helps the utilization of calcium in plants. The uptake of calcium was found to be promoted by boron addition to the medium (Warington, 1934) and which was accompanied by increase in the growth of the plants (Minarik and Shive, 1939, Handerson and Veal, 1948). Talibi (1935) observed depression in calcium uptake in flax supplied with boron. Boron deficiency is known to slightly increase the total calcium in tomato but it inhibited calcium translocation and brought about abnormal changes in calcium metabolism of the cell wall (Yamanouchi *et al.*, 1986). No effect of boron deficiency was observed on tissue calcium by Reeve and Shive (1944) and Hernandez-Medina and Shive (1946) in tomato and corn respectively. In siberian millet, tissue calcium was found to be depressed by excess boron (McIlrath and de Bruyn, 1956). Blamey and Chapman (1979) observed an increase in calcium content in spanish peanuts as the plants developed boron toxicity symptoms, whereas, total as well as soluble boron content decreased at excess B-excess Ca. Neales and Hinde (1962) observed that increase in boron supply neither increased the calcium content nor counteracted the effect of calcium deficiency in bean radicle.

Boron application decreased the calcium content in *Brassica napus* (Myers *et al.*, 1983), barley (Singh and Singh, 1984) and groundnut (Patel and Golakiya, 1986). Boron uptake in groundnut increased by the application of boron but when calcium carbonate was also applied with boron it was reduced (Patel and Golakiya, 1986). Leaf tip burn caused by the deficiency of either boron or calcium in strawberry was corrected by the application of aqueous salt of calcium and not by boric acid to the buds and emerging leaves (Mason and Guttridge, 1974). In maize, boron deficiency effects were accentuated by calcium deficiency and alleviated by excess calcium. Low B-low Ca supply increased the activities of starch phosphorylase, peroxidase, ribonuclease and polyphenoloxidase and decreased the biomass, leaf boron and calcium, sugars and starch contents. Excess calcium increased tissue Ca and B and reduced the concentrations of sugars, starch, nucleic acids, total and inorganic phosphorus and the acitivites of starch phosphorylase and ATPase at low and normal levels of boron (Chatterjee *et al.*, 1987).

Excess calcium has been reported to decrease total and soluble boron in a number of plant species (Reeve and Shive, 1944; Hernandez-Medina and Shive, 1946; McIlrath and de Bruyn, 1956; Yamanouchi, 1980; Sherrel, 1983). Reports are also there, that high levels of lime applied to the soil accentuated boron deficiency symptoms (Gupta and Cutcliffe, 1972, Adams and Winsor, 1974). Application of high calcium reduced the development of internal brown flecks in potatoes grown in boron deficiency in sand culture and decrease in calcium level caused increased incidents of brown flecks (Combrink and Hammes, 1972). Hemphill (1982) reported a decrease in leaf boron and increase in incidence of canker in *Beta vulgaris* by the application of lime in the absence of applied boron.

Calcium when applied as gypsum to the soils lowered the effects of boron deficiency (Gupta and Chandra, 1972; Morrill *et al.*, 1977). Cooper and Thompson (1972) reported that increasing calcium supply could correct the internal bark necrosis in delicious apple trees. Hill and Morrill (1975) observed significant interaction between boron and calcium in green house condition. On the other hand, the negative effects of boron deficiency on pod yield in pea could not be corrected by applying high calcium, which also caused decrease in other yield parameter (Salinas *et al.*, 1985). de Bruyn (1966) reported that calcium promotes pollen germination and also counteracted the toxic effects of high boron. Pollen germination is enhanced in the absence of calcium (Samaranayake *et al.*, 1979).

Magnesium

Yield of many crops, viz., clover, alfalfa, peas, barley, flax, sugarbeet, forages and grasses was increased when boron was supplied along with magnesium rather than when supplied alone (Anspok, 1971). Almost similar effects of interaction of Mg with boron was observed in potato grown in sal podzolic sandy loam soil of Polesic (Al'Shevskii *et al.,* 1986). Yamanouchi (1980) observed that boron conent in the leaves of tomato, sunflower, sugar-beet, radish, cucumber and soybean decreased by increasing magnesium content in culture solution. Application of boron was reported to decrease the magnesium content in barley (Singh and Singh, 1984).

Iron

Iron is known to accumulate in older leaf blades in boron deficiency (Ohki, 1975). An excess supply of boron (10 ppm) caused a decrease in the availability and distribution of iron in groundnut (Gopal, 1971). In rice and soybean, boron concentration decreased with an increase in iron supply (Yamanouchi, 1980).

Manganese

Boron-Manganese interaction was observed by Ohki (1973) in *Gossypium hirsutum* grown in nutrient solution in a green house. He (Ohki, 1974, 1975) further added that low boron supply increased the critical levels of manganese in old leaves and induced low manganese concentration in young newly formed leaf tissue. Enhanced manganese uptake in boron deficient tomatoes has been observed and it was further reported that manganese becomes relatively immobile in the roots at excess boron (Garate, 1984). Leaves of spanish groundnut showing boron toxicity symptoms contained high concentration of manganese (Blamey and Chapman, 1979). In cauliflower, Chatterjee *et al.* (1989) reported that manganese deficiency symptoms were aggravated by low boron under manganese deficiency. Low Mn - low boron decreased biomass, curd formation, leaf Mn, sugars, starch, chlorophyll, Hill reaction, specific activity of aldolase and increased the concentration of proetin and inorganic phosphorus. At excess Mn - excess B, plants showed boron toxicity symptoms and increased the concentration of leaf Mn, sugars, nucleic acids, phosphorus fractions and leaf boron. In boron deficiency, excess manganese decreased the concentration of leaf boron, leaf Mn, DNA, protein nitrogen, chlorophyll and activities of peroxidase and aldolase.

Copper

Translocation of copper has been reported to be stimulated by boron excess but its uptake was not affected by boron levels (Alvarez-Tinaut *et al.,* 1979). Gomez *et al.* (1981) did not find any correlation between boron level of the medium and copper concentration in leaves of sunflower.

Zinc

Boron deficiency has been reported to decrease the zinc concentration in cotton (Ohki, 1975) and its uptake in tomato (Alvarez-Tinaut *et al.*, 1979) and ***Phaseolus vulgaris*** (Pilipenko and Solovyeva, 1979) but the translocation of zinc was stimulated by excess boron (Alvarez-Tinaut *et al.*, 1979). Blamey and Chapman (1979) reported an increase in zinc concentration with the development of boron toxicity symptoms in the leaves of spanish groundnut. Chandel *et al.* (1985) made a radioactive study on the zinc and boron interaction in green gram and maize. They observed maximum uptake of ^{65}Zn at 0.1 ppm boron supply. Zinc application decreased the boron content and high rates of boron decreased the zinc content of the shoot.

8. CHLORINE

Chloride as an essential micronutrient occurs in plants as a simple anion (Cl^-), in concentration of 30 to 40 μmol g^{-1} fresh weight, which is similar to that of macronutrients. However, the requirement of Cl^- for optimal growth is considerably lower (< 0.3 μmol g^{-1} f. wt.) (Marschner, 1986; Mengel and Kirkby, 1987). The work of Nobbe and Siegert (1862), Lipman (1938) and Eaton (1942) indicated that perhaps Cl is an essential element for plants. Broyer *et al.* (1954), for the first time, demonstrated the essentiality of chloride as a micronutrient in tomato. Ozanne (1958) has observed chlorine deficiency under field conditions in California and West Australia. Chloride is known to play an important role in the growth and metabolism of plants as it is associated with chloroplast activities specifically with PS II (Hind *et al.*, 1949), ATPase activity (Hager and Helmle, 1981) and turgor regulation (Marschner, 1986). In chloride deficiency, leaf wilting might be due to improper water uptake or inhibition of transpiration process (Marschner, 1986). In sensitive crops, chloride concentration (0.5 to 2%) is associated with reduction in yield and quality while for tolerant species (Reisenauer *et al.*, 1973) it is more than 4 % in dry matter. In most plant species, chloride is taken up very rapidly and its uptake is metabolically controlled (Mengel and Kirkby, 1987). Gerson and Poole (1972) have reported that the uptake of chloride occurs against concentration gradient. Chloride uptake is an active process (MacDonald *et al.*, 1975) required ATP formed during photosynthetic phosphorylation, as energy source. In corn root protoplasts, Lin (1981) suggested that its uptake is a carrier mediated process.

Chloride is not only taken up by roots, it may also be absorbed by aerial plant parts as chloride or chlorine gas (Johnson *et al.*, 1957).

Marschner (1986) reported critical deficiency level of chloride in plants in the range of 2-20 μ mol Kg^{-1} in dry matter.

In plants normally chloride accumulated in old leaves but under deficient conditions it is translocated to younger leaves (Woolley *et al.*, 1958). Chloride plays an important role in photosynthesis as it influences the O_2 evolution and water splitting system at photosystem II (PS II) (Arnon and Whatley, 1949; Hind *et al.*, 1949; Bove *et al.*, 1963; Kelley and Izawa, 1978; Ball *et al.*, 1984). It has been shown by Kelley and Izawa (1978) that chlorine acts as a cofactor in Mn containing O_2 evolving system :

$$H_2O \xrightarrow[Mn+Cl]{O_2} PS\ II \xrightarrow{e^-} PS\ I \xrightarrow{e^-}$$

In sugarbeet, Terry (1977) reported that growth reduction in chloride deficiency might be due to reduced rate of cell multiplication in leaf blades which in turn reduced the leaf surface area and consequently the plant growth but not due to net photosynthesis per unit of chlorophyll. Chloride deficiency is known to inhibit stomatal opening and impair the water uptake, thus, indirectly influencing the rate of photosynthesis in onion (Schnabl, 1980). Inhibition of synthesis or degradation of proteins in chloride deficiency resulted in accumulation of certain amino acids and amides (Frency *et al.*, 1959). Application of chloride decreased both total N and nitrate nitrogen in tops although protein nitrogen was not affected (Muraka *et al.*, 1973).

Clarkson and Hanson (1980) have suggested that Cl_2 is known to play an important role in cation neutralisation.

Application of chloride specifically stimulated the activity of H^+ - pumping ATPase located in membranes of cytoplasm vesicles or in the tonoplast or both (Hager and Helmle, 1981; Mettler *et al.*, 1982; Churchill and Sze, 1983). This chloride stimulated H^+ - ATPase operates as an electrogenic H^+ - pump at the tonoplast that transports H^+ from the cytoplasm into the vacuole (Churchill and Sze, 1983) thereby creating the typical pH gradient between the two compartments (cytosol, $pH > 7$; vacuole $pH << 6$).

Chloride acts as an specific activator of an ATPase activity at 10 to 50 mM concentration in the gland of *Limonium vulgare* (Hill and Hill, 1976). Chloride also stimulates the activity of asparagine synthetase which used glutamine as a substrate :

$$\begin{matrix}\text{Glutamine +} \\ \text{aspartic acid}\end{matrix} \xrightarrow[\text{asparagine synthetase}]{NH_3} \begin{matrix}\text{asparagine +} \\ \text{Glutamic acid}\end{matrix}$$

Chloride also increases the affinity of the enzyme asparagine synthetase by a factor of 50 (Rognes, 1980), therefore it plays a specific role in nitrogen metabolism (Marschner, 1986).

Chloride is one of the major osmotically active solutes in the vacuoles thus affecting the turgor potential of leaves or of specialised tissues such as the pulvini of *Mimosa* (Kumon and Tsurumi, 1984).

9. COBALT

The essentiality of cobalt was established in 1948 when Rickes *et al.* showed it to be a constituent of B_{12} or cyanocobalamin which is required by all animals. Cobalt has been shown to be essential for many lower plants but not for all higher plants (Bollard, 1983). Holm-Hansen *et al.* (1954) found cobalt to be essential for the growth of fresh water blue green algae. Cobalt is also known to be essential for symbiotic nitrogen fixation in soybean, lucerne, subterranean clover in presence of *Rhizobium* (Ahmad and Evans, 1960; Reisenauer, 1960), in Azolla in the presence of *Anabaena azollae* (Johnson *et al.*, 1966) and in non-leguminous plants *Alnus* and *Casurina* (Hewitt and Bond, 1966). There was apparently no requirement for cobalt when nitrate was supplied as a nitrogen source. In root nodules the metal requirement is usually associated with the production of cobamide compounds by the *Rhizobia* (Nicholas, 1975). For the first time Kliewer and Evans (1963a, b) isolated the cobalamin coenzyme B_{12} from the root nodules of legumes and non-legume plants and established a close correlation between cobalt supply, the B_{12} co-enzyme content of *Rhizobium*, the formation of leghaemoglobin and N_2-fixing micro-organisms have an absolute cobalt requirement.

In co-enzyme cobalamin (Vitamin B_{12} and its derivaties) Co (III) is the metal component which is chelated to four nitrogen atoms at the centre of a porphyrin structure similar to that of iron in hemin. This Co complex provides a prosthetic group which is associated with a nucleotide in the B_{12} co-enzyme.

In *Rhizobium* three specific cobalamin dependant enzyme system function which influence the nodule formation and nitrogen fixation in legumes (Dilworth *et al.*, 1978). These enzymes are :

(1) Methylemalonyl - coenzyme A mutase - it is involved in the synthesis of haem in the bacteroids and needed for leghaemoglobin production in nodules. This might account for the correlation between Co deficiency and a decrease in leghaemoglobin concentration in the nodules (Gladstone *et al.*, 1977)

(2) Ribonucleotide reductase is involved in the reduction of ribonucleotides to deoxyribonucleotides. Suggestions indicate that in Co deficiency not only defective DNA synthesis occurs but the cell division is depressed and cell size increases in *Rhizobia* (Chatel *et al.*, 1978).

(3) Methionine synthetase which is involved in protein synthesis and which is impaired in Co deficiency.

Evidence put forward by Hallsworth and co-workers (1965) suggest that Co is required by higher plants because by adding very low concentration of Co, a significant increase in the dry weight yield of non-nodulated subterranean clover was obtained. Some observations on Co requirement has also been made on the detection of B_{12} on the aseptic cultures of pea, wheat and lupins (Fries, 1962). But the work of Wilson and Nicholas (1967) is the first report of Co deficiency in higher plants. It is known that cobamide compounds are involved in several enzymatic reactions involving group transfer. Wilson and Nicholas (1967) provided evidence of cobalt requirement for subterranean clover and wheat in which optimum requirement was found to be 0.006 ppm. Vitamin B_{12} controls the biosynthesis of leghaemoglobin in plants. Leghaemoglobin can not be synthesized either by bacteria or by the host alone in absence of cobalt (Robson *et al.*, 1979; Dilworth *et al.*, 1979).

Yamada (1958) reported that cobalt is essential for germinating pollen of *Lilium* and optimum concentration is 30 ppm which is toxic for most plant species.

9.1. Toxicity

Cobalt (even in very small amounts) is known to be highly toxic to plants. It has been reported that in nutrient solutions a level of 0.5 ppm cobalt decreases the plant growth (Brenchley, 1938, Vergnano and Hunter, 1952; Nicholas and Thomas, 1953; De Kock, 1956; Rauser, 1978). Less than 0.1 ppm coblat produces toxicity symptoms in a number of crop plants. Most of the reports on cobalt toxicity are available from culture experiments (Hewitt, 1948, 1953; Nicholas, 1950; Nicholas and Thomas, 1953; De Kock, 1956; Agarwala and Kumar, 1962; Bisht, 1972; Bisht *et al.*, 1976; Agarwala *et al.*, 1977; Rauser, 1978; Bisht and Agarwala, 1982). Excess cobalt has not been so far reported to be harmful to animals and human beings (Peterson and Girling, 1981). But excess amounts of cobalt is reported to cause accumulation of such high concentrations of the element in plants which upon entering the food chain, may pose health problems. Young (1979) reported numerous diseases in man and animals caused by high concentration of cobalt compounds. Excess Co decrease the ability of thyroid to accumulate iodine and a result of which goiter may occur in man (Martin, 1986). But it is not known why cobalt is toxic to plants.

10. Nickel

Nickel has been shown to be an essential element for bacteria (Thauer *et al.*, 1980), and certain strains of blue green algae (Van Baalen and O'Donell, 1978). It is not generally regarded as essential for plants or for animals. Some reports are available on the favourable role of Ni in low concentration on the germination and growth of various crop species (Misra and Kar, 1974). Dixon *et al.* (1975) reported Ni to be a component of urease from seeds of jackbean plants, the enzyme catalyses the reaction.

$$CO(NH_2)_2 + H_2O \text{ ---------------> } 2NH_3 + CO_2$$

The enzyme urease consists of six subunits with 2 nickel atoms in each subunit (Alagna *et al.*, 1984). Ni is not required for the synthesis of the enzyme protein (Winkler *et al.*, 1983) but the metal is essential for the enzyme structure and its function (Klucas *et al.*, 1983). In soybean callus culture, Ni is essential if urea is the sole source of nitrogen but not when other nitrogen sources are used (Polaceo, 1976). Some suggestions support the essentiality of Ni for plants supplied with ureides (eg., Allantoin) which play a significant role in nitrogen metabolism (Mengel and Kirkby, 1986). In *Clostridium pasteurianum* Ni influences the activity of carbon monoxide dehydrogenase (Diekert *et al.*, 1979). Ni appears to be particularly mobile in the phloem and after uptake considerable amounts of Ni are transferred to seeds and fruits (Cataldo *et al.*, 1978; Mitchell *et al.*, 1978).

10.1. Toxicity of Nickel

Toxic effects of nickel in bean and corn in solution culture was first reported by Haselhoff (1893). After that, nickel toxicity has been reported in barley (Wolfe, 1913), buck wheat (Cotton, 1930), tobacco (Spencer, 1937), bean (Brenchley, 1938), flax (Millikan, 1949) and oat (Hunter and Vergnano, 1953; Vergnano and Hunter, 1952). Nickel accumulates as citrate complexes in the ratio of 1:2 or low as Ni $(H_2O)_6$ (Lee *et al.*, 1977).

11. SODIUM

Sodium has long been known to play a beneficial role in the growth of higher plants specially when potassium was deficient. Several investigations on possible sodium requirements for plants have been made since the time of Hellriegel and Wilfarth (1898). *Atriplex vesicaria* when grown in nutrient culture condition showed specific requirement for sodium in the presence of potassium (Brownell and Wood, 1957; Brownell, 1965). Allen and Arnon (1955) found that sodium was necessary in addition to potassium for the growth of *Anabaena cylindrica.* The essentiality of sodium (Na) as a micronutrient has been established for the growth of C_4 and CAM plants by Brownell and Crossland (1972 and 1974) where sodium has been suggested to play an important role in photosynthesis as it influences the C_4 photosynthetic pathway and CAM pathway. Sodium influences the cell expansion and water balance by substituting potassium and it also stimulates growth of plants (Mengel and Kirkby, 1987).

In sugarbeet, there is an increase in the expansion of leaf segments, leaf area, thickness and succulence when high K^+ is replaced by sodium (Marschner and Possingham, 1975; Milford *et al.,* 1977).

Application of high sodium increases not only leaf area but also stomata number per unit leaf area and growth rate while rate of net photosynthesis decreases and this decrease in photosynthetic rate is caused by lower chlorophyll concentration at high sodium and low K concentration (Lawlor and Milford, 1973).

Durrant *et al.* (1978) have concluded that the application of sodium fertilizers increased the leaf area index which in turn induces light interception, thus improves the water use efficiency under moderate water stress conditions.

Marschner *et al.* (1981) have reported that sodium improves the water balance of plants under water stress conditions through stomatal regulation.

Although Na is present in the phloem tissue of higher plants but it is much less mobile than K (Raven, 1977). In different plant species, the replacement of K^+ by Na^+ differs considerably (Marschner, 1971) and it depends on the uptake potential of Na^+.

In sugarbeet, Na^+ is readily transported to the shoots where it replaces most of the K^+ and increases the growth of plant, while in beans, exclusion mechanism exists for the blocking of Na^+ transport to the shoots thus it showed no growth response (Hawker *et al.,* 1974) and this difference in growth response to sodium by different plant species is related to difference in its translocation to the shoots. In natrophobic plants, e.g., beans, sodium is retained mainly in the roots and lower stem whereas in natrophilic species, e.g., in sugarbeet it is translocated readily into the leaves (Marschner, 1986). In beans, this restricted upward Na^+ transport is caused by selective accumulation of Na^+ in the xylem parenchyma cells of the roots (Kramer *et al.,* 1977) and stems (Rains, 1969).

Sodium is involved in the shuttle of metabolites between the mesophyll and bundle sheath chloroplasts as C_4 plants are more efficient in utilizing low ambient CO_2 concentration than C_3 plants (Johnston *et al.,* 1984).

In sodium deficient *Amaranthus tricolar,* Nable and Brownell (1984) have reported accumulation of high levels of alanine.

Sodium also affects the enzyme activities which are sensitive to K^+, as it replaces the higher proportion of K^+ at the cellular level. The enzyme starch synthase, is stimulated by K^+ is also stimulated by Na^+ (Hawker *et al.,* 1974).

$$\text{Glucose - 1 P} \xrightarrow[\text{ADP-glucose phosphorylase}]{} \text{ADP - Glucose} \xrightarrow[K^+ >> Na^+]{\text{starch synthase}} \text{starch}$$

Hawker *et al.* (1974) have also reported that in sugarbeet leaves, the content of starch was lowered but sucrose content was much higher which in turn stimulated the cell expansion in the leaf tissue and phloem transport to the sinks such as storage roots of sugarbeet. Sodium is known to cause further accumulation of sucrose in storage roots and this accumulation of sucrose is related to ATPase enzyme located at the tonoplast of beet root cells (Willenbrink, 1983) and also K^+, Na^+ - AT Pase which requires both K^+ and Na^+ for their maximal activity in roots of natrophilic species (Kylin and Hansson, 1971).

In *Aerobacter aerogenes* and *Salmonella typhimurium* oxaloacetic acid decarboxylase is activated by sodium (Stern, 1967; O'Brien *et al.* 1969). At present, it is not known why Na is specifically required by C_4 plants in photosynthesis but in some halo-tolerant bacteria some enzymes of citric acid cycle are activated by sodium (Hewitt, 1983). The importance of some enzymes of carboxylic acid metabolism in photosynthesis of C_4 plants might have a bearing on this problem. The activity of phosphoenol pyruvate carboxylase is activated by sodium (Shomer-Ilan and Waisel, 1973). The sodium directs dark fixation of CO_2 into amino acids rather than carboxylic acids in cell free systems (Joshi *et al.*, 1962).

12. Iodine

Iodine (I) is an essential element for animal nutrition as it is a component of the hormone thyroxine (Nason and McElroy, 1963; Mengel and Kirkby, 1987). Hsiao (1969) reported essentiality of iodine for the growth of marine brown alga *Patalonia fascia.* For the growth of excised roots of tomato, iodine is known to be essential when supplied at a concentration of 0. 6 ppm in culture medium (Street, 1957).

Marine algae are particularly rich in iodine (Grimm, 1952). In red and brown algae, iodine is accumulated in particular cells or group of cells (Kylin, 1929; Roche and Yagi, 1952). In most of the marine algae, the level of iodine is about 10 ppm (Shaw, 1962). Response to iodine was determined by the species of red algae, some responded upto 1 ppm iodine, while others were inhibited by as low as 0.07 ppm iodine concentration (Fries, 1966). In *Laminaria digitata*, it can constitute about 1% of the tissue dry weight while in other species, iodine is found in the range of 0.3 to 0.7 % of dry weight (Shaw, 1959).

The toxicity effects of iodine could be reduced by high chloride levels which suggested that these two elements showed competitive effect (Lewis and Powers, 1941).

Application of iodide causes leaf abscission in beans and reduced ascorbic acid content in tomato (Hageman *et al.*, 1942; Herrett *et al.*, 1962).

In higher plants, most of the iodide remained in inorganic form but a small proportion is incorporated into organic compounds such as tyrosine (Boszormenyi and Cseh, 1959; Cowan and Esfahani, 1967). Many tyrosine containing compounds are known diidotyrosine in barley and mungbean (Fowden, 1959) and 3-iodotyrosine, 3, 5-diidotyrosine and 3', 5, 3'-triidotyrosine in several other plants which are closely related to thyroxine (Fowden, 1959; Mengel and Kirkby, 1987). In algae, similar compounds have also been reported by several other workers (Tong and Chaikoff, 1955; Fenical, 1975). The idotyrosines are protein bound (Bollard, 1983). In animal systems chloroperoxidase system oxidises iodide ions at some stage of its incorporation into iodotyrosine derivatives which are essential for thyroid action. Perhaps a similar system operates in the presence of plant peroxidases (Bollard and Butler, 1966).

13. ALUMINIUM

Eisenmenger (1935) for the first time reported growth stimulation by low concentration of aluminium (Al) in the culture solution. Later, similar results have been obtained in several plant species by various other workers, viz., legumes (Klimashevskii *et al.*, 1970; MacLeod and Jackson, 1965), sugarbeet (Keser *et al.*, 1975), tea (Matsumoto *et al.*, 1976a), cereals (Howeler and Cadavid, 1976; Clark, 1977) and in *Aspergillus niger* and *Chlorella* (Bertrand, 1963; Bertrand and DeWolf, 1968; Foy and Gerloff, 1972). There is some evidence that low levels of Al can have a beneficial effect on plant growth although the mechanism is not clear (Foy, 1974). But aluminium is not considered to be an essential element for plants.

Higher plants usually contain Al concentration in about 200 ppm in dry matter (Mengel and Kirkby, 1987). In tea, Chenery (1955) has reported levels of Al as high as 2000-5000 ppm and it is considered necessary for the normal growth of the tea bush.

Some other workers (Humphreys and Truman, 1964; Sivasubramanian and Talibudeen, 1971) have reported Al content in tea foliage upto 16000 ppm. Seeds generally contain very low level of Al content (Bertrand and Levy, 1931). Generally in roots Al content is much higher than in upper parts as it is fixed in the free space of root (Mengel and Kirkby, 1987; Rahardjo *et al.*, 1987). Al toxicity is associated with acid soils where the availability of both Mn and Al is very high (Schmehl *et al.*, 1950; Foy and Wheeler, 1979; Mengel and Kirkby, 1987), Al is more soluble in soil solution at low pH. This high levels of Al in soil solution is harmful to plants and especially to roots (Fox, 1979). Addition of lime to soil increases the soil pH and exchangeable Ca but it counteracted the toxicity of Al by decreasing extractable Al in soil solution (Foy *et al.*,1965; Haynes, 1984; Gupta and Singh, 1990; Shuman *et al.*, 1990; Singleton *et al.*, 1987).

Tolerance to excess Al depends on the ability of plants to modify the pH of soil root interface (Foy *et al.*, 1965; 1967). An Al exclusion mechanism also exists in Al tolerant wheat cultivars (Henning, 1975), which protects the plants from Al toxicity. In cowpea (*Vigna unguiculata)*, Horst *et al.*, (1982) indicated a mucilage layer around the root tips which prevents the young root tissues from Al toxicity by adsorbing Al. On the other hand, in tea large amount of Al stored in the older leaves is de-toxified by organic compounds (Sivasubramanian and Talibudeen, 1972).

There is also some evidence that plants supplied with NH_4 - nitrogen in culture solution contain less Al than when supplied with NO_3 as nitrogen source (Kotze *et al.*, 1977).

13.1. Interaction

The symptoms of Al toxicity often resemble those of P deficiency (Wright, 1937; Munns, 1965) or of Ca deficiency (Johnson and Jackson, 1964; Edwards and Horton, 1977) and this may be due to shortage of these elements rather than direct Al toxicity (Ligon and Pierre, 1932). Within the root tissues, Al precipitated phosphate thereby reducing the phosphate translocation (Pierre and Stuart, 1933; Lee, 1971; Mugwira, 1979). The addition of superphosphates to acid soils often improves plant growth by alleviating Al toxicity (Bollard, 1983). Aluminium application also retards the uptake of Mg^{2+} (Grimme, 1983). As it affects the plasmamembrane (Hecht - Buchholz and Foy, 1981) it may have a detrimental effect on ion uptake. High Al concentration decreases shoot and root Ca, Zn, Cu, Mn, Fe and Mo (Hiatt *et al.*, 1963; Mugwira, 1979; Spiers, 1990; Chibiliti and Byrne, 1990; Tang *et al.*, 1989) in several plant species. The reduction in level of Ca and other elements due to high Al level is the result of reduced root surface area rather than a specific Al effect (Clarkson, 1966).

Al toxicity inhibits cell division and cell wall extension (Morimura and Matsumoto, 1978; Klimashevskii and Dedov, 1975) there by resulting in cessation of root growth (Bollard, 1983).

Al is known to be adsorbed on DNA double helix thus inhibiting the separation of the strands (Matsumoto and Morimura, 1980).

Hampp and Schnabl (1974) reported inhibitory effect of Al ions on photosynthetic CO_2 fixation in *Spinach* chloroplasts. Aluminium also binds to calmodulin; a Ca^{2+} containing regulator protein thus affecting the various enzyme activities which are promoted by calmodulin (Mengel and Kirkby, 1987).

14. SILICON

In diatoms, essentiality of silicon is well established (Werner and Roth, 1983). In higher plants, the essentiality of silicon has been proved for several species of silicon accumulators such as *Equisetum arvense* (Chen and Lewin, 1969), certain wet land grasses (Takahashi and Miyake, 1977) and in rice (Lewin and Reimann, 1969).

Silicon acts as a structural component of cell (Marschner, 1986). In cell walls, silicon was deposited in the outer epidermal cells as amorphous silica or opal phytoliths with distinct three dimensional shapes (Wynn parry and Smithson, 1964; Schwarz, 1973), thus it prevents the cuticular water loss Vormvan der, 1980) and decreased the susceptibility to lodging, fungal attack and Mn and Fe toxicities (Volk *et al.,* 1958; Miyake and Takahashi, 1983; Coors, 1986). It also reduced transpiration coefficients at higher levels of Si supply (Mengel and Kirkby, 1987).

Monosilicic acid (Si $(OH)_4$) is the form of Si in which it is taken up by plants (Mengel and Kirkby, 1987). Jones and Handreck (1965) concluded that its uptake is a passive process. While Vorm van der (1980) indicated that its uptake, whether it is passive or active, depends on plant species. It is transported through xylem tissue, therefore its distribution within the plant system depends on the rate of transpiration of different plant parts (Mengel and Kirkby, 1987). In xylem sap it occurs in the form of mono silicic acid (Jones and Handreck, 1967). The rate of transpiration is 30% higher in silicon deficient plants than in plants supplied adequate silicon (Mitsui and Takatoh, 1963; Yoshida *et al.,* 1969).

As most of silicon is associated with cell wall component, its metabolic effect may be via an effect on cell wall metabolism. Silicon deficiency is known to decrease the incorporation of Pi into ATP, ADP and sugar phosphates in sugarcane (Wong You Cheong and Chan, 1973) and lignin content in cell walls of wheat root (Jones *et al.,* 1978). They also have reported accumulation of phenolic compounds in Si deficiency. Silicon occurs in plants as ester like derivatives of silicic acid which acts as bridge in the structural organization of polyuronides (Jones, 1978). It reacts with O-diphenols such as caffeic acid and corresponding esters to form mono, di or polymeric si-complexes (Weiss and Herzog, 1978; Jones *et al.,* 1978), thus it affects the stability of plants not only as an inert deposition in lignified cell walls but by modulating lignin biosynthesis. In sugarcane, Ayres (1966) reported increased cane yield and sugar content by the application of silicon.

In beans, Horst and Marschner (1978) have reported a beneficial effect of Si in alleviating the detrimental effect of Mn toxicity as it ensures even distribution of Mn within the leaf tissue but does not affect its uptake. Similar results had been observed by Vlamis and Williams (1967) and Horiguchi

(1988). Vorm van der and Diestvan (1979) have observed that, the application of silicon can also protect the plant from high concentration of Fe and Mn by reducing their uptake. In 1929, Fisher reported increased availability of phosphate in barley by the application of Si grown in soil culture. Russell (1973), Scheffer *et al.* (1982) and Fox (1978) have observed similar results when silicon was supplied as sodium silicate.

15. VANADIUM

The essentiality of vanadium for *Scenedesmus obliguus* was established by Arnon and Wessel (1953) and for *Chlorella pyrenoidosa* by Meisch *et al.*, (1977). Growth stimulation in *Chlorella pyrenoidosa* was fourfold when the iron source was $FeCl_3$ but it was negligible when iron was supplied as Fe-EDTA or ferric citrate. But absorption of iron was not involved in the process. Basiouny (1984) indicated the importance of vanadium for the growth of tomato. In aerial parts of higher plants, Bertrand (1941) determined the vanadium content and he found it is in the range of 0.2 to 4 ppm with an average of 1 ppm in dry matter. For algae and fungi it is about 1 ppm in dry matter (Bertrand, 1943; Cannon, 1963; Byrne *et al.*, 1976). The content of vanadium is much higher in *Amanita muscaria* and it is between 60 to 325 ppm (Bertrand, 1943; 1950; Byrne *et al.*, 1976; Meisch *et al.*, 1978, 1979), but it is always not so high in other species of *Amanita* (Bayer and Kneifel, 1972; Kneifel and Bayer, 1973).

In *Azotobacter,* Horner *et al.* (1942) have reported the essentiality of vanadium (or Mo) for N_2 fixation where it could substitute for Mo. On the other hand, Esposito and Wilson (1956) suggested that vanadium cannot replace Mo in *Azotobacter vinelandii.* Allen (1956) indicated that vanadium (V) is not essential for N_2 fixation in *Anabaena.*

Addition of sodium vanadate externally to *Nitella transluscense* inhibited the H+ ATPase activity in plasmalemma and depolarised strongly and permanently the membrane potential both in light and dark (Cruz-Mireles and Ortega-Blake, 1991), thus affecting P state of the alga. Vanadate supplied in growth medium markedly reduced the activity of nitrate reductase in both tomato seedlings and *Chlorella vulgaris* (Buczek, 1973; Ramadoss, 1979) where it formed a complex with the enzyme nitrate reductase (Ramadoss, 1979, 1980).

Vanadium (to be 100 μM) is known to effect adversely several physiological and biochemical processes, viz., rhythmic leaflet movement in *Albizzia julibrissin* (Satter and Fries, 1981), stomatal movement and leaf unrolling (Saxe and Rajagopal, 1981), H^+ secretion and K^+ uptake by plant cells (Cocucci *et al.*, 1980) and auxin enhanced elongation of pea epicotyls (Jacobs and Taiz, 1980).

In green algae, an increase in vanadium concentration in growth medium increased the chlorophyll content and cell production and showed maximum affect at 0.5 ppm V supply whereas toxicity of the element was obtained at 100 ppm (Meisch *et al.*, 1977). Meisch and Bauer (1978) suggested that vanadium affects the synthesis of δ-aminolevulinic acid, a precursor of porphyrins and chlorophyll, but the concentration at which vanadium promotes the synthesis of porphyrin, is found to be inhibitory for cell division (Meisch and Benzschawel, 1978). Radioactive studies have cast a doubt on the synthesis of δ-aminolevulinic acid by condensation of glycine with succinyl CoA, which is the process of synthesis of porphyrin in animal systems. It has been suggested that δ-aminolevulinic acid is synthesised by transamination of 4,5 dioxovaleric acid (Hewitt, 1983). Gassman *et al.* (1968) have reported that δ-aminolevulinic acid transaminase is present in *Chlorella vulgaris.* May be, vanadium is involved also at some step of the synthesis of porphyrin in higher plants. Stimulated photosynthetic gaseous exchange has been reported in *Chlorella* in presence of vanadium in culture solutions (Warburg *et al.*, 1955).

It influences the accumulation of intracellular starch, the content of P700 and cytochrome 'f' and carotenoids (Meisch *et al.*, 1980; Wilhelm and Wild, 1980; Becker and Meisch, 1981).

In *Neurospora crassa*, the activity of ATPase contained in the plasma membrane is inhibited in presence of micromolar vanadium quantities but not in the mitochondria (Bowman *et al.*, 1978).

In *Phaseolus vulgaris*, the activity of phosphatases were inhibited in all fractions of leaf by 10 to 100 μM concentration of vanadate (Saxe and Rajagopal, 1981). Vanadate even in micromolar concentration is a plant inhibitor of several ATPases (Cantley *et al.*, 1977; Saxe and Satter, 1979; Cocucci *et al.*, 1980) in both plants and animals. In addition to ATPases, vanadate has also been reported to affect a range of other enzymes (Simons, 1979). Vanadate is not a specific inhibitor of (Na^+, K^+) ATPase as was believed earlier (Cantley *et al.*, 1977).

16. CADMIUM

Cadmium is known to act as a cumulative poison for animals as it causes chronic disorders, emplysema and disturbances in kidney functioning. Cd is chemically very similar to zinc in its metabolic functions and uptake but Cd is however toxic to both animals and plants. It disturbs the enzyme activities as it has much higher affinity for thiol groups (SH) in enzymes and other proteins (Mengel and Kirkby, 1987).

Cadmium is one of the heavy metals which in general is most readily taken and translocated into the shoots (Diez and Rosopulo, 1976). In higher plants Cd transport into the shoots is usually directly related to its concentration in the external soil solution (Jones *et al.*, 1975; Petterson, 1976; Gotoh and Iriye, 1989; Anderson and Siman, 1991).

Cd uptake depends much on soil pH and presence of other cation species in soil solution. It was observed that its availability increases with the lowering of soil pH (Hasselbach, 1940; Oberlander *et al.*, 1989) and decreases at higher soil pH levels and lower cation exchange capacity of the soil (Herms and Brumnier, 1979).

In Oat, increased grain Cd content is associated with the pH of soil solution in the range of 4.7 to 7.4 (Eriksson *et al.*, 1990), various heavy metal ions influenced the availability of Cd in wheat, some (Fe^{3+}) enhanced its uptake while others (Fe^{2+}, Zn, Mn) reduced it. The enhancement of Cd uptake was considered to be mainly the result of displacement processes in the soil and in the root free space whereas inhibition was attributed to competition for ion carriers in the plasmamembrane and reduction in passive membrane permeability (Oberlander *et al.*, 1989).

Several plants of the Zawar zinc deposits of Rajasthan (India) accumulate very high levels of cadmium and the sequence of its accumulation in different plant organs of the same plant species was roots, stem and leaves (Aery and Tiagi, 1988). A similar pattern of accumulation of Cd have been also observed in several other plant species by other workers (Estan *et al.*, 1988; Dang *et al.*, 1990).

The translocation of Cd from the leaves into seeds is generally very low and the content of Cd in cereal grains from highly contaminated soils exceeds only 1 ppm (Sommer, 1979). On the other hand, Sela *et al.* (1988) reported similar concentration of Cd in both roots and shoots of *Azolla filiculoides*. This indicated a high mobility of Cd within the plants. Frossard *et al.* (1989) reported the highest Cd accumulation in the shoots of ryegrass. Nishizeno *et al.* (1989) examined the distribution and accumulation of heavy metal ions (Cu^{2+}, Zn^{2+} and Cd^{2+}) in the roots of *Polygonum cuspidatum* and they found that about 90% of the metal ions was located in the cell wall fraction of the roots.

Inhibition of chlorophyll biosynthesis on exposure of etiolated cowpea leaf segments to Cd ions for 24 hrs has been suggested by Mutchuchelian *et al.* (1988). In plants, excess Cd may also disturb Fe metabolism and cause chlorosis (Mengel and Kirkby, 1987).

In germinating seeds of maize plants, Cd^{2+} (10-5 mol/litre) inhibited alcohol dehydrogenase activity and Cd^{2+} was considered to interact with imidazol ring of histidine (Leblova and Zatioukalova, 1988).

The activity of ATPase of the plasmamembrane from roots of *Zea mays* was stimulated by Cd in the presence of Mg^{2+} and it is further stimulated by K^+ (50mM) addition. Here Cd-ATP is an alternative substrate for the enzyme ATPase (Kennedy and Gonsalves, 1989).

Induction of peroxidase and enzymes of intermediary metabolism of higher plants is considered as an indirect effect of metal (Cd) toxicity (Assche and Clijsters, 1990). Burzynski and Buczek (1989) have suggested antagonistic effect of Cd and Mo on chlorophyll concentration, δ-amino-levulinic acid, dehydratase activity, number of chloroplasts in the mesophyll cells and translocation of heavy metals Fe, Cu and Zn into the leaf of *Cucumis sativus* L.

Corn (*Zea mays*) grown in soil culture showed reduced biomass at the rate of 30 mg Cd/Kg soil and Cd toxicity effect on plant yield was reversed by the addition of 5mg Zn/Kg soil, thus increased the corn yield and suppressed the Cd toxicity (Singh and Nayyar, 1989). Cadmium application in growing medium reduced the plant yield in *Capsicum annuum* (4.36 Cd/1) (Estan *et al.*, 1989), in barley (30 mg Cd/Kg) (Singh and Rakipov, 1988), lettuce (1 mg Cd/Kg) (Yang and Lee, 1990), in onion and fenugreek (50 mg Cd/Kg) (Dang *et al.*, 1990).

Rani and Paliwal (1988) have reported increased germination rate of *Echinochlea colona* seeds with upto 50 μM Cd^{2+} treatment and shoot elongation with 25 μM Cd^{2+} concentration and these were inhibited with further increase in Cd concentration.

ACKNOWLEDGEMENT

The authors would like to thank Dr (Mrs) Nirmala Nautiyal, Dr (Mrs) Neena Khurana, Dr (Miss) Pratima Sinha and Dr (Miss) Radha Jain for their help in the preparation of the manuscript.

REFERENCES

Abadia, A., Lemoine, Y., Tremolieras, A., Ambard - Bretteville, F. and Remy, R. 1989. Iron deficiency in pea : effects on pigment, lipid and pigment - protein complex composition in thylakoids. *Plant Physiol. Biochem.*, **27** : 679-687.

Abruna, F. 1980. Response of soybean to liming on acid tropical soils. In *Proc. World Soybean Research Conference II.* (F.T. Corbin, ed). PP. 34-46. Westview Press, Boulder, Colorado.

Abutalybov. M.G. and Gummatov, M.R. 1966. The effect of various nutritional conditions on the nature of iron distribution in plants *Izv. Akad. Nauk. Azerb. SSR. Ser. Biol.*, 5:21-27.

Adams, P., Graves, C.T. and Winsor, G.W. 1980. Some effects of micronutrient and liming on the production and quality of glass house carnations grown in a peat-sand substrate. *J. Hort. Sci* ., **5**:89-96.

Adams, P. and Winsor, G.S. 1994. Some responses of glass house tomatoes to boron. *J. Hort. Sci* ., **49**: 355-363.

Adriano. D.C. 1969. Phosphorous-zinc Interaction in *Zea mays* and *Phaseolus vulgaris*.Ph.D. Dissertation-Kansas State University 89p.

Adriano, D.C., Paulsen, G.M. and Murphy, L.S. 1971. Phosphorus- iron and phsophorus-zinc relationship in corn (*Zea mays* L.) seedling as affected by mineral nutrition. *Agron. J.*, **63**:36-39.

Aery, N.C. and Tiagi, Y.D. 1988. Accumulation of cadmium by plants of Zawar mines, Rajasthan, India. *Acta Biol. Hung.,* **39**: 87-88.

Agarwala, S.C. 1952. The effect of molybdenum and nitrate status on the carbohydrate and nitrogen metabolism of cauliflower plant in sand culture. Ann. Rept. Long-Ashton Res. Sta. p. 70.

Agarwala, S.C., Abidi, A. and Sharma, C.P. 1991. Variable boron supply and sugarbeet metabolism. *Proc. Natl. Acad. Sci. India Sect B,* **61**:109-114.

Agarwala, S.C., Bisht, S.S. and Sharma C.P. 1977. Relative effectiveness of certain heavy metals by producing toxicity and symptoms of iron deficiency in barley. *Can. J. Bot.,* **55**: 1299-1307.

Agarwala, S.C., Bisht, S.S., Sharma, C.P. and Afzal, A. 1976. Effect of deficiency of certain micronutrients on the activity of aldolase in radish plants grown in sand culture. *Can. J. Biochem.,* **54**. 76-78.

Agarwala, S.C., Chatterjee, C. and Gupta, S. 1988. Sulphur-iron Interaction in mustard. *Proc. Int. Congr. Plant Physiol.* vol. 2 (S.K. Sinha, P.V. Sane, S.C. Bhargava and P.K. Agarwal, eds.),pp. 1116-1121. New Delhi.

Agarwala, S.C., Chatterjee, C., Gupta, S. and Nautiyal, N. 1988. Iron-manganese-magnesium interaction in cauliflower. *J. Plant Nutr.*, **11**: 1005-1014.

Agarwala, S.C., Chatterjee, C., Nautiyal, B.D., Dube, B.K. and Nautiyal, N. 1985 Induction of deficiency of zinc, copper and molybdenum in sugar-cane and its effect on growth, sugar concentration and some enzyme activities. *Sugar Cane,* 6: 1-[illegible]

Agarwala, S.C., Chatterjee, C. and Nautiyal, N. 1986. Effect of manganese supply on the physiological availability of iron in rice plants grown in sand culture. *Soil Sci. Plant Nutr.* (TOKYO), **32** : 169-178.

Agarwala, S.C., Chatterjee, C. and Nautiyal, N. 1986. Copper- boron interaction in sugarcane. *Taiwan Sugar,* **33**: 10- 16.

Agarwala, S.C., Chatterjee, C. and Nautiyal, N. 1988. Effect of induced molybdenum deficiency on growth and enzyme activity in sorghum. *Trop. Agric.,* **65**: 333-336.

Agarwala, S.C., Chatterjee, C. and Nautiyal, N. 1989. Copper molybdenum interaction in maize. *Soil Sci. Plant Nutr.,* **35**: 435-442.

Agarwala, S.C., Chatterjee, C., Nautiyal, N. and Sharma. C.P. 1986. Molybdenum interaction in maize. *Soil Sci. Plant Nutr.,* **35**:435-442.

Agarwala, S.C., Chatterjee, C., Nautiyal, N. and Sharma, C.P. 1986. Molybdenum nutrition of isolates of four *Aspergillus* species. *Can. J. Microbiol.,* **32**: 557-561.

Agarwala,S.C.,Chatterjee, C., Sharma, C.P. and Nautiyal, N. 1985. Copper nutrition of sugarbeet. *J. Expt. Bot.,* **36**: 881-888.

Agarwala, S.C., Chatterjee, C., Sharma, P.N., Sharma, C.P. and Nautiyal, N. 1979. Pollen development in maize plants subjected to molybdenum deficiency, *Can. J. Bot.,* **57**: 1946-1950.

Agarwala, S.C., Farooq, S. and Sharma, S.C. 1977. Interaction of boron supply and nitrogen sources in growth, boron uptake, nitrogen fractions and sugars in radish. *Proc, national Symposium: Nitrogen Assimilation and Crop Productivity.* pp. 185-194.

Agarwala, S.C. and Hewitt, E.J. 1954a. Molybdenum as a plant nutrient III. The interrelationships of molybdenum and nitrate supply in the growth and molybdenum content of cauliflower plants grown in sand culture. *J. Hort. Sci.,* **29**: 278-290.

Agarwala, S.C. and Hewitt, E.J. 1954b. Molybdenum as a plant nutrient. IV. The interrelationships of molybdenum and nitrate supply in chlorophyll and ascorbic acid fractions in cauliflower plants grown in sand culture. *J. Hort. Sci.,* **29**: 291-300.

Agarwala, S.C. and Hewitt, E.J. 1955a. Molybdenum as a plant nutrient V. The interrelationships of molybdenum and nitrate supply in the concentration of sugars, nitrate and organic nitrogen in cauliflower plants grown in sand culture. *J. Hort. Sci.,* **30**. 151- 162.

Agarwala. S.C. and Hewitt, E.J. 1955b. Molybdenum as a plant nutrient VI. Effect of molybdenum supply on the growth and composition of cauliflower plants given different sources of nitrogen supply in sand culture. *J. Hort. Sci.,* **30**: 163- 180.

Agarwala, S.C., Khurana, N., Chatterjee, C. and Sharma, C.P. 1988. Manganese-sulphur interaction in cauliflower. International symposium on manganese in soils and plants (M.J. Webb, R.O. Nable, R.D. Grahama and R.J. Hannam, eds.) Manganese Symposicim Inc., Adelaide.

Agarwala, S.C. and Kumar, A. 1962. The effect of heavy metals and bicarbonate excess on sunflower plants grown in sand culture with special reference to catalase and peroxidase. *J. Indian Bot. Sci.,* **41**:77-92.

Agarwala, S.C., Kumar, A. and Sharma, C.P. 1961. The effect of excess supply of heavy metals on barley during germination, with special reference to catalase and peroxidase. *Nature,* **191**: 726-727.

Agarwala, S.C. and Mehrotra, S.C. 1984. Iron-magnesium antagonism in growth and metabolism of radish. *Plant Soil,* **80**: 355-361.

Agarwala, S.C., Mehrotra, S.C., Farooq, S. and Sharma, C.P. 1982. Effect of iron supply on sugars and nitrogen in two varieties of *Brassica oleracea* that differ in iron nutrition. *Indian J. Plant Nutr.,* **1**: 101-106.

Agarwala, S.C., Mehrotra, N.K. and Sharma, C.P. 1981. Iron- nutrition of wheat I. Growth and metabolic changes in seven varieties of wheat (*Triticum aestivum L. (D)* grown at graded levels of iron supply in sand culture. *J. Indian Bot. Soc.,* **60**: 191-196.

Agarwala, S.C., Mehrotra, P.S. and Bisht, S.S. 1977. Effect of iron deficiency on growth, chlorophyll, oxygen uptake, tissue iron and activity of certain enzymes in three varieties of chickpea. *J. Indian Bot. Soc.,* **56**: 207-216.

Agarwala, S.C., Nautiyal, B.D. and Chatterjee, C. 1986. Manganese, copper and molybenum nutrition of papaya. *J. Hort. Sci.,* **61**: 397-405

Agarwala, S.C. and Sharma, C.P. 1961b. The relation of iron supply to the tissue concentration of iron, chlorophyll and catalase in barley plants grown in sand culture. *Physiol. Plant.,* **14**: 275-283.

Agarwala, S.C. and Sharma, C.P. 1976. Pot and sand culture technique for the study of the mineral nutrient deficiencies under Indian condition. *Geophytol.,* **6**: 356-367.

Agarwala, S.C., Sharma, C.P. and Farooq, S. 1965. Effect of iron supply on growth, chlorophyll, tissue iron and activity of certain enzymes in maize and radish. *Plant Physiol.,* **40**: 493-499.

Agarwala, S.C., Sharma, C.P., Farooq, S and Chatterjee, C. 1978. Effect of molybdenum deficiency on the growth and metabolism of corn plants raised in sand culture. *Can. J. Bot.,* **56**: 1905-1908.

Agarwala, S.C., Sharma, C.P. and Kumar, A. 1964. Inter- relationship of iron and manganese supply in growth, chlorophyll and porphyrin enzymes in barley plants. *Plant Physiol.,* **39**: 603-609.

Agarwala, S.C., Sharma, P.N., Chatterjee, C. and Sharma, C.P. 1980. Copper deficiency induced changes in wheat anthers. *Proc. Indian Natl. Sci. Acad. Part B.,* **46**: 172-176.

Agarwala, S.C., Sharma, P.N., Chatterjee, C. and Sharma, C.P. 1981. Development and enzyme changes during pollen development in boron deficient maize plants. *J. Plant Nutr.,* **3**: 329- 336.

Aghulon, H. 1910. (Role of boron in plants). Ph.D. thesis Univ. of Paris.

Ahmed, S. and Evans, H.J. 1960. Cobalt-a micronutrient element for the growth of soybean plants under, symbiotic condition. *Soil Sci.,* **90**: 205-210.

Ahmed, A.M., Zidan, M.A. and Adam, M.S. 1988. Effect of high boron levels of growth and some metabolic activities of the halotolerant *Duneliella tertiolecta. Biol. Plant.,* **30**: 357-361.

Akulova, E.A. and Mukhin, E.N. 1968. The role of plastocyanin from *Pisum sativum* L. leaves in the reduction of NADP by chloroplast fragments. Dorlady Akad. Nauk. SSR, 180A.

Alagna, L., Hasnain, S.S., Piggptt, B. and Williams, D.J. 1984. The nickel ion environment in jack bean urease. *Biochem. J.,* **220**: 591-595.

Alam, S.M. 1983. Effects of copper and phosphorus on growth and nutrient content of rice. *Pakistan J. Sci. Ind. Res.,* **26**: 370-373.

Alam, S.M. 1983. Effects of flooded and unflooded soil conditions, Fe and Mn application on growth and nutrient content by rice plants. *J. Sci. Tech.,* **7**: 1-2.

Albert, L.S. 1965. Ribonucleic acid content, boron deficiency symptoms, and elongation of tomato root tips. *Plant Physiol.,* **40**: 649-659.

Aleshin, E.P., Sheudzhen, A.Kh. and Rymar, V.T. 1989. Copper in mineral nutrition of rice. *Vestnik Sel'Skokhozyaistvennoi Nauk. Moscow, USSR,* **7**: 107-111.

Alexander, T.R. 1942. Anatomical and physiological responses of squash to various levels of boron supply. *Bot. Gaz.,* **103**: 475-491.

Alexander, A.G. and Samuels, G. 1967. Effect of variable calcium and manganese of the enzyme, nucleotide and nutrient constituents of sugarcane grown in sand culture, *J. Agric. Univ. P.R.,* **51**: 238-251.

Allakhverdiev, S.I., Shafiev, M.A. and Klimov, V.V. 1986. Effect of extraction and subsequent addition of manganese ions on photo-oxidation of chlorophyll P680 in the photosystem II preparation. *Biofizica.,* **31**: 223-226.

Allen, M.B. 1956. Photosynthetic nitrogen fixation by blue-green algae. *Sci. Monthly.,* **83**: 100-106.

Alemela, L., Garcia, A.L. and Madrid, R. 1983. Iron deficiency in maize plants. Effects on the chlorophyllase activity, chlorophylls, pheophytins and proteins in the leaf. *Agrochimica.,* **27**: 123-131.

Al-Molla, R.M.M. 1986. Some physiological aspects of soybean development and yield as affected by boron fertilizer ion. *Sciences and Engineering.,* **46:** 3268B.

Al-Shevskii, N.G., Zhitomir, Skhi, Zhitomir 1986. Effect of magnesium and boron on metabolism & productivity of potato on sod- podzolic sandy loam soil of polesic. *Agrokhim., 7: 93-102.*

Altman, Nfelter, F. and Kaindl, K. 1968. Untersuchungen uper den Einfluss Von Zn-Ionen and dis m RNA. Synthese in *Chlorella zellen. Z. Naturforsch.,* **236:** 395-396.

Alvarez-Tinaut, M.C., Leal, A., Gomez, M. and Recalde, L. 1979. Physiological effect of boron-manganese interaction in tomato plants III. Absorption and distribution of micro-elements, manganese, zinc and copper. *An. Edafol. Agrobiol.,* **38**: 1013-1029.

Alvarez-Tinaut, M.C., Leal, A. and Recalde, L. 1981. Physiological effects of the boron-manganese interaction in tomato plants. *Agrochim.,* **25**: 501-514.

Amberger, A. 1973. *Die Rolle des Magans in Stoffwechsel der plfanzen. Agrochim.,* **17**: 165-183.

Amberger, A. 1974. Micronutrient dynamics in the soil and function in plant metabolism. *Proc. Egyptian Bot. Soc. Workshop.,* **1**: 91-101.

Amberger, A., Gutser, R. and Wunsch, A. 1982. Iron chlorosis induced by high copper and manganese supply. *J. Plant Nutr.,* **5**: 715-720.

Ambler, J.E., Brown, J.C. and Gauch, H.G. 1971. Sites of iron reduction in soybean. *Agron. J.,* **63**: 95-97.

Anderson, I. and Evans, H.J. (1956). Effect of manganese and certain other metal cations on isocitric dehydrogenase and malic enzyme activities in *Phaseolus vulgaris. Plant Physiol.,* **31**: 21- 28.

Anderson, A. and Siman, G. 1991. Levels of Cd and some other trace elements in soils and crops as influenced by lime and fertilizer level. *Acta Agric. Scandinavica.,* **41**: 3-11.

Anspok, P.L. 1971. Effect of magnesium and boron on crop yield. *Agrokhim.,* **9**: 122-124.

Antil, R.S., Yadav, D.S., Kumar, V. and Singh, M. 1988. Nitrogen- copper relationship in raya (*Brassica juncea* Coss). *Indian Soc. Soil Sci.,* **36**: 704-708.

Antonov, I. 1970. Results of boron, manganese and zinc dressing of sugarbeets grown on calcareous Chernozem. *Pochvozn. Agrokhim.,* **5**: 61-69.

Arnon, D.I. 1950. Functional aspects of copper in plants. In: Copper metabolism (W.D.Mc Elroy and H.B. Glass, eds.)pp. 89-110. Johns Hopkins press, Baltimore, Maryland.

Arnon, D.I. 1954. The role of micronutrients in plant nutrition with special reference to photosynthesis and nitrogen assimilation. In : *Trace Elements* (C.A. Lamb, O.G. Bentley and J.M. Beattie, eds.). pp. 1-32. Academic Press, New York.

Arnon, D.I. 1958. The role of micronutrients in plant nutrient with special reference to photosynthesis and nitrogen assimilation, In : *Trace Elements.* Vol. I. (C.A. Lamb, D.P. Bentley and J.M. Beattie, eds.), 1-32, Academic Press, New York.

Arnon, D.I. 1965. Ferredoxin and photosythesis. *Science,* **149**:1460-1470.

Arnon, D.I. and Stout, P.R. 1939. Molybdenum as an essential element for higher plants. *Plant Physiol.,* **14**: 599-602.

Arnon, D.I. and Wessel, G. 1953. Vanadium as an essential element for green plants. *Nature,* **172**:1039-1040.

Arnon, D.I. and Whatley, F.R. 1949. Is chloride a co-enzyme of photosynthesis ? *Science,* **110** :554-556.

Arora, S.K., Pandita, M.L. and Singh, K. 1982 Effect of P chlorophenoxyacetic acid, boron, copper, molybdenum and zinc and on the fruit quality of tomato varieties in summer and the rainy season. *Indian J.Agric. Sci.,* **52**:648-652.

Ashour, N.I. and Thalooth, A.T. 1974. Effect of foliar application with microelements and gibberellic acid on photosynthetic pigments and nitrogen content in leaves of horse bean plant. *Egypt. J. Physiol. Sci.,* **1**:39-45.

Askew, H.O. 1935. The boron status of fruit and leaves in relation to 'internal cork' of apple in Nelson district. *Newzealand J. Sci. Technol.,* **A17**: 388-391.

Assche, F. Van.,Ceulemans, R. and Clijsters, H. 1980. Zinc mediated effect on leaf CO_2 diffusion conductances and net photosynthesis in *Phaseolus vulgaris* L. *Photosynthesis Res.*, **1**: 171-180.

Assche, F. Van and Clijsters, H. 1983. Multiple effects of heavy metal toxicity on photosynthesis. Effects of stress on photosynthesis. (R. Marcelle, H. Clijsters, and M. Van. Poucks eds.) Department SBM, The Hague, Netherlands.

Assche, F. Van and Clijsters, H. 1990. Effects of metals and enzyme activity in plants. *Plant Cell Environ.*, **13**: 195- 206.

Assche, F. Van; Clijsters, H. and Marcellc, R. 1979. Photosynthesis in *Phaseolus vulgaris* L. as influenced by supraoptimal zinc nutrition. *Photosynthesis and development.*, 175-184.

Augsten, H. and Hundt, I. 1970. Investigation into the influence of boron on the respiratory metabolism of *Helianthus annuus* L. *Biol. Zentralbl.*, **89**: 497-505.

Aulakh, M.S. et.al., 1977. Effect of sulphur fertilization on the yield and quality of potato (*Solanum tuberosum* L.)*J. Indian Soc. Soil Sci.*, **25**: 182-185.

Avila-Rincon, M.J., Gomez-Rodriguez, M.V. and Alvarez-Tinaut, M.C. 1981. Effect of boron, GA_3 treatments on growth and boron, copper , manganse and zinc distribution in dwarf bean (*Phaseolus vulgaris* L.) plants. I. Vegetative stage of development*An. Edafol. Agrobiol.*, **40**: 1303-1313.

Avakyan, Z.A. 1971. Comparative toxicity of free ions and complexes of copper with amino acid for *Candida utilis. Mikrobiol.*, **40**: 417-423.

Ayres, A.S. 1966. Calcium silicate slag as a growth stimulant for sugarcane on low silicon-soils. *Soil Sci.*, **101**: 216-227.

Badawi, A.M., Sweidan, A.M., Fayck, M.A. and ElHawary, A.H.I. 1981. Effect of boron, zinc and calcium on growth, fruit quality and storage ability of Le-conte pear. *AIN SHAMS Univ. Fac. Agric. Res. Bull. O* : 1-10.

Bahl, G.S. et. al. 1986. Sulphur and zinc nutrition of groundnut grown on Tolewal loamy sand soil. *Indian J. Agric. Sci.*, **56**: 421-433.

Bailey, L.F. and McHargue, J.S. 1944. Effect of boron, copper,manganese and zinc on the enzyme activity of tomato and alfalfa plant in the green house. *Plant Physiol.*, **19**: 105-116.

Bajescu, I., Chiriac, A. and Tiganas, L. 1983. Effect of nitrogen and phosphorus on manganese nutrition of some plants. *An. Inst. Cercet. Pedol. Agrochim.*, **45**: 67-80.

Baker, J.E., Gauch, H.G. and Dugger, W.M.Jr. 1956. Effect of boron on the water relations of higher plants. *Plant Physiol.*, **31**: 89-94.

Baker, N.R., Fernyhough, P. and Meek, I.T. 1982. Light independent inhibition of photosynthetic electron transport by zinc. *Physiol. Plant.*, **56**: 217-222.

Ball, M.C., Taylor, S.E. and Terry,N. 1984. Properties of thylakoid membranes of the mangrove *Avicennia graminanis* and *Avicennia marina* and sugarbeet, *Beta vulgaris* grown under different salinity conditions. *Plant Physiol.*, **76**: 531- 535.

Barakat, M.Z., Bassiouni, M., El-Wakil, M. 1972. Evaluation of iodine content of certain vegetables. *Bull. Acad. Pol. Sci. Ser. Sci. Biol.*, **20**: 531-533.

Bar-Akiva, A. 1961. Biochemical indications as a means of distinguishing between iron and manganese deficiency symptoms in cirtrs plants. *Nature*, **190**: 647-648.

Bar-Akiva, A. 1964. Visual symptoms and chemical analysis vs. biochemical indicators as a means of diagnosing iron and manganese deficiencies in citrus plants. In : Plant Analysis and Fertilizer Problems (C. Bould, P. Prevot and J.R. Mangness, eds.). pp. 9-25. 4th Am. Soc. Hort. Sci.

Bar-akiva, A. and Lavon, R. 1967. Visible symptoms and metabolic patterns in micronutrient deficient Eureka lemon leaves. *Israel J. Agric. Res.*, **17**: 7-16.

Bar-Akiva, A. and Lavon, R. 1968. Peroxidase activity as an indicator of the iron requirement of citrus plants. *ISR.J. Agric. Res.*, **18**: 145-153.

Bar-Akiva, A. and Lavon, R. 1969. Carbonic anhydrase activity as an indicator of zinc deficiency in citrus leaves. *J. Hort. Sci.*, **44: 359-362.**

Bar-Akiva, A., Sagiv, J. and Hasdai, D. 1971. Effect of mineral deficiencies and other cofactors on the aldolase enzyme activity of citrus leaves. *Physiol. Plant.*, **25**: 386-390.

Baranov, V.I. 1980. Combined effects of the trace elements copper and iron, phenols, amino acids and their oxidation products on the decomposition of indolylacetic acid and growth processes. Mikroelem. Okruzhayushchei Srede, 107-111.

Barbaria, C.J. and Patel, C.L. 1980. Effect of application of iron, farmyard manure and sulphur on the availability of iron in medium black calcareous soil at different moisture regimes. *J. Indian Soc. Soil.*, **28**: 302-306.

Bard, R.C. and Gunsalus, I.C. 1950. Glucose metabolism of *Clostridium prefringens* existence of a metallo-aldolase. *J. Bacteriol.*, **59**: 387-400.

Barnette, R.M., Camp, J.P., Warner, J.D. and Gall, O.E. 1936. Use of zinc sulphate under corn and other field crops. *Fla. Agric. Exp. Sta. Bull.*, **293** : 3

Baran, M., Lachica, M., Chueca, A. and Sandmann, G. 1990. The role of copper in the structural organisation of photosystem II in chloroplast membranes. *Curr. Res. Photosynth. Proc. Int. Conf. Photosynth. 8,th* **1**: 303-306.

Barshad, I. 1951. Factors affecting the molybdenum content of pasture plants. 2. Effect of soluble phosphate, available nitrogen and soluble sulphates. *Soil Sci.*, **71**: 387-398.

Baser, B.L. and Somani, L.L. 1982. Effect of soil application of manganese on dry matter yield and uptake of manganese and iron by maize. *Am. Edafol. Agrobiol.*, **41:** 2211-2220.

Basiouny, F.M. 1984. Distribution of polyphenols in plants and cell suspension cultures. *Physiol. Plant.*, **15**: 261-277.

Baszynaki, T., Ruszkowka, M., Krol. M. and Tukendrof, A. 1978. The effect of copper deficiency on the photosynthetic apparatus in higher plants. *Z. Pflanzenphysiol.*, **89:** 207-216.

Bayer, E. and Kneifel, H. 1972. Isolation of amavadion, a vanadium compound occurring inAmanita muscaria. *Z. Naturforsch.,* 27b : 207.

Becker, L.J.M. and Meisch, H.U. 1981. Effect of vanadate and iron stress on the pigment composition of *Chlorella fusca. Z. Naturforsch.,* **36c**: 207-209.

Bell, P.F., Chaney., R.L. and Angle, J.S. 1988. Staining localization of ferric reduction in roots. *J. Plant Nutr.,* **11**: 1237-1252.

Bell, R.W., Edwards, D.G. and Asher, C.J. 1989. Effects of calcium supply on uptake of calcium and selected mineral nutrients by tropical food legumes in solution culture. *Aust. J. Agric. Res.,* **40:** 1003-1013.

Bell, R.W., Mclay, L., Plaskett, D., Dell, B. and Loneragan, J.R. 1990. Internal boron requirement of green gram (*Vigna radiata*). *Dev. Plant Soil Sci.,* **41:** 275-280.

Belokopylsky, A.M., Ginturi, E.N., Mosulishvili, L.T. and Kharabadze, N.E. 1968. Measurement of trace amounts of heavy metal in DNA and RNA by activation analysis. *Biofizika.,* **13**: 950-953.

Belver, A. and Donaire, J.P. 1983. Partial purification of soluble lipoxygenase of sunflower cotyledons, action of boron on the enzyme and lipid constituents. *Z. Pflanzen. Physiol.,* **109**: 309-317.

Bendall, D.S. 1958. Cytochromes and some respiratory enzymes in mitochondria from the spadix of *Arum maculatum. Biochem. J.,* **70**: 381-390.

Bennett, J.P. 1945. Iron in leaves. *Soil Sci.,* **60:** 91- 105.

Bennett, J.H., Krizek, D.T., Wergin, W.P., Cleming, A.L., Mirecki, R.M. and Wyse, R.E. 1984. Physiological and ultrastructural changes in chloroplasts of snap bean plants under LPS lamps during chlorosis and regreening. *J. Plant Nutr.,* **7**: 819-832.

Ben-Yehoshua, S. and Biggs, R.H. 1970. Effects of iron and copper ions in promotion of selective abscission and ethylene production by citrus plant fruit and the inactivation of indoleacetic acid. *Plant Physiol.,* **45:** 604-607.

Berry, J.A. and Reisenaur, H.M. 1967. The influence of molybdenum on iron nutrition of tomato. *Plant Soil,* **27**: 303-313.

Berry, W.F. 1971. The nutrient status of Zn in lettuce evaluated by plant analysis. *J. Am. Soc. Hort. Sci.,* **96**: 412-414.

Bertrand, D. 1943. Le Vanadium chez les champinous et plus specialment chez les amanites. *Bull Soc. Chim. Biol.,* **25**: 194-197.

Bertrand, D. 1941. Distribution of vanadium in plants. *Compt. Rend. Acad. Sci.,* **212**: 1170-1172.

Bertrand, G. 1897 Sur l' intervention bu manganese dans les oxidation provoguces parla laccase. *Compt Rend.Acad.Sci., Paris,* **124**: 1032-1035.

Bentrand, D. 1950. Survey of contemporary knowledge of biogeochemistry. 2. The biogeochemistry of vanadium. *Bull. Am. Museum Nat. His.,* **94:** 40/-466.

Bertrand, D and Wolf, A. 1961. Sur la necessite' dzinc comme oligoelement pour pour la glucose-6-phospate dehydrogenase at la 6-phosphate gluconique-dehydrogenase de L' *Aspergillus niger. C.R. Acad. Sci., Paris.,* **245**: 1179-1182.

Bertrand, D. and Wolf, A. 1961. Sur la necessite' du zinc comme oligoelement pour la synthese due quelgnes aminoacid, chez 1. *Aspergillus niger. C.R. Acad. Sci. Paris,* **253**: 1342-1344.

Besong, B.E. and Lawanson, A.O. 1991. Boron stress and stress and mitochondrial quinone accumulation in *Zea mays* seedlings. *J. Plant Physiol.,* **138:** 80-88.

Bienfait, H.F., Bino, R.J., Vander Blick, A.M., Duivenvoorden, J.F. and Fontaine, J.M. 1983. Characterization of ferric reducing activity in roots of iron-deficient *Phaseolus vulgaris. Physiol. Plant.,* **59**: 196-202.

Bienfait, H.F., Duivenvoorden, J. and Verkerke, W. 1982. Ferric reduction by roots of chlorotic bean plants. Indications for an enzyme process. *J. Plant Nutr.,* **5**: 451-456.

Bilteanu, G., Dumitru, G. and Rodica, P. 1973. ^{65}Zn absorption in the pea as a function of continuous variation of Mo together with N, P and K. *Stud. Corect Biol. Ser. Bot.,* 497-501.

Bingham, F.T. 1963. Relation between phosphorus and micronutrient in plants. *Soil Sci. Soc. Am. Proc.,* **27**: 389-391.

Bingham, F.T. and Garber, M.J. 1960. Solubility and availability of micronutrients in relation to phosphorus fertilization. *Soil Sci. Soc. Am. Proc.,* **24**: 209-213.

Birnbaum, E.H., Dugger, W.M. and Beasley Bud, C.A. 1977. Interaction of boron with components *in vitro. Plant Physiol.,* **56**: 1034-1038.

Bishop, W.B.S. 1928. The distribution of manganese in plants and its importance in plant metabolism. *Aust. J. Expt. Biol. Med. Sci.,* **5**: 125-141.

Bishop, N.I. 1964. Site of action of copper in photosynthesis. *Nature,* London, **204**: 401-402.

Bishop, N.I. 1966. Partial reactions of photosynthesis and photo reduction. *Rev. Plant Physiol.,* **17**: 185-208.

Bisht, S.S. 1972. Effect of heavy metals on plant metabolism. Ph.D. Thesis Bot. Deptt. Lucknow University, Lucknow.

Bisht, S.S. and Agarwala, S.C. 1982. Effect of excess supply of certain heavy metals on growth and phosphorus fractions in corn. *Indian J. Plant Nutr.,* **1**: 73-79.

Bisht, S.S., Sharma, C.P. and Kumar, A. 1976. Plant response to excess concentration to heavy metals. *Geophytol.,* **6:** 296- 307.

Biswas, C.R. et al. 1977. Zinc availability to maize and wheat in relation to P and K status of the soil. In : A long term fertilizer experiment. *J. Indian Soc. Soil Sci.,* **25:** 414-421.

Blamey, P.P.C. and Chapman. J. 1979. Boron toxicity in spanish groundnuts. *Agrochemo physics. Agron., J.,* **71** : 57-59.

Blumberg, W.W., Peisach, J. and Levine, W. 1967. Structural properties of stellacyanin a copper mucoprotein from *Rhusvernicifera. J. Biol. Chem.,* **242:** 2847-2858.

Boardman, N.K. 1975. Trace elements in photosynthesis. In : Trace Elements in Soil-Plant Animal System. (N.J.D. Nicholas and A.B. Ezan, eds). pp. 199-212. Acad. Press, New York.

Boawn, L.C. and Brown, J.C. 1968. Further evidence for a P-Zn imbalance in plants. *Soil Sci. Soc. Amer. Proc.*, **32**: 94-97.

Boawn, L.C., Rasmussen, P.E. and Brown, J.W. 1959. Relationship between tissue zinc levels and maturity period of field beans. *Agron. J.*, **61**: 49-51.

Boawn, L.C., Viets, F.G., Crawford, C.L. and Nelson, J.L. 1960. Effect of nitrogen carrier, nitrogen rate, zinc rate and soil pH on zinc uptake by sorghum, potatoes and sugarbeets. *Soil Sci.*, **90**: 329-337.

Boawn, L.C., Jr. Viets, F.G. and Croawford, C.L. 1954. Effect of phosphate fertilizer on zinc nutrition of field beans. *Soil Sci.*, **78**: 1-7.

Bollard, E.G. 1983. Involvement of unusual elements in plant growth and nutrition. In : *Inorganic Plant Nutrition, Encyclopaedia, Plant Physiol.* New Series Vol. 158. (A. Lauchli and R.L. Bieleski, eds.) pp. 695-744. Springer, Verlag, Berlin, Heidelberg, New York.

Bollard, E.G. and Butler, G.W. 1966. Mineral nutrition of plants. *Ann. Rev. Plant Physiol.*, **17**: 77-112.

Bolle-Jones, E.W. 1955. The interrelationship of iron and potassium in the potato plant. *Plant Soil.*, **6**: 129-173.

Bonilla, I., Cadahia, C., Carpena, O. and Hernado, V. 1980. Effect of boron on nitrogen metabolism and sugar levels of sugarbeet. *Plant Soil*, **57**: 3-9.

Bordier, H.P. and Lindberg, B. 1986. The 110-KDa reaction centre protein of photosystem I, P700-chlorophyll *a*-protein I, is an iron sulfur protein. *J.Biol. Chem.*, **261**: 14292-14300.

Bortels, H. 1927. Uber die Bedeutung von Eisen zink und Kupper fur mikroorganismen. *Biochim. Z.*, **182**: 301-358.

Bortels, H. 1930. Molybdal als katalysater bei der biologischen stickstoffbindung. *Arch. Mikrobiol.*, **1**: 333.

Bothe, H., Yates, M.G. and Cannon, F.C. 1983. Physiology, biochemistry and genetic dinitrogen fixation. In: *Encyclopaedia of Plant Physiol.* New Series Vol. 15A. (A Lauchli and R.L. Bieleski, eds.), pp. 241-285. Springer, Verlag, Berlin, New York.

Bove, J.M., Bove C., Whatley, F.R. and Arnon, D.J. 1963. Chloride requirement for oxygen evolution in photosynthesis. *Z. Naturforsch. B: Inorg. Chem. Org. Chem. Biochem. Biophys. Biol.*, **18b** : 683-688.

Bowen, H.J.M. 1966. Trace elements in biochemistry Vol. 24. Academic press, London and New York.

Bowen, J.E. 1968. Borate absorption in excised sugarcane leaves. *Plant Cell Physiol.*, **36**: 420-424.

Bowen, J.E. 1969. Absorption of borate ionic species by *Saccharum officinarum* L. *Plant Cell Physiol.*, **10**: 227-230.

Bowen, J.E. 1972. Effect of environment factors on water utilization and boron accumulation and translocation in sugarcane. *Plant Cell Physiol.*, **13**: 703-714.

Bowen, J.E. 1981. Microelement nutrition of sugarcane 11. Interactions in microelement accumulation. *Trop. Agric.*, **58**: 215-220.

Bowen, J.E. 1981. Kinetics of active uptake of boron, zinc, copper and manganese in barley and sugarcane. *J. Plant Nutr.,* **3:** 215-223.

Bowen, J.E., and Nissen, P. 1976. Boron uptake by excised barley roots. I. Uptake into the free space. *Plant. Physiol.,* **57**: 353-357.

Bowen, M.R., Howarth J. and Loughman, K.A. 1975. Effect of auxin and other factors on the rooting of *Pinus cantorta* Dough cuttings. *Ann. Bot.* **39:** 647-656.

Bowman, B.J., Mainzer, S.E., Allen, K.E., and Slayman. 1978. Effects of inhibitors on the plasma membrane and mitochondrial adenosine triphosphates of *Neurospora crassa. Biochim. Biophys. Acta,* **512:** 13-28.

Bozhenko, V.P., Belyaeva, V.N. and Shkol' nik, M. Ya. 1973. Nucleotide composition of total RNA and its individual fractions in the apical cane and roots of sunflower under the effect of high temperatures and boron deficiency. *Fiziol. Biokhim. Kul. Rast.,* **5:** 125-130.

Bradfield, J.R.C. 1947. Plant carbonic anhydrase. *Nature.,* **159:** 467-468.

Branden, R. 1978. Ribulose-1, 5-diphosphate carboxylase and oxygenase from green plants are two different enzymes. *Biochem. Biophys.Res. Commun.,* **81**: 539-546.

Brenchley, W.E. 1938. Comparative effects of cobalt, nickel and copper on plant growth. *Ann. Appl. Biol.,* **25**: 671-694.

Brenchley, W.E. and Warington, K. 1927. The role of boron in the growth of plants. *Ann. Bot.,* **41:** 167-187.

Breslow, E. and Girotti, A.W. 1966. The interaction of ribonuclease with metal ions. *J. Biol. Chem.,* **241:** 5651-5660.

Brigg's G.O. 1943. Effect of boron in the substrate on the rate of nitrate absorption and on nitrogen distribution in *Nasturtium. Plant Physiol.,* **18**: 415-432.

Brown, A.L., Krantz, B.A. and Eddings, J.L. 1970. Zinc-phosphorus interaction as measured by plant response and soil analysis. *Soil Sci.,* **110**: 415-420.

Brown, A.L. Yamaguchi, S. and Leal Diaz, L. 1965. Evidence for translocation of iron in plants. *Plant Physiol.,* **40**: 35-38.

Brown, J.C. 1960. An evaluation of bicarbonate- induced iron chlorosis. *Soil Sci.,* **89**: 246-247.

Brown, J.C. 1961. Iron chlorosis in plants. *Adv. Agron.,* **13**: 329-369.

Brown, J.C. (1966). Fe and Ca uptake as related to root-sap and stem-exudate citrate in soybeans. *Physoil. Plant,* **19**: 968-976.

Brown, J.C. 1978. Mechanism of iron uptake in plants. *Plant Cell Environ.,* **1**: 221-257.

Brown, J.C. 1979. Effect of boron stress on copper enzyme activity in tomato. *J. Plant Nutr.,* **1**: 39-53.

Brown, J.C. and Ambler, J.E. 1969. Characterisation of boron deficiency in soybeans. *Physiol. Plant.,* **22**: 177-185.

Brown J.C. and Chaney, R.L. 1971. Effect of iron on the transport of citrate into the xylem of soybeans and tomatoes. *Plant Physiol.*, **47**: 836-840.

Brown, J.C., Holmes, R.S. and Tiffin, L.O. 1959a. Hypothesis concerning iron chlorosis. *Soil Sci. Soc. Am. Proc.,* **23**: 231-234.

Brown, J.C., Holmes, R.S. and Tiffin, L.O. 1961. Iron chlorosis in soybeans as related to the genotypes of root stock. 3: Chlorosis susceptibility and reductive capacity at the root. *Soil Sci.,* **91**: 127-132.

Brown, J.C. and Steinberg. R.A. 1953. Iron and copper enzymes in leaf lamina of tobacco when deficient in micronutrients or grown in calcareous and organic soils. *Plant Physiol.,* **28:** 488- 494.

Brown, J.C. and Tiffin, L.O. 1960. Iron chlorosis in soybeans as related to the genotypes of root stock. II A relationship between susceptibility to chlorosis and capacity to absorb iron frcm iron chelate. *Soil Sci.,* **89**: 8-15.

Brown, J.C., Tiffin, L.O.Holmes, R.S., Specht, A.W. and Resnicky, J.W. 1959c. Internal inactivation of iron in soybeans as affected by root growth medium. *Soil Sci.,* **87**: 89-94.

Brown, P.H., Graham, R.D. and Nicholas, D.J.D 1984. Physiological effects of manganese deficiency. In: *Trace Elements* (C.A. Lamb, O.G. Bentley and J.M. Beattie, eds.) pp. 135-155. Academic Press, New York.

Brown, P.H., Graham R.D. and Nicholas, D.J.D. 1984. The effects of manganese and nitrate supply on the levels of phenolics and lignin in young wheat plants. *Plant Soil.,* **81**: 437-440.

Brown, R. and Possinghamm J.V. 1957. Iron deficiency and growth of pea roots. *Roy. Soc. Sci. B. Biol. Sci.,* **147**: 145- 166.

Brown, T.E., Eyster, H.C. and Tanner, H.A. 1958. Physiological effects of manganese deficiency. In: *Trace Elements* (C.A. Lamb, O.G. Bentley and J.M. Beattie, eds.) pp. 135-155. Academic Press, New York.

Brownell, P.F. 1965. Sodium as a essential micronutrient element for a higher plant (*Atriplex vesicaria*). *Plant Physiol.,* **40**: 460-463.

Brownell, P.F. and Crossland, C.J. 1972. The requirement for sodium as a micronutrient by species having the C_4 dicarboxylic photosynthetic pathway. *Plant Physiol.,* **49**: 794-797.

Brownell, P.F. and Crossland, C.J. 1974. Growth responses to sodium by *Bryophyllum tibuflorum* under conditions inducing crassulacean acid metabolism. *Plant Physiol.,* **54**: 416- 417.

Brownell, P.F. and Wood, J.G. 1957. Sodium as an essential element for *Atriplex vesicaria. Nature,* **179:** 635-636.

Broyer, T.C., Carlton, A.B., Johnson, C.M. and Stout, P.R.C. 1954. Chlorine - a micro nutrient element for higher plants. *Plant Physiol.,* **29**: 526-532.

Buczek, J. 1973. Effect of vanadium on nitrate reductase activity in tomato leaves. *Acta Soc. Bot. Pol.,* **42**: 223-232.

Bulen, W.A. and LeCompte, J.R. 1966. The nitrogenase system from *Azotobacter*, two enzyme requirement for N_2 reduction, ATP - dependent H_2 evolution and ATP hydrolysis. *Proc. Natn. Acad. Sci., U.S.A.,* **56:** 979-986.

Burleson, C.A., Dacus, A.D. and Gerard, C.J. 1961. The effect of phosphorus fertilisation in the zinc nutrition of several irrigated crops. *Soil Sci. Soc. Am. Proc.*, **25**: 365-368.

Burnell, J.N. 1986. Purification and properties of phosphoenol pyruvate carboxykinase from C4 plants. *Aust. J. Plant Physiol.*, **13**: 577-587.

Burnell, J.N. 1987. Photosynthesis in phosphoenol pyruvate carboxykinase type C4 plants properties of NAD malic enzyme from Urochloa panicoides. *Aust. J. Plant Physiol.*, **14**: 517- 525.

Burnell, J.N. 1988. The biochemistry of manganese in plants. In : *Manganese in Soils and Plants* (R.D. Graham, R.J. Hannam and N.C. Urens, eds.) pp. 125-137. Kluwer Academic Publishers.

Burris, R.H. 1971. Fixation by free-living micro-organism enzymology. In : *The Chemistry and Biochemistry of Nitrogen Fixation.* (J.R. Postgate, ed.) pp. 105-106. London and New York, Plenum P.

Burstrom, H. 1939. Uber die Schwermetallkatalyseder Nitrate- assimilation. *Planta.*, **29**: 292-305.

Burton, K.W., Morgan, E. and Roig, A. 1985. Interactive effects of cadmium, copper and nickel on the growth of Sitka Spruce and studies of metal uptake from nutrient solutions. *New Phytol.*, **10**: 546-557.

Burzynski, M. and Buczek, J. 1989. Interaction between cadmium and molybdenum affecting the chlorophyll content and accumulation of some heavy metals in the second leaf of *Cucumis sativus* L. *Acta Physiol. Plant.*, **11**: 137-145.

Bussler, W. 1958. Manganvergiftung bei hoheren. *Pflanzen. Z. Pflanzenernahr. Dung. Bodenk.*, **81**: 256-265.

Bussler, W. 1960. Relationship between root formation and boron in sunflower. *Z. Pflanzen. Dung. Bodenk.*, **92**: 1-14.

Bussler, W. 1981. Physiological functions and utilization of copper. Copper Soils and Plants. *Proc. Golden Jubilee Int. Symp.* 213-234.

Bussler, W. and Doering, H.W. 1979. Distribution and solubility of boron in sunflower plants in relation to the boron concentration of the nutrient solution. *Z. Pfl. Ernahr. Bodenk.*, **142**: 719-730.

Butorac, K. 1981. Response of lucerne (*Medicago sativa* L.) to nitrogen, molybdenum and copper fertilization. *Poljopr. Znan. Smotra.*, **55**: 269-287.

Butorac, A. and Uscumlic, L. 1977. Some results of copper concentration in lucerne. *Poljopr. Znan. Smotra.*, **41**· 85- 94.

Buzetti, S., Muraoka, T. and Sa, M.E. de. 1990. Boron rates for soybeans under different soil acidity conditions. II Critical levels in plants and seeds. *Revista Brasileira de Ciencia do Solo.*, **14**: 163-166.

Buzover, F.Y. 1951. Effect of boron on accumulation of carbohydrates and enzymatic activity of potato. *Dokl. Akad. Nauk. U.S.S.R.*, **78**: 1239-1242.

Byrne, A.L.. Ravnik, V. and Kosta, L. 1976. Trace element concentration in higher fungi. *Sci. Total Environ.*, **6**: 65-78.

Camp, A.F. and Fudge, B.R. 1939. Some symptoms of citrus malnutrition in Florida. *Florida Agric. Expt. Sta. Bull.,* 335.

Camp, A.F. and Fudge, B.R. 1945. Zinc as a nutrient in plant growth. *Soil Sci.,* **60:** 157-164.

Camprubi, P., Aldrufen, A., Pages, M., Bargallo, R. and Lopez. J. 1981. Effects of sodium borohydride on ethylens production and carbohydrate status in petals of carnation (*Dianthus cryophyllus* L.) cut flowers. *Acta Hort.,* **113:** 51-57.

Cannon, H.L. 1963. The Biogeochemistry of vanadium. *Soil. Sci.,* **96:** 196-204.

Cantley, L.C. Jr., Josephson, L., Warner, R., Yanagisawa, M., Lechene, C. and Guidotti, G. 1977. Vanadate is a potent (Na, K), ATPase inhibitor found in ATP derived from muscle. *J. Biol. Chem.,* **252:** 7421-7423.

Carpena, A.D., Carpena, R.R., Zornoza, P. and Collado, G. 1984. A possible role for boron in higher plants. *J. Plant. Nutr.,* **7:** 1341-1354.

Carpena, O., Costa, F. and Navarro, S. 1968. Effect of deficiencies and excesses of iron and manganese on various nitrogen fractions in *Capsicum annuum. Atti. Simp. Int. Agrochim.,* **7**: 233- 244.

Carpena, O., Mataix, J.J. and Carpena, R.O. 1979. Effect of boron on amino acid and flavanoid level in tomato plants. *Agrochim.,* **23:** 47-50.

Carpena, O., Zornoza, P. and Mataix, J. 1982. Incidence of P, Mn and B deficiencies on the level of the whole and individual flavanoid groups in tomato leaves. *J. Plant Nutr.,* **5**: 197-208.

Carpena, P., Hernando, V., Cadahia, C. and Bonillo, I. 1978. Effect of boron on the activity of nitrate reductase in sugarbeet (*Beta vulgaris*). *Int. Ser. N.Z. Dep. Sci. Ind. Res.,* **134** (Plant Nutr. VI) : 83-90.

Cartwright, B. and Hallsworth, E.G. 1970. Effects of copper deficiency on root nodules of *Subterranean clover. Plant Soil,* **33**: 685-698.

Casimiro, A. 1987. Effect of copper on water relation and growth of *Triticum.* Berlin, Springer-Verlag 454-465, ISBN 3- 540.- 16082-5 ISBNO - 387-16082-5.

Casimiro, A., Barroso, J.P. and Pais, M.S. 1990. Effect of copper deficiency on photosynthetic electron transport in wheat plants. *Plant Physiol.,* **79:** 459-464.

Cataldo, D.A., Garland, T.R., Wildung, R.E. and Drucker, H. 1978. Nickel in plants. II Distribution and chemical form in soyben plants. *Plant Physiol.,* **62**: 566-570.

Chahal, D.S. et. al. 1979. Response of sorghum to phosphorus as influenced by iron application. *Bull. Indian Soc. Soil. Sci.,* **12**: 421-427.

Chandel, A.S., Sharma, D. and Rathore, V.S. 1978. Zinc and copper interaction in green gram (*Phaseolus aureus* Roxb.) and maize (Zea mays L.). *J. Nucl. Agric. Biol.,* 7: 18-22.

Chandel, A.S., Sharma, D. and Rathore, V.S. 1985. Radio active studies on zinc and boron interaction in green gram and maize. *Indian J. Agri. Sci.,* **55:** 371-374.

Chandel, A.S., Tiwari, S.K. and Saxena, S.C. 1989. Effect of micro-nutrient application on soybean (*Glycine max*) grown in Uttar Pradesh foothills. *Indian J. Agric. Sci.,* **59:** 62- 63.

Chandler, W.H., Hoagland, D.R. and Hibbard, P.L. 1931. Little leaf or rosette of fruit trees. *Proc. Am. Soc. Hort. Sci.*, **28:** 556-560.

Chaney, R.L. 1988. Recent progress and needed research in plant Fe nutrition. *J. Plant Nutr.*, **11**: 1589-1603.

Chaney R.L. and Bell, P.F. 1987. Complexity of iron nutrition. Lessons for plant soil interaction research. *J. Plant Nutr.*, **10:** 963-994.

Chaney, R.L., Brown, J.C. and Tiffin, L.O. 1972. Obligatory reduction of ferric chelates in iron uptake by soybeans. *Plant Physiol.*, **50:** 208-213.

Chapman, H.D. 1966. Diagnostic Criteria for Plants and Soils. Div. Agr. Sci. Univ. California, Berkeley, 793p.

Chapman, H.D. 1973. visual symptoms of mineral deficiencies and excesses on citrus. *Indian Phytopath.*, **26**: 385-423.

Chapman, H.D., Leibig, G.F. Jr., and Vanselow, A.P. 1940. Some nutritional relationships as revealed by a study of mineral deficiency and excess symptoms on citrus. *Soil. Sci. Soc. Am. Proc.*, **4**: 196-200.

Chapman, K.S.R. and Jackson, J.F. 1974. Increased RNA labelling in boron deficient root tip segments. *Phytochemistry,* **13:** 1311-1318.

Chatel, D.L., Robson, A.D., Gartell, J.W. and Dilworth, M.L. 1978. The effect of inoculation and cobalt application on the growth and nitrogen fixation by sweet lupins. *Aust. J. Agric. Res.*, **29:** 1191-1202.

Chatterjee, C. and Agarwala, S.C. 1979. Effect of molybdenum and iron supply in molybdenum (99Mo) and iron (59 Fe) uptake and activity of certain enzymes in tomato plants grown in sand culture. *J. Nucl. Agric. Biol.*, **8:** 21-23.

Chatterjee, C., Nautiyal, N. and Agarwala, S.C. 1983. Interaction of iron source and nitrogen supply in rice. *J. Indian Bot. Soc.*, **62**: 54-58.

Chatterjee, C., Nautiyal, N. and Agarwala, S.C. 1984. Iron molybdenum interaction in mustard. *Indian J. Plant Nutr.*, **3**: 79-86.

Chatterjee, C. Nautiyal, N. and Agarwala, S.C. 1985. Metabolic changes in mustard plants associated with molybdenum deficiency. *New Phytol.*, **100:** 511-518.

Chatterjee, C., Nautiyal, N. and Agarwala, S.C. 1992. Excess sulphur partially alleviates copper deficiency effects in mustard. *Soil Sci. Plant Nutr.*, **38**: 92-99.

Chatterjee, C., Sinha, P. and Agarwala, S.C. 1990. Boron nutrition of cowpea. *Proc. Indian Acad. Sci. (Plant Sci.)*, **100**: 311-318.

Chatterjee, C., Sinha, P. and Agarwala, S.C. 1990. Interactive effect of boron and phoshporus on growth and metabolism of maize grown in refined sand. *Can. J. Plant Sci.*, **70**: 455-460.

Chatterjee, C., Sinha, P., Nautiyal, N., Agarwala, S.C. and Sharma, C.P. 1987. Metabolic changes associated with boron-calcium interaction in maize. *Soil Sci. Plant Nutr.*, **33:** 607-617.

Chatterjee, C., Sinha, P., Nautiyal, N., Khurana, N. and Agarwala, S.C. 1989. Physiological response of cauliflower in relation to manganese - boron interaction. *J. Hort. Sci.*, **64:** 591-596.

Chaturvedi, G.S. and Singh, A. 1981. Effect of iron and manganese on major biochemical constituents, growth and yield in wheat, *Triticum aestivum* L. *Indian J. Exptl. Biol.*, **19**: 1135-1139.

Chaudhry, F.M. and Loneragan, J.F. 1972. Zinc absorption by wheat seedlings : II. Inhibition by hydrogen ions and by micronutrient cations. *Soil Sci. Soc. Am. Proc.*, **36**: 327-331.

Chaudhry, F.M. and Wallace, A. 1976. Response of Earlirose rice to iron, zinc and BPOS when grown on calcareous soil. *Commun. Soil Sci. Plant Anal.*, **7**: 89-91.

Chen, C.H. and Lewin, J. 1969. Silicon as a nutrient element for *Equisetum arvense. Can. J. Bot.*, **47**: 174-200.

Chenery, M. 1955. A preliminary study of aluminium and the tea bush. *Plant Soil.*, **6**: 174-200.

Cheniae, G.M. and Martin, I.F. 1968. *Biochim Biophys. Acta.*, **153**: 819.

Cheniae, G.M. and Martin, I.F. 1971. Photoactivation of the manganese catalyst of O_2 evolution. I. Biochemical and kinetic aspects. *Biochem. Biophys. Acta.*, **253**: 167-181.

Chernavina, I.A., Zhogova, E.P. and Nikiforova, T.A. 1973. Plastocyanin and cytochrome f in electron transport chain of photosynthesis in oat chloroplast. *Fiziol. Rast.*, **20**: 988-994.

Chesters, G.G. and Ralison, G.N. 1951. The role of zinc in plant metabolism. *Biol. Revs. Cambridge Phil. Soc.*, **26**: 239- 252.

Chibiliti, G.B. and Byrne, D.H. 1990. Interaction of aluminium and calcium on "Nemaguard" peach seedling nutrient contents and grown in sand culture. *Sci. Hort.*, **43**: 29-36.

Churchill, K.A. and Sze, H. 1983. Anion-sensitive H-pumping ATPase in membrane vesicles from oat roots. *Plant Physiol.*, **71**: 610-617.

Clark, R.B. 1977. Effect of aluminium on growth and mineral elements of Al-tolerant and Al-intolerant corn. *Plant Soil*, **47**: 653-662.

Clarkson, D.T. 1966. Effect of aluminium on the uptake and metabolism of phosphorus by barley seedlings. *Plant Physiol.*, **41**: 165-172.

Clarkson, D.T. and Hanson, J.B. 1980. The mineral nutrition of higher plants. *Ann. Rev. Plant Physiol.*, **31**: 279-298.

Cocucci, M., Ballarin-Denti, A. and Marre, M.T. 1980. Effects of orthovanadate on H+ secretion, K uptake, electric potential difference and membrane ATPase activities of higher plant tissues. *Plant Sci. Lett.* **17**: 391-400.

Cohen, M.S. and Albert, L.S. 1974. Autoradiographic examination of meristem of intact boron deficient squash roots treated with thymidine. *Plant Physiol.*, **54**: 766-768.

Coke, L. and Whittington, W.J. 1968. The role of boron in plant growth. IV. Interrelationships between boron and indole-3-acetic acid in the metabolism of bean radicles. *J. Exp. Bot.*, **19**: 295-308.

Collander, R. 1941. Selective absorption of cations by higher plants. *Plant Physiol.*, **16**: 691-720.

Combrink, N.J.J. and Hammes, P.S. 1972. The effect of calcium, phosphorus and boron on the occurrence of internal brown fleck in potatoes. *Agroplantae.*, **4:** 81-85.

Cooper, R.E. and Girton, R.E. 1963. Physiological effect of manganese deficiency related to age in soybeans (*Glycine max.*). *Am. J. Bot.*, **50**: 105-110.

Cooper, R.E. and Thompson, A.H. 1972. Solution culture investigations of the influence of manganese, calcium, boron and pH on the internal bark necrosis of delicious apple trees. *J. Am. Soc. Hort. Sci.*, **97:** 138-141.

Coors, J.G. 1986. Resistance to European corn borer, *Ostrinia nubialis* (Hubner), in maize *Zea mays* L., as affected by soil silica, plant silica, structural carbohydrates, and lignin. *Plant Soil* (In press, 1987).

Cory, S., Finch, L.R. and Hinde, R.W. 1966. The incorporation of P^{32} phosphate into nucleic acids of normal and boron deficient bean roots. *Phytochemistry*, **5**: 625-634.

Cotton, M. 1930. Toxic effects of iodine and nickel on buck wheat grown in solution cultures. *Torry Bot. Club. Bull.*, **57**: 127-140.

Court, R.D., Williams, W.T. and Hegarty, M.P. 1972. The effect of mineral nutrient deficiency on the content of free amino acids in *Setaria sphacelata. Aust. J. Biol. Sci.*, **25:** 77-87.

Cowan, J.W. and Esfahani, M. 1967. Chemical form and physiological availability of iodine in lettuce (*Lactuca sativa*). *Nature,* **213**: 289-290.

Cresswell, C.F. and Nelson, H. 1973. The influence of boron on the RNA level, ∝-amylase activity and level of sugars of germinating *Themeda triandra.* Forsk, seeds. *Ann. Bot.*, **37:** 427-438.

Cripps, E.G. 1956. Boron nutrition of the hop. *J. Hort. Sci.*, **31**: 24-34.

Crisp, P., Collier, G.F. and Thomas, T.H. 1976. The effect of boron on tipburn and auxin activity in lettuce. *Sci. Hort.*, **5**: 215-226.

Cruz-Mireles, R.M. and Ortega-Blake, I. 1991. Effect of Na_3VO_4 on the P state of *Nitella translucens. Plant Physiol.*, **96:** 91-97.

Cutcliffe, J.A. and Gupta, U.C. 1980. Effect of added nitrogen, phosphorus and potassium on leaf tissue boron concentration of three vegetable crops. *Can. J. Plant Sci.*, **60:** 570-576.

Dami, H.M., Paul, G.S., Kakkar, V.K., Saini, S.J. and Sareen, K. 1970. Effect of boron on protein and nucleic acid contents of wheat plant. *Res. Bull. Punjab Univ. Sci.*, **21**: 425-430.

Dang, Y.P., Chhabra, R. and Verma, K.S. 1990. Effect of cadmium, nickel, lead and zinc on growth and chemical composition of onion and genugreek. *Commun. Soil Sci. Plant Anal.*, **21**: 717- 735.

D'Arrigo, C.M., Radaelli, L., Ioppolo, A. and Sambuco, G. 1983. Sodium in mineral nutrition. 11. The influence of NaCl on the uptake of some trace elements in bean plants *(Phaseolus vulgaris L.). Agrochim.*, **27**: 165-172.

Das, N. and Srivastava, R.S.L. 1970. Studies on the effect of copper, manganese and zinc in the growth and amino acids metabolism in *Raphanus sativus.* L. *Proc. Nat Acad. Sci. Indian Sec. B. Biol. Sci.*, **39**: 213-216.

Datir, M.S., Shinde, P.H. and Shengte, A.K. 1976. The effect of liming of an acid laterite soil of Maharashtra on the availability of manganese and molybdenum. I. *Maharashtra Agric. Univ.*, **1:** 179-181.

Davis, T.D., Jolley, V.D., Walser, R.H., Brown, J.C. and Blaylock, A.D. 1986. Net photosynthesis of iron-efficient and iron-inefficient soybean cultivars grown under varying iron levels. *J. Plant Nutr.*, **9:** 671-681.

Davydova, V.M. 1967. The effect of zinc and vitamins B and B on nitrogen metabolism in tomatoes. Authors Abstracts of Dissertation. Botanical Institute, Leningrad.

Dawson, C.R. 1950. The copper–protein, ascorbic acid oxidase. In : *Copper Metabolism* (W.D. McElroy and H.B. Glass, eds.) pp. 18-47. John Hopkins press, Baltimore, Maryland.

Dawson, C.R. and Tarpley, W.B. 1951. Copper oxidase. In :*The Enzymes* (J.B. Summer and K. Myrback, eds.) 2: 454-498. Academic press, New York.

Day, R. and Franklin, J. 1946. Plant carbonic anhydrase. *Science,* **104:** 363-365.

De Bruyn, J.A. 1966. The *in vitro* germination of pollen of *Sataria spchacelata.* 2. Relationship between boron and certain cations. *Plant Physiol.,* **19:** 322-327.

Debus, R.J., Fehér, G. and Okamura, M.Y. 1986. Iron depleted reaction centres from *Rhodoseudomonas sphaeroides* R-26. 1: Characterization and reconstitution with iron (2+), manganese (2+),cobalt (2+),nickel(2+),copper(2+), zinc (2+). *Biochemistry,* **25** : 2276-2287.

Dechen, A.R., Haag, H.P., Sarruge, J.R. and Oliveira, G.D. 1980. Mineral nutrition of vegetable crops. XXXVI. Effects of calcium levels in nutrient solution on the concentration of boron, copper, iron, manganese and zinc in tomato plants *(Lycopersicon esculentum Mill.). An. Esc. Super. Agric. "Luir de Queiroz", Univ. Sao. Paulo.,* **37**: 1077-1103.

Dekock, P.C. 1955. Iron nutrition of plants at high pH. *Soil Sci.,* **79**: 167-175.

DeKock, P.C. 1956. Heavy metal toxicity and iron chlorosis. *Am. Bot.,* **20:** 133-141.

DeKock, P.C., Hall, A. and McDonald, M. 1960. A relation between the ratios of phosphorus to iron and potassium to calcium in mustard leaves. *Plant Soil,* **12:** 128-142.

DeKock, P.C. and Morrison, R.I. 1956. The effect of iron status on the pattern of free amino-acids and organic acids in plant leaves. *Biochem. J.,* **63:** 13.

DeKock, P.C. and Morrison, R.I. 1958. The metabolism of chlorotic leaves. I. Amino acids. *Biochem. J.,* **70:** 266-272.

DeKock, P.C., Rutherford, M. and Cheshire, M.V. 1971. The fine structure of leaf cells of copper deficient oats. *Ann. Bot. (London),* **35:** 193-199.

Delhaize, E., Loneragan, J.F. and Webbs, J. 1982. Enzyme diagnosis of copper deficiency in *Subterranean clover.* II. A simple field test. *Aust. J. Agric. Res.,* **33:** 981-987.

Demetri, J.L., Ellis, R., Jr., and Paulson, G.M. 1972. Nodulation and nitrogen fixation by two soybean varieties as affected by phosphorus and zinc nutrition. *Agron. J.,* **64:** 566-569.

Demetriades, S.D. 1955. On the abnormal metabolism observed in iron deficient plant of *Hibiscus esculentus*. *Ann. Inst. Phytopath. Benaki,* **9**: 10-29.

Demetriades, S.D. 1956. Sur'l accumulation d' asparagine Chez l. *Hibiscus esculentus* L. Privedefer. *Ann. Inst. Phytopath. Benaki,* **10**: 13-21.

Dev, S. and Shukla, U.C. 1980. Nitrogen zinc content in maize as affected by their different sources. *J. Indian Soc. Soil Sci.,* **28**: 336-341.

Diekert, G.B., Graf, E.G. and Thauer, R.K. 1979. Nickel requirement for carbon monoxide dehydrogenase formation in *Clostridium pasteurianum. Arch. Microbiol.,* **122**:177-120.

Diez, Th. and Rosopulo, A. 1976. Schwermetallgehalte in Boden and Pflanzen nach extrem hohenKlanchlammgaben. *Landwirtsch Forsch Sonderheft,* **22**: 236-248.

Dilworth, M.J., Robson, A.D. and Chatel, D.L. 1979. Cobalt and nitrogen fixation in *Lupinus angustifolius* L. 11. Nodule formation and function. *New Phytol.,* **83**: 63-79.

Dittrich, P., Campbell, W.H. and Black, C.C. Jr. (1973). Phosphoenolpyruvate carboxykinase in plants exhibiting crassulacean acid metabolism. *Plant Physiol.* **52**: 351-361.

Dixon, N.E., Gazzola, C., Blakeley, R.L. and Zerner, B. 1975. Jack bean urease (E.C.3.5.1.5.). A metalloenzyme. A simple biological role for nickel. *J. Am. Chem. Soc.,* **97**: 4131-4133.

Djendov, C. 1971. L'influence des diverses concentration du fer surles plantes d' helianthe. *Rev. Roum. Biol. Ser. Bot.,* **16**: 413-419.

Domska, O., Benedycka, Z. and Krauze, A. 1989. Residual effect of molybdenum fertilization on the content of nitrogen compounds in beans. *Acta Acad. Agric. ac Technicae olstenensis Agricultura* No. **48**: 99-106.

Dove, W.F. and Davidson, N. 1962. Cation effect on the denaturation of DNA. *J.Mol. Biol.,* **5**: 467-472.

Downes, G.M. and Turoey, N.D. 1990. The effect of nitrogen and copper on characteristics of wood tissue in Pinus radiata, *Can. J. For. Res.,* **20**: 1369-1377.

Draycott, A.P. and Farley, F.R. 1973. Response by sugarbeet to soil dressing and foliar spray of manganese. *J. Sci. Food Agric.,* **24**: 675-683.

Dressler, H. and Dawson, C.R. 1960. On the nature and mode of action of the copper protein, tyrosinase. I. Exchange experiments with radioactive copper and the functioning enzyme. *Biochem. et Biophys. Acta.,* **45**: 515-524.

Duda, C.T. and Cherry, J.H. 1971. *Plant Physiol.,* **47**: 262.

Dugger, W.M. 1983. Boron in Plant Metabolism. In : *Encyclopaedia of Plant Physiol.* New Ser. 15B pp. 626-6[illegible]0. Springer-Verlag, Berlin, New York.

Dugger, W.M. Jr. and Hymphreys, T.E. 1960. Influence of boron on enzymatic reactions associated wit[illegible] biosynthesis of sucrose. *Plant Physiol.,* **35**: 523-530.

Dugger, W.M. Jr., Humphre[illegible]s, T.E. and Calhoun, B. (1957). The influence of boron on starch phosphorylas[illegible] [illegible]d its significance in translocation of sugars in plants. *Plant Physiol.,* **32**: 364-[illegible].

Duke, M.V. and Salin, M.L. 1983. Iso-enzymes of cuprozinc superoxide dismutase from *Pisum sativum*. *Phytochemistry*, **22:** 2369-73.

Durrant, M.J., Draycott, A.P. and Milford, G.F.J. 1978. Effect of sodium fertilizer on water status and yield of sugarbeet. *Ann. Appl. Biol.*, **88**: 321-328.

Dutta, T.R. and McIlarth, W.J. 1964. Effect of boron on growth and lignification in sunflower tissue and organ cultures. *Bot. Gaz.*, **125:** 89-96.

Dvornic, V., Moles, I., Gaina, R. and Dumitresue, A. 1972. Effect of molybdenum and boron microelements on some physiological and biochemical processes in the bean - 416 in the yield experiment. *Rev. Roum. Biol. Ser. Bot.*, **17**: 257-262.

Dwivedi, R.S. and Randhawa, N.S. 1974. Evaluation of a rapid test for the hidden hunger of zinc in plants. *Plant Soil*, **40:** 445-451.

Dwivedi, R.S. et al. 1975. Phosphorus-zinc interaction I. sites of immobilization of the zinc in maize at a high level of phosphorus. *Plant Soil.*, **43:** 639-648.

Dyar, J.J. and Webb, K.L. 1961. A relationship between boronauxin in C translocation in bean plants. *Plant Physiol.*, **36**: 672-676.

Eaton, F.M. 1942. Toxicity and accumulation of chloride and sulphate in plants. *J. Agric. Res.*, **64:** 357-399.

Echevarria, M., Soikes, R., Beeson, K.C. and Kalinowski, J. 1970. Soil plant nutrition inter-relationship : Chemical composition of the forage species of Junin. *Ancient* (La Molina) **8**: 179-194.

Edward, G.E. and Mohamad, A.K. 1973, Reduction in carbonic anhydrase activity in zinc deficient leaves of *Phaseolus vulgaris*, L. *Crop. Sci.*, **13:** 351-354.

Edwards, H. and Walker, D. 1983. C_3 C_4: Mechanisms and Cellular and Environmental Regulation of Photosynthesis. Black well, London.

Edwards, J.H. and Horton, B.D. 1977. Aluminium induced calcium deficiency in peach seedlings. *J. Am. Soc. Hortic. Sci.*, **102**: 459-461.

Eisenmenger, W.S. 1935. Toxicity of aluminium on seedlings and action of certain ions in the elimination of the toxic effect. *Plant Physiol.*, **10:** 1-25

Elamin, O.M. and Wilcox, G.E. 1986. Manganese toxicity in watermelon plants as influenced by nitrogen form. *J. Am. Soc. Hort. Sci.*, **111**: 765-768.

Elamin, O.M. and Wilcox, G.E. 1986a. Manganese toxicity development in muskmelons as influenced by nitrogen form. *J. Am. Soc. Hort. Sci.*, **111:** 323 327.

Elamin, O.M. and Wilcox, G.E. 1986b. Effect of magnesium and manganese nutrition on muskmelon growth and manganese toxicity. *J. Am. Soc. Hort. Sci.*, **111:** 582-587.

El-Gharably, G.A. and Bussler, W. 1986. Critical levels of boron in cotton plants. *Egypt J. Bot.*, **26:** 81-90.

El-Hac [illegible], Z.A., Francis, R.R. and Shehata, M.A. 1980. Effect of spraying some micronutrients on the pigments in wheat blades, yield and carbohydrate fractions of grains. Research Bulletin, Faculty of Agriculture, Ain Shams University, No. 13[illegible] [illegible]pp.

Elmstrom, G.W. and Howard, F.D. 1969. Iron accumulation, root peroxidase activity and varietal interactions in soybean genotypes that differ in iron nutrition. *Plant Physiol.,* **44**: 1108-1114.

Elstner, E.I., Pistorius, E., Boger, P. and Trebst, A. 1968. Zur Rolle on plastocyanin and cytochrome in photosynthestis ehein Electronenentransport. *Planta,* **79:** 146-161.

Eltinge, E.T. 1941. Effect of manganese deficiency upon the histology of *Lycopersicon esculentum. Plant Physiol.,* **161**: 189-195.

Emel'yanova, V.M., Novikov, A.A. and Nelyubova, G.L. 1971. Effect of boron and manganese on the yield of sugarbeet on grey forest soil of a nonchernozen zone. *Agrochim.,* **10**: 117- 125.

Ender, C., Li,. M.Q., Marin, B., Nobiling, R., Povh, B., Reiss, O. and Traxel, K. 1983. Trace element distribution at the subcellular level-possible correlation between zinc and mRNA. *IFEE Trans. Nucl. Sc. No*-**30** : 1332-4.

Epstein, E. 1960. Spaces, barriers and iron carriers : ion absorption by plants. *Am. J. Bot.,* **47**: 393-399.

Eriksson, J.E., Anderson, A. and Wenblad, A.C.H. 1990 Cd, Ni and Zn contents of oat grain as related to soil factors and precipitation. *Swedish J. Agric. Res.,* **20**: 81-87.

Eriksson, M. 1979. The effect of boron on nectar production and seed setting on red clover *(Trifolium pratense L.).Swedish J. Agric. Res.,* **9**: 137-141.

Esposito, R.G. and Wilson, P.W. 1956. Trace metal requirements of *Azotobacter. Proc. Soc. Exptl. Biol. Med.,* **93:** 564-567.

Estan, M.T., Romero, M. and Guillen, M.G. 1988. Cadmium toxicity in capsicums (*Capsicum annuum* L.). *Agrochim.,* **32:** 141- 150.

Esteban, R.M., Collado, J.G., Lopez, A.F.J. and Fernandez, H.M. 1985. Effect of boron on soluble protein and sugar contents of tomato roots. *Plant Soil,* **88:** 149-151.

Eun, M.Y. 1981. Influence of flooding, soil pH and zinc on growth and chemical composition of rice plants. *Dissertation Abstracts International B.,* **41:** 3971 ord No. 8110413.

Evans, H.J. 1956. Role of molybdenum in plant nutrition. *Soil Sci.,* **81**: 199-208.

Evans, H.J. and Hall, N.S. 1955. Association of molybdenum with nitrate reductase from soybean leaves. *Science,* **22**: 922- 923.

Evans, H.J. and Russel, S.A. 1971. Physiological chemistry of symbiotic nitrogen fixation by legumes In: *The Chemistry and Biochemistry of Nitrogen Fixation.* (J.R. post-gate,ed.) pp. 191-244. Plenum Publishing Co.

Evans, H.J. and Sorger, G.J. 1966. Role of mineral elements with emphasis on the univalent cation. *Ann. Rev. Plant Physiol.,* **17**: 47-76.

Evans, M.C.W. and Ford, R.C. 1986. Evidence for two tightly bound iron-quinones in the electron acceptor complex I Photosystem II. *FEBS Lett.,* **195:** 290-294.

Eyster, C., Brown, T.E. Tanner, H.A. and Hood, S.L. 1958. Manganese requirement with respect to growth, Hill reaction and photosynthesis. *Plant Physiol.,* **33**: 235-241.

Fackler, U., Goldbach, H., Weiler, E.W. and Amberger, A. 1985. Influence of boron deficiency on indole-3 acetic acid and abscisic acid levels in roots and shoot tips. *J.Plant Physiol.*, **119**: 295-299.

Falke, H., Podlesak, W. and Daetz, F. 1988. Effect of pH value on plant availability of soil molybdenum. Tagunsber. Akad. Landwirtschaftwiss D D R. 267 (Charakt, Pflanzenverfuegbarer Naehrst. Boden.) 351-356.

Fenical, W. 1975. Halogenation in the Rhodophyta - a review. *J. Phycol.*, **11**: 245-259.

Fido, R.J., Gundry, C.S., Hewitt, E.J. and Notton. B.A. 1977. Ultra-structural features of molybdenum deficiency and whiptail of cauliflower leaves; effects of nitrogen source and tungsten substitution for molybdenum. *Aust. Plant Physiol.*, **4**: 675-689.

Firgany, A.H., El-Hindi, M.H. and El-Moursy, S.A. 1983. Effect of seed soaking in manganese and copper solution on onion crop. *Egypt. J. Agron.*, **5**: 119-125.

Fisher, R.A. 1929. A preliminary note on the effect of sodium silicate in increasing the yield of barley. *J. Agric. Sci.*, **19**: 132-139.

Fomasieri, F.D. Malheiros, E.B., Vitti, G.C., Massari C.A. and Fornasieri, J.L. 1988. Effect of inoculation with ***Rhizobium phaseoli*** and of molybdenum fertilizer on *Phaseolus vulgaris* cv. carioca 80: *Cientifica.*, **16**: 271-279.

Fork, D.F. and Urbach, W. 1965. Evidence for the localization of plastocyanin in the electron-transport chain of photosynthesis. *Proc. Natl. Acad. Sci. USA*, **53**: 1307-1315.

Fowden, L. 1959. Radioactive iodine incorporation into organic compounds of various angiosperms. *Physiol. Plant.*, **12**: 657-664.

Fowler, J.W. and Morgan, P.W. 1972. The relationship of the peroxidative indoleacetic acid oxidase system to *in vivo* ethylene synthesis in cotton. *Plant Physiol.*, **49**: 555-559.

Fox, R.H. 1978. Selection for phosphorus efficiency in corn. *Commun. Soil Sci. Plant Anal.*, **9**: 13-37.

Fox, R.H. 1979. Soil pH, aluminium saturation and corn grain yield. *Soil Sci.*, **127**: 330-334.

Foy, C.D. 1974. Effect of aluminium on plant growth. In : The Plant Roots and Its Environment (E.W. Carson, ed). pp. 601-642. Univ. Press of Virginia, Charlottesville.

Foy, C.D. 1984. Physiological effect of hydrogen aluminium and manganese toxicities in acid soil. In: *Soil Acidity and Liming*- Agronomy Monography No. 12 2nd ed. (F. Adams, ed.) pp. 57-97. Am. Soc. Agron. Madison, Wisconsin.

Foy, C.D. Armiger, W.H., Flerring, A.L. and Lewis, C.F. 1967. Differential tolerance of cotton varieties to an acid soil high in exchangeable aluminium. *Agron. J.*, **59**: 415-418.

Foy, C.D., Burns, G.R., Brown, J.C., and Fleming, A.L. 1965. Differential aluminium tolerance of two wheat varieties associated with plant-induced pH changes around their roots. *Proc. Soil Sci. Soc. Am.*, **29**: 64-67.

Foy, C.D. and Gerloff, G.C. 1972. Responses of *Chlorella pyrenoidosa* to aluminium and low pH. *J. Phycol.,* **8:** 268- 271.

Foy, C.D., Webb, H.W. and Jones J.E. 1981. Adaptation of cotton genotypes on an acid, manganese toxic soils. *Agron. J.,* **73:** 107-111.

Foy, C.D. and Wheeler, N.C. 1979. Adaptation of ornamental species to an acid soil high in exchangeable aluminium. *J. Am. Soc. Hortic. Sci.,* **104:** 762-767.

Freney, J.R., Delwiche, C.C. and Johnson, C.M. 1959. The effect of chloride on the free amino acid of cabbage and cauliflower plants. *Aust. J. Soil Sci.,* **12:** 160-167.

Fridovich, J. 1975. Superoxide dismutase. *Ann. Rev. Biochem.,* **44:** 147-159.

Fridovich. I. 1983. Superoxide radical : an endogenous toxicant. *Annu. Rev. Pharmacol. Toxicol.,* **23:** 239-259.

Fries, L. 1962. Vitamim B_{12} in *Pisum sativum* L. *Physiol. Plant.,* **15:** 566-571.

Fries, L. 1966. Influence of iodine and bromine on growth of some red algae in axenic culture. *Physiol. Plant.,* **19:** 800-808.

Frossard, R., Stadelmann, F.X. and Niederhauser, J. 1989. Effects of different heavy metals on fructan, sugar and starch content of rye grass. *J. Plant Physiol.,* **134:** 182-185.

Fujimoto, S., Nakagawa, I., Ishimitsu, S. and O'hara, A. 1977. Purification and some properties of violet-coloured acid phosphatase from spinach leaves. *Chem Pharm. Bull.,* **25:** 1459-1462.

Fujiwara, A. and Tsutsumi, M. 1957. Biochemical studies of microelements in green plants. II. The chemical composition of rice and barley plants affected by iron deficiency. *Tohoku J. Agric. Rec.,* **7**: 397-408.

Fujiwara, A. and Tsutsumi, M. 1962. Biochemical studies of microelements deficiencies on the rate of photosynthesis and rate of respiration of rice and barley plants. *Tohoku J. Agric. Res.,* **13**: 389-395.

Gaina, R. and Silli, E. 1973.. Cu and B influence on some physiological processes in corn plant. *Stud Cerect. Biol. Ser. Bot.,* **25**: 151-158.

Gajarova, K.M. 1973. Effect of micronutrient fertilizer on protein accumulation in barley. *IZV. Akad. Nauk. Kaz. SSR. Ser. Biol.,* **4:** 87.

Galrao, E.Z. 1988. Wheat response to copper in an organic paddy soil. *Revista Brasileriade Ciencia do Solo,* **12:** 275-279.

Gangadhara, G.A. et al. 1990. Effect of sulphur on yield, oil content of sunflower and uptake of micronutrients by plants. *J.Indian Soc. Soil. Sci.,* **38:** 692-695.

Gangwar, M.S. and Mann, J.S. 1972. Zinc nutrition of rice in relation to iron and manganese uptake under different water regimes. *Indian J. Agric. Sci.,* **42:** 1032-1035.

Ganiron, R.B., Adriano, D.C., Paulson, G.M. and Murphy, L.S. 1969. Effect of phosphorus carriers and zinc sources on phosphorus-zinc interaction in corn. *Soil, Soc. Proc.,* **33:** 205-309.

Garate, A., Carpena-Ruiz, R.O. and Ramon, A.M. 1984. Influence of boron on manganese and other nutrient in juices of vascular tissue. *An. Edafol. Agrobiol.,* **43:** 1464-1478.

Garner, C.D., Arber, J.M., Harvey, I., Hasnain, S.S., Eady, R.R., Smith, B.E., De Boer, E. and Weaver, R. 1989. Characterisation of molybdenum and vanadium centers in enzymes by X-ray absorption spectroscopy. *Polyhedron*, **8:** 1649-1652.

Gassman, M., Pluscec, J. and Bogorad, L. 1968. δ-Aminolevulinic acid transaminase in *Chlorella vulgaris*. *Plant Physiol.*, **43**: 1411-1414.

Gauch, H.G. and Dugger, W.M. Jr. 1954. The physiological role of boron in higher plants : a review and interpretation. *Univ. Mryland. Agric. Expt. Sta. Tech. Bull.*, A.80.

Gerloff, G.C., Stout, P.R. and Jones, L.P.H. 1959. Molybdenum manganese - iron antagonism in the nutrition of tomato plants. *Plant Physiol.*, **84:** 608-613.

Gerretsen, F.C. 1950. Manganese in relation to photosynthesis II. Redox potentials of illuminated crude chloroplast suspensions. *Plant Soil*, **2:** 159-193.

Gerretsen, F.C. 1950. Manganese in relation to photosynthesis II. Uptake of oxygen by illuminated crude chloroplast suspensions. *Plant Soil.*, **2:** 323-343.

Gerson. D.F. and Poole, R.J. 1972. Chloride accumulation by mung bean root tips. A low affinity active transport system at the plasmalemma. *Plant Physiol.*, **50:** 603-607.

Ghalaie, K.R. 1985. Effect of potassium, sulphur, boron and molybdenum fertilisation on alfalfa production and herbage production. *DissertationAbstracts International B (Science and Engineering)*. **46**: 1754B.

Ghosh, A.K., Rai, R.K., Saxena, Y.R., and Srivastava, A.K. 1990. Effect of sulfur application on the nutritional status, yield, and quality of sugarcane. *J. Indian Soc. Soil Sci.*, **38**: 73-76.

Gibson, T.S. and Leece, D.R. 1981. Estimations of physiologically active zinc in maize by biochemical assay. *Plant Soil*, **63:** 395-406.

Gilbert, F.A. 1952. Copper in nutrition. *Adv. Agron.*, **4**: 147-173.

Gilbert, S.G., Sell, H.M. and Drosdoff, M. 1946. The effect of copper deficiency on the metabolism and oil synthesis of the tung tree. *Plant Physiol.*, **21**: 290-303.

Giordano, P.M., Koontz, H.V., and Rubins, E.J. 1966. ^{14}C distribution in photosynthesis of tomato as influenced by substrate copper and molybdenum level and nitrogen source. *Plant Soil*, **24**: 437-446.

Giordano, P.M. and Mortvedt. J.J. 1974. Response of several rice cultivars to zinc. *Agron. J.*, **66:** 220-223.

Giordano, P.M., Noggle, J.C. and Mortvedt, J.J. 1974. Zinc uptake by rice as affected by metabolic inhibitors and competing cations. *Plant Soil*, **41**: 637-646.

Gladstones, J.S., Loneragan, J.F. and Goodchild, N.A. 1977. Field responses to cobalt and molybdenum by different legume species with inferences on the role of cobalt in legume growth. *Aust. J. Agric. Res.*, **28**: 619-628.

Glikman, T.S. and Zavgorodnyaya, L.N. 1973. Possibility of water photo-oxidation by manganese containing derivatives oxidation of reduced products by quinones. *Biokhimiya.*, **38**: 101-105.

Goldstein, B.I. and Gerasimova, V.V. 1963. The denaturation and fragmentation of deoxyribonucleic acid in animal cells. *Ukrainsky Biokhimicheskij Zhurnal.*,**35:** 3-17.

Golov, V.I. and Kazachkov. Yu. N. 1979. Accumulation and distribution of essential trace elements in soybean plants. *Vladivostok. USSR* pp. 60-70.

Gomez, R.M.V., Alvarez-Tinaut, M.C. and Lune del Castillo, J. 1981. Catalase, peroxidase and IAA oxidase activities and O diphenolic content in sunflower leaves grown with different boron levels. *J. Plant Nutr.*, **4:** 375-383.

Gomez, R.M.V., Gomez, O.M. and Alvarez, T.M.C. 1981. Boron, copper, iron, manganese and zinc content in leaves of flowering sunflower plants (*Helianthus annuus* L.) grown with different boron supplies. *Plant Soil,* **62:** 461-464.

Goor, J.V. and Lune, P.V. 1980. Redistribution of potassium, boron, iron, magnesium and calcium in apple tress determined by an indirect method. *Physiologia Plant.*, **48:** 21-26.

Gopal, N.H. 1971. Effect of excess boron supply on accumulation of boron and nitrogen metabolism in groundnut plants. *Proc. Indian Acad. Sci. Soc. B.*, **73:** 192-201.

Gotoh, C.D. and Iriye, F.A.N. 1989. The action of divalent zinc, cadmium, mercury, copper and leaf ions on the ATPase activity of a plasma membrane fraction from roots of *Zea mays. Plant Soil.*, **117:** 167-175.

Grafflin, A.L. and Ochoa, S. 1950. Partial purification of isocitric dehydrogenase and oxalosuccinic carboxylase. *Biochem. Biophys. Acta.*, **4:** 205-210.

Graham. R.D. 1975. Male sterility in wheat plants deficient in copper. *Nature* (London), **254:** 514-515.

Graham, R.D. 1976. Anamolous water relations in copper deficient wheat plants. *Aust. J. Plant Physiol.*, **3:** 229-236.

Graham, R.D. 1979. Transport of copper and manganese to the xylem exudate of sunflower. *Plant Cell Env.*, **2:** 139-143.

Graham, R.D. 1980. The distribution of copper and soluble carbohydrates in wheat plants grown at high and low levels of copper supply. *Zeitschrift fur Pflanzenerna chrung Und Boden Kunde.*, **143:** 161-169.

Grant, E.A. and Gupta, U.C. 1972. Effects of rate of placement of phosphorus, pH, and temperature on the early growth, phosphorous, boron and zinc content of corn. *Can. J. Plant Sci.*, **52:** 35- 40.

Graupe, B. 1966. Molybdenum metabolism of reminants. In : Mineralsto off-versorgung von Pflanze Und Tier, Tagungsberichte Nr. 85, Dt. Akad. d. Landw. Wiss., Berlin. pp. 309-317.

Graves, C.J. Adams, P. and Winsor, G.W. 1979. Some effects of copper deficiency on the flowering, copper status and phenolase activity of different cultivars of *Chrysanthemum morifolium* L. *Sci. Food and Agric.*, **30:** 751-758.

Greenwood, E.A.N. and Hallsworth, F.G. 1960. Studies on the nutrition of forage legumes. II. Some interactions of calcium, phosphorus, copper and molybdenum on the growth and chemical composition of *Trifolium subterraneum* (L). *Plant Soil*, **12:** 97-127.

Gribanov, V.M. 1954. A study on the effects of various doses of copper and manganese on nitrogen fixation in *Azotobacter chrooccum. Mikrobiol. Zhurnal.*, **16:** 3-12.

Griffiths. D.E. and Wharton, D.C. 1961. Copper in cytochrome oxidase. *Biochem. Biophys. Research Communs.*, **4:** 199-204.

Grimm, M.R. 1952. Iodine content of some marine algae. *Pac. Sci.*, **6:** 318-323.

Grimme, H. 1983. Aluminium induced magnesium deficiency in oats. *Z. Pflanzenernahr. Bodenk.*, **146:** 666-676.

Grinkevich, N.I.,.Borovkova, L.I. and Gribovskaya, I.F. 1970.The effect of trace elements on the alkaloid content of *Atropa belladona* L. *Farmatsiya* : 41-45.

Gris, E. 1844. Nauvelles experiences sur l'action des composes ferrugineux solubles applique a la vegetation at specialement on triatment de la chlorose et da la de'bilite des plantes. *2 C.R. Acad. Sci. (Paris).*, **19:** 1118-1119.

Gross, G.G. 1980. The biochemistry of lignification. *Adv. Bot. Res.*, **8:** 26-63.

Gudkova, G.V. 1969. Effect of some copper containing fungicides on the carbohydrates metabolism in leaves and tubers of potatoes. *Za. Leningrad Sel' Skokhoz Inst.*, **134**: 29-32.

Guerinot, M.L. 1991. Iron uptake and metabolism in the rhizobia/legume symbioses. *Plant Soil.*, **130:** 199-209.

Guikema, J.A. and Sherman, L.A. 1983. Organisation and function of chlorophyll in membranes of cyanobacteria during iron starvation. *Plant Physiol.*, **73**: 250-256.

Gunsalus, I.C., Pederseon, T.C. and Sligar, S.G. 1975. Oxygenase catalysed biological hydroxylations. *Ann. Rev. Biochem.*, **44**: 317-340.

Gupta, I.C. and Chandra, H. 1972. Effect of gypsum in reducing boron hazard of saline water and soil. *Ann. Arid Zone,* **11**: 228-230.

Gupta, R.K. and Singh, R.D. 1990. Studies on the antagonistic effects of lime on iron and aluminium. *J. Indian Soc. Soil Sci.*, **38**: 130-134.

Gupta, S.P., and Gupta, V.K. 1984. Influence of zinc on calcium, magnesium, sodium, potassium and phosphorus nutrition of soybean (*Glycine max*) in sodic soils. *Indian J. Ecol.*, **11:** 236- 242.

Gupta, U.C. 1972. Effects of manganese and lime on yield and on the concentration of manganese, molybdenum, boron, copper and iron in the root storage tissue of barley. *Soil Sci.*, **114:** 131-136.

Gupta, U.C. and Cutcliffe, J.A. 1972. Effect of lime and boron on brown-heart, leaf tissue calcium/boron ratios, and boron concentrations of rutabaga. *Soil Sci. Am. Proc.*, **36:** 936- 939.

Gupta, U.C. and Macleod, L.B. 1975. Effects of sulphur and molybdenum on the molybdenum, copper and sulphur concentrations of forage crops. *Soil Sci.*, **120:** 441-447.

Gupta, U.C. and Munro, D.C. 1969. Influence of sulphur, molybdenum and phosphorus on chemical composition and fields of Brussels sprout and of molybdenum on sulphur contents of several plant species grown in the green house. *Soil Sci.*, **107:** 114-118.

Gupta, V.K. and Potalia, B.S. 1991. Influence of nitrogen carriers and zinc levels on copper, manganese and iron nutrition of maize (*Zea mays* L.). *Indian J. Agric. Res.*, **25:** 1-6.

Gupta, V.K. and Rajhans 1980. A note on effect of zinc application on the yield and phosphorus nutrition of coriander. *Haryana J. Hort.*, **9:** 82-83.

Gupta, V.K. and Rajhans 1983. Zinc,potassium relationship in wheat. *Haryana Agric. Univ. J. Res.*, **13**: 140-146.

Gupta, V.K. and Ram. K. 1980. Effect of copper and molybdenum on the copper/molybdenum ratio and their concentration in different organs of cowpea (*Vigna sinensis* L.). *Plant Soil*, **56:** 235-241.

Gupta, V.K. and Singh, B. 1984. Growth and nutrient composition of peanut (*Arachis hypogaea* L.) as influenced by zinc and molybdenum application. *Indian J. Plant Nutr.*, **3:** 125- 135.

Gupta, V.K. and Singh, B. 1985. Residual effect of zinc and magnesium on maize crop. *J. Indian Soc. Soil Sci.*, **33:** 204-207.

Gupta, V.K. and Singh, S. 1981. Influence of molybdenum and iron interaction on copper, manganese and zinc nutrition of paddy (*Oryza sativa*). *Indian J. Plant Physiol.*, **24**: 137-144.

Gurevich, A.S. and Yakushkina, N.I. 1989. Cytokinins as one of the factors involved in the effect of copper on the photosynthetic apparatus of potato. *Fiziol. Biokhim. Kul't Rast.*, **21:** 156-161.

Haas, A.R. 1936. Zinc relation in mottle leaf of citrus. *Bot. Gaz.*, **98**: 65-86.

Haas, A.R.C. and Klotz, L.J. 1931b. Further evidence for the necessity of boron for health in citrus. *Bot. Gaz.*, **92**: 94-100.

Habermann, H.M. and Shoemaker, F.W. 1983. Copper deficiency in *Helianthus annuus* L. developmental and metabolic consequences, stomatal function and senescence. *Growth*, **47**: 184-200.

Hageman, R.H., Hodge, E.S., McHargue, J.S. 1942. Effect of potassium iodide on the ascorbic acid content and growth of tomato plants. *Plant Physiol.*, **17:** 465-470.

Hageman, R.H. and Reed, A.J. 1980. Nitrate reductase from higher plants. In : *Methods in enzymolgy*, **69:** 270-281.

Hager, A. and Helmle, M. 1981. Properties of an ATP fueled, ct- dependent proton pump localized in membranes of microsomal vesicles from maize coleoptiles. *Z. Naturforsch. Bio. Sci., C* : **36:** 997-1008.

Hallsworth, E.G., Wilson, S.B., and Adams, W.A. 1965. Effect of cobalt on the non nodulated legume. *Nature.*, **205:** 307.

Hampp, R. and Schnabl, H. 1975. Effect of aluminium ions on $14CO_2$-fixation and membrane system of isolated spinach chloroplasta. *Z. Pflanzenphysiol.*, **76:** 300-306.

Hamzah, S. and Gomej, J.B. 1980. Ultrastructure of mineral deficient leaves of *Hevea* II. Effect of micronutrient deficiencies. Rubber *Res. Inst. Malayasia*, **28**: 17-25.

Handerson, J.H.M. and Veal. M.P. 1948. The effect of interactionship of boron and manganese on the growth and calcium uptake of blue lupine (*Lupinus angustifolius*) in solution culture. *Plant Physiol.,* **23**: 609-620.

Hanger, B.C. 1965. The influence of iron upon the toxicity of manganese, molybdenum, copper and boron in red clover (*Trifolium pratense* L.). *J. Aust. Inst. Agric. Sci.,* **31**: 315-317.

Hannam, R.J., Davies, W.J., Graham, R.D. and Riggs, J.L. (1984). The effect of soil and foliar applied manganese in preventing the onset of manganese deficiency in *Lupinus angustifolius. Aust. J. Agric. Res.,* **35**: 529-538.

Hannam, R.J., Graham, R.D. and Riggs, J.L. 1985. Redistribution of manganese in maturing *Lupinus angustifolius* cv. Illyarrie in relation to levels of previous accumulation. *Ann. Bot.,* **56**: 821-834.

Harrison, S.J., Lepp, N.W. and Phipps, D.A. 1983. Copper uptake by excised roots : 3. Effect of manganese on copper uptake. *Z. Pflanzenphysiol.,* **109**: 285-289.

Hartt, C.E. 1934. Some effects of potassium on the growth of sugarcane and upon the absorption and migration of ash constituents. *Plant Physiol.,* **9**: 399-452.

Hasselbach, G. 1940. Effects of long term application of domestic wastes on the heavy metal transfer from soil to plant. III. Soil specific effects on the uptake and extraction of heavy metals. Verband Deutscher Landwirtschaftlicher. Unter suchungs-Und Forschurg sanstalten. *Reihe Kongressberschte.,* **30**: 281- 286.

Hatch, M.D. and Mau, S.L. 1977. Properties of phosphoenolpyruvate carboxykinase operative in C4 pathway photosynthesis. *Aust. J. Plant Physiol.,* **4**: 207-216.

Hatch, M.D. and Slack, C.R. 1970. Photosynthetic CO_2-fixation pathways. *Ann. Rev. Plant Physiol.,* **21**: 141-162.

Hawker, J.S., Marschner, H. and Downton, W.J.S. 1974. Effect of sodium and potassium on starch synthesis in leaves. *Aust. J. Plant Physiol.,* **1**: 491-501.

Haynes, R.J. and Swift, R.S. 1985. Effects of soil acidification on the chemical extractability of Fe, Mn, Zn and Cu and the growth and micronutrient uptake of high bush blueberry plants. *Plant Soil,* **84**: 201-212.

Haynes, R.J. 1984. Effect of lime, silicate and phosphate applications on the concentrations of extractable aluminium and phosphate in a spodasol. *Soil Sci.,* **138**: 8-14.

Healy, W.B., Cheng, S. and McElroy, W.D. 1955. Metal toxicity and iron deficiency effects on enzymes in *Neurospora. Arch. Biochem. Biophys.,* **54**: 206-214.

Heath, R.L. and Hinde, G. 1969. On the functional site of manganese in photosynthetic oxygen evolution. *Biochem. Biophys. Acta,* **189**: 223-233.

Hecht-Buchholz, C. and Foy, C.D. 1981. Effect of aluminium toxicity on root morphology of barley. *Plant Soil.,* **63**: 93-95.

Hecht-Buchholz, C.C., Jorns, A. and Keil, P. 1987. Effect of excess aluminium and manganese on Norway spruce seedlings as related to magnesium nutrition. *J. Plant Nutr.,* **10**: 1103- 1110.

Heenan, D.P. and Campbell, L.C. 1980. Transport and distribution of manganese in two cultivars of soybean (*Glycine max* (L). Merr.) *Aust. J. Agric. Res.,* **31** : 943-949.

Heenan, D.P. and Campbell, L.C. 1981. Influence of potassium and manganese on growth and uptake of magnesium by soybeans (*Glycine max*) cultivar. Bragg. *Plant Soil.,* **61:** 447-456.

Heenan, D.P. and Campbell, L.C. 1990. The influence of temperature on the accumulation and distribution of manganese in two cultivars of soybeans (*Glycine max*). *Aust. J. Agric. Res.,* **41**: 835-843.

Hellriegel, H. and Wilfarth, H. 1898. Vegetations versucha uber der Kalibedarf einiger. *Pflazen. Azab. deut. Landw. G.E.,* 34: Cited by Dorph Petersen and Steenbjerg (669).

Henning, S.J. 1975. Aluminium toxicity in the primary meristem of wheat roots. Ph.D. Thesis. Oregon State Univ. Corvallis, Oregon, USA.

Henriques, F.S. 1989. Effects of copper deficiency on the photosynthetic apparatus of sugarbeet (*Beta vulgaris* L.). *J. Plant Physiol.,* **135**: 453-458.

Herms, U. and Brumnier, G. 1979. Einflub der Redoxbedingungen auf die Loslichkeit von Schwermetallen in Boden und Sedimenten. *Milt Dtsch. Bodenkdl. Ges.,* **29:** 533-544.

Hernandez-Medina, E. and Shive, J.W. 1946. Calcium-boron relationship in the nutrition of corn and distribution of these elements in the plants. *J. Agric. Univ. Puerto. Rico.,* : 253-291.

Herrett, R.A., Hatfield, H.H., Crosby, D.G., and Vlitos. A.J. 1962. Leaf abscission induced by the iodide plant. *Plant Physiol.,* **37:** 358-363.

Heuterman, G.A., Woodbridge, C.G. and Kalin, E.W. 1973. Boron levels in stems, leaves and flower part of carnation. *Hort. Sci.,* **8:** 126-127.

Hewitt, E.J. 1948. Relation of manganese and other metal toxicities to the iron status of plants. *Nature,* **161**: 489- 490.

Hewitt, E.J. 1949. Experiments on iron metabolism in plants. I. some effects of metal induced iron deficiency. *Long. Ashton Research Sta. Ann. Rept.,* **1948** :66-80.

Hewitt, E.J. 1951. Experiments on iron metabolism in plants III. The relation of molybdenum and nitrogen supply to metal induced iron deficiency in sugarbeet. *Long Ashton Res. Sta. Ann. Rept.,* **1950:** 64-70.

Hewitt, E.J. 1952. Sand and Water Culture Methods used in the Study of Plant Nutrition. *Commonwealth Bur. Hort. Plantation Crops. (Gr. Brit.) Tech. Commun.* No.22.

Hewitt, E.J. 1953. Metal interrelationship on plant nutrition. *J. Exptl. Bot.,* **4:** 59-64.

Hewitt, E.J. 1954. Metal interrelationships in plant nutrition I. Effects of some metal toxicities on sugarbeet, tomato, oat, potato, and marrowstem Kali in sand culture. *J. Exptl. Bot.,* **4**: 59-64.

Hewitt, E.J. 1955. Metal interrelationships in plant nutrition II. The relation of metal toxicity, molybdenum and nitrogen source to chlorophyll and magnesium content of sugarbeet in sand culture. *J. Exptl. Bot.,* **5:** 110-118.

Hewitt, E.J. 1963. Essential nutrient elements. Requirements and interactions in plants. In: *Plant Physiology Vol. III. (F.C. Steward, ed.) pp 136-160. Academic Press Inc., New York.*

Hewitt, E.J. 1966. Sand and Water Culture Method used in the Study of Plants Nutrition. Rev. 2nd ed. Common W. Agric. Bureaux. England.

Hewitt, E.J. 1983. The Essential and Functional Mineral Elements. In: *Diagnosis of Mineral Disorders in Plants* Vol. I. Principles, (Late C. Bould, E.J. Hewitt, P. Needham and J.B.D. Robinson, Eds.). pp.7-53. H.M. S.O. London.

Hewitt, E.J. 1983. The Effects of Mineral Deficiencies and Excesses on Growth and Composition. In: *Diagnosis of Mineral Disorders in Plants.* Vol. I. Principles (Late C. Bould, E.J. Hewitt, P. Needham and J.B.D. Robinson, eds.) pp. 54-110. H.M.S.O., London.

Hewitt, E.J. and Agarwala, S.C. 1952. Reduction of triphenyltetra - zolium chloride by plant tissues and its relation to molybdenum status. *Nature* (London), **169:** 545-546.

Hewitt, E.J., Agarwala, S.C. and Williams, A.H. 1957. Molybdenum as a plant nutrient. VIII. The effect of different molybdenum levels and nitrogen supplies on the nitrogen fractions in cauliflower plants grown in sand culture. *J. Hort. Sci.,* **32** : 34-48.

Hewitt, E.J. and Bolle-Hones, E.W. 1952. Molybdenum as a plant nutrient. II. The effects of molybdenum deficiency on some horticultural and agricultural crops grown in sand culture. *J. Hort. Sci.,* **27:** 257-265.

Hewitt, E.J. and Bolle-Jones, E.W. 1953. Studies in iron deficiency of crops. II. The interrelationships of iron and potassium in the potato plant. *J. Hort. Sci.,* **28**: 185-195.

Hewitt, E.J. and Bond, G. 1966. The cobalt requirement on non-legume root nodule plants. *J. Exp. Bot.,* **17:** 480-491.

Hewitt, E.J. and Gundry, C.S. 1970. The molybdenum requirement of plants in relation to nitrogen supply. *J. Hort. Sci.,* **45:** 351-358.

Hewitt, E.J., Hucklesby, D.P., Mann, A.F., Notton, B.A. and Ruckledge, G.J. 1979. Regulation of nitrogen assimilation in plants. In: Nitrogen Assimilation of Plants (E.J. Hewitt and V. Cutting. eds.).(6th Long Ashton Symp.) pp.255-287. Academic Press, Inc., New York.

Hewitt, E.J. and Jones, E.W. 1947. The production of molybdenum deficiency in plants grown in sand culture with special reference to tomato and brassica crops. *J. Pomil. Hort. Sci.,* **23**: 254-262.

Hewitt, E.J., Jones, E.W. and Williams, A.H. 1949. Relation of molybdenum and manganese to the free amino acid content of the cauliflower. *Nature,* **163:** 681-682.

Hewitt, E.J. and McCready, C.C. 1956. Molybdenum as a plant nutrient. VII. The effects of different molybdenum and nitrogen supplies on yields and composition of tomato plants grown in sand culture. *J. Hort. Sci.,* **31:** 284-290.

Hewitt, E.J. and Notton, B.A. 1980. Nitrate reductase systems in eukaryotic and prokaryotic organisms. In: Molybdenum Containing Enzymes (M. Coughlah, ed.). pp. 275-325. Pergamon Press.

Hewitt, E.J. and Tatham, P. 1960. Interaction of mineral deficiency, and nitrogen supply on acid phosphatase activity in leaf extracts. *J. Exptl. Bot.,* **11**: 367-376.

Hiatt, A.J., Amos, D.F. and Massey, H.F. 1963. Effect of aluminium on copper absorption by wheat. *Agron. J.*, **55:** 284-287.

Hill, A.E. and Hill, B.S. 1976. Mineral ions. In : Transport in Plants II. *Encyclopaedia of Plant Physiology, New Ser, Vol. 2. Part B: Tissues and Organs. (U. Luttge, M.G. Pitman, eds.)pp. 225-243. Springer, Berlin, Heidelberg, New York.*

Hill, J., Robson, A.D. and Loneragan, J.F. 1979. The effect of copper supply and shading on Cu retranslocation of nutrients of the oldest leaf of wheat. *Ann. Bot.*, **44:** 279-287.

Hill, J.M. 1973. The changes with age in the distribution of copper and some copper containing oxidase in red clover (*Trifolium pratense* L. Cv. Dorset Malgrass). *J. Exp. Bot.*, **24:** 525-533.

Hill, W.L. and Morrill, L.G. 1975. Boron, calcium, and potassium interaction in spanish peanuts. *Soil Sci. Soc. Am. Proc.*, **39:** 80-83.

Hind, G., Nakatini, N.Y. and Izawa, U. 1949. The role of Cl in photosyntyhesis. The Cl requirement of electron transport. *Biochem. Biophys. Acta.*, **172:** 277-289.

Hirasawa, M., Sung, J.D., Malkin, R., Zilber, A., Droux,M.and Knaff, D.B. 1988. Evidence for the presence of a (2Fe-2S) ferredoxin in bean sprouts. *Biochem. Biophys. Acta.*, **934:** 169-176.

Hirsch, M., Pengelly, W.L. and Torrey, J.G. 1982. Endogenous IAA levels in boron deficient and control root tips of sunflowers. *Bot. Gaz.*, (Chicago) **143:** 15-19.

Hirsch, M. and Torrey, J.G. 1980. Ultrastructural changes in sunflowers root cells in relation to boron deficiency and added auxin. *Can. J. Bot.*, **58:** 856-866.

Hiyama, T. 1987. Iron-sulfur clusters bound to photosynthetic membranes. *Iron Sulfur Protein Res.*, (Pap. Int. Symp. Front). (H. Matsubara, Y. Kutsube and K. Wada. eds.). pp. 254- 259. Japan Sci. Soc. Press : Tokyo, Japan.

Hoagland, D.R. 1944. Lectures on the Inorganic Nutrition of Plants. Chronica Botanica, Waltham, Massachusetts.

Hoagland, D.R., Chandler, W.H. and Hibbard, P.L. 1936. Little leaf or rossette of fruit trees. V. Effect of zinc on the growth of plants of various types of controlled soil and water culture experiments. *Proc. Am. Soc. Hort. Sci.*, **33:** 131-141.

Hoch, F.L. and Vallee, B.L. 1958. The metabolic role of zinc. In : Trace Elements (C.A. Lamb, O.G. Bentley and J.M. Beattie, eds.) pp. 337-363.

Hoffer, G.N. 1930. Testing corn stalks chemically to aid in determining their plant food needs. Purdue Univ. Agric. Expt. Sta. Ext. Bull. 298.

Holden, M. and Pirie, N.W. 1955. A comparison of leaf and pancreatic ribonuclease. *Biochem. J.*, **60:** 53-56.

Holm-Hansen, O., Gerloff, G.C. and Koog, F. 1954. Cobalt as an essential element for blue green algae. *Physiol. Plant.*, 7: 665-675.

Homann, P. 1967. Studies on the manganese of the chloroplast. *Plant Physiol.*, **42:** 997-1007.

Horiguchi, T. 1988. Mechanism of manganese toxicity and tolerance of plants. IV. Effects of silicon and deviation of manganese toxicity of rice plants. *Soil Sci. Plant Nutr.*, **34:** 65- 73.

Horner, C.K., Burk., D., Allison, F.E. and Sherman, M.S. 1942. Nitrogen fixation by *Azotobacter* as influenced by molybdenum and vanadium. *J. Agric. Res.,* **65:** 173-193.

Horst, W.J. 1988. The physiology of manganese toxicity. In : Manganese in Soils and Plants. (R.D. Graham, R.J. Hannam and N.C. Uren, eds.) pp. 175-188. Kluwer Academic Publishers.

Horst. W.J. and Marschner, H. 1978. Effect of silicon on manganese tolerance of bean plants. (*Phaseolus vulgaris* L.). *Plant Soil.,* **50:** 287-303.

Horst, W.J. and Marschner, H. 1978a. Effect of excessive manganese supply on uptake and translocation of calcium in bean plants (*Phaseolue vulgaris* L.). *Z. Pflanzen-Physiol.,* **87:** 137- 148.

Horst, W.J., Wagner, A. and Marschner, H. 1982. Mucilage protects root meristem from aluminium injury. *Z. Pflanzenphysiol.,* **105:** 435-444.

Horvath, G., Droppa, M. and Terry, N. 1983. Changes in photosynthetic attributes in response to copper dapletion in sugarbeet (*Beta vulgaris* L.). *J. Plant Nutr.,* **6:** 971-982.

Howeler, R.H. and Cadavid, L.R. 1976. Screening of rice cultivar for tolerance to Al-toxicity in nutrient solutions as compared with a field-screening method. *Agron, J.,* **68:** 551-555.

Hsiao, S.I.C. 1969. Life history and iodine nutrition of the marine brown alga, *Petalonia fascia* (O.F. Mull, eds.), *Kuntze, Can. J. Bot.,* **47:** 1611.1616.

Hsu, W.P. and Miller, G.W. 1968. Iron in relation to aconitase hydratase activity in *Glycine max* Merr. *Biochim. Biophys. Acta,* **151**: 711-715.

Hucklesby, D.P. 1961. A histogical study of whiptail in cauliflower plants. M.Sc. thesis, Univ. London.

Hudak, J. 1973. Effect of boric acid on the division and elongation of cells in root tips of *Vicia faba* (1) Cv. Povozaky. Acta. Fac. Rerum Nat. Univ. Comenianc. *Plant Physiol.,* **6:** 119-126.

Hulagur, B.P. and Dangarwala, R.T. 1982. Effect of zinc, copper, and phosphorus fertilization on the uptake of iron, manganese and molybdenum by hybrid maize. *Madras Agric. J.,* **69:** 11-16.

Hulagur, B.F., Dangarwala, R.T. 1983. Effect of zinc, copper and phosphorus fertilization on the content and uptake of introgen and secondary nutrients by hybrid maize. *Madras Agric. J.,* **70:** 88-91.

Humphreys, F. and Truman, R. 1964. Aluminium and the phosphorus requirements of *Pinus radiata. Plant Soil,* **20:** 131-134.

Hunter, J.G. and Vergnano, O. 1953. Trace-element toxicites in oat plants. *Ann. Appl. Biol.,* **40:** 761-777.

Hutchinson, T.C. 1970a. Lime chlorosis as a factor in seedling establishment on calcareous soils. II. The development of leaf water deficits in plants showing lime chlorosis. *New Phytol.,* **69:** 143-157.

Hylmo, B. 1958. Passive components in the ion absorption of plant. II. The zonal water flow, ion passage and pore size in roots of *Vicia faba*. *Physiol. Plant.*, **11:** 382-400.

Ichioka, P.S. and Arnon, D.I. 1955. Molybdenum in relation to nitrogen metabolism. II. Assimilation of ammonia and urea without molybdenum by *Scenedesmus*. *Physiol. Plant.*, **8:** 552-560.

Igoshina, T.I. 1970. A study of the physicochemical state of zinc in tomato chloroplasts in relation to its physiological role. Authors Abstracts of Dissertation. Botanical Institute, Leningrad.

Iqtidar, A. and Rehman, S.P. 1984. Effect of boron, on the protein and amino acid composition of wheat grain. *J. Agric. Sci.*, **103:** 75-80.

Isaev, B.M. and Rustamov, K. 1975. Effect of the trace element copper on some physiological and biochemical processes and productivity of cotton. *Uzb. Biol.*, **19:** 27-30.

Ivanova, S.M. 1964. Changes in the metabolism and the content of mobile and bound iron in apple with chlorosis. In : Voprosobmena veschestv prodovykh lovoschnykh. Kortya molydevenyaske Kishinev. 3-12.

Ivanova, E.V. 1970. Effect of various ratios of phosphorus and iron in the medium on the absorption and distribution of these elements in kidney beans (*Phaseolus*). Vestn. Mosk. Univ. Ser. 6. *Biol. Pochvoved.*, **25:** 87-91.

Iyenger, B.R.V. and Seshagiri Rao, T. 1971. Studies on manganese induced iron chlorosis in coffee (*Coffea arabica* L.) II. Effect on chlorophyll content and activities of catalase and peroxidase enzymes. *Mysore J. Agric. Sci.*, **5:** 167-173.

Jacob, V.J., Engbe, N.E. and Okoloko, G.E. 1972. Abnormalities in vegetative and floral organs of *Cola nitida* caused by boron deficiency. *Can. J. Bot.*, **50**: 381-383.

Jacobs, M. and Taiz, L. 1980. Vanadate inhibition of auxin enhanced H^+ secretion and elongation in pea epicotyls and oat coleoptiles. *Proc. Natl. Acad. Sci. USA*, **77:** 7242-7246.

Jacobson, L. 1945. Iron in the leaves and chloroplasts of some plants in relation to their chlorophyll contents. *Plant Physiol.*, **20:** 233-245.

Jacobson, L. and Oertli, J.J. 1956. The relationship between iron and chlorophyll content in chlorotic sunflower leaves. *Plant Physiol.*, **31:** 199-204.

Jawaharlal, M. et. al. 1988. Influence of the method of application of zinc and iron on the major nutrient content of onion (*Allium cepa* var cepa). *South Indian Hort.*, **36:** 308-312.

Jefferey, R.A. and Wallace, A. 1968. Detection of iron ethylenediamine-bis (O-hydroxy-phenylacetate) in plant tissue. *Agron. J.*, **60:** 613-616.

Ji, Z.H., Korcak, R.F., Wergin, W.P., Fan, F. and Faust, M. 1984. Cellular ultrastructure and net photosynthesis of apple seedlings under iron stress. *J. Plant Nutr.*, **7:** 911-928.

Johnson, A.D.W. and Simons, J.G. 1979. Diagnostic indices of zinc deficiency in tropical legumes. *J. Plant Nutr.*, **1:** 123-149.

Johnson, C.M., Stout, P.R., Broyer, T.C. and Carlton, A.B. 1957. Comparative chlorine requirements of different plant species. *Plant Soil.,* **8:** 337-353.

Johnson, G.V., Mayeux, P.A. and Evans, H.J. 1966. A cobalt requirement for symbiotic growth of *Azolla filiculoides* in absence of combined nitrogen. *Plant Physiol.,* **41:** 852-855.

Johnson, H.S. and Hatch, M.D. 1970. Properties and regulation of leaf nicotinamide-adenine dinucleotide phosphate-malate dehydrogenase and malic enzyme in plants with the C-4 dicarboxylic acid pathway of photosynthesis. *Biochem. J.,* **119:** 273-280.

Johnson, J.L. 1980. The molybdenum cofactor common to nitrate reductase, xanthine dehydrogenase and sulphite oxidase. In Molybdenum and Molybdenum-containing Ezymes (M.P. Coughlan, ed.) pp. 347-383. Pergamon.

Johnson, J.L., Hainline, B.E. and Rajagopalan, K.V. 1980. Characterization of the molybdenum cofactor of sulphite oxidase, xathine oxidase and nitrate reductase. Identification of a pteridine as a structural component. *J. Biol.Chem.,* **255:** 1783-1786.

Johnson, R.E. and Jackson, W.A. 1964. Calcium uptake and transport by wheat seedlings as affected by aluminium. *Proc. Soil Sci. Soc. Am.,* **28:** 381-386.

Johnston, E.S. and Fischer, P.L. 1930. The essential nature of boron to the growth and fruiting of the tomato. *Plant Physiol.,* **5:** 387-392.

Johnston, M., Grof, C.P.L. and Brownell, P.F. 1984. Responses to ambient CO_2 concentrations by sodium deficient C_4 plants. *Aust. J. Plant Physiol.,* **11:** 137-141.

Jones, E.W. and Hewitt, E.J. 1949. Experiments on iron metabolism in plants. II. The interrelationship of iron and potassium in the metabolism of the potato plant. *Long Ashton Res. Sta. Ann. Rept.,* 49-57.

Jones, J.B. Jr. 1970. Distribution of 15 elements in corn leaves. *Commun. Soil Sci. Plant Anal.,* **1:** 27-34.

Jones, L.H., Shepardson, W.B. and Peters, C.A. 1949. The function of manganese in the assimilation of nitrates. *Plant Physiol.,* **24:** 300-306.

Jones, L.H.P. 1978. Mineral components of plant cell walls. *Am. J. Clin. Nutr.,* **31:** 94-98.

Jones, L.H.P. and Handreck, K.A. 1967. Silica in soils, plants and animals. *Adv. Agron.,* **19:** 107-149.

Jones, L.H.P. and Handreck, K.A. 1965. Studies of silica in the oat plant III. Uptake of silica from soils by the plant. *Plant Soil.,* **23:** 79-96.

Jones, L.H.P., Hartley, R.D. and Jarvis, S.C. 1978. Mineral content of forage plants in relation to nutritional quality silicon. In Annual Report of Grassland Research Institute, Hurley, pp. 25-26.

Jones, O.T.G. 1968. Ferrochelatase in spinach chloroplasts. *Biochem. J.,* **107:** 113-119.

Jones, R.L., Hinsely, T.D. Ziegler, E.L. and Tyler, J.J. 1975. Cadmium and zinc contents of corn leaf and grain produced by sludge amended. *Soil J. Environ. Qeial.,* **4:** 509-514.

Joselow, M. and Dawson, C.R. 1951. The copper of ascorbic acid oxidase; exchange studies with radioactive copper. *J. Biol. Chem.*, **191**: 11-20.

Joshi, G., Dolan, T., Gee, R. and Saltman, P. 1962. Sodium chloride effect on dark fixation of CO_2 by marine and terrestial plants. *Plant Physiol.*, **37**: 446-449.

Jyung, W.H., Fhamann, A., Schlender, K.K. and Scala, J. 1975. Zinc nutrition and starch metabolism in *Phaseolus vulgaris* L. *Plant Physiol.*, **55**: 414-420.

Jyung, W.H. and Camp, M.E. 1976. The effect of zinc on the formation of ribulose diphosphate carboxylase in *Phaseolus vulgaris*. *Physiol. Plant.*, **36**: 350-355.

Kaleya, P.M., Manolov, P., Ignatov, G. and Lilova, I. 1989. Changes in the structure and function of the photosynthetic apparatus in peach leaves under iron deficiency. *Dokl. Bolg. Akad. Nauk.*, **42**: 105-108.

Kalia, B.D. and Sharma, C.M. 1988. Effect of micronutrients on yield of soybean. *Indian J. Agron.*, **33**: 199-200.

Kanematsu, S. and Asada, K. 1989. Copper, zinc-superoxide dismutase in rice : occurrence of an active, monomeric enzyme and two types of isoenzyme in leaf and non-photosynthetic tissues. *Plant Cell Physiol.*, **30**: 381-391.

Karakis, K.D. and Rudakova, E.V. 1971. Effect of zinc on the activity of enzyme system of IAA formation from tryptophan. *Fiziol. Biokhim. Kul't Rast.*, **3**: 512-516.

Karamanos, R.E., Hodge, N. and Stewart, J.W.B. 1989. The effect of sulphur on manganese and copper nutrition of canola. *Can. J. Soil Sci.*, **69**: 119-125.

Karki, A.B., Pradhan, S.B. and Maharjan, P.L. 1974. Influence of molybdenum, lime and phosphorus and their interactions on the yield and chemical composition of soybean. *Rev. Ecol. Biol. Sol.*, **114**: 249-256.

Karzhasova, A.V. and Kelekenov, D.K. 1982. Effect of trace elements on the amino acid composition and heterogeneity of tobacco leaf cytoplasmic proteins. *F. Zv. Akad. Nauk. Kaz, SSR. Ser. Biol.*, **2**: 25-30.

Kashyap, J. et. al. 1988. Influence of zinc levels on dry matter yield, Zn and P concentration in parts of cotton genotypes at varying growth stages. *J. Indian Soc. Soil Sci.*, **36**: 386- 388.

Katoh, S. and Takamiya, A. 1961. A new leaf copper protein plastocyanin-a natural Hill oxidant. *Nature (London)*, **189**: 665-666.

Kaur, N.P. Arora, C.L. and Sadana, U.S. 1991. Effect of manganese stress on the reproductive phase of wheat. *New Trends in Plant Physiology*. (K.K. Dhir, I.S. Dua and K.S. Chark, eds.). pp. 269-272.

Kazak, M., Biczok, G. and Szemes, I. 1984. Possibilities and necessity of the placement of calcium fertilizers on acidic sandy soils. *Zesz. Probl. Postepow Nauk. Roln.*, **285**: 55-60.

Keilen, D. and Mann. T. 1938. Polyphenol oxidase, purification, nature and properties, *Proc. Roy. Soc. (London) B.*, **125**: 187-204.

Kelley, P.M. and Izawa, S. 1978. The role of chloride ion in photosystem. II. I. Effects of chloride ion on photosystem II : electron transport and on hydroxylamine inhibition. *Biochem. Biophys. Acta*, **502**: 198-201.

Kene, D.R. and Deshpande, T.L. 1980. Effect of application of zinc, manganese and iron on the biological quality ofgrain of hybrid sorghum grown in black soils. *J. Indian Soc. Soil Sci.*, **28:** 199-202.

Kennedy, C.D. and Gonsalves, F.A.N. 1989. The action of divalent zinc, cadmium, mercury, copper and leaf ions on the ATPase activity of a plasma membrane fraction from roots of *Zea mays. Plant Soil.*, **117**: 167-175.

Kenten, R.H. 1955. The oxidation in indolyl-3-acetic acid by waxpod bean root sap and peroxidase systems. *Biochem. J.*, **59**: 110-121.

Kenten, R.H. and Mann, P.J.G. 1949. The oxidation of manganese by plant extracts in the presence of hydrogen peroxide. *Biochem. J.*, **45**: 255-263.

Kenten, R.H. and Mann, P.J.G. 1955. The oxidation of manganese by illuminated chloroplast suspension. *Biochem. J.*, 61: 279- 286.

Kentworthy, A.L. 1961. Interpreting the balance of nutrient elements in leaves of fruit tree. In : *Plant Analysis and Fertilizers Problems* (W. Reuther, ed.) pp. 28-43. *Am. Inst. Biol. Sci.*, Washington, D.C.

Keser, M., Neubauer, B.F. and Hutchinson, F.E. 1975. Influences of aluminium ions on developmental morphology of sugarbeet roots. *Agron. J.*, **67**: 84-88.

Kessler, B. 1957. Effect of certain nucleic acid components upon the status of iron, deoxyribonucleic acid and lime induced chlorosis in fruit trees. *Nature.*, **179:** 1015-1016.

Khatiashvili, R.J., Abulashvili, J.G., Burdzhanadze, M.B. and Gogoladze, N.T. 1985. The effect of trace elements on the uptake of anions by the grapevine leaves. *Soobshch Akad. Nauk. Gruz. SSR.*, **120:** 161-164.

Khatiashvili, R.M. and Abulashvili, J.G. 1984. Effect of iron, phosphorus and potassium ions on the uptake of manganese by grape vines. *Soobshch Akad. Nauk. Gruz. SSR.*, **115:** 381-384.

Khovrychev, M.P., Fedrova, T.A. and Robotnova, J.L. 1974. Effect of copper ions on growth of *Candida ultilis* continuous culture. *Mikrobiol.*, **43:** 99-102.

Kibalenko, A.P., Komlyak, M.M. and Velyka, S.L. 1973. Significance of boron and nitrogen metabolism and protein synthesis in plants. *Dopov. Acad. Nauk. Urk. Rerser B Hool Hoofiz Khim Biol.*, **35:** 457-463.

Kim, K.R., Byun, J.R. and Kim, C.C. 1970. Studies on boron deficiency in a peach orchard. *Res. Rep. Office Rural Devel.*, **12:** 29-34.

Kirkpatric, J.C., Thompson, J.M. and Edwards, J.H. 1974. Effect of aluminium concentration of growth and chemical composition of peach seedlings. *Hort. Sci.*, **10**: 132-134.

Kleinkopf, G.E., Wallace, A. and Hartstock, T. 1976. Lime chlorosis on photosynthesis and transpiration of iron-inefficient soybeans. *Commun. Soil Sci. Plant Anal.*, **7:** 97-99.

Kliewer, M. and Evans, H.J. 1963a. Identification of cobamide coenzyme in nodules of symbionts and isolation of the B12 coenzyme from *Rhizobium meliloti. Plant Physiol.*, **38**: 55- 59.

Klimashevskii, E.L., Marleava, Yu. A., Seregina, M.L., Grodzinskii, D.M. and Kozarenko, T.D. 1970. Specifics of the physiological activity of pea plants in connection with unequal resistance of different varieties to mobile aluminium. *Sov. Plant Physiol.,* **17:** 372-378.

Klimovitskya, Z.M. and Lobanova, Z.I. 1975. Nature of UV spectra of native DNA and RNA after acidic and alkaline denaturation depending on level of the plant supply of Mn++ ions. *Fiziol. Biokhim. Kult Rast.,* **7**: 456-458.

Klimovitskya, Z.M. Lobanova, Z.I., Prokopivnyuk, L.N. 1969. The physiochemical properties of plant RNA and DNA in manganese deficiency. *Fiziol. Biokhim. Kul't Rast.,* **1:** 269-275.

Klucas, R.V., Hanus, F.J., and Russell, S.A. 1983. Nickel. A micronutrient element for hydrogen - dependent growth of *Rhizobium japonicum* and for expression of urease activity in soybean leaves. *Proc., Natl. Acad. Sci. USA,* **90:** 2253-2257.

Klucas, R.V., Koch, B., Russell, S.A. and Evans, H.J. 1968. Purification and some properties of the nitrogenase from soybean (*Glycine max* Merr.) nodules. *Plant Physiol.,* **43**: 1906- 1912.

Kluthcouski, J. and Nelson, L.E. 1979. Variations in the manganese concentration in soybean trifoliates. *Commun. Soil Sci. Plant Anal.,* **10:** 1299-1310.

Kneifel, H. and Bayer, E. 1973. Determination of the structure of the vanadium compound, amavadines from fly agaric. *Angeco. Chem.,* **85**: 508.

Knezek, B.D. and Maier, R.H. 1966. Fractionation of iron in chlorotic and non-chlorotic tomato leaf tissue. *J. Agric. Food Chem.,* **14:** 266-268.

Kobakhidzie, K.V., Chekoniya, T.A. and Dzhavshana-Shvili, E.Z. 1971. The effect of manganese on some physiological processes occurring in the maize plant and accumulation of chemical substances in the kernels. *Soobashch Akad. Nauk. Gruz-SSR.,* **62:** 153-155.

Kobowitz, F. 1938. Spaltung und Resynthese der Poly-phenoloxydase und des hamocyanins. *Biochim.,* **2:** 299-32-57.

Kohno, Y. and Foy, C.D. 1983. Differential tolerance of bush bean cultivars to excess manganese in solution and sand culture. *J. Plant Nutr.,* **6:** 877-893.

Kohno, Y., Roy, C.D., Fleming, A.L. and Krizek, D.T. 1984. Effect of manganese on concentration of zinc, potassium, calcium and mangenese in 2 bush bean (*Phaseolous vulgaris*) cultivars grown in solution culture. *J. Plant Nutr.,* **7:** 897-910.

Kok, B. and Rurainski, H.I. 1965. Plastocyanin photooxidation by detergent treated chloroplast. *Biochem. Biophys. Acta.,* **94:** 588-590.

Komai, H. and Neilands, J.B. 1968. Effect of zinc ions on δ-amino levulinate dehydratase in *Ustilago sphaerogena. Arch. Biochem. Biophys.,* **124:** 456-461.

Kositsyn, A.V. and Igoshina, T.I. 1971. The association of zinc with the soluble proteins of tomato chloroplasts. *Fiziol. Rastenii.,* **18:** 1113-1118.

Kotze, W.A.G., Shear, C.B. and Faust, M. 1977. Effect of nitrogen source and the presence or absence of aluminium on the growth and calcium nutrition of apple seedlings. *J. Am. Soc. Hortic. Sci.,* **101**: 305-309.

Kotze, W.A. and Villiers, J. de 1989. Seasonal uptake and distribution of nutrient elements by Kiwifruit vines. 2. Micronutrients. *South African J. Plant Soil.* **6 :** 265- 270.

Koula, V. and Cingrosova, K. 1975. The effect of cuprous oxide on some vital processes in potato plants. *Sbornik Vysoki Skoly Zemedelske V Praze, Fakulta Agronomicka A.*, **1:** 177-201.

Kovalsky, V.V. and Rish, M.A. 1970. A copper deficient biogeo chemical province. In: *Biologicheskaya Rol Medi.* (The Biological Role of Copper) (V.V. Kovalsky, ed.) pp. 299-314. Nauka, Moscow.

Kowalewska, G., Folkowski, L., Hoffman, S.K. and Szozepanaik, L.S. 1987. Replacement of magnesium by copper (II) in the chlorophyll porphyrin ring of planktonic algae. *Acta. Physiol. Plant.*, **9:** 43-52.

Kramer, D., Lauchli, A., Yeo, A.R. and Gullasch, J. 1977. Transfer cells in roots of *Phaseolus coccineus* : Ultrastructure and possible function in exclusion of sodium from the short. *Ann. Bot.*, **41:** 1031-1040.

Kreutzer, K. 1970. The influence of manganese application on the concentration of pigments and nutrients in manganese deficient needles of *Picia abies* (Karst). *Z. Pflanzenernaehr Bodenkd.*, **127:** 84-91.

Krivoshei, I.D. and Kononovich, A.D. 1978. Effect of trace elements on seed yield of lucerne. Botanika Issledovaniya No. **20** : 206-210.

Krueger, R.W., Carol, J.L. and Luck, S.A. 1987. Metabolic requirement of *Cucurbita pepo* for boron. *Plant Physiol.*, (BETHSDA) **83:** 254-258.

Krupnikova, T.A. and Smirnov, Yu. S. 1981. Content of phenolic compounds in plants in relation to boron availability. *Bot. Zh. Leningred.*, **66**: 536-542.

Krupnikova, T.A., Smirnov, Yu. S., and Shkol'nik, M.Ya. 1975. Boron effect on auxin oxidase activity in plants differing in the requirement for this element. *Fiziol. Biochem. Kul't Rast.*, **7**: 569-573.

Kumar, A. and Alexander, A.G. 1972. Effect of manganese deficiency on leaf chlorophyll and certain enzymes in sugarcane raise in sand culture. *J. Agric. Univ. P.R.*, **56:** 75-78.

Kumar, V. and Singh, M. 1981. Zinc and sulphur interaction in relation to uptake and utilization of iron in soybean (*Glycine max* (L) Merr.). *Indian J. Plant Physiol.*, **24:** 53-60.

Kumar, V. et. al. 1981. Magnesium and zinc relationship in relation to dry matter yield and the concentration and uptake of nutrients in wheat. *Soil Sci.*, **131**: 151-155.

Kumon, K. and Tsurumi, S. 1984. Ion efflux from pulvinar cells during slow downward movement of the petiole of *Mimosa pudica* L. induced by photo, stimulation. *J. Plant Physiol.*, **115**: 439-443.

Kuo, S. and Mikkelsen, D.S. 1981. Effect of phosphorus and manganese on growth response and uptake of iron, manganese and phosphorus by sorghum (*S. vulgare*). *Plant Soil.*, **62:** 15- 22.

Kvariani, L.D. 1971. Nitrogen exchange in soybean leaves during treatment of seeds with copper. *Soobshch Akad. Nauk. Gruz. SSR.*, **61**: 661-663.

Kylin, H. 1929. Uber das Vorkommen Von Jodiden, Bromiden Und Jodidoxydasen bei den Meeresalgen. *Hoppe-Seyler's Z. Physiol. Chem.*, **186:** 50-84.

Kylin, A. and Hansson, G. 1971. Transport of sodium and potassium and properties of (sodium + potassium) activated adenosine triphosphatase : Possible connection with salt tolerance in plants. *Proc. 8th Colloq. Int. Potash Inst. Bern.*, pp. 64- 68.

Laine, T. 1934. On the absorption of electrolytes by the cut roots of plants and the chemistry of plant exudation sap. *Acta Bot. Fennica.*, **16:** 1-64.

Landi, S. and Fagioli, F. 1983. Efficiency of manganese and copper by excised roots of maize genotypes. *J. Plant Nutr.*, **6:** 957-970.

Lauchli, A. 1972. Translocation of inorganic solutes. *Ann. Rev. Plant Physiol.*, **23:** 197-218.

Lawlor, D.W. and Milford, G.F.J. 1973. The effect of sodium on growth of water-stressed sugarbeet. *Ann. Bot.*, **37:** 597- 604.

Leaf, G.L. 1953. Boron in relation to water balance in plants. *Proc. Iowa. Sci.*, **60:** 176-191.

Leblova, S. and Zatioukalova, S. 1988. Interaction of cadmium, copper, cobalt, nickel and lead ions with alcohol dehydrogenase isolated from germinating maize (Zea mays) seeds. *Biologia Czechoslovakia,* **43**: 1105-1135.

Lee, C.M.C. 1978. The effect of zinc nutrition on protein and ribosome metabolism in the leaves of *Phaseolus vulgaris* L. *Dissertation Abstracts International,* **8:** 396, 2598.

Lee, C.R. 1971. Influence of aluminium of plant growth and mineral nutrition of potatoes. *Agron. J.*, **63:** 604-608.

Lee, J., Reeves, R.D., Brooks, R.P. and Jaffre, T. 1977. Isolation and identification of a citrato-complex of nickel from nickel-accumulating plants. *Phytochemistry,* **16**: 1503- 1505.

Lee, S.G. and Aronoff, S. 1967. Boron in plants : A biochemical role. *Science,* **158:** 789-799.

Lee, T.S., Kim, M.S. and Hong, S.K. 1990. Effect of some lining material application on the alleviation of the symptoms of Mn toxicity including marginal leaf chlorosis. *Korean J. Ginseng Sci.*, **14**: 44-49.

Leela, D. and Pathak, C.H. 1968. Effect of iron, magnesium and calcium deficiency on the growth of a enzyme activities of the excised root of *Trigonella foenum. Indian J. Exptl. Biol.*, **6:** 127-128.

Leeper, G.W. 1941. Manganese deficiency and accumulation of nitrates in plants. *J. Aust. Inst. Agric. Sci.,* **1:** 161- 162.

Leidi, E.O., Gomez, M. and De la Guardia, M.D. 1986. Evaluation of catalase and peroxidase activity as indicators of iron and manganese nutrition for soybean. *J. Plant Nutr.,* **9:** 1239- 1249.

Leidi, E.O., Gomez, M. and Del Rio, L.A. 1987. Evaluation of biochemical indicators of iron and manganese nutrition for soybean plants. II. Superoxide dismutase, chlorophyll content and photosystem II activity. *J. Plant Nutr.*, **10:** 261-271.

Lerer, M. and Bar-Akiva, A. 1976. Nitrogen constituents in manganese deficient lemon leaves. *Physiol. Plant.*, **38:** 13-18.

Lerer, M. and Bar-Akiva, A. 1979. Effect of manganese deficiency on chloroplast of lemon leaves. *Physiol Plant.*, **47:** 163- 166.

Lessman, G.M. and Ellis, B.G. 1971. Response of *Phasoeolus vulgaris*, to zinc as influenced by phosphorus level and source. *Soil Sci. Soc. Am. Proc.*, **35:** 935-938.

Levesque, M.P. and Mathur, S.P. 1983. The effects of using copper for mitigating histosol subsidence on : I The yield and nutrition of oats and lettuce grown on histosols, mineral sublayers and their mixtures. *Soil Sci.*, **135:** 88-100.

Lewis, D.H. 1980a. Are there interrelationship between the metabolic role of boron, synthesis of phenolic phytoalexine and the germination of pollen. *New Phytol.*, **84:** 261-270.

Lewis, J. and Reimann, B.E.F. 1969. Silicon and plant growth *Ann. Rev. Plant Physiol.*, **20:** 289-304.

Lewis, J.C. and Powers, W.L. 1941. Antagonistic action of chlorides on the toxicity of Iodides to corn. *Plant Physiol.*, **16:** 393-398.

Ligon, W.S. and Pierre, W.H. 1932. Soluble aluminium studies. II Minimum concentration of aluminium found to be toxic to corn, sorghum and barley in cluture solutions. *Soil Sci.*, **34**: 307-317.

Lim, S.U. and Kim, J.H. 1983. The effect of boron application on the yield and quality of potatoes (*Solanum tuberosum* L.) *Hanguk Nonghwa Hakroechi.*, **26:** 191-197.

Lin, C.H. and Stocking, C.R. 1978. Influence of leaf age, light, dark and iron deficiency on polyribosome levels in maize leaves. *Plant Cell Physiol.*, **19:** 461-470.

Lin, W. 1981. Inhibition of anion transport in corn root protoplasts. *Plant Physiol.*, **68:** 435-438.

Lindsay, W.L. 1972. Zinc in soils and plant nutriton. *Adv. Agron.*, **24**: 47-186.

Lindskog, S. and Malmstrom, B.G. 1960. A reversible dissociation of zinc in bovine carbonic anhydrase. *Biochem. Biophys. Research Communs.*, **2**: 213-217.

Lipman, C.B. 1938. Importance of silicon, aluminium and chlorine for higher plants. *Soil Sci.*, **45:** 189-198.

Lipman, C.B. and Mackinney, G. 1931. Proof of the essential nature of copper for higher green plants. *Plant Physiol.*, **6:** 593-599.

Liu, W., Meimci, PI. and Lishu, Wu. 1987. Interaction between boron and potassium in cotton nutrition. *Acta, Pedot. Sin.*, **24:** 45-50.

Livdane, B.A. 1978. Accumulation of starch and proteins in maturing barley seeds under copper deficiency. Pulb. Diga, Latvian SSR : Zinatne, 92-99.

Lohins, M.P. 1937. Plant development in the absence of boron. *Mededel. Landbouwhooge School Wageningen*, **41:** Part 3.

Lohnis, M.P. 1940. Histology of symptoms of boron deficiency in plants. Mededel. Landbouwhooge School. Wageningen. **44:** 3-36.

Lohnis, M.P. 1960. Effect of magnesium and calcium supply on the uptake of manganese by various crop plants. *Plant Soil,* **12:** 339-376.

Loneragan, J.F. 1951. The effect of applied phosphate on the uptake of zinc by flax. *Aust. J. Sci. Res., B.* **4:** 108- 114.

Loneragan, J.F. 1988. Distribution and movement of manganese in plants. In: *Manganese in Soils and Plants.* (R.D. Graham, R.J. Hannam and N.C. Uren, eds.) pp. 113-124. Kluwer Academic Publishers.

Loomis, W.E. and Shull, I.A. 1937. *Methods in Plant Physiology,* McGraw Hill Book Co. Inc. New York.

Lopez-Andreu, F.J., Esteban, R.M., Molla, E. and Carpena, O. 1988a Effect of nutrition on tomato fruit quality. I. Yield, physical parameters, nitrogen fractions, acidity and sugar contents. *An. Edafol. Agrobiol.,* **47:** 1181-1190.

Lopez-Andreu, F.J., Esteban, R.M., Molla, E. and Carpena, O. 1988b. Effect of nutrition on tomato fruit quality. II Carotenoids, ascorbic acid, pectic substances and flavonoids. *An. Edafol. Agrobiol.,* **47:** 1191-1198.

Lotti, G., Saviozzi, A. and Baizini. S. 1989. The distribution of boron and major mineral elements in plant organs in relation to the content of their leaves. *Agrochim.,* **33:** 129-142.

Loughman, B.C. 1963. Effect of boric acid on phosphoglucomutase of pea seed. *Nature,* **191:** 1399-1400.

Lousialot, A.J., Burrows, F.W., Gilbert, S.G., and Nason, A. 1945. Effect of copper and zinc deficiencies on the photosynthesis activity of the foliage of young tung trees. *Plant Physiol.,,* **20**: 283-288.

Lovatt, C.J., Albert, L.S. and Tremblay, G.C. 1981. Synthesis, Salvage and catabolism of uridine nucleotide in boron - deficient squash roots. *Plant Physiol.,* **68:** 1389-1394.

Low, C., Homma, S. and Kuno, K. 1991. Effect of copper and manganese application on nickel toxicity in mulberry seedlings. *Nippon Sanshigaku Zasshi,* **60**: 200-207.

Lundegardh, H. 1939. Mangan als Katalysator der Pflanzenatmung *Planta,* **29:** 419-426.

Luo, C.M. 1982. Effect of several mineral elements and growth regulators on stomatal differentiation and density in leaves of wheat. *Plant Physiol. Communication.* (Zhiwu shenglixue Tongxun). No. **6:** 22-25.

Lutz, J.A. Jr., Genter, C.F. and Hawkins, G.W. 1972. Effect of soil pH on element concentration and uptake by maize. II Cu, B, Zn, Mn, Mo, Al and Fe. *Agron. J.,* **64:** 583-585.

Lyttleton, J.W. 1960. Stabilisation by manganous ions of ribosomes from embryonic plant tissue. *Nature,* **187:** 1026- 1027.

Ma, W.Q., Liu, W.D. and Pi, M.M. 1990. Effect of boron on nitrogen metabolism in red clover. *Plant Physiol. Commun.,* **4:** 43-46.

Maas, E.V., Moore, D.P., and Mason, B.J. 1968. Manganese absorption by excised barley roots. *Plant Physiol.*, **43**: 527-530.

Mac Donald, I.R., Macklon, A.E.S. and MacLeod, R.W.G. 1975. Energy supply and light enhanced chloride uptake in wheat laminae. *Plant Physiol.*, **56**: 699-702.

Mackay, D.C., Chipman, E.W. and Gupta, U.G. 1966. Copper and molybdenum nutrition of crops grown on acid sphagnum peat soil. *Soil Sci. Soc. An. Proc.*, **30**: 755-759.

Maclachlan, G.A. and Waygood, E.R. 1956a. Catalysis of indoleacetic acid oxidation by manganic ions. *Physiol. Plant.*, **9**: 321-330.

Maclachlan, G.A. and Waygood, E.R. 1956b. Kinetics of the enzymically catalyzed oxidation of indoleacetic acid. *Can. J. Biochem. Physiol.*, **34**: 1233-1250.

MacLeod, L.B. and Jackson, L.P. 1965. Effect of concentration of the aluminium ion on root development and establishment of legume seedlings. *Can. J. Soil Sci.*, **45**: 221-234.

Maeno, H. and Feigelson, P.H. 1965. The purification of copper in tryptophan pyrrolase action. *Biochem. Biophys. Res. Commun.*, **21**: 297-302.

Mac Vicar, R. and Burris, R.H. 1948. Relation of boron to certain plant oxidases. *Arch. Biochem.*, **17**: 31-39.

Maevskaya, A.N., Seregeichik, A.A. and Yakovleva, N.S. 1981. Effect of boron excess and deficiency in a nutritional medium of B-glucosidase activity and growth of *Cucumis sativa L. Fiziol. Rast.*, **28**: 536-540.

Magdolna, D., Terry, N. and Horvath, G. 1984. Variation in photosynthetic pigments and plastoquinone contents in sugar beet *(Beta vulgaris*, cultivars. F. 58-554 HI) chloroplasts with changes in leaf copper content. *Plant Physiol.*, **74**: 717-720.

Mahmoud, K., Filsoof, F. and Rezai-Nejad, Y. 1988. Effect of sulphur treatment on yield and uptake of Fe, Zn and Mn by corn, sorghum and soybeans. *J. Plant Nutr.*, **11**: 1353-1360.

Malewar, G.V., Jadhav, D.K. and Ghonsikar, C.P. 1982. Absorption and transport of Fe in genotypes of chickpea. *International Chick pea Newslett.*, **6**: 13-14.

Malik, C.P. and Sharma, R. 1978. Effect of boron on the activity of some enzymes in the germinating pollen of *Sesbania aegyptica*, Pers. Physiol. Sex. Reprod. Flowering plants Int. Symp. I[st]. 89- 96.

Malkin, R. 1987. Iron-sulfur centres of chloroplast membranes. Iron-sulfur protein Res. (pap. Int. Symp. Front). I (H. Matsubara, Y. Katsube and K. Wada, eds.) pp. 260-269. Japan Sci. Soc. Press : Tokyo, Japan.

Malkin, R. and Bearden, A.J. 1973. Light induced changes of bound chloroplast plastocyanin as studied by EPR spectroscopy : The role of plastocyanin in non-cyclic photosynthetic electron transport. *Biochem. Biophys. Acta.*, **292**: 169-185.

Mandley, S.L. 1981. Iron uptake and translocation by *Macrocystis pyrifera. Plant Physiol.*, **68**: 914-918.

Mann, M.S. and Takkar, P.N. 1982. Effect of rates of $Ca(OH)_2$ and $ZnSO_4$ on soluble zinc concentration of spray solution and absorption of Zn^{65} by sweet orange leaves. *Sci. Hort.*, **18:** 263-268.

Marschner, H. 1971. Why can sodium replace potassium in plants ? *Proc. 8th Colloq. Int. Potash Inst. Bern.*, pp. 50-63.

Marschner, H. 1986. Mineral Nutrition of Higher Plants. Academic Press, New York.

Marschner, H., Kulper, P.J.C. and Kylin, A. 1981. Genotypic differences in the response of sugarbeet plants to replacement of potassium by sodium. *Physiol. Plant.*, **51:** 239-244.

Marschner, H. and Possingham, J.V. 1975. Effect of K^+ and Na^+ on growth of leaf discs of sugar beet and spinach. *Z. Pflazenphysiol.*, **75**: 6-16.

Marschner, H., Romheld, V. and Kissel, M. 1986. Different strategies in higher plants in mobilization and uptake of iron. *J. Plant Nutr.*, **9**: 695-713.

Marsh, R.P. 1942. Comparative study of calcium-boron metabolism of representative dicots and monocots. *Soil Sci.*, **53**: 75-78.

Marsh, H.V. Jr., Evans, H.J. and Matrone, G. 1963a. Investigations of the role of iron in chlorophyll metabolism. I. Effect of iron deficiency on chlorophyll and heme content and on the activities of certain enzymes in leaves. *Plant Physiol.*, **38**: 632-637.

Martin, M.S., Saco, M.D. and Alvarez, M. 1988. Effect of molybdenum on the first stage of growth and alkaloid content of *Nicotiana rustica*, L. *An. R. Acad. Farm.*, **54:** 340-351.

Martynov, E.G. 1980. Accumulation of flavonoids in the fruits of *Amelanchia melanocarpa*, under the effect of minor elements. Khimya Priradnykh soedinenii. No. **3:** 415-417.

Mason, H.S. 1957. Mechanism of oxygen metabolism. *Adv. Enzymol.*, **19**: 79-233.

Mason, G.F. and Guttridge, C.G. 1974. The role of calcium, boron and some divalent ions in leaf tipburn of strawberry. *Sci. Hort.*, **2:** 299-308.

Mathan, K.K. 1980. Effect of various levels of aluminium on the dry matter yield, content and uptake of phosphorus, aluminium, manganese, magnesium and iron in maize. *Madras Agric. J.*, **67**: 751-757.

Matsumoto, H. and Morimura, S. 1980. Repressed few plate activity of chromatin of pea roots treated by aluminium. *Plant Cell Physiol.*, **21**: 951-959.

Matsumoto, H., Hirasawa, E., Morimura, S. and Takahashi, E. 1976a. Localization of aluminium in tea leaves. *Plant Cell Physiol.*, **17:** 627-631.

Maze', P. 1914. Influences respectives des elements de la solution minerals sur le development dumiis. *Ann. Inst. Pasteur.*, **28:** 21-46.

Maze', P. 1915. Determination des elements mineraux rares; necessaries an development du mais. *Compt. Rend. Acad. Sci.*, **160:** 211-214.

McCord, J.M., Keele, B.B. and Fridovich, J. 1971. An enzyme basis theory of obligate as scrobisis The physiological function of superoxide.

McDermott, A.E., Yachandra, V.K., Guiles, R.D., Britt, R.D., Dexhelmer, S.L., Sauer, K. and Klein, M.P. 1987. Iron x-ray absorption spectra of acceptors in PSI. *Prog. Photosynth. Res. Proc. Int. Congr. Photosynth.*, 7th, **1:** 249-252.

McElroy, W.D. and Nason, A. 1954. Mechanism of action of micronutrient elements in exzyme systems. *Ann. Rev. Plant Physiol.,* **5:** 1-30.

McGeorge, W.T. 1949. Lime induced chlorosis; relation between active iron and citric and oxalic acids. *Soil Sci.,* **68:** 381-390.

McHargue, J.S. 1919. The effect of manganese on the growth of wheat : A source of manganese for agricultural purposes. *Ind. Eng. Chem.,* **11:** 332-335.

McIlrath, W.J. and Palser, B.F. 1956. Responses of tomato, turnip and cotton to variation in boron nutrition. I. Physiological responses. *Bot. Gaz.,* **118:** 43-52.

McIlrath, W.J. and de Bruyn, J.A. 1956. Calcium-boron relationship in siberian millet. *Soil Sci.,* **81:** 301-310.

Mc Leod, K.A. 1975. The control of growth of tomato pollen. *Ann. Bot. (Lond.).,* **39:** 591-596.

Medina, A. and Nicholas, D.J.D. 1957. Metalloenzymes in the reduction of nitrite to ammonia in *Neurospora. Biochim. Biophys. Acta.,* **25:** 138-141.

Meerow, A.W. and Broschat, T.K. 1991. Anatomical aspects of K, Mn and Fe deficiencies in queen palm. *Hort. Sci.,* **26**: 581- 583.

Mehrotra, S.C. 1988. Effect of Mn deficiency on uptake, transport and distribution of ^{32}P in cauliflower and tomato. *Indian J. Expt. Biol.,* **26:** 806-807.

Mehrotra, S.C. and Bisht, S.S. 1990. Genotypic differences in concentration of free amino acids in susceptible and tolerant cultivars of chick-pea under iron deficiency stress. *Ind. J. Expt. Biol.,* **28:** 893-894.

Meisch, H.U. and Bauer, J. 1978. The role of vanadium in green plants. IV: Influence on the formation of δ- aminolevulinic acid in *Chlorella. Arch. Microbiol.,* **117:** 49-53.

Meisch, H.U. and Becker, L., J.M. and Schwab, D. 1980. Ultrastructural changes in *Chlorella fusca,* during iron deficiency and vanadium treatment. *Protoplasma,* **103:** 273-280.

Meisch, H.U. and Benzschawel, H. 1978. The role of Vanadium in green plants. III Influence on cell division of *Chlorella. Arch.Microbiol.,* **116**: 91-95.

Meisch, H.U., Benzschawel, H. and Bielig, H.J. 1977. The role of vanadium in green plants. II. Vanadium in green algae - two sites of action. *Arch. Microbiol.,* **144:** 67-70.

Meisch, H.U., Reinle, W. and Schmilt, J.A. 1979. High vanadium content in mushrooms is not restricted to the fly agaric. *(Amanita muscaria). Naturwitsen schaften.,* **66**: 620-621.

Meisch, H.U., Schniett, J.A. and Reinle, W. 1978. Schwermetalle in haheren Pilzen III. *Vanadium und Molybdan. Z. Naturforsch.,* **33**: C : 1-6.

Mendel, R.R. and Muller, A.J. 1976. A common genetic determinant of xanthine dehydrogenase and nitrate reductase in *Nicotina tabacum. Biochem. Physiol. P. fl.,* **170:** 538-541.

Mengel, K. and Kirkby, E.A. 1982. Principles of Plant Nutrition 3rd. edn. Int. Potash Inst., Bern, Switzerland.

Mengel, K. and Kirkby, E.A. 1987. Principles of Plant Nutrition. 4th edn. Int. Potash Inst., Bern, Switzerland.

Mercer, F.V., Nittim, M. and Possingham, J.V. 1962. The effect of manganese deficiency on the structure of spinach chloroplasts. *J. Cell Biol.*, **15**: 379-381.

Mercer, E.R. and Richmond, J.L. 1970. Fate of nutrients in soils. Copper, pp. 9. In: Letcombe Laboratory Annual Report.

Mettler, I.J., Mandala, S. and Taiz, L. 1982. Characterization of *in vitro*, proton pumping by microsomal vesicles isolated from corn coleoptiles. *Plant Physiol.*, **70:** 1738-1742.

Michalski, W.P. and Kaninga, Z. 1982. Photosynthetic apparatus of chilling sensitive plants XI. Reversibility by light of cold and dark - induced inactivation of cyanide senstive superoxide dismutase activity in tomato leaf chloroplast. *Biochem. Biophys. Acta.*, **680:** 250-257.

Michelson, A.M., McCord, J. and Fridovich, I. 1977. Superoxide and superoxide dismutases. Academic Press, New York.

Milford, G.F.J., Cormack, W.F. and Durrant, M.J. 1977. Effects of sodium chloride on water status and growth of sugar beet. *J. Exptl. Bot.*, **28**: 1380-1388.

Milivojevic, D. and Rastovic, A. 1983. Ultrastructure and photosynthetic activity of chloroplasts in manganese and iron deficient maize leaves. *Arh. Poljopr. Nauke.*, **44:** 43-49.

Miller, L.F. 1933. Effect of manganese deficiency on the growth and sugar content of plants. *Am. J. Bot.*, **20:** 621-631.

Millikan, C.R. 1949. Effect of flax of a toxic concentration of boron, iron, molybdenum, aluminium, copper zinc, manganese, cobalt or nickel in the nutrient solutions. *Proc. Roy. Soc. Vict.*, **61**: 25-42.

Millikan, C.R. 1963. Effect of different levels of zinc and phosphorus on the growth of subterranean clover *(Trifolium subterraneum L.). Aust. J. Agric. Res.*, **14:** 180-205.

Milosavljevic, M. and Popovic, R. 1970. The effect of boron and manganese on the intensity of photosynthesis in grape vines. *Arch, Poljopr. Nauk.*, **23**: 15-34.

Minarik, C.E. and Shive, J.W. 1939. The effect of boron in the substrate on calcium accumulation by soybean plants. *Am. J. Bot.*, **26:** 827-831.

Mirchev, S. and Gezenchova, L. 1986. Uptake of molybdenum and manganese by some crops grown on chernozems and grey forest soil. *Pochvonanic Agrokhimya i Rastitelna Zashchita*, **21:** 49- 57.

Misra, S.G. and Dwivedi, R.S. 1976. Studies on residual effect of micronutrients on nutrient elements of soil : I. *Vijnana Parishad Anusandhan Patrika.*, **19:** 381-389.

Mishra, D. and Kar, M. 1974. Nickel in plant growth and metabolism. *Bot. Rev.*, **40:** 395-452.

Misra, S.M. and Patil, B.D. 1987. Effect of boron on seed yield in lucerne (*Medicago sativa* L.). *J. Agron, Crop Sci.*, **158:** 34-37.

Mishra, S.M., Sinha, N.C. and Tomex, P.S. 1991. Seed yield of sorghum in an Alfisol in relation to boron nutrition at the reproductive phase. *Plant Soil,* **132**: 293-296.

Mitchell, G.A., Bingham, F.T. and Page, A.L. 1978. Yield and metal composition of lettuce and wheat grown on soils amended with sewage sludge enriched with cadmium, copper, nickel, and zinc. *J. Environ. Qual.,* **7**: 165-171.

Mitchell, J.W., Dugger, W.M., Jr. and Gauch, H.G. 1953. Increased translocation of plant growth - modifying substance due to application of boron, *Science,* **118**: 354-355.

Mitsui, S. and Takatoh, H. 1963. Nutritional study of silicon in graminaceous crops. Plant I. *Soil Sci. Plant Nutr.,* **9**: 49-53.

Miyake, Y. and Takahashi, E. 1978. Silicon deficiency of tomato plant. *Soil Sci. Plant Nutr.,* **24**: 175-189.

Miyake, Y. and Takahashi, E. 1983. Effect of silicon on the growth of solution cultured cucumber plant. *Soil Sci. Plant Nutr.,* **29**: 71-83.

Mizuno, N. and Dobashi, K. 1982. Composition of mineral nutrient, moisture, soluble solids and transformation of copper deficient wheat plant. *Nippon Dojo Hiryogaku Zasshi.,* **53**: 503-506.

Mizuno, N., Inazu, O. and Kamada, K. 1982. Characteristics of concentration of copper deficient wheat plants. *9th Intt. Plant Nutr. Colloq.,* **2**: 396-399.

Mizuno, N., Inazu, O. and Dobashi, K. 1983. Composition of carbohydrates and soluble solids, moisture and morphological changes in wheat plants grown on copper deficient soil. *Soil Sci. Plant Nutri.,* **29**: 1-6.

Mohapatra, A.R. and Kibe, M.M. 1971. Zinc deficiency in tomato. *Indian J. Exp. Biol.,* **9**: 283-285.

Mondy, N.I. and Munshi, C.B. 1988. Effect of molybdenum, fertilization on the nitrogenous constituents of potatoes. *J. Food Sci.,* **53**: 1909-1910.

Moore, A.L. and Cottingham, I.R. 1983. Characteristics of the higher plant respiratory chain. In: Metals and micronutrients uptake and utilization by plants. (D.A. Robb. and W.S. Pierpoint, eds.), Academic Press, New York, London.

Moore, D.P. 1972. Mechanism of micronutrient uptake by plants. In: *Micronutrients in Agriculture.* pp. 171-198. *Soil Sci. Soc. Am.* Inc. Madison.

Moore, D.P. 1974. Physiological effects of pH on roots. In: The Plant Root and Its Environment (E.W. Carson, ed.). pp. 135-151. University Press of Virginia, Charlottesvile.

Moore, H.M. and Hirsch, A.M. 1983. Effects of boron deficiency on mitosis and incorporation of tritiated thymidine into nuclei of sunflower root tips. *Am. J. Bot.,* **70**: 165-172.

Morgan, P.W. and Hall, W.C. 1963. Indoleacetic acid oxidising enzyme and inhibitors from light grown cotton, *Plant Physiol.,* **38**: 365-370.

Morgan, P.W., Joham, H.E. and Amin, J.V. 1966. Effect of manganese toxicity on the indole-acetic acid oxidase system of cotton. *Plant Physiol.,* **41**: 718-724.

Morgan, P.W., Taylor, D.M. and Joham, H.E. 1976. Manipulation of 1AA oxidase activity and auxin-deficiency symptoms in intact cotton plants with manganese nutrition. *Physiol. Plant.,* **34:** 149-156.

Morimura, S. and Matsumoto, H. 1978. Effect of aluminium on some properties and template activity of purified pea - DNA. *Plant Cell Physiol.,* **19:** 429-436.

Morrill, L.G., Hill, W.E., Chrudimsky, W.W., Ashlock, L.O., Tripp, L.D., Tucker, B.B. and Weatherley, L. 1977. Boron requirements of spanish peanuts in Oklahoma : Effect on yield quality and interaction with other nutrients. Pulb. OKlahoma State Univ. Agric. Exp. Stn. Np. MP - 99, pp. 20.

Mugwira, L.M. 1979. Aluminium effects on the growth and mineral levels of triticale heat and rye. *J. Plant Nutr.,* **1:** 214- 240.

Mukhi, A.K., and Shukla, U.C. 1987. Iron and zinc relationship in rice grown in submerged soils. *J. Indian Soc. Soil Sci.,* **35:** 685-689.

Mukhopadhyay, M.S., and Sharma, A. 1991. Manganese in cell metabolism of higher plants. *Bot. Rev.,* **57:** 117-149.

Mulder, E.G. 1948. Importance of molybdenum in the nitrogen metabolism of micro-organisms and higher plants. *Plant Soil.,* **1**: 94-119.

Mulder, E.G. 1954. Molybdenum in relation to growth of higher plants and microorganisms. *Plant Soil.,* **5**: 368-415.

Mulder, E.G. and Gerretsen, F.C. 1952. Soil manganese in relation to plant growth. *Adv. Agron.,* **4:** 222-227.

Munns, D.N. 1965. Soil acidity and growth of a legume II. Reactions of aluminium and phosphate in solution and effects of aluminium, phosphate, calcium and pH on *Medicago sativa,* L. and *Trifolium subterraneum,* L. in solution culture. *Aust. J. Agric. Res.,* **16:** 743-755.

Munns, D.N., Jacobson, L. and Johnson, C.M. 1963. Uptake and distribution of manganese in oat plants. II. A kinetic model. *Plant Soil,* **19:** 193-204.

Munshi, C.B. and Mondy, N.I. 1988. Effect of soil application of molybdenum on the biochemical composition of katahdin potatoes : Nitrate nitrogen and total glycoalkaloids. *J. Agric. Food Chem.,* **36**: 688-690.

Muraka, I.P., Jackson, T.L. and Moore, D.P. 1973. Effects of N, K and Cl on N components of Russet Burbank potatoes. *Agron, J.,* **65:** 868.

Murlidharan, A. and George, C.M. 1971. Effect of phosphorus and molybdenum on the growth and nodulation of groundnut. *Agric. Res. J. Kerala.,* **9**: 48-53.

Murthy, K.S.N., Rao, J.A.V.P. and Prasad, S.S. 1988. Nutritional disorders, magnesium and iron deficiencies in light soil tobacco nurseries. *Tob. Res.,* **14:** 136-139.

Mutchuchelian, K., Rani, S.M.V. and Paliwal, K. 1988. Differential action of copper and cadmium on chlorophyll biosynthesis and nitrate reductase activity in *Vigna sinensis L. Indian J. Plant Physiol.,* **31**: 169-173.

Myers, L.F., Lipsett, J. and Kirchner, R. 1983. Response of rapeseed (*Brassica napus*) to phosphorus, boron and lime on an acid soil near Canberra. *Aust. J. Exp. Agric. Anim. Husb.,* **121**: 172-177.

Nable, R.O. 1983. Translocation of manganese in subterranean clover and its relation to deficiency diagnosis. Ph.D. Thesis Murdoch University, West Australia.

Nable, R.O., Bar-Akiva, A. and Loneragan, J.F. 1984. Functional nutrient requirement and its use as a critical value for diagnosis of manganese deficiency in subterranean clover *(Trifolium subterraneum,* L.cv. Seaton Park). *Ann. Bot.,* **54**: 39-49.

Nable, R.O. and Brownell, P.F. 1984. Effect of sodium and light upon the concentrations of alanine in leaves of C_4 plants. Aust. *J. Plant Physiol.,* **11**: 319-324.

Nable, R.O., Houtz, R.L. and Cheniae, G.M. 1988. Early inhibition of photosynthesis during development of Mn toxicity in tobacco. *Plant Physiol.,* **86**: 1136-1142.

Nable, R.O. and Loneragan, J.F. 1984. Translocation of manganese in subterranean clover *(Trifolium subterraneum,* L. cv. Seaton Park). I. Redistribution during vegetative growth. *Aust. J. Plant Physiol.,* **11**: 101-111.

Nable, R.O. and Loneragan, J.F. 1984. Translocation of manganese in subterranean clover *(Trifolium subterraneum,* L. cv. Seaton Park) II. Effect of leaf senescence and of restricting supply of manganese to part of a split root system. *Aust. J. Plant Physiol.,* **11**: 113-118.

Nagarajah, S. and Ulrich, A. 1966. Iron nutrition of the sugarbeet plant in relation to growth, mineral balance and riboflavin formation. *Soil Sci.,* **102**: 399-407.

Naik, G.R., Patil, T.M. and Hedge, B.A. 1985. Shift in the photosynthetic carboxylation pattern by iron chlorosis in sugarcane leaves. *Photosynthetica,* **19**: 581-585.

Nason, A. 1950. Effect of zinc deficiency on the synthesis of tryptophan by *Neurospora extract. Science,* **112**: 111- 122.

Nason, A. 1958. The metabolic role of vanadium and molybdenum in plants and animals. In: Trace Elements (C.A. Lamb, O.G. Bentley and J.M. Beattie, eds.) pp. 269-296. Academic Press.

Nason, A., Kaplan, N.O. and Colowick, S.P. 1951. Changes in enzymatic constitution in zinc-deficient *Neurospora. J. Biol. Chem.,* **188**: 397-406.

Nason, A. and McElroy, W.O. 1963. Modes of action of the essential mineral elements. In: Plant Physiology (F.C. Steward, ed.) Vol. III. pp. 451-536. Academic Press, New York.

Nason, A., Oldewurtel. H.A. and Propst, L.M. 1952. Role of micronutrient elements in the metabolism of higher plants. I. Change in oxidative enzyme constitution of tomato leaves deficient in micronutrient elements. *Arch. Biochem. Biophys.,* **38**: 1-13.

Nautiyal, N., Chatterjee, C. and Agarwala, S.C. 1988. Iron molybdenum interaction in *Aspergillus niger. Trans. Br. Mycol. Soc.,* **90**: 91-96.

Navarro, S., Carpena, O., Garcia, A.L. and Costa, F. 1972. Amino acids and other nitrogen fractions in normal citrus leaves and leaves deficient in iron and manganese. In IX Simposium Internazioale di Agrochimica, La Fitonutrizine oligominerable. 208-220.

Nayyar, V.K. *et. al.,* 1990. Micronutrients in soils and crops of Punjab Res. Bull. Dept. of Soils Punjab Agric. Univ. pp. 148.

Nazrul-Islam, A.K.M. 1986. Effect of interaction of calcium and manganese on the growth and nutrition of *Epilobium hirsutum*, (L.). *Soil Sci. Plant Nutr.*, **32:** 161-168.

Neales, T.F. 1959. Effect of boron supply on the sugars soluble in 80 per cent ethanol in flax seedling. *Nature,* **183**: 4830.

Neales. T.F. 1960. Some effects of boron on root growth. *Aust. J. Biol. Sci.,* **13**: 232-278.

Neales. T.F. and Hinde, R.W. 1962. A text of calcium/boron interaction hypothesis using growth of the bean radicle. *Plant Physiol.,* **15:** 212-228.

Nelson, L.E. 1983. Tolerance of 20 rice cultivars to excess Al and Mn. *Agron. J.,* **75:** 134-138.

Nelyubova, G.L. and Sychov, Y.P. 1969. Efficiency of boron in relation to the level of phosphorus. *Dokh. Mosk. S-kh Akad. In K.A. Timiryazev.,* **154:** 135-139.

Ness, P.J. and Woolhouse, H.W. 1980. RNA synthesis in *Phaseolus* chloroplasts. I. Ribonucleic acid synthesis and senescing leaves. *J. Exp. Bot.,* **31:** 223-233.

Newman, K.H. and Steward, F.C. 1969. Investigations on the growth and metabolism of culture explants of *Daucus carota.*, I. Effects of iron, molybdenum and manganese on growth. *Plant,* **81**: 333-350.

Nguyen, J. and Feierabend. J. 1978. Some properties and subcellular localization of xanthine dehydrogenase in pea leaves. *Plant Sci. Letts.,* **13**: 125-132.

Nicholas, D.J.D. 1950. Some effects of metal excess on crop plants grown in soil culture. III. Effect of cobalt, nickel and zinc on the growth, metal and chlorophyll content of tomato. *Long Ashton Res. Sta. Ann. Rept.,* 87-102.

Nicholas, D.J.D. 1961. Minor mineral elements. *Ann. Rev. Plant Physiol.,* **12:** 63-90.

Nicholas, D.J.D. 1975. The functions of trace elements in plants. In : Trace Elements in Soil-Plant-Animal Systems. (D.J.D. Nicholas and A.R. Egan, eds.), pp. 181-198. Academic Press. New York.

Nicholas, D.J.D. and Commissiong, K. 1957. Effects of molybdenum, copper and iron on some enzymes in *Neurospora crassa. J. Gen. Microbiol.,* **17:** 699-707.

Nicholas, D.J.D. and Goodman, T. 1958. The effects of deficiencies of zinc and iron on some enzyme system in *Neurospora. J. Exptl. Bot., 9:* 83-97.

Nicholas, D.J.D. and Nason, A. 1954. Molybdenum and nitrate reductase. II. Molybdenum as a constitutent of nitrate reductase. *J. Biol. Chem.,* **207:** 353-360.

Nicholas, D.J.D. and Nason, A. 1955. Role of molybdenum as a constituent of nitrate reductase from soybean leaves. *Plant Physiol.,* **30:** 135-138.

Nicholas, D.J.D. and Thomas, W.D.E. 1953. Some effects of heavy metals on plants grown in soil culture. I. The effect of cobalt on fertiliser and soil phosphate uptakes and the iron and cobalt status of tomato. *Plant Soil,* **5:** 67-80.

Nielsen, E.S., Nielsen, L.K. and Anderson, S.W. 1969. The effect of deleterious concentration of copper on the photosynthesis of *Chlorella pyrenoidosa. Physiol. Plant.,* **22**: 1121-1133.

Nikolov. B.A. 1967/68. Leaf feeding of corn with molybdenum and manganese. *God. Sofilskiya Univ. Biol. Fak. Kn. Bot. Microbiol. Fiziol. Biokhim. Rast.*, **62**: 209-218.

Ninh, T. 1971. The effects of copper ions on transpiration. *Bot. Kozl.*, **58:** 99-102.

Nishio. J.N. 1986. Iron nutrition - mediated chloroplast development. *Dissertation Abstracts International.* B. (Sciences and Engineering). **46:** 2904B-2905B.

Nishio, J.N., Taylor, S.E. and Terry, N. 1985b. Changes in thylakoid galactolipids and proteins during iron nutrition mediated chloroplast development. *Plant Physiol.*, **77**: 705-711.

Nishizeno, H., Kubota, K., Suzuki, S. and Ishii, F. 1989. Accumulation of heavy metals in cell walls of *Polygonum cuspidatum*, roots from metalliferous habitats. *Plant Cell Physiol.*, **30**: 595-598.

Nissen, P. 1974. Uptake mechanisms : Inorganic and organic. *Ann. Rev. Plant Physiol.*, **25:** 53-79.

Nobbe, F. and Siegert, T. 1862. Uber das Chlor als Spezifischer Nahrstoff des Buchweizen plfanze. *Landwirtsch. Ver. Sta.*, **4:** 318-340.

Noort, D.V. and Wellace, A. 1966. Role of iron in chlorophyll synthesis. In : Current Topics in Plant Nutrition. (A. Wallace, ed.) pp. 27-28. Pamell, Ave Los, Angles, California.

Notton, B.A. 1972. Molybdenum and tungsten in nitrate reductase formation. Ph.D. thesis, Univ. Bristol.

Notton, B.A. and Hewitt, E.J. 1979. Structure and properties of higher plant nitrate reductase especially *Spinacea oleracea.*, In : Nitrogen Assimilation of Plants (E.J. Hewitt and C.V. Cutting, eds.) (6th Long Ashton Symp. Academic Press.

Oberlander, H.E., Piatti-Funfkirchen, M. and Roth, K. 1989. The influence of various heavy metals on cadmium uptake. Uptake by wheat. *Bodenkultur.*, **40:** 295-303.

O'Brien, R.W., Frost, G.M. and Stern, J.R. 1969. Enzymatic analysis of the requirement for sodium in aerobic growth of *Salmonella typhimurium* on citrate. *J. Bact.*, **99**: 395-400.

O'Connor, G.A., Lindsay, W.L. and Olsen, S.R. 1975. Iron diffusion to plant roots. *Soil Sci.*, **119:** 285-289.

Odhnoff, C. 1957. Boron deficiency and growth. *Physiol. Plant.*, **10**: 984-1000.

Oertli, J.J. and Grgurcvic, E. 1975. Effect of pH on the absorption of boron by excised barley roots. *Agron. J.*, **67**: 278-280.

Oertli, J.J. and Jacobson, L. 1960. Some quantitative consideration in iron nutrition, of higher plants. *Plant Physiol.*, **35:** 683-688.

Ohki, K. 1974. Manganese nutrition of cotton under two boron levels. II. Critical Mn level. *Agron. J.*, **66**: 572-575.

Ohki, K. 1975. Mn and B effects on micronutrients and P in cotton. *Agron. J.*, **67:** 204-207.

Ohki, K. 1975a. Lower and upper critical zinc levels in relation to cotton growth and development. *Physiol. Plant.*, **35:** 96-100.

Ohki, K. 1976. Effect of zinc nutrition on photosynthesis and carbonic anhydrase activity on cotton. *Physiol. Plant.*, **38**: 300-304.

Ohki, K. 1978. Zinc concentration in soybean as related to growth, photosynthesis and carbonic anhydrase activity. *Crop Sci.*, **18**: 79-82.

Ohki, K. 1981. Manganese critical levels for soybean growth and physiological process. *J. Plant Nutr.*, **3**: 271-284.

Ohki, K. 1984. Manganese deficiency and toxicity effects on growth, development and nutrient composition in wheat. *Agron. J.*, **76**: 213-218.

Ohki, K. 1985. Manganese deficiency and toxicity effects on photosynthesis, chlorophyll, and transpiration in wheat. *Crop Sci.*, **25**: 187-191.

Ohki, K. and Young, C.T. 1978. Photosynthesis, carbonic andhydrase and amino acids related to zinc status in cotton. *Phytosynth. Plant Dev. Proc. Conf.*, 161-173.

Okada, S. and Asada, K. 1983. A manganese protein from thylakoids of a cyanobacterium, *Plectonema boryanum*, : inhibition of oxygen evolution with its antibody. *Plant Cell Physiol.*, **24**: 163-172.

Okanenko, A.S. and Ostrovskaya, L.K. 1950. The physiological role of copper in plants growing on peatbog soils. In: *Mikroelementry V. Zhini Rasteniji Zhivotnykh (Trace Elements in the Living Processes of Plants and Animals).* (A.P. Vinogradov, ed.) pp.400-409. Izdatelstov Akademii Nauk. SSR. Moscow.

Olsen, S.R., 1972. Micronutrient Interactions. In: Micronutrients in Agriculture (J.J. Mortvedt, P.M. Giordano and W.L. Lindsay, eds.) pp. 243-261. Soil Soc. Am. Inc. Madison, Wisconsin, USA.

Olsen, S.R. and Brown, J.C. 1980. Factors related to iron uptake by dicotyledonous and monocotyledonous plants. I. pH and reductant II. The reduction of Fe^{3+} as influenced by roots and inhibitors III. Competition between roots and external factors for iron. *J. Plant Nutr.*, **2**: 629-682.

Olsen, R.A. and Miller, R.O. 1986. Absorption of ferric iron by plants. *J. Plant Nutr.*, **9**: 751-757.

Omholt, T.E. and Boyer, G.L. 1988. Reduction of iron in squash callus cultures. *J. Plant Nutr.*, **11**: 1227-1235.

Oswa, T. and Ikeda, H. 1980. Heavy metal toxicities in vegetable crops. VIII. The effect of form of nitrogen supplied and pH level of the nutrient solution on manganese toxicities in vegetable crops. *Engei. Gakkai Zasshi.*, **49**: 197-202.

Osawa, T. and Tazuke, A. 1990. The effect of potassium and calcium concentration in the nutrient solution on copper toxicities in vegetable crops. *J. Japanese Soc. Hort. Sci.*, **59**: 365-370.

Ostrovskaya, L.K. and Geller, B.A. 1955. The influence of micro doses of copper on nitrogen metabolism in oat leaves. *Doklady Akademi Nauk SSSR.*, **103**: 727-730.

O'sullivan, M. 1970. Aldolase activity in plants as an indicator of zinc deficiency. *J. Sci. Fd. Agric.*, **21**: 607-609.

Ozanne, P.G. 1955. The effect of nitrogen on zinc deficiency in subterranean clover. *Aust. J. Biol. Sci.*, **8**: 47-55.

Ozanne, P.G. 1958. Chlorine deficiency in soils. *Nautre* (London), **182**: 1172-1173.

Ozolinya, G.R. and Klavina, D. 1981. Regulation of the activity of copper containing enzymes at various copper levels in plant tissues. *Somoregue. Meta. Rast.*, 66-73.

Ozolinya, G.R. and Lapinya, L.P. 1965. Dynamics of nucleic acid metabolism in maize and flax plants as a function of copper and nitrogen supply. In : *Mikroelementry i Produktivnost Rastenij (Trace Elements and Plant Productivity.).* (V.A.V. Peive, ed.) pp. 75-102. Izdatelstvo Akademi Nauk Latvijskoj SSR, Riga.

Ozolinya, G.R. and Livdane, B. 1970. Copper induced changes in the composition of soluble protein in barley seed endosperm. In: *Biologi-cheskaya Rol' Mikroelementovi Ikh Primeseniye V Selskom Khozyaistve i Meditsine. (The Biological Role of Trace Elements and their use in Agriculture and Medicine).* (V. Peive and M. Ya Shkolnik, eds.) Nauka, Leningrad **1**: 345.

Ozolinya, G.R., Zarin, V.E., Klyavinya, D.R., Lapinya, L.P. and Livdane, B.A. 1979. Activity of copper containing enzymes depending on providing plants with copper. In : Makro i Mikro elementy. V. Mineralnom Pitanii Rastenij (Macro and Micro elements in Mineral Nutrition of Plants). (G.Ya. Rin'kis, ed.) pp. 169-211. Znatne, Riga.

Paivoke, A. 1983. The long term effects of zinc on the growth and development, chlorophyll content and nitrogen fixation of the garden pea. *Ann. Bot. Fenn.*, **20**: 205-213.

Palaniyandi, R. and Smith, C.B. 1979. Effects of nitrogen sources on growth response and magnesium and manganese leaf concentration. *Commun. Soil Sci. Plant Anal.*, **10**: 869- 882.

Pandev, S., Stanev, V. and Kudrev, T. 1981. Effect of trace elements iron, manganese, copper, zinc, molybdenum and boron deficiency on some photosynthetic indexes of pepper. *Fiziol. Rast.*, (Sofia). **7**: 46-54.

Pardo M.B., Gomez-Moreno, C. and Peleato, M.L. 1990. Effect of iron deficiency on ferredoxin levels in *Anabaena variabilis*, PCC 6309. *Arch. Microbiol.*, **153**: 528-530.

Paribok, T.A. and Alekseyeva - Popova, N.V. 1965. The effect of zinc on the uptake and utilization of phosphorus by plants. *Fiziol. Rast.*, **12**: 591-595.

Parisi, A.F. and Vallee, B.L. 1969. Zinc metalloenzymes : Characteristics and significances in biology and medicine. *Am. J. Clin. Nutr.*, **22**: 1222-1239.

Parmar, A.S. and Hammond, H.D. 1971. Effect of boron and GA on the senescence of leaf discs from *Lycopersicon esculentum*, cultivar Rutgers. *Physiol. Plant.*, **25**: 169-177.

Patel, M.S. and Golakiya, B.A. 1986. Effect of calcium carbonate and boron application on yield and nutrient uptake by groundnut. *J. Indian Soc. Soil Sci.*, **34**: 815-820.

Patil, B.A. and Joshi, G.V. 1972. Effect of Mg^{++} ions and Mn^{++} ions on sucrose-synthetase in sugarcane leaves. *Proc. Indian Nat. Sci. Acad. part B Biol. Sci.*, **38**: 50-54.

Patnaik, S. and Bhadrachalam, A. 1965. Effect of increased concentration of iron and manganese in the growth medium on grain yield and composition of indica rice. *Indian J. Exptl. Biol.*, **3**: 149-203.

Patner, A., Gur, A. and Snir, Y. 1983. Nutrition of apple trees with zinc. *Alan. Hanotes.*, **38:** 137-155.

Pattanaik, S. 1950. The effect of boron on the catalase activity of rice plant. *Curr. Sci.*, **19**: 153-154.

Pedash, F.I. and Sereda, V.P. 1979. Effect of boron on amino acid composition in main organs of soybean plants. *Vestn. Khar'k Univ. No.*, **189:** 62-65.

Pedash, F.I. and Sereda, V.P. 1983. Effect of boron on the composition of free amino acids in main organs of sunflower plants. *Vestn. Khar'k Univ.*, **250**: 58-62.

Peive, Ya. V. 1970. Copper oxidoreductases in plants. In : Biologicheskaya Rol Medi (The Biological Role of copper) (V.V. Kovalsky, ed.) Nauka, Moscow. pp. 22-30.

Peive, Ya. V., Zhiznevskaya, G. Ya., Yagodin, B.A. and Dubrovo, P.N. 1964. A study of the activity of dehydrogenases in the root nodules of broad beans in relation to the involvement of trace elements in nitrogen fixation. Abstracts 2nd Biochem. Conf. Beltic Countries and Bylorussian SSR. (Ya. V. Peive. ed.) pp. 175-176. Nauka, Tiza.

Peive, Ya. V, Zhiznevskaya, G. Ya., and Tenisone, I.V. 1965. Comparative investigation of the activity of nitrate reductase and dehydrogenase in the root nodules of lupins and broad beans. In : Ya. V. Peive, ed.) pp. 27-48. Mikroelementy i Produktivnost Raste nif. Znatne, Riga.

Penalosa, J.M., Zornoza, P. and Carpena, O. 1987. Study of boron and manganese deficiencies in tomato plants. *An. Edafol. Agrobiol.*, **46:** 749-758.

Peneva, N. and Velinova, I. 1981. Effect of phosphorus and calcium carbonate on root uptake and mobility of some micro elements. *Prochvozn. Agrokhim.*, **16:** 25-29.

Perez-Lopez, A. and Reyes-Jurado, R.D. 1983. Effect of nitrogen and boron application in *Carica papaya*, L. I. Growth and yield. *J. Agric. Univ. P.R.*, **67**: 181-187.

Perkins, M.Y. and Aronoff, S. 1956. Identification of the blue fluroescent compounds in B deficient plants. *Arch. Biochem. Biophys.*, **64:** 506-582.

Perur, N.G., Smith, R.L. and Wisbe, H.H. 1961. Effect of iron chlorosis on protein fractions of corn leaf tissue. *Plant Physiol.*, **36**: 736-739.

Peterson, P.J. and Girling, 1981. Other trace metals, Lepp. Vol.1. Petranyi, P., Jendrisak, J.J. and Burgess, R.R. (1977). RNA Polymerase II from wheat germ contains tightly bound zinc. *Biochem. Biophys. Res. Commun.*, **74:** 1031-1038.

Petrovic, N. 1987. Effect of iron deficiency at different nitrogen concentrations on activity of RuBP Carboxylase in some cultivated plants. *Archiv Za poljoprivredne Nauke*, **48:** 385-393.

Pettersson, A. 1976. Heavy metal ion uptake by plants from solutions with metal ion, plant species and growth period variations. *Plant Soil*, **95**: 445-459.

Pienkos, P.T., Shah, V.K. and Brill, W.J. 1980. Molybdenum in nitrogenase. In : Molybdenum and molybdenum - containing Enzymes (M.P. Coughlan, ed.) pp. 385-401. Pergamon, P.

Pierre, W.H. and Stuart, A.D. 1933. Soluble aluminium studies. IV. The effects of phosphorus in reducing the detrimental effects of soil acidity on plant growth. *Soil Sci.*, **36:** 211-225.

Pilipenko, T.I. 1983. Elongation growth and IAA content in tips of plant roots with varying growth responses to boron. *Vestn. Khark's Univ.*, **250**: 55-58.

Pillar, J. 1969. The effects of trace elements, boron, molybdenum and manganese on the creation of free amino acid. *Ved. Pr. Vyak. Ustavu Rast Linne. J. Vyroby V. Plestanoch.*, **7**: 183-194.

Pillar, J. 1977. Study of the effect of copper, molybdenum and boron on the content of amino acids in lucerne. *Vedecke Prace Vyskumneho Ustavu Rastlinni J. Vyroby V Piestanoch.*, **14**: 155- 165.

Pinero, M.T., Cadahia, C. and Hernando, V. 1977. Study of the action of boron on the absorption and metabolism of essential elements, yields and quality index of tomato II. Effect on yields and nutritive quality indeces. *An. Edafol. Agrobiol.*, **36**: 1223-1231.

Pirie, N.W. 1962. Plant amine oxidase - a copper protein enzyme. Reports of Pothamst. Exp. St. for 1961. pp.96.

Pirson, A. 1937. Ernahrungs and Stoffwechselphysiolo-gische untersuchungen an *Foutinalis und Chlorella. Z. Bot.*, **31**: 193-267.

Pissarek, H.P. 1974. Investigation of the anatomical changes in oats and sunflower caused by copper deficiency. *Z. Pflanzenernaehr. Bodenkd.*, **137**: 224-234.

Platt-Aloia, K.A., Thomson, W.W. and Terry, N. 1983. Changes in plastid ultrastructure during iron nutrition mediated chloroplast development. *Protoplasma.*, **114**: 95-99.

Plesnicar, M. and Bendall, D.S. 1970. The plastocyanin content of chloroplast from some higher plants estimated by a sensitive enzyme assay. *Biochem. Biophys. Acta.*, **216**: 192-199.

Polacco, J.C. 1976. Nitrogen metabolism in soybean tissue culture. I. Assimilation of urea. *Plant. Physiol.*, **58**: 350-357.

Polikarpochikina, R.T. and Khavbin, E.E. 1972. Role of zinc in nitrogen metabolism of growing cells. *Fiziol. Rast.*, **19**: 597-603.

Popovic, R., Martinovic, B., Vuckovic, M. and Rastovic, A.1977. The effect of manganese and iron on photosynthetic activity. *Biol. Vest.*, **25**: 190.

Possingham, J.V. 1956. Effect of mineral nutrition on the content of free amino acids and amides in tomato plants. I. Comparison of the deficiencies of Cu, Zn, Mn, Fe and Mo. *Aust. J. Biol. Sci.*, **9**: 539-551.

Possingham, J.V., Vesk, M. and Mercer, F.V. 1964. The firm structure of leaf cells manganese deficient spinach. *J. Ultrastructure Res.*, **11**: 68-83.

Powers, W.H., Lewis, S. and Dawson, C.R. 1944. The preparation and properties of highly purified ascorbic acid oxidase. *J. Gen. Physiol.*, **27**: 167-180.

Prasad, K.G. *et. al.*, 1971. Effect of zinc application on phosphorus concentration and uptake in maize (*Zea mays* L.). *Indian J. Agric. Sci.*, **41**: 1068-1073.

Praske, J.A. and Plocke, D.J. 1971. A role of zinc in the structural integrity of the cytoplasmic ribosome of *Euglena gracilis. Plant Physiol.*, **48**: 150-155.

Prasolova, M.F., Mambe/tkulov, Sh., Favlinova, O.A. and Pechnov, V.A.. 1976. Effect of inorganic ions on sucrose synthetase of the sugarbeet root. *Fiziol. Rast.* (NOso)., **23**: 292-299.

Price, C.A. 1962. RNA synthesis, zinc deficiency and the kinetics of growth. *Plant Physiol.,* **37**: Supply. XXI.3.

Price, C.A. 1968. Iron compounds and plant nutrition. *Ann. Rev. Plant Physiol.,* **19**: 239-248.

Price, C.A., Clark, H.E. and Funkhouser, R.E. 1972. Functions of micronutrients in plants. In: *Micronutrient in Agriculture,* pp. 731-742. Soil Sci. Soc. Am. Madison, Wisconsin.

Prince, R.C. 1986. Manganese at the active site of the chloroplast oxygen-evolving complex. *Trends in Biochem. Sci.,* **132**: 491-492.

Prince, R.C., Bonner, W.D. Jr. and Bershok, P.A. 1981. *Fed. Proc.,* **40**: 166-167.

Pudova, R.A. 1970. Effect of trace elements on the transpiration of green ash seedlings. *Fiziol. Rast.,* **17**: 1086-1088.

Pushnik, J.C. and Miller, G.W. 1989. Iron regulation of chloroplast photosynthetic function. Mediation of PSI development. *J. Plant Nutr.,* **12**: 407-421.

Qian, L., Ye, J. and Hong Y. 1986. The role of manganese and calcium in oxidizing water in spinach PS II particles. *Zhiwu Shengli Yuebao.,* **12**: 237-243.

Qiu, Z.Q., Liu, Y.Q., Tan, C.J., Hu, J.C., Zhang, M.Z. and Wang, F.W. 1990. Application of manganese in combination with phosphorus in cultivation of soybeans. *J. Shenyang Agric. Univ.,* **21**: 243-246.

Quinland-Watson, T.A.F. 1953. The effect of zinc deficiency on the aldolase activity in the leaves of oats and clover. *Biochem. J.,* **53**: 457-460.

Radeva, V., Kichukova, A. and Georgieva, V. 1991. Foliar application of boron to sainfoin grown for seed production. *Fiziol Rast.,* **1**: 55-59.

Radeva, V. and Kudrev, T. 1984. Foliar boron spraying of alfalfa for seed production. *Fiziol. Rast.,* **10**: 99-105.

Radmer, R. and Cheniae, G.M. 1971. Photoactivation of the manganese catalyst of oxygen evolution II. A 2 - Quantum mechanism. *Biochem. Biophys. Acta.,* **253**: 182-186.

Ragimova, A.M. 1987. The effect of micro fertilizers on yield, quality and NO_3-N accumulation in cucurbitaceous crop fruits. Baku, Azerbaifjan, SSR 186-190.

Rahardjo, M., Akao, S. and Ichiki, H. 1987. A elementary study of the physiological disorder in acid soil due to aluminium and manganese toxicity in corn. *Panelitian Pertanian,* **7**: 5- 8.

Rahimi, A. and Bussler, W. 1973. The effect of copper deficiency on the tissue structure of higher plants. *Z. Pflanzenernaehr. Bodenkd.,* **135**: 183-195.

Rahimi, A. and Bussler, W. 1974. Kuperfermangel bei hoheren Pflanzen und sein histochemischer Nachweis. *Landwirtsch. Forsch. Sondesh.,* **30**: 101-111.

Rahimi, A. and Bussler, W. 1979. The development and the Zn, Fe and P content of higher plants in relation to the zinc supply. *Z. Pflanzenernaehrung Bodenkd.,* **142**: 15-27.

Rains, D.W. 1969. Cation absorption by slices of stem tissues of bean and cotton. *Experientia*, **25**: 215-216.

Rains, D.W. 1971. Mineral Metabolism. In : Plant Biochemistry. 3rd edn. (J. Bonner and I.E. Varner, eds.) Academic Press, New York.

Rajagopalan, K.V. 1980. Sulphite oxidase. In : Molybdenum and molybdenum-containing Enzymes. (M.P. Coughlan, ed.) pp. 243-272. Pergamon P.

Raju, R.A. and Reddy, M.N. 1989. Response of rice *(Oryza sativa*, L.) to algalization in combination with chemical nutrients. *Trop. Agri.*, (U.K.), **66**: 334-336.

Ramadoss, C.S. 1979. The effect of vanadium on nitrate reductase of *Chlorella vulgaris. Planta*, **146:** 539-544.

Ramadoss, C.S. 1980. Effects of vanadate on the molybdo-proteins xanthine oxidase and nitrate reductase : Kinetic evidence for multiple site extraction. *Z. Naturforsch.*, **350**: 702-707.

Ramani, S. and Kannan, S. 1974. Effects of certain cations on manganese absorption by excised rice roots. *Commun. Soil Sci. Plant Anal.*, **5:** 427-436.

Ramon, A.M., Carpena Ruiz, R.O. and Garate, A. 1989. *In vitro* stabilization and distribution of nitrate reductase in tomato plants. Incidence of boron deficiency. *J. Plant Physiol.*, **1**: 126-128.

Ramshaw, J.A.M., Seawen, M.D. and Boulter, D. 1974. The amino acid sequence of plastocyanin from *Vicia faba*, L. *Biochem. J.*, **141:** 935-943.

Randall, P.J. and Bouma, D. 1973. Zinc deficiency, carbonic anhydrase and photosynthesis in leaves of spinach. *Plant Physiol.*, **52**: 229-232.

Rani, S.M.V. and Paliwal, K. 1988. Germination as a decreasing index of heavy metal tolerance in a forage crop. *Indian J. Range Management.*, **9:** 113-118.

Rao, G.S.C. and Gupta, A.P. 1979. Influence of manganese on the accumulation of free amino acids in cane leaves. *International Sugar J.*, **81:** 99-102.

Raper, H.S. 1928. The serobic oxidase. *Physiol. Rev.*, **8:** 245-282.

Repota, V.V. 1970. The autoradiographic method in the study of the nucleic acids and proteins of pea roots in the presence of boron deficiency. *Fiziol. Biokhim. Kul'T Rast.*, **2** : 210- 214.

Raulin, J. 1869. Etudes Chimiques sur la Ve'ge'tation. *Ann. Sci. Nat. Botanique, Ser. S.*, **11:** 292-299.

Rauser, W.E. 1978. Early effects of phytotoxic burdens of cadmium, cobalt, nickel and zinc in white beans. *Can. J. Bot.*, **56**: 1744-1749.

Raven, J.A. 1977. H+ and Ca^{2+} in phloem and symplast : relation of relative immobility of the ions to the cytoplasmic nature of the transport paths. *New Phytol.*, **79:** 465-480.

Rayment, A.F. and Gupta, U.C. 1984. Copper-molybdenum fertilising of sheep pasture on new Found Land Peat Soils. Publ. Dublin; Ireland; Irish National Pest Committee. 511-521.

Reed, H.S. 1929. The relation of copper and zinc salts to leaf structure. *Am. J. Bot.*, **26:** 29-38.

Reed, H.S. 1946. Effect of zinc deficiency on phosphorus metabolism of tomato plants. *Am. J. Bot.*, **33**: 778-784.

Reed, H.S. 1947. A physiological study of boron deficiency in plants. *Hilgardia.*, **17:** 377-411.

Reddy, G.D. *et. al.*, 1973. Zinc phosphate interaction in maize. *J. Indian Soc. Soil. Sci.*, **21**: 433-445.

Reddy, K.J. and Rao. K.V.N. 1979. Effect of zinc on growth and metabolism in two varieties of *Cicer arietinum*, L. *Indian J. Plant Physiol.*, **22:** 245-266.

Reddy, K.S. and Mills, H.A. 1991. Interspecific responses of marigold to manganese as influenced by nitrogen source. *Hort. Sci.*, **26**: 1281-1282.

Reeve, E. and Shive, J.W. 1944. Potassium-boron and calcium-boron relationship in plant nutrition. *Soil Sci.*, **57**: 1-14.

Reilly, C. 1969. The uptake and accumulation of copper by *Becium homblei* (De Wild) Duvign et Plancke. *New Phytol.*, **68**: 1081-1087.

Reisenauer, H.M. 1960. Cobalt in nitrogen fixation by a legume. *Nature*, **186**: 375-376.

Reisenauer, H.M. 1963. The effect of sulphur on the absorption and utilization of molybdenum by peas. *Soil Sci. Soc. Am. Proc.*, **27:** 553-555.

Reisenauer, H.M., Walsh, L.M. and Hoeft, R.G. 1973. Testing soils for sulphur, boron, molybdenum and chlorine. In : Soil Testing and Plant Analysis (L.M. Walsh and J.D. Beaton, ed.) pp. 173-200. *Soil Sci. Soc. of Am. Inc. Madison*, Wisconsin.

Rengel, Z. and Robinson, D.L. 1959. Aluminium effects on micronutrient uptake by annual ryegrass. *Commun. Soil Sci. Plant Anal.*, **20**: 253-269.

Rickes, E.L., Brink, N.G., Koniuszy, S.R., Wood, T.R. and Folkers, K. 1948. Crystalline Vitamin B_{12}. *Science.*, **107:** 396-397.

Robertson, G.A. and Loughman, B.C. 1973. Rubidium uptake and boron deficiency in *Vicia faba*, L. *J. Exp. Bot.*, **24:** 1046- 1052.

Robertson, G.A. and Loughman, B.C. 1974. Reversible effect of boron on the absorption and incorporation of phosphate in *Vicia faba*. *New Phytol.*, **73**: 291-298.

Robinson, J.M., Smith, M.G. and Gibbs, M. 1980. Influence of hydrogen peroxidase photoassimilation in the spinach chloroplast. *Plant Physiol.*, **65:** 755-759.

Robson, A.D., Dilworth, M.J. and Chatel, D.L. 1979. Cobalt and nitrogen fixation in *Lupinus angustifolius*, L. I Growth, nitrogen concentrations and cobalt distribution. *New phytol.*, **83**: 53-62.

Robson, A.D. and Loneragan, J.F. 1970. Sensitivity of annual *Medicago* species to manganese toxicity as affected by calcium and pH. *Aust. J. Agric. Res.*, **21:** 223-232.

Roche, J. and Yagi, Y. 1952. Sur la fixation de l'iode radioactif par ler algues et sur les constituants iode's des laminaires. *C.R. Soc. Biol.*, **146:** 642-645.

Roden, E.G. and Allen. J.R. 1982. Effect of boron, manganese and zinc on nodulation and nitrogen fixation in Southern peas. *Commun. Soil Sci. Plant Anal.*, **13:** 243-258.

Rognes, S.E. 1980. Anion regulation of lupin asparagine synthetase : Chloride activation of the glutamine utilizing reaction. *Phytochemistry.*, **19:** 2287-2293.

Romheld, V. 1987. Existence of two different strategies for the acquisition of iron in higher plants. In : *Iron transport in microbes. plants and animals.* (G. Winkelmann, D.vander Helm and J.B. Neilands, eds.), pp. 353-374. VCH VerlagsGesellschaft, Weinheim, FRG.

Roemheld, V. 1991. The role of phytosiderophores in acquisition of iron and other micronutrients in graminaceous species : an ecological approach. *Plant Soil,* **130**: 127-134.

Romheld, V. and Marschner, H. 1986a. Evidence for a specific uptake system for iron phytosiderophores in roots of grasses. *Plant Physiol.,* **80:** 175-180.

Romheld, V. and Marschner, H. 1986b. Mobilization of iron in the rhizosphere of different plant species. In : *Advances in plant nutrition.* Vol. II. (B. Tinker and A. Lauchli, eds.) pp. 151:204. Praegor, New York.

Romheld, V. and Marschner, H. 1990. Genotypical differences among graminaceous species in release of phytosiderophores. *Plant Soil,* **123:** 147-153.

Rorison, I.H. 1986. The response of plants to acid soil. *Experientia,* **42**: 357-362.

Rosenfield, C.L., Reed, D.W. and Kent, M.W. 1991. Dependency of iron reduction on development of unique root morphology in *Ficus benjamina,* L. *Plant Physiol.,* **95:** 1120-1124.

Roy, T.T., Hille, F.K. and Clerk, H.E. 1966. Requirement of Ginkgo : pollen dereived tissue cultures for boron and effects of boron deficiency. *Plant Physiol.,* **41:** 815-820.

Rubin, B.A. and Loginova, L.N. 1968. Metalloproteins as catalysts of terminal stages of biological oxidation. *Uspekhi Sovremenoj Biologii.,* **65:** 442-465.

Ruck, H.C. and Bolas, B.D. 1954. The effect of manganese on the assimilation and respiration rate of isolated rooted leaves. *Am. J. Bot.,* **18**: 267-297.

Rudgers, L.A., Demetero, J.L., Paulsen, G.M. and Ellis, R. Jr. 1970. Interaction among atrazine, temperature and phosphorus induced zinc deficiency in corn *(Zea mays,* L.). *Soil Sci. Soc. Am. Proc.,* **34:** 240-244.

Rueter, J.G. 1988. Iron stimulation of photosynthesis and nitrogen fixation in *Anabaena,* 7120 and *Trichodesmium (Cyanophyceae). J. Phycol.,* **24:** 249-254.

Ruszkowska, M. and Lyszez, S. 1975. Activity of copper protein oxidases as an indicator of copper supply of plants. *Pamiet Pulawski.,* **62**: 251-263.

Russel, E.W. 1973. Conditions and plant growth. 10th Edition, Longman.

Ruszkowska, M., Lyszez, S. and Sykut, S. 1975. The activity of catechol oxidase in sunflower leaves as indicator of copper supply in plants. *Pol. J. Soil Sci.,* **8:** 67-74.

Ryan, F.J. and Tolbert, N.E. 1975. Ribulose diphosphate carboxylase/oxygenase : III. Isolation and properties. *J. Biol. Chem.,* **250:** 4229-4233.

Sachs, J. 1860. Baricht uber die physiologische Thatigkeit an der versuchsstation in Tharanot. IV. Vegetations ersuchemit Ausschluss der Bodens ubder die Nahrstoffe und sonstigen Ernahrungsbedingungen von mais, Bohnen und anderan pflanzen. Landwirft. Vers. Sta., **2:** 219-268.

Saeed, M. and Woodbridge, C.G. 1981. Effect of boron on carbohydrate and aminoacids in grape vines. *Pak. J. Agric. Res.,* **2**: 108-114.

Sakal, R. et. al. 1988. Effect of different soil fertility levels on response of wheat to zinc application on calciorthents. *J. Indian Soc. Soil Sci.,* **36**: 125-127.

Salin, M.L. and Bridges, S.M. 1980. Isolation and characterization of an iron containing superoxide dismutase from a eukaryote, *Brassica campesteris. Arch.Biochem. Biophys.,* **201:** 369-374.

Salinas, M.R., Caro, M., Cerda, A. and Santacruz, F. 1985. Interactive effect of boron and potassium on pea. *(Pisum sativum)*, growth. *An. Edafol. Agrobiol.,* **44:** 1727-1736.

Salinas, R., Cerda, A. and Martinez, V. 1986. The interactive effect of boron and macronutrients (Phosphorus, potassium, calcium and magnesium) on pod yield and chemical composition of pea *(Pisum sativum). J. Hort. Sci.,* **6:** 343-348.

Samaranayake, P., Rupatunga, K.W., Fernando, D.M. and Jayaeskera, N.E.M. 1979. *In vitro* germination of pollen and pollen tube growth of *Hevea brasiliensis,* in the presence of calcium and boron. *J. Rubber Res. Inst. Sri Lanka,* **56:** 29-33.

Samuels, G. and Alexander, A.G. 1969. Influence of variable manganese and silicon on the nutrition, sugar production and enzyme activity of immature sugarcane. *J. Agric. Univ. P.R.,* **53**: 14-27.

Sandmann, G. 1985. Consequences of iron deficiency on photosynthetic and respiratory electron transport in blue green algae. *Photosynth. Res.,* **6:** 261-271.

Sandmann, G. and Boger, P. 1983. The enzymatological function of heavy metals and their role in electron transfer processes of plants. In : Encyclopaedia of Plant Physiology. New Series, Vol. 15A. (A. Lauchli and R.L. Bieleski, eds.) pp. 563-596. Springer- Verlag, Berlin and New York.

Satter, R.L. and Fries. 1981. Vanadate inhibition of phytocrome effects on leaflet movement in *Albizzia julibrissin. Plant Sci. Lett.,* **21:** 175-180.

Saxe, H. and Rajagopal, R. 1981. Effect of vanadate of bean leaf movement, stomatal conductance, barley leaf uprolling, respiration and phosphatase activity. *Plant Physiol.,* **58**: 880-884.

Saxe, H. and Satter, R.L. 1979. Effect of vanadate on rhythmic leaf movement in *Albizzia julibrissin. Plant Physiol.,* **64:** 905-907.

Schantz, A. and Hunter, S.H. 1949. An inert metal carrier for culture media. Abst. of papers, Soc. Am. Bact. 34.

Scheffer, K., Schreiber, A. and Kickuth, R. 1982. Die sorptive Bindung von Dunger-phosphaten in Boden und die phosphatmobilisierene Wirkung der Kieselsaure 2. Mitt Die phosphatmobilisierende Wirkung der Kieselsaure. *Arch. Acker-Pflanzenbau Badenkd.,* **26**: 143-152.

Schmehl, W.R., Peech, M. and Bradfield, R. 1950. Causes of poor growth of plants on acid soils and beneficial effects of liming. I. Evolution of factors responsible for acid-soil injury. *Soil Sci.,* **70:** 393-410.

Schmid, W.E., Haag, H.P. and Epstein, E. 1965. Absorption of zinc by excised barley roots. *Physiol. Plant.,* **18:** 860- 869.

Schnabl, H. 1980. Der Anionemetabalismus in starkehaltigen und starkefreien Schlie Bzellen-protoplasten. *Ber. Dtsch. Bot. Ges.,* **93:** 595-605.

Schneider, E. and Price, C.A. 1962. Decreased RNA level : a possible cause of growth inhibition in zinc deficiency. *Biochem. Biophys. Acta.,* **55:** 406-411.

Scholz, G., Becker, R., Stephen, U.N., Rudolph, A. and Pich, A. 1988. The regulation of iron uptake and possible functions of nicotianamine in higher plants. *Biochem Physiol. Pflanz.,* **183**: 257-269.

Schwarz, K. 1973. A bound form of silicon in glycosaminoglycans and polyuronides. *Proc. Nat. Acad. Sci.,* **70:** 1608-1612.

Scienza, A., Miravalle, R., Coselli, M. and Deroten, G. 1981. The effect of boron deficiency on the development and chemical composition of Barbera groups. *Vignevin.,* **8:** 37-42.

Scott, L.E. 1944. Boron nutrition of the grape. *Soil Sci.,* **57:** 55-65.

Scripture, P.N. and McHargue, J.S. 1943. Effect of boron deficiency on the soluble nitrogen and carbohydrate content of alfalfa. *J. Am. Soc. Agron.,* **35:** 988-992.

Scripture, P.N. and McHargue, J.S. 1945. Boron supply in relation to carbohydrate metabolism and distribution in radish. *J. Am. Soc. Agron.,* **37**: 360-364.

Seatz, L.F. 1960. Zinc availability and uptake by plants as affected by the calcium and magnesium saturation and phosphorus content of the soil. *Int. Congr. Soil Sci. Trans.* 7th (Madison, Wis.) II, 271-280.

Sedberry, J.E. Jr., Eun, M.Y., Wilson, F.E., Brandon, D.M. and Bligh, D.P. 1980. Effects of application of copper and zinc on yield of Saturn rice grown on crowley silt loam and on chemical composition of rice leaf tissue. Louisiana State University, 91- 95.

Sekiya, J., Mizuno, K., Kimura, O. and Shimose, N. 1990. Ascorbate oxidase in cucumber calli and enhancement of enzyme activity by copper sulfate. *Soil Sci. Plant Nutr.,* (Tokyo). **36:**43-51.

Sela, M., Tel. Jr. E., Fritz, E. and Hattermann, A. 1988. Localization and toxic effects of cadmium, copper and uranium in *Azolla. Plant Physiol.,* **88**: 30-36.

Sevilla, F., Lopez-George, J., Gomez, M. and Del Rio, L.A. 1980. Manganese superoxide dismutase from higher plant. Purification of a new Mn - containing enzyme. *Planta.,* **150**: 153-157.

Sevini, I., D'Alessio, S., Murpurgo, L. and Avigliano, L. 1990. The role of Cu in the stability of ascorbic acid oxidase towards denaturing agents. *Eur. J. Biochem.,* **190:** 491-495.

Shah, V.K. and Brill, W.S. 1977. Isolation of an iron - molybdenum cofactor from nitrogenase. *Proc. Natn. Acad. Sci., USA,* **74:** 3249-3253.

Shaltenene, E. 1965. Changes in the phosphorus - carbohydrate metabolism of plants brought about by the action of copper and zinc. Authors Abstract of Dissertation, University, Vilnius.

Shàrma, C.P. 1961. The relation of iron supply to the growth and ascorbic acid content of barley plants grown in sand culture. *J. Indian Bot. Soc.*, **40:** 548-554.

Sharma, C.P., Agarwala, S.C., Chatterjee, C. and Srivastava, S.C. 1986. Changes in sugarcane metabolism associated with variation in manganese supply. *Sugarcane,* **1:** 10-15.

Sharma, C.P., Gupta, J.P. and Agarwala, S.C. 1981. Metabolic changes in *Citrullus*, subjected to zinc stress. *J. Plant Nutri.*, **3:** 337-344.

Sharma, C.P., Khurana, N., Chatterjee, C. and Agarwala, S.C. 1988. Response of pigeon pea to variable levels of manganese. *Proc. Indian Acad. Sci.* (Plant Sci.), **98:** 283-290.

Sharma, C.P. and Sharma P.N. 1987. Mineral nutrient deficiencies affect plant water relations. *J. Plant Nutr.*, **10**: 1637- 1643.

Sharma, C.P., Sharma, P.N. and Bisht, S.S. 1984. Induction of water stress by boron deficiency. *Actes Proc.*, **4:** 1303- 1309.

Sharma, C.P., Sharma, P.N., Bisht, S.S. and Nautiyal, B.D. 1982. Zinc deficiency induced changes in cabbage. *Plant Nutr. Proc. Int. Plant Nutr. Colloq.* 9th (A. Scaife, ed.), **2**: 601-606.

Sharma, C.P., Sharma, P.N. and Singh, S. 1985. Potassium and iron interrelationship in water relations of cauliflower. *Indian J. Plant Nutr.*, **4:** 167-175.

Sharma, C.P., Sharma, P.N., Chatterjee, C. and Agarwala, S.C. 1991. Manganese deficiency in maize affects pollen viability. *Plant Soil,* **138**: 139-142.

Sharma, P.N. 1992. Pollen fertility in manganese - deficient wheat. *Trop. Agric.*, **69:** 21-24.

Sharma, P.N., Chatterjee, C., Nautiyal, N. and Agarwala, S.C. 1984. Effect of iron deficiency on the development and physiology of maize pollen. *J. Indian Bot. Soc.*, **63**: 377-380.

Sharma, P.N., Chatterjee, C., Sharma, C.P. and Agarwala, S.C. 1987. Zinc deficiency and anther development in maize. *Plant Cell Physiol.*, **28**: 11-18.

Sharma, P.N., Chatterjee, C., Sharma, C.P., Nautiyal, N. and Agarwala, S.C. 1979. Effect of zinc deficiency on the development and physiology of wheat pollen. *J. Indian Bot. Soc.*, **58:** 330-334.

Sharma, P.N. and Ramchandra, T. 1989. Water relations of sunflower plants subjected to deficiency and toxicity of boron. *Indian J. Exp. Biol.*, **27**: 650-652.

Sharma, P.N. and Ramchandra, T. 1990. Water relations and photosynthesis in mustard plants subjected to boron deficiency. *Indian J. Plant Physiol.*, **33:** 150-154.

Shaw, T.I. 1959. The mechanism of iodide accumulation by the brown seaweed *Laminaria digitata*,. The uptake of 131 I. *Proc. R. Soc. London. Ser. B.*, **150:** 356-371.

Shaw, T.I. 1962. Halogens. In : Physiology and biochemistry of algae (Lewin R.A., ed.). Academic Press, London, New York.

Shchitaeva, V., Agakishiev, D., Lisyuk, O., Yakovlw, T., Akmamedova, O. and Kuvushinova, O. 1984. Effect of zinc and manganese on phosphorus metabolism and activity of cation-stimulated ATPase in fine filament cotton. *IZV. Akad, Nauk, Turk. SSR., Ser. Biol. Nauk.,* **6:** 22-29.

Sheat, D.C., Fletcher, B.H. and Street, H.E. 1959. *New Phytol.,* **58**: 128.

Shelp, B.J., Shattuck, V.I. and Proctor, J.T.A. 1987. Boron nutrition and mobility and its relation to the elemental composition of greenhouse grown root crops : II Radish. *Commun. Soil Plant Anal.,* **18**: 203-220.

Sherrel, C.G. 1983. Boron deficiency and response in white clover *(Trifolium repens),* and red clover *(Trifolium pertense),* and lucerne *(Medicago sativa). N.Z.J. Agric. Res.,* **26:** 197- 204.

Shetty, A. and Miller, G.W. 1966. Influence of iron chlorosis on pigment and protein metabolism in leaves of *Nicotiana tabacum* L. Ph.D. dissertation. Utah State University.

Shiralipour, A., Harris, H.C. and West, S.H. 1969. Boron deficiency and amino acid and protein contents of peanut leaves. *Crop Sci.,* **9:** 455-456.

Shkol'nik, M.Ya. 1974. General conception of the physiological role of boron in plants. *Fiziol. Rast.,* **21:** 174-186.

Shkol'nik, M.Ya. and Abdarashitov, S.A. 1958. Influence of micronutrients on synthesis and translocation of carbohydrates. *Plant Physiol.* (USSR), **5:** 393-399.

Shkol'nik, M.Ya. and Abysheva, L.N. 1971. The effect of boron deficiency on the content of catcehina, leucoanthocyanins and flavanol in buck wheat, *(Fagopyrum esculentum Moench). Bot. Zh.,* **56:** 543-548.

Shkol'nik, M.Ya. and Abysheva, L.N. 1982. Effect of high concentration of chromium, nickel and boron of flavanol of *Lycopersicon esculentum*, (Solanaceae) leaves. *Bot. Zh.,* **67**: 771-777

Shkol'nik, M.Ya. and Davydova, V.N. 1965. The effect of zinc and vitamins B_1 and B_6 on the incorporation of ^{14}C tyrosin into proteins. *Doklady6 Akademii SSSR,* **161:** 1465-1466.

Shkol'nik, M.Ya. and Greshishcheva, V.N. 1958. The effect of trace elements on photosynthesis, carbohydrate levels and the translocation of assimilates in plants supplied with either nitrate or ammonium nitrogen. *Trady Botanicheskogo Institute Akademii Nauk, SSR : Ser. IV Experimentalnaya Botanika, Izdatelstvo Akademii Nauk, SSR Leningrad.,* **12**: 154-168.

Shkol'nik, M.Ya. and Krupnikova, T.A. 1981. Formation of quinone formation from exogenous polyphenols in leaf homogenate of sunflower grown under conditions of boron deficiency. *Fiziol. Rast.* (Moscow), **28:** 541-546.

Shkol'nik, M.Ya., Krupnikova, T.A. and Smirnov, Yu. S. 1981. Activity of polyphenol oxidase and sensitivity to boron deficiency of some monocotyledonous and dicotyledonous plants. *Fiziol. Rast.,* **28**: 391-397.

Shkol'nik, M.Ya. and Maevskaya, A.N. 1962. Significance of boron in nucleic acid metabolism. *Plant Physiol.* (USSR), **9:** 270-278.

Shkol'nik, M.Ya., Maevskaya, A.N. and Davydova, V.N. 1972. Effect of boron deficiency on the content of hydroxycinnamic acid in corn leaves. *Fiziol. Rast.*, **19:** 1240-1244.

Shkol'nik, M.Ya., Maevskaya, A.N. and Sergeichik, A.A. 1980. Activity of phenylalanine - ammonialyases and sensitivity to boron deficiency in some monocotyledon and dicotyledon plants. *Fiziol. Rast.*, **27**: 773-777.

Shkol'nik, M.Ya., Soloviyova-Troitzkaya, E.A. 1962. On the physiological role of boron. 2. The significance of temperature in the eliminating of the boron deficiency by supplying RNA. *Bot. J.*, **47:** 626-635.

Shomer - Ilan, A. and Waisel, Y. 1973. The effect of sodium chloride on the balance between the C_3- and C_4- carbon fixation pathyways. *Physiol. Plant.*, **29**: 190-193.

Shrotri, C.k., Mohanty, P., Rathore, V.S. and Tewari M.N. 1983. Zinc deficiency limits and photosynthetic enzyme activities in *Zea mays*. L. *Biochem. Physiol. Pflanzen*, **178**: 213-217.

Shrotri, C.K., Rathore, V.S. and Mohanty, P. 1989. Zinc deficinecy effects on photosynthetic pigments contents, CO_2 fixation and photosynthetic enzyme activity in bajra *(Pennisetum typhoides). Nat. Acad. Sci. Lett.* (India), 12: 1-4.

Shrotri, C.K., Tewari, M.N. and Rathore, V.S. 1980. Effect of zinc nutrition on sucrose biosynthesis in maize. *Phytochemistry*, **19**: 139-140.

Shukla, P. and Pathak, A.N. 1973. Effect of molybdenum, phosphorus and sulphur on the yield and composition of berseem in acid soil. *J. Indian Soc. Soil Sci.*, **21:** 187-192.

Shukla, M.P. *et.al.*, 1981. Sulphur, zinc and boron nutrition of rai *(Brassica juncea). J. Indian Soc. Soil Sci.*, **31:** 517- 520.

Shuman, L.M., Ramscur, E.L. and Duncan, R.R. 1990. Soil aluminium effects on the growth and aluminium concentration of sorghum. *Agron. J.*, **82:** 313-318.

Shuvalov, V.A. and Krasnovsky, A.A. 1971. The luminescence of zinc protoporphyrine in microorganisms and plants. Phosphorescence and delayed fluorescence. *Molekulyarnaya Biologiya*, **5:** 698-703.

Sideris, C.P. and Young, H.Y. 194[illegible] Growth and chemical composition of *Ananas comosus*, (L.) Merr, in solution culture with different iron manganese ratio. *Plant Physiol.*, **24:** 416- 440.

Sidhu, B.S. 1989. Movement and availability of manganese in coarse textured soils under rice, wheat system. Ph.D. Thesis, Punjab Agric. Univ., Ludhiana.

Simons, T.J.B. 1979. Vanadate, a new tool for biologist. *Nature*, **281**: 337-338.

Simpson, D.J. and Robinson, S.P. 1984. Freeze fracture ultrastructure of the thylakoid membranes in chloroplast from manganese deficient plants. *Plant Physiol.*, **74**: 735-741.

Sims, J.L., Laggett, J.E. and Pal, U.R. 1979. Molybdenum - sulphur interaction effects on growth, yield and selected chemical constituents of burley tobacco. *Agron. J.*, **71:** 75-78.

Sing, J., Sing, A., Addy, S.K., Sing, S.B., Awasthi, C.P. and Srivastava, S. 1988. Effect of boron and zinc on the activity of oxidative and hydrolytic enzymes in cauliflower *(Brassica oleracea,* Botrytis group). *J. Recent Adv. Appl. Sci.,* **31**: 400-405.

Singh, B.K. and Singh, R.P. 1985. Zinc deficiency symptoms in lowland rice as induced by modified urea materials applied at different rates of nitrogen in calcareous soils. *Plant Soil,* **87:** 439-440.

Singh, B.R. and Steenberg, K. 1975. Interaction of micronutrients in barley grown on zinc - polluted soils. *Soil Sci. Soc. Am. Proc.,* **39**: 674-679.

Singh, D. and Singh, V. 1989. Effect of potassium, iron and liming on the growth and utilization of potassium and iron by barley. *Trans. Indian Soc. Desert Technol.,* **14**: 15-21.

Singh, M. and Singh, S.P. 1983. Effects of zinc and phosphorus on absorption of iron and nitrogen by submerged paddy. *Soil Sci.,* **135:** 71-78.

Singh, R.R. 1977. Effect of various concentrations of boron on growth characters and chemical composition of leaves of mango *(Mangifera indica,* L.) cv Langra. *Bangladesh Hort.,* **5**: 30- 34.

Singh, S.P. and Nayyar, V.K. 1989. Effect of cadmium and zinc on growth of corn in a coarse texture soil (Typic Ustipsamments). *Int. J. Environ. Stud.,* **34:** 57-63.

Singh, S.P. and Rakipov. N.G. 1988. Effects of Cd, Cu and Ni on barley and their removal from the soil. *Int. J. Environ, Stud.,* **31**: 291-295.

Singh, V. and Singh, S.P. 1984. Effect of applied boron on nutrients and uptake by barley crop. *Curr. Agric.,* **8:** 86- 90.

Singleton, P.L., Edmeades, D.C., Smart, R.E. and Wheeler, D.M. 1987. Soil acidity and aluminium and manganese toxicity in the Te Kauwhata area, North Island, New Zealand. *N.Z.J. Agric. Res.,* **30:** 517-522.

Sirkar, S. and Amin, J.V. 1974. The manganese toxicity of cotton. *Plant Physiol.,* **54:** 539-543.

Sisler, E.C., Dugger, W.M.Jr. and Gauch, H.G. 1956. The role of boron in the translocation of organic compounds in plants. *Plant Physiol.,* **31**: 11-17.

Sisodia, A.K., Sawarkar, N.J. and Rai, M.M. 1975. Effect of sulphur and molybdenum on yield and nutrient uptake in berseem *(Trifolium alexandrinum). J. Indian Soc. Soil Sci.,* **23:** 96- 102.

Sivasubramanian, S. and Talfbudeen, O. 1971. Effects of aluminium on growth of tea *(Camellia sinensis),* and its uptake of potassium and phosphorus. *J. Sci. Food Agric.,* **22:** 325-329.

Skok, J. 1956. Relationship of boron nutrient to radiosensitivity of sunflower plant. *Plant Physiol.,* **3:** XIII.

Skok, J. 1958. The role of boron in the plant cell. In: "Trace Elements" (C.A. Lamb, O.G., Bentley and J.B. Beattie, eds.) pp. 227-273. Acad. Press, New York.

Skok, J. 1961. Effect of G A and light on the utilization of boron and the development of boron deficiency symptoms. *Plant Physiol.,* 36 (Suppl.) : XV-XVI.

Skok, J. and McIlrath, W.J. 1958. Distribution of boron in cells of dicotyledonous plants in relation to growth. *Plant Physiol.,* **33**: 428-431.

Smirnov, Yu.S. 1978. The activity of polyphenoloxidase in *Helianthus annuus,* grown in the environment enriched with trace elements. *Botanicheskii Zh.,* **63**: 1636-1639.

Smith, P.F. 1953. Heavy metal accumulation by citrus roots. *Bot. Gaz.,* **114**: 426-436.

Snaith, S.M. 1975. Characterization of jack-bean D-mannosidase as zinc metalloenzyme. *Biochem. J.,* **147**: 83-90.

Snir, I. 1983. Carbonic anhydrase activity as an indicator of zinc deficiency in pecan leaves. *Plant Soil,* **74**: 287- 289.

Soboleva, G.A. and Bokuchava, M.L. 1969. Plant phenoloxidase. *Uskekha Biologicheskoj Khimii.,* **10**: 120-137.

Soltanpour, P.N. 1969. Effect of nitrogen, phosphorus and zinc placement on yield and composition of potatoes. *Agron. J.,* **61**: 288-289

Somers, I.I. and Shive, J.W. 1942. The iron-manganese relation in plant metabolism. *Plant Physiol.,* **17**: 582-602.

Sommer, A.L. 1928. Further evidence of the essential nature of zinc for the growth of higher green plants. *Plant Physiol.,* **3**: 217-221.

Sommer, A.L. and Lipman, C.B. 1926. Evidence of the indispensable nature of zinc and boron for higher green plants. *Plant Physiol.,* **1**: 231-249.

Sommer, G. 1979. Pot experiments to establish the danger levels of cadmium, copper, lead and zinc in relation to the application of refuge materials in agriculture. *Landw. Forschung. Sonderhsth.,* **35**: 350-364.

Spencer, D. 1954. The effect of molybdate on the activity of tomato and phosphatases. *Aust. J. Biol. Sci.,* **7**: 151-160.

Spencer, W.F. 1966. Effect of copper on yield and uptake of phosphorus and iron by citrus seedlings grown at various phosphorus levels. *Soil Sci.,* **102**: 296-299.

Spencer, D. and Possingham, J.V. 1960. The effect of nutrient deficiencies on the Hill Reaction of isolated chloroplasts of tomato. *Aust. J. Biol. Sci.,* **13**: 441-455.

Spiers, J.M. 1990. Influence of aluminium and manganese on rabitteye blueberries. *Hort. Sci.,* **25**: 515-516.

Spiller, S.C. 1980. The influence of iron stress on the photosynthetic apparatus of *Beta vulgaris. Dissertation Abstracts International B.* **41**: 1-17.

Spiller, S.C., Kaufman, L.S., Thompson, W.F. and Briggs, W.R. 1987. Specific mRNA and rRNA levels in greening pea leaves during recovery from iron stress. *Plant Physiol.,* **84**: 409-414.

Spiller, S.C. and Terry, N. 1980. Limiting factors in photosynthesis. II. Iron stress diminishes photochemical capacity by reducing the number of photosynthetic limits. *Plant Physiol.,* **65**: 121-125.

Spurr, A.R. 1952. Fluorescence in ultraviolet light in the study of boron deficiency in celery. *Science,* **116**: 421-423.

Srivastava, S.K., Ratnam, N.V. and Tiwari, D.K. 1982. Response of soybeans to micronutrients. *Indian J. Bot.*, **5**: 99-101.

Stanley, R.C. 1961. Effects of various forms of boron on germinating pollen. *Plant Physiol.*, **36:** (Suppl.) XVII.

Stanley, R.C. and Lichtenberg, E.A. 1963. The effect of various boron compound on *in vitro*, germination of pollen. *Physiol. Plant.*, **16**: 337-346.

Starck, K. 1963. Effect of boron on cell walls structure of sunflower. *Acta Bot. Palon.*, **32:** 3-15.

Starovoitova, V.P. 1983. Effectiveness of boron in flax production under condition of nitrogen fertilizer. *Izv. Timiryazev S.Kh. Akad.*, **1:** 71-77.

Stashauskaite, S.A., Navaitene, G.S. and Zhlabene, R.I. 1972. A study of the involvement of zinc in the process of phosphorus assimilation in maize roots. *Fiziol. Rast.*, **19:** 119-124.

Steinberg, R.A. 1955. Effect of boron deficiency on nicotine formation in tobacco. *Plant Physiol.*, **30:** 84-86.

Steinberg, R.A. 1956. Metabolism of inorganic nitrogen by plants. In: Inorganic Nitrogen Metabolism. (W.D. McElroy and D. Glassed, eds.) pp. 153-158. John Hopkins Univ. Press Baltimore, Maryland.

Stephen, U.W. and Gruen, M. 1989. Physiological disorder of the nicotianamine auxotroph tomato mutant *chloronerva* at different levels of iron nutrition. II. Iron deficiency response and heavy metal metabolism. *Biochem. Physiol. Pflanz.*, **185:** 189-200.

Stern, J.R. 1967. Oxaloacetate decarboxylase of *Aerobacter aerogenes.*, I. Inhibition by aviden and requirement of sodium ion. *Biochemistry,* **6:** 3545-3551.

Steward, F.C., Crane, F., Miller, K., Zacharias, R.M., Robson, R. and Margolis, D. 1959. Nutritional and environmental effects on the nitrogen metabolism of plants. In: Utilization of Nitrogen and its Compounds by Plants. *Symposia Soc. Exptl. Biol.*, **13**: 148-176.

Steward, F.C. and Margolis, D. 1962. The effect of manganese upon the free amino acids and amides of the tomato plant. Contreb. *Boyce. Thompson Inst.*, **21**: 393-410.

Steward, F.C., Newman, F.H. and Rao, K.V.N. 1968. Investigations on the growth and metabolism of cultured explants of *Daucus carota.*, II. Effects of iron, molybdenum and manganese on metabolism. *Planta.*, **81**: 351-371.

Stiles, W. and Skelding, A.D. 1940. The salt relations of plant tissues. II. The absorption of manganese salts by storage tissue. *Ann. Bot.*, **4:** 673-700.

Stocking, C.R. 1975. Iron deficiency and the structure and physiology of maize chloroplast. *Plant Physiol.*, **55:** 626- 631.

Stout, P.R. and Meagher, W.R. 1948. Studies of the molybdenum nutrition of plants with radioactive molybdenum. *Science, N.Y.*, **108:** 471-473.

Stout, P.R., Meagher, W.R., Pearson, G.A. and Johnson, C.M. 1951. Molybdenum nutrition of crop plants. I. The influence of phosphate and sulfate on the absorption of molybdenum from soils and from solution cultures. *Plant Soil,* **3**: 51-87.

Stoyanov, I. and Kha, Zul. Tkh. 1981. Relationships between growth and contents of gibberellin like substances and abscisic acid in maize under the influence of different iron concentrations in the nutrient medium. *Fiziol. Rast.*, **7**: 70-77.

Street, H.R. 1957. Excised root culture. *Biol. Rev.* **32**: 117- 155.

Stukenholtz, D.D., Olsen, R.J., Gogan, G. and Olson, R.A. 1966. On the mechanism of phosphorus - zinc interaction in corn nutrition. *Soil Sci. Soc. Am. Proc.*, **30**: 759-763.

Swamy, N.N. 1980. Effect of boron on the nutritive value of rice. *Oryza*, **17**: 142-143.

Syworotkin, G.S. 1958. The boron contents of plants with a latex system. Spurenelements in der Landwirsts Chaff. Academic Verlag, Berlin, pp. 283-288.

Szlovak, S. and Zoltanne, M. 1981. The relation between iron, copper and zinc content of maize leaf blades and transpiration. *Novenytermeles.*, **30**: 61-73.

Takahashi, H. and Ito, S. 1976. Physiological studies on zinc deficiency on maize *(Zea mays,* L.) I. Influence of zinc concentration in the nutrient solution. *J. Agric. Sci.*, **21**: 164-176.

Takahashi, M. and Asada, K. 1977. Manganese binding to sodium cyanide treated chloroplasts. Effects of light and redoxpotentials on the binding. *Plant Cell Physiol.*, **18**: 807-814.

Takahashi, E. and Miyaka, Y. 1977. Silica and plant growth. In : *Proc. Intern. Seminar on Soil Environment and Fertility Management in Intensive Agriculture.* pp. 603-611. Nippon Dojohiryo Gakkai, Tokyo.

Takkar, P.N. *et. al.*, 1976. Yield and uptake response of corn to zinc as influenced by phosphorus fertilization. *Agron. J.*, **68**: 942-946.

Tal, M. 1969. Metal ions and ribosomal conformation. *Biochem. Biophys. Acta.*, **195**: 76-86.

Talibi, G.A. 1935. Bodentung von Mikroeelementov und des verhalt nisses von Ca/Mg fur das Pflanzenwachstum bei Kalkungen Saurer Boden Z. Pfl-Ernahr. *Dung. Bodenk.*, **39**: 257-264.

Tang, C., Robson, A.D. and Dilworth, M.J. 1990. A split root experiment shows that iron is required in nodule initiation in *Lupinus angustifolius*, L. *New Phytol.*, **115**: 61-67.

Tang, Van H., Truong, T.N. and Landelout, H. 1989. Effect of aluminium on the mineral nutrition of rice. *Plant Soil.*, **114**: 173-185.

Tanner, P.D., 1977. Toxic effects of maize on manganese as affected by calcium and molybdenum application in sand culture. *Rhod. J. Agric. Res.*, **15**: 32-35.

Tanukov, G.R. 1981. Effect of the trace elements zinc and manganese on the winter hardiness and productivity of grapes. *Sadovod. Vinograd. Vinodel. Hold.*, **36**: 39-41.

Taylor, D.M., Morgan, P.W., Joham, H.E. and Amin, J.V. 1968. Influence of substrate and tissue manganese on the 1AA oxidase system in cotton. *Plant Physiol.*, **43**: 243-247.

Taylor, G.J. and Stodt, K.J. 1990. Interactive effects of cadmium, copper, manganese, nickel, and zinc on root growth of wheat (*Triticum aestivum*) in solution culture. *Dev. Plant Soil Sci.*, **4:** 317-322.

Taylor, O.A., Fetuga, B.L. and Oyenuga, V.A. 1983. Accumulation of mineral elements in 5 tropical leafy vegetables as influenced by nitrogen fertilization and age. *Sci. Hortic.*, (AMST). **18:** 313-322.

Taylor, S.E., Terry, N. and Huston, R.P. 1982. Limiting factors in photosynthesis. III. Effects of iron nutrition on the activities of 3 regulatory enzymes of photosynthetic carbon metabolism. *Plant Physiol.*, **70:** 1541-1543.

Teare, I.D. 1974. Boron nutrition and acid soluble phosphorus compounds in bean roots. *Hort. Sci.*, **9:** 236-238.

Teichler-Zallen, D. 1969. The effect of manganese on chloroplast structure and photosynthetic ability of *Chlamydomonas reinhardi.*, *Plant Physiol.*, **44:** 701-710.

Terry, M. and Ulrich, A. 1974. Photosynthetic and respiratory CO_2 exchange of sugarbeet leaves as influenced by manganese deficiency. *Crop Sci.*, **14:** 502-504.

Terry, N. 1977. Photosynthesis, growth and the role of chloride. *Plant Physiol.*, 60: 69-75.

Terry, N. 1980. Limiting factors in photosynthesis. 1. use of iron stress to control photochemical capacity *in vivo.*, *Plant Physiol.*, **65:** 114-120.

Terry, N. 1983. Limiting factors in photosynthesis IV. Iron stress mediated changes in light harvesting and electron transport capacity and its effects on photosynthesis *in vivo. Plant Physiol.*, **71**: 855-860.

Terry, N. and Abadia, J. 1986. Function of iron in chloroplasts. *J. Plant Nutr.*, **9:** 609-696.

Terry, N. and Low, G. 1982. Leaf chlorophyll content and its relation to the intracellular localization of iron. *J. Plant Nutr.*, **5:** 301-310.

Tesu, V., Toma, L.D. and Grease, E. 1980. The effect of major and minor elements applied via the foliage on photosynthesis, respiration and the water regime in the grape vine cultivar Altiqote. Institutal Agronomic, *"Ion Ionescu del Le Brad"*, Lalsi, Romania **24:** 75-78.

Teuteberg, W. 1979. Boron for rape. *Proc. Inst. Rape seed Conf.* 5th., **1:** 260-265.

Thomas, M.D. 1965. Photosynthesis (carbon assimilation). Environmental and metabolic relationships. In : Plant Physiology Vol. IV A. (F.C. Steward, ed.), 9-202. Academic Press, New York.

Thorne, D.W. 1957. Zinc deficiency and its control. *Adv. Agron.*, **9:** 31-65.

Tiffin, L.O. 1967. Translocation of manganese, iron, cobalt and zinc in tomato. *Plant Physiol.*, **42:** 1427-1432.

Tiffin, L.O. 1972. Translocation of micronutrients in plants. In: *Micronutrients in Agriculture.* (J.J. Mortvedt, D.M. Giordano and W.L. Lindsay, eds.) pp. 199-229. Soil Sci. Soc. Am. Madison, Wisconsin.

Tiffin, L.O. and Brown, J.C. 1959. Absorption of iron from chelate by sunflower roots. *Science.*, **130:** 274-275.

Tiffin, L.O. and Brown, J.C. 1961. Selective absorption of iron from iron chelates by soybean plants. *Plant Physiol.,* **36:** 710-714.

Timashov, N.D. 1970. Effect of B deficiency on respiration and esterification of P32 in the intact sunflower tissue. *Fiziol. Rast.,* **17**: 1004-1011.

Titu, H. and Djendov, C. 1972. The influence of iron deficiency on the ultrastructure of cellular organelles in the leaf mesophyll of sunflower *(Helianthus annuus,* L.). *Rev. Roum. Biol. Ser. Bot.,* **17**: 3-8.

Tiwari, K.N. *et. al.,* 1976. Studies on Fe and Zn nutrition of rice at varying moisture regimes in a black clay soil of Uttar Pradesh. *J. Indian Soc. Soil Sci.,* **24**: 303-307.

Tobin, A.J. 1970. Carbonic anhydrase from parsley leaves. *J. Biol. Chem.,* **245:** 2656-2666.

Tong, W. and Chaikoff, I.L. 1955. Metabolism of I 131 by the marine algae, *Nereocystis luetheana. J. Biol. Chem.,* **215:** 473-484.

Tripathi, B.C. and Mohanty, P. 1980. Zinc inhibited electron transport of photosynthesis in initiated barley chloroplasts. *Plant Physiol.,* **16:** 1174-1176.

Troitskaya, E.A., Dranik, L.I. and Shkol'nik, M.Ya. 1971. The phenol composition of boron deficient sunflower leaves. *Fiziol. Rast.,* **18:** 393-396.

Tso, T.C. and McMurtrey, J.E. Jr. 1960. Mineral deficiency and organic constituents in tobacco plants. II. Amino acids. *Plant Physiol.,* **35:** 865-870.

Tso, T.C. and Engelhaupt, M.E. 1963. Amino acid content related to gradual development of calcium and boron deficiency symptoms in tobacco plants. *Tobacco Sci.,* **7:** 12-16.

Tsutsumi, M. and Fujiwara, A. 1964. Biochemical studies of micro elements in green plants. VI. Effects of micro elements deficiencies on the status of phosphorus fraction in barley plants. *Tohoku. J. Agric Res.,* **15**: 169-174.

Turner, D.W. and Barkus, B. 1983. The uptake and distribution of mineral nutrients in bananas in response to supply of K, Mg and Mn. *Fertilizer Res.,* **4:** 89-99.

Twyman, E.S. 1951. The iron-manganese requirements of plants. *New Phytol.,* **50:** 210-226.

Twyman, E.S. 1959. The effect of iron supply on the yield and composition of leaves of tomato plants. *Plant Soil,* **10**: 375-388.

Urvas, L. 1989. Effects of molybdenum fertilization on timothy *(Phleum Pratense,* L.) growth in peat soil. *Ann. Agric. Fenn.,* **28:** 151-180.

Valenzuela, J.L. and Romero, L. 1988. Biochemical indicators and iron index for the appraisal of the mineral status in leaves of cucumber and tomato. *J. Plant Nutr.,* **11**: 1177-1184.

Vallee, B.L. 1960. Metal and enzyme interaction. In : *The Enzyme* (P.B. Broyer, H. Lardy and K. Myroback, eds.). 3: 225- 276. Acad. Press, New York.

Vallee, B.L. and Wacker, W.E.C. 1976. Metalloprotein. In : The Proteins 2nd edn. Vol. 5. CH.Hevrath, ed.) pp. 192. Acad. Press, New York, London.

Van Schrevan, D.V. 1937. Zinc as an essential element for sugarbeet and potato plants. *Meded. Inst. Suikerbieten.*, **7:** 18-21.

Vasil, I.K. 1964. Effect of boron on pollen germination and pollen tube growth. In: *Pollen Physiology and Fertilization*, (H.F. Linskens, ed.). North Holand Amsterdem.

Vekirchik, K.N. 1981. Effect of zinc on the nucleic acid content in spinach plants (*Spinacia oleracea* L.). *Nukleinovye Kisloty. Khromatin, Rast.*, 25-27.

Veltrup, W. 1977. The uptake of copper by barley roots in the presence of zinc. *Z. Pflanzen. Physiol.*, **83:** 201-206.

Verma, T.S. and Tripathi, B.R. 1983. Zinc and iron interaction in submerged paddy. *Plant Soil*, **72**: 107-116.

Verma, T.S. and Tripathi, B.R. 1986. Interaction effect of P-Zn and P-Cu on dry matter yield and micronutrient availability to rice in water logged alfisols. *Acta Agronomica Hung.*, **35**: 83-90.

Verstreate, D.R., Stroch, T.A. and Dunham, V.L. 1980. A comparison of the influence of iron on the growth and nitrate metabolism of *Anabaena* and *Scenedesmus*. *Physiol. Plant*, **50:** 47-51.

Vesk, M., Possingham, J.V. and Mercer, F.V. 1966. The effect of mineral nutrient deficiencies on the structure of the leaf cells of tomato, spinach and maize. *Aust. J. Bot.*, **14:** 1-18.

Vielemeyer, H.P., Fisher, F. and Bergmann, W. 1966. Uber den Einfluss des Eisenund Manganernahrung auf die peroxidase - Und Katalaseakftivitat. *Sowie den Gehalt an loslichen Kohlenhydraten in den Blattern einiger land wirtschaftlicher kulturflanzen Albrecht-Thaer-Arch.*, **10:** 727-745.

Vlamis, J. and Williams, D.E. 1967. Manganese and silicon interactions in the graminae. *Plant Soil*, **27**: 131-140.

Vlasyuk, P.A. 1980. Significance of the trace elements, manganese, molybdenum, zinc, boron and lithium in plant metabolism and productivity. Karaboammohoski Mikroelem. PP. 5-41.

Vlasyuk, P.A., Karakis, K.D. and Rudakova, E.V. 1972. Enzyme system IAA synthesis from tryptophan in maize under different condition of zinc supply. *Fiziol. Rast.*, **18:** 1243-1247.

Vlasyuk, P.A., Klimovitskaya, Z.M., Lobanova, Z.I. and Prokoplvnyak, L.M. 1973. Composition of plant chromatin as a function of the presence of manganese ions in the medium. *Fiziol. Biokhim., Kul't Rast.*, **5:** 227-231.

Vlasyuk, P.A. and Koval'chuk, M.I. 1971. Change in activity of nucleases with manganese deficiency in plants. *Fiziol. Biokhim. Kul't Rast.*, **3:** 456-459.

Vlasyuk, P.A., Rudakova, E.V. and Karakis, K.D. 1969. The effect of zinc on the biosynthesis of nucleic acids, protein and physiologically active substances in maize leaves. *Izvestiya Akademii Nauk SSSR, Ser. Biol.*, **1:** 122-126.

Vlasyuk, P.A. and Zorya, V.T. 1970. The biosynthesis of proteins with different concentrations of manganese in plants. *Fiziol. Biokhim. Kul't Rast.*, **2**: 142-147.

Voica, C. and Mihaila, M. 1982. The effect of deficiencies in certain trace elements on the total quantity of nitrogen and protein substances in the organs of oat and buck wheat plants. *An Univ. Bucur. Biol.,* **31:** 17-20.

Volk, R.J., Kahn, R.P. and Weintraub, R.L. 1958. Silicon content of the rice plant as a factor influencing its resistance to infection by the blast fungus, *Piricularia oryzae. Phytopathol.,* **48:** 179-184.

Vorm. P.D.J. van der and Diest, A. van. 1979. Aspects of the Fe and Mn nutrition of rice plants. II. Iron and manganese uptake by rice plants, grown on aerobic water cultures. *Plant Soil,* **52:** 19-29.

Vorm, P.D.J. van der 1980. Uptake of Si by five plant species as influenced by variations in Si-supply. *Plant Soil,* **56**: 153-156.

Wacker, W.E. 1962. Nucleic acid and metals. II. Changes in nucleic acid protein and metal content as a consequence of zinc deficiency in *Euglena gracilis. Biochem.,* **1:** 859-862.

Wadleigh, C.H. and Brown, J.W. 1952. The chemical status of bean plants affected with bicarbonate induced chlorosis. *Bot. Gaz.,* **113:** 272-392.

Wadleigh, C.H. and Shiva, J.W. 1939. A micro chemical study of the effect of boron deficiency in cotton seedlings. *Soil Sci.,* **47:** 33-36.

Wagner, H. and Michael, G. 1971. The influence of varied nitrogen supply on the production of cytokinins in sunflower roots. *Biochem. Physiol. Pfl. (BPP),* 162· 147-158.

Wallace, A. 1962. A decade of synthetic chelating agents. In: Inorganic plant nutrition. (A. Wallace, ed.) pp. 2278, Parnell Avenue, Los Angeles, 24, California.

Wallace, A. 1963. Role of chelating agents on the availability of nutrients to plants. *Soil Sci. Soc. Am. Proc.,* **27:** 176- 179.

Wallace, A., Frolich, E. and Lunt, O.R. 1966. Calcium requirements of higher plants. *Nature,* **209**: 634.

Wallace, A. and Lunt, O.R. 1960. Iron chlorosis in horticultural plants. *A review ; Proc. Am. Soc. Hort. Sci.,* **75:** 819-841.

Wallace, A., North, C.P., Mueller, R.T., Shannon, L.M. and Hemaidan, N. 1955. Behaviour of chelating agents in plants. *Proc. Am. Soc. Hort. Sci.,* **65:** 9-16.

Wallace, A., Shannon, L.M., Lunt., O.R. and Impey, R.L. 1957. Some aspects of the use of metal chelates as micronutrient fertilizer sources. *Soil Sci.,* **84:** 27-42.

Wallace, T. 1961. The Diagnosis of Mineral Deficiencies in Plants by Visual Symptoms. A Colour Atlas and Guide. Her Majesty's Stationery Office, London.

Warburg, O. and Christian, W. 1943. Isolierung und Kristallisation des Garungs ferments Zymohexase. *Biochem. Z.,* **314:** 149-170.

Warburg, O., Krippahl, G., and Buchhole, W. 1955. Wirkung von Vanadium auf die Photosynthese. *Z. Naturforsch.,* **10b** : 422.

Warington, K. 1926. The changes induced in the anatomical structure of *Vicia faba,* by the absence of boron from the nutrient solution. *Ann. Bot.,* **40:** 27-42.

Warington, K. 1923. The effect of boric acid and borax on the broad bean and certain other plants. *Ann. Bot.*, **37**: 629- 672.

Warington, K. 1940. The growth and anatomical structure of the carrot (*Daucus carota*) as affected by boron deficiency. *Ann. Appl. Biol.*, **27**: 176-183.

Watkins, C.B. and Ferguson, I.B. 1982. The interaction of copper and zinc with calcium in apple fruit. *Sci. Hortic. (Amsterdum).*, **17**: 319-325.

Watanabe, R., Chorney, W., Skok, J. and Wender, S.H. 1964. Effect of boron deficiency on polyphenol production in the sunflower. *Phytochemistry*, **3**: 391-393.

Waygood, E.R. 1955. Carbonic anhydrase (Plant and animal). *Meth Enzymol.*, **2**: 836-846.

Waygood, E.R., Oaks, A. and Maclachlan, G.A. 1956. The enzymically catalysed oxidation of indoleacetic acid. *Can. J. Bot.*, **34**: 905-926.

Wenerer, W.S. and Romano, A.H. 1963. Zinc stimulation of RNA and protein synthesis in *Rhizopus nigricans. Science*, **142**: 1668-1669.

Weiland, R.T., Noble, R.D. and Crang, R.E. 1975. Photosynthesis and chloroplast ultrastructural consequences of manganese deficiency in soybean. *Am. J. Bot.*, **62**: 501-508.

Weinstein, L.M. and Robbins, W.H. 1955. The effect of different iron and manganese nutrient levels on the catalase and cytochrome oxidase activity on green and albino sunflower leaf tissue. *Plant Physiol.*, **30**: 27-32.

Weiss, A. and Herzog, A. 1978. Isolation and characterization of a silicon-organic complex from plants. In : *Biochemistry of Silicon and Related problems.* (G. Banz and I. Lindovist, eds.) pp. 109-127, Plenum, New York.

Welkie, G.W. and Miller, G.W. 1960. Iron nutrition of *Nicotiana tabacum*, L. in relation to riboflavin, riboflavin-5-phosphate and flavin adenine dinucleotide content. *Plant Physiol.*, **35**: 516-520.

Warner, D. and Roth, R. 1983. Silica metabolism In : *Inorganic Plant Nutrition, Encyclopaedia Plant Physiol.* New Series, Vol. 15B (A. Lauchli and R.L. Bieleski, eds.) pp. 682- 694. Springer-Verlag, Berlin, Heidelberg, New York.

West, E.S. 1938. Zinc cured mottle leaf in citrus induced by excess phosphate. *J. Counc. Sci. Ind. Res.*, **11**: 182-184.

Whittington, W.J. 1957. The role of boron in plant growth behaviour. I. The effect of general growth, seed production and cytological behaviour *J. Exp. Bot.*, **8**: 353-367.

Whittington, W.J. 1959. The role of boron in plant growth II. The effect on growth of the radicle. *J. Exp. Bot.*, **10**: 93- 103.

Wilhelm, C. and Wild, A. 1980. The effect of vanadium on the content of chlorophyll, P700 and cytochrome f at different light intensities in *Chlorella fusca. Biochem. Physiol. Pflanz.*, **175**: 163-171.

Wilkinson, B. 1957. Boron in glasshouse tomato plant. *Nature*, **180**: 666.

Willenbrink, J. 1983. Mechanism des Z uckertransports durch Membranen bei *Beta vulgaria. Kal-Briefe.*, **16**: 585-594.

Willet, I.R., Jakobsen, P. and Zarcinas, B.A. 1985. Nitrogen induced boron deficiency in lucerne. *Plant Soil,* **86:** 443-446.

Williams, D.E. and Vlamis, J. 1957. The effect of silicon on yield and manganese - 54 uptake and distribution in the leaves of barley plants grown in culture solutions. *Plant Physiol.,* **32:** 404-408.

Wilson, D.O., Boswell, F.C., Ohki, K., Parker, M.B., Shuman, L.M. and Jellum, M.D. 1982. Changes in soybean (*Glycine max.*) seed oil and protein as influenced by manganese nutrition. *Crop Sci.,* **22**: 948-952.

Wilson, R.D. 1948. Some responses of lettuce to the application of molybdenum. *J. Aust. Inst. Agric. Sci.,* **14:** 180-187.

Wilson, R.D. and Waring, E.J. 1948. Some observations and experiments concerning the role of molybdenum in nutrition of the cauliflower plant. *J. Aust. Inst. Agric. Sci.,* **14**: 141- 145.

Winder, F.G. and Denney, J.M. 1959. Effect of iron and zinc on nucleic acids and protein synthesis in *Myobacterium smeqmatis. Nature,* London, **184**: 742-743.

Winfield, M.E. 1945. The role of boron in plant metabolism. III. The influence of boron on certain enzyme system. *Aust. J. Exp. Biol. Med. Sci.,* **23**: 267-272.

Wittwer, S.J. and Teubner, F.G. 1959. Foliar absorption of mineral nutrients. *Ann. Rev.. Plant Physiol.,* **10:** 13-32.

Wojnar, R. and Roth, J.S. 1964. Metal ions and ribonucleic acid : their nature and interference with the assay for ribonuclease and ribonuclease inhibitor. *Biochem. Biophys. Acta,* **87**: 17- 24.

Wolley, J.T., Broyer, T.C. and Johnson, G.V. 1958. Movement of chlorine within plants. *Plant Physiol.,* **33**: 1-8.

Wong You Cheong, Y. and Chan, P.Y. 1973. Incorporation of ^{32}P in phosphate esters of the sugar cane plant and the effect of Si and Al on the distribution of these esters. *Plant Soil,* **38:** 113-123.

Wood, J.G. and Silby, P.M. 1952. Carbonic anhydrase activity in plants in relation to zinc content. *Aust. J. Sci. Res.,(B)* **5:** 244-255.

Woodbridge, C.G., Venegas, A. and Crandall, P.C. 1971. The boron content of developing pear, apple and cherry flower buds. *J. Am. Soc. Hort. Sci.,* **96**: 613-615.

Wright, K.E. 1937. Effects of phosphorus and lime in reducing aluminium toxicity of acid soils. *Plant Physiol.,* **12:** 173-181.

Wu, Z. and Wu, Y. 1990. Effect of copper and zinc on growth and superoxide dismutase activity of rice seedlings. *Zhiwu Shengli Xuebao.,* **16**: 139-146.

Wynn Parry, D. and Smithson, E. 1964. Types of opaline silica depositions in the leaves of British grasses. *Ann. Bot.,* **28**: 169-185.

Wynn, R.M. and Malkin, R. 1988. Characterization of an isolated chloroplast membrane iron-sulphur protein and its identification as the photosystem I iron-shulphur A/ iron-sulphur B binding protein. *FEBS Lett.,* **229:** 293-297.

Xie, Y.Z., Qiu, R.L., Lin, C.P. and Li, Z.M. 1989. Effects of fertilizer application on biomass yield and quality of sweet potatoes Cv. Sushu NO.2. *Jiangsu Agric. Sci.*, **7**: 7-19.

Yachandra, V.K., Guiles, R.D., McDermott, A., Britt, R.D., Dexheimer, S.L., Sauer, K. and Klein, M.P. 1986. The state of manganese in the photosynthetic apparatus and structure of the manganese complex in photosystem II studied using EXAFS spectroscopy. The SI State of the oxygen-evolving photosystem complex for spinach. *Biochem. Biophys. Acta,* **850:** 324- 332.

Yagodin, B.A., Zadorozhnii, G.P. and Khalidov, K. Kh. 1979. Effect of zinc, copper and coblalt on the absorption of nutrient elements. *Khlopkovodstov.,* **11:** 31-32.

Yakunin, A.F., Ch'AngWang, N. and Gogotov, I.N. 1990. Formation of alternative nitrogenases of the heterocystic cyanobacterium *Anabaena variabilis. Doklady. Biochemistry,* **307:** 239- 240.

Yamada, Y., Tang, X.S., Itah, S. and Satoh, K. 1987. Purification and properties of an oxygen-evolving photosystem II reaction- center complex from spinach. *Biokhim. Biophys. Acta,* **891:** 129-137.

Yamanouchi, M. 1980. Effect of phosphorus, potassium, calcium, magnesium and iron treatment on the absorption and translocation of boron in several crops grown in high concentration of boron. *Nippon Dojo. Hiryogaku Zasshi.,* **51**: 126-130.

Yamanouchi, T., Hara, T. and Sonoda, Y. 1986. Effect of boron deficiency and calcium supply on the calcium metabolism in tomato plant. *Plant Soil,* **93**: 223-230.

Yang, Y.J. and Lee, B.Y. 1990. Effect of heavy metal treatments on the growth and uptake in hydroponically cultured lettuce. *J. Korean Soc. Horti. Sci.,* **31**: 37-41.

Yoshida, S., Navasero, S.A. and Ramirez, E.A. 1969. Effects of silica and nitrogen supply on some leaf characters of the rice plant. *Plant Soil,* **31**: 48-56.

Young, H. and Wallace, A. 1976. Iron deficiency on corn root mitochondrial adenosine triphosphatase activity. *Commun. Soil Sci. Plant Anal.,* **7:** 47-49.

Zaitseva, N.I., Bychovski, V. Ya. and Bukin, V.N. 1970. A study of the biosynthesis of vitamin B_{12} and porphyrins in *Propionibacterium shermanii. Doklady Akad. Nauk. SSSR.,* **190:** 1476-1479.

Zakharchisina, V.A. and Klyuchko, A.N. 1970. Changes in the activity of mitochondria in zinc deficient barley. In : Biologicheskaya Rol' Mikroelementov i Ikh Primeneniya V Selskom Khozhyaistve i Meditsine (The Biological role of Trace Elements and Their use in Agriculture and Medicine) (Ya. V. Peive and M. Ya. Shkolnik, eds.) *Nauka. Leningrad.,* **1**: 314.

Zakharieva, T., Kasabov, D. and Roemheld, V. 1988. Response of peanuts to iron-manganese interaction in calcareous soils. *J. Plant Nutr.,* **11**: 1015-1024.

Zaki, M., El-Saied, Helal, R.M., Abdall, I.M. and Amer, A.S. 1981. Preliminary investigation in improving yield and quality of tomato *(Lycopersicon esculentum),* by aid of micronutrients foliar sprays. *Ain. Shams. Univ. Fac. Agric. Res. Bull.,* **1623**: 1-15.

Zhu, Q. and Zhao, G. 1990. Response of repeseed to boron fertilizers at different levels of nitrogen application. *Turang (Nanjing)* **22:** 92-94.

Zornaza, P. and Esteban, R.M. 1984. Flavonoid contents of tomato plants for the study of the nutritional status. *Plant Soil,* **82:** 269-271.

Zumft, W.G. 1972. Ferredoxin : Nitric oxido-reductase from *Chlorella*, purification and properties. *Biochem. Biophys. Acta.,* **276**: 363-375.

Advancements In Micronutrient Research
Editor : A. Hemantaranjan
1996, Scientific Publishers, Jodhpur, pp. 367–382

10

VESICULAR-ARBUSCULAR MYCORRHIZAE AND MICRONUTRIENT NUTRITION

SUSAN B. PERSAD-CHINNERY AND LOUIS E. CHINNERY
Department of Biology, Faculty of Natural Sciences,
The University of the West Indies, Cave Hill Campus,
P.O. Box 64, Bridgetown, Barbados, WEST INDIES

CONTENTS

1. INTRODUCTION

Zygomycetous fungi of the order Glomales (Morton and Benny, 1990) form mutualistic symbioses with the roots of about 80% of plant species (Gianinazzi, 1991), including most major crop species. This symbiosis is the commonest form of mycorrhiza (fungus root, from the Greek: *mykes* = mushroom or fungus and *rhiza* = root) and is termed vesicular-arbuscular (VA) because of the formation of arbuscules, highly branched, intracellular hyphal termini, that form the interface between plant host and fungal endophyte, and, in most species, terminal swellings of hyphal tips called vesicles that have storage and perennating functions. Since species of the Gigasporaceae do not form vesicles within the roots (Gerdemann and Trappe, 1974; Morton and Benny, 1990; Walker and Sanders, 1986), some authors have suggested that arbuscular mycorrhiza would be a better name (Morton and Benny, 1990). Despite this argument, most workers in the field continue to use the term VA mycorrhiza and that name will be used here.

VA mycorrhizal fungi are obligate symbionts that have been shown to increase plant growth and nutrient uptake (Harley and Smith, 1983; Smith, 1980), plant reproduction (Koide *et al.*, 1988), plant drought resistance (Michelsen and Rosendahl, 1990) and plant resistance to pathogens (Caron, 1989). VA mycorrhizae exhibit increased nutrient uptake because of the extraradical hyphae that extend the surface area of the roots and the volume of soil from which nutrients can be extracted. VA mycorrhizal hyphae may extend 12cm from the roots (Marschner and George, 1991), traversing beyond the depletion zones created by root activity. In addition, the small diameters of the extraradical hyphae enable them to extract nutrients from pore spaces which roots and root hairs are unable to penetrate, and it is believed that VA mycorrhizal hyphae can alter unusable forms of nutrients to make them available to their host plant (Brundrett, 1991).

Most studies of the VA mycorrhizal symbiosis have focused on the nutritional benefits to the host plant in terms of phosphorus, and to a lesser extent nitrogen nutrition (Harley and Smith, 1983; Jackson and Mason, 1984). Research on the uptake of micronutrients by VA mycorrhizae are not as prevalent and a significant portion of the observations presented in the literature have been made on greenhouse pot cultures. Extrapolating results from greenhouse experiments to the situation in the field may not necessarily lead to an accurate assessment of the field situation. Thus, more studies of VA mycorrhizae in the field environment are needed to determine their agronomic potential. It should also be noted that it is extremely difficult to draw conclusions on any one nutrient from field experiments due to complex chemical interactions that naturally occur in the soil (Chinnery and Harding, 1980; Taper and Leach, 1957). Although there are conflicting reports on the effects of micronutrients on host plants and VA mycorrhizal fungi, several studies suggest that the existence of the VA mycorrhizal mutualism enhances the uptake of various micronutrients, when these are in poor supply, and provides protection when they occur at levels that could lead to plant toxicity.

Enhancement of nutrient uptake by VA mycorrhizae can be attributed to direct hyphal uptake and indirect effects caused by morphological and physiological changes in the host roots (Li *et al.,* 1991). VA mycorrhizal hyphae have been shown to actively take up micronutrients and translocate them to the roots (Cooper and Tinker, 1978; Rhodes and Gerdemann, 1975; Swaminathan and Verma, 1979). VA mycorrhizal plants have more B, Cu, Ca, Fe, K, Mg, Mn, P, and S than non-mycorrhizal plants (Gerdemann, 1968; Mosse, 1973).

2. Boron

Dixon *et al.* (1989) proposed that B may indirectly influence the plant-fungus symbiosis by enhancing root exudation which results from the effect

B has on root phyto-hormone levels and on the membrane permeability of root cortical cells. The rate of root exudation is directly related to fundamental changes in root membrane permeability and membrane mediated root exudation results in increases in VA mycorrhizal colonisation (Graham *et al.*, 1981). Low B levels may stimulate root colonisation by VA mycorrhizal fungi by increasing root exudation and B may have varied effects on different species of VA mycorrhizal fungi (Dixon *et al.*, 1989).

Dixon *et al.* (1989) studied the effects of B on the growth and colonisation of *Citrus jambhiri* Lush. seedlings inoculated with *Glomus fasciculatum* (Thaxter *sensu* Gerd.) Gerd. & Trappe and *Gl. etunicatum* Beck. & Gerd. Boron at a concentration of 25 $\mu g\ ml^{-1}$ applied to leaves, soil or both promoted root exudation of reducing sugars and amino acids, 2wks after germination but not after 4wks and B fertilized plants inoculated with either of the *Glomus* species had higher levels of VA mycorrhizal colonisation when compared to *Glomus* colonised plants that had not received B fertilization. Root colonisation was greater in seedlings inoculated with *Gl. fasciculatum* than with *Gl. etunicatum* after 16wks. It has been suggested that colonisation of roots by VA mycorrhizal fungi is influenced by several factors, one of these being root exudates that act as an attractant to these fungi.

3. ZINC

It is thought that Zn is taken up by plants as Zn^{2+} and/or $ZnOH^{+}$, with higher plants being able to absorb the latter (Marschner, 1986). VA mycorrhizal plants may more efficiently absorb Zn from the soil than non-colonised plants, such that mycorrhizal plants have increased Zn uptake when Zn is not available in phytotoxic amounts. Excessive Zn and soil pH appear to have an inhibitory effect on VA mycorrhizae (Christie and Kilpatrick, 1992).

Sharma and Srivastava (1991) studied the effects of VA mycorrhizae and Zn application on greengram [*Vigna radiata* (L.) Wilczek] grown in a sandy loam (Zn-sufficient) and a clay loam (Zn-deficient). VA mycorrhizal plants had increased Zn uptake in both soil types when compared to non-mycorrhizal plants. In the sandy loam soil, VA mycorrhizal plants of *Gl. macrocarpum* Tul. & Tul. applied with 2.5mg Zn kg^{-1} soil had about 8% greater increase in total plant dry weight than mycorrhizal plants without supplemental Zn. No significant increases in total dry weights of mycorrhizal plants, which had received 5.0 and 10.0mg Zn kg^{-1}, occurred, suggesting that deviations from "optimum" concentrations of Zn may affect VA mycorrhizae. Less pronounced increases in total dry weight were observed in the clay loam soil for the different Zn applications. In both soil types, VA mycorrhizal colonisation decreased with increasing Zn concentration yet increased water uptake was reported at all levels of Zn applications. This suggests that mycorrhizal effectiveness may not be related to root colonisation. Mohandas (1989) also reported increased uptake of Zn when mycorrhizal papaya (*Carica papaya* L. cv. Coorg Honey Dew.) was colonised by *Gl. fasciculatum* and *Gl. mosseae* (Nicol. & Gerd.) Gerd. & Trappe. Sorghum [*Sorghum bicolor* (L.) Moench] colonised by the VA mycorrhizal fungus *Gl. fasiculatum* had higher shoot concentrations of Zn than non-mycorrhizal sorghum plants when the soil was supplemented with 0.0273mg kg^{-1} Zn (Raju *et al.*, 1987). Compared with non-mycorrhizal controls, increases of 16 and 14% in the concentration of Zn in the roots and shoots respectively were observed in apple [*Malus pumila* var. *domestica*(= *M. domestica* Borkh.?) cv. M26] colonised with *Gl. macrocarpum* and grown in a P sufficient soil (Gnekow and Marschner, 1989). Mycorrhizal (*Glomus* spp.) oats (*Avena sativa* L.) grown in a sewage-sludge treated soil had increased root Zn and decreased shoot Zn. Alfalfa (*Medicago sativa* L.) grown in the same soil and colonised by the same fungal isolates showed decreases in Zn in both roots and shoots (Ricken and Höfner, 1993).

Faber *et al.* (1990) studied the uptake of Zn by VA mycorrhizal corn (*Zea mays* L. cv. Jubilee) colonised by *Gl. clairodeum* Schenck & Smith in an attempt to resolve four concerns. Firstly, they tried to determine if mycorrhizal colonisation worsened Zn deficiency when adequate amounts of P and low extractable soil Zn were present. No interaction between Zn and VA mycorrhizal colonisation was observed. Secondly, they tried to determine if the mycorrhizal association could increase Zn uptake from a soil low in this micronutrient. It was found that mycorrhizal plants not receiving Zn could initially show symptoms of alleviated Zn deficiency and mycorrhizal plants supplemented with Zn had heavier weights and higher Zn tissue concentrations than non-supplemented and non-mycorrhizal plants. However, by the time of the last harvest all plants attained similar weights irrespective of their Zn treatment. Thirdly, they tried to determine if VA mycorrhizal root colonisation could be correlated to tissue Zn concentrations. It was concluded that percentage colonisation was not an accurate assessment since different root sizes having the same percentage colonisation may have significantly different quantities of hyphae. An alternative method of recording the presence or absence of fungal structures such as arbuscules and vesicles was also deemed an inaccurate evaluation of mycorrhizal effectiveness. Faber *et al.* (1990) suggested that the measurement of total hyphal biomass should be a better measure of mycorrhizal effectiveness, however, the limitation of identifying viable from non-viable hyphae makes this method unapplicable. Therefore, a correlation between VA mycorrhizal colonisation with tissue Zn concentrations could not be determined. Fourthly, Faber *et al.* (1990) tried to determine if water extractable Zn is a better indicator of Zn deficiency in corn than total Zn. It was concluded that either water extractable or total Zn are adequate to identify Zn deficient plants. This study suggests that Zn at a concentration of 3.5 $\mu M\ l^{-1}$ does not have any adverse effects on VA mycorrhizal fungi and VA mycorrhizae.

Inconsistent foliar Zn accumulation was observed in mycorrhizal cassava (*Manihot esculenta* Crantz) colonised by *Gl. microcarpum* (Tul. & Tul.) Gerd. & Trappe (Potty, 1989). Zinc was not added to the soil and plants were fertilized with P. Dehn and Schuepp (1989) reported that VA mycorrhizal colonisation enhanced the uptake of Zn and Cd to the roots but not to the shoots. However, in soils containing phytotoxic quantities of these metals, increased root uptake accompanied by decreased shoot concentrations were observed.

Shetty *et al.* (1993) compared Zn tolerance of two grasses - *Andropogon gerardii* Vitman, an obligate mycotroph, and *Festuca arundinacea* Schreber, a facultative mycotroph. Inoculation with VA mycorrhizal fungi was found to improve survival and growth of *A. gerardii* but not of *F. arundinacea* in a high Zn soil. More Zn was retained in the roots than shoots of mycorrhizal *A. gerardii*, a trend not seen in non-mycorrhizal plants of this species or in mycorrhizal or non-mycorrhizal *F. arundinacea*.

It is well documented that the benefits of VA mycorrhizae decrease with increasing P fertilization (Menge *et al.*, 1978). When P deficiency occurs, plants may release larger amounts of sugars and amino acids that may stimulate colonisation by VA mycorrhizal fungi, partially explaining why colonisation is more prominent at low P levels. Limited studies have been conducted to observe the interaction between P and Zn and the effect this may have on VA mycorrhizae and plant nutrition. The effects of the interaction of P and Zn on VA mycorrhizae were studied by Lambert *et al.* (1979) using soybean (*Glycine max* (L.) Merr.) and corn (*Z. mays*) colonised by *Gigaspora gigantea* (Nicol. & Gerd.) Gerd. & Trappe. As expected, as P fertilization increased, VA mycorrhizal colonisation decreased. With no correction for differences in plant size, when P was added at concentrations of 0, 25, 75 and 200 $\mu g\ g^{-1}$ to soil, the concentration of Zn was decreased in mycorrhizal soybean and both mycorrhizal and non-mycorrhizal corn but was not decreased in non-mycorrhizal soybean plants. Two mechanisms by which P could reduce Zn concentrations in mycorrhizal plants were proposed. Firstly, P may suppress the mycorrhizal uptake of Zn (and Cu, see below) and, secondly, larger P sufficient plants may not concentrate Zn as much as deficient plants. Similar results were obtained by Gardiner and Christensen (1991) who found that Zn uptake of pear (*Pyrus communis* L.) seedlings colonised by *Gl. intraradices* Schenck & Smith or *Gl. mosseae* was enhanced when the concentration of P added to the soil was between 0.25 and 0.40mg l^{-1}. In their study of the P-Zn interaction, Shetty *et al.* (1993) found that mycorrhizal and non-mycorrhizal plants of two species of grass (*Andopogon gerardii* and *Festuca arundinacea*) had no differences in shoot Zn concentration. They also found that mycorrhizal plants had significantly higher shoot P concentrations than non-mycorrhizal plants at all Zn levels.

These studies make it clear that further understanding of how macro-and micro-nutrients interact with each other in the soil at deficient, optimal and supraoptimal levels and how this in turn may affect plant nutrition and the efficiency of mycorrhizae is needed. Such information is necessary if the agronomic potential of VA mycorrhizal fungi is to be realised. Evidence from studies conducted on Zn nutrition of mycorrhizal plants suggest that when P is not limiting to plant growth, VA mycorrhizal fungi become more functional as suppliers of this micronutrient. In soils low in P, however, P transfer to the plant becomes the major benefit of VA mycorrhizal fungi.

Plants that grow in toxic micronutrient concentrations are more tolerant to these environments than plants growing under non-toxic conditions (Ernst *et al.*, 1992). Several mechanisms have been proposed to explain the tolerance of plants to heavy metals (e.g., Cd, Cu, Pb, Zn); metals may become bound to the plant cell wall, the plasma membrane, uptake kinetics may be altered such that plants growing in heavy metal soils take up less of these heavy metals, cytosolic buffers may be present in the plant cell (Ernst *et al.*, 1992) and VA mycorrhizal fungi may provide protection to the plant.

VA mycorrhizal fungi are increasingly being shown to be significant in our understanding of heavy metal tolerance of plants that grow in soils containing toxic or very high levels of these metals. Soils contaminated by heavy metals may contain tolerant strains of VA mycorrhizal fungi (Leyval *et al.*, 1993; Shetty *et al.*, 1993; Weissenhorn *et al.*, 1992). Leyval *et al.* (1993), for example, found that *Gl. mosseae* isolated from contaminated soils were more resistant to Zn and Cd than a reference strain of this species.

In recent years, interest has been expressed in the revegetation of severely disturbed soils such as mine spoils in which toxic levels of micronutrients may be present. Arines and Vilarino (1991) studied the growth, micronutrient content and mycorrhizal status of fescue (*Festuca pratensis* L.) and red clover (*Trifolium pratense* L.) grown in a lignite mine soil. Species of *Glomus*, *Gigaspora* and *Acaulospora* were used to colonise the fescue and red clover plants grown in this substrate that was rich in Al, Fe, Mg and Mn and poor in P and K. It was found that all plants had toxic levels of Al and Mn, with red clover having higher concentrations of these elements and Zn than fescue. Only *Gl. deserticola* Trappe, Bloss & Menge was able to consistently colonise the plant roots in spite of the toxic levels of several elements present within the plant. However, this species of VA mycorrhizal fungi did not significantly increase plant growth or nutrition except when compared to a native inoculum containing an unidentified *Gigaspora* sp. Excessive amounts of micronutrients may be toxic to both plants and VA mycorrhizal fungi and as such the benefits of the mycorrhizal association can be lost. This study also suggests that some species of VA mycorrhizal fungi and plants may have different tolerances to the same level of micronutrient toxicity.

In a study of mycorrhizal plants grown in soils containing toxic levels of micronutrients, Heggo *et al.* (1990) found that mycorrhizal soybean plants (*Glycine max* cv. Essex) colonised by several VA mycorrhizal fungi and grown in soils with a pH range of 5.0 to 7.5 that contained varying levels of Zn, Cd, Cu, Mn and Fe had reduced foliar Zn, Cd, and Mn when the soils had high toxic levels of these elements. However, when plants were grown in soils containing low to moderate levels of Zn, Cd, and Mn increased foliar concentrations of these elements were found. These results, and those reported by other authors (Cooper and Tinker, 1978; Swaminathan and Verma, 1979), suggest that VA mycorrhizae may take up nutrients selectively and that this may be influenced by the availability of the specific nutrient in the soil.

Denny and Wilkins (1987) proposed a mechanism by which mycorrhizal fungi may ameliorate the severity of Zn toxicity. The mechanism was deduced from work done on ectomycorrhizal fungi but since it involves extraradical hyphae it may be applicable to other types of mycorrhizae. It was proposed that Zn may become adsorbed to electronegative polysaccharides in the hyphal cell wall and in the slime surrounding the hyphae. As Zn is highly immobile in soil, additional Zn does not rapidly diffuse to replace the adsorbed Zn and as such this would result in lower Zn concentrations in the soil, especially around the mycorrhizae. Extraradical VA mycorrhizal hyphae are capable of indeterminate growth in the soil and as such new binding sites for Zn are continuously being produced. It was also suggested that Zn may become bound at the plant-fungus interface as another means of reducing the severity of Zn toxicity in the plant. Bradley *et al.* (1982) found that ericoid mycorrhizae could reduce the accumulation of metallic elements in the shoots when they were present in toxic levels in the soil. They proposed that the observed phenomenon may be accomplished by the binding of metallic elements to the cell walls of the fungal hyphae and that hyphae may selectively adsorb metals to their cell walls.

Zinc has also been shown to inhibit the germination of some VA mycorrhizal spores or reduce the rate at which they germinate (Hepper, 1979).

4. COPPER

Increased Cu concentrations have been observed in several mycorrhizal plants, e.g., cassava colonised by *Gl. microcarpum* (Potty, 1989), sorghum with *Gl. fasiculatum* (Raju *et al.*, 1987) and oats (Ricken and Höfner, 1993).

In the soil, Cu is highly immobile and depletion zones readily develop around roots. Although, VA mycorrhizal colonisation may increase Cu concentrations in the roots, shoots or the entire plant when Cu is at low or moderate levels in the soil (Li *et al.*, 1991), the mechanism(s) involved is not known. The rate of Cu transfer to the plant may depend on the plant's demand as a result of growth. Li *et al.* (1991) calculated that between 53 and 62% of the total Cu uptake is delivered to the host plant by VA mycorrhizal hyphae. This can be compared with 34 to 90% of P uptake delivered in this way (Li *et al.*, 1991). These results help explain the reduction in the concentration of Cu in plants with increasing P since VA mycorrhizal development is inversely proportional to the availability of soil P (Gildon and Tinker, 1983; Timmer and Leyden, 1980).

Gnekow and Marschner (1989) observed increases of Cu concentration of 13 and 126% in stems and roots, respectively, of VA mycorrhizal apple plants. (Ricken and Höfner, 1993) found that Cu was enhanced in shoots but decreased in roots of oats mycorrhizal with *Glomus* spp. Graham *et al.* (1986) observed that the growth of Carrizo citrange [*Poncirus trifoliata* (L.) Raf. x *Citrus sinensis* (L.) Osbeck] seedlings colonised with *Gl. intraradices* and non-colonised seedlings was logarithmically reduced with increasing Cu concentrations from 3 to 248μg g^{-1} soil. However, plant dry weight of mycorrhizal citrange decreased more rapidly with increasing Cu concentration than in non-mycorrhizal plants. The minimum toxic concentration of Cu for colonised plants was found to be 19μg g^{-1} soil and the corresponding value for uncolonised plants was 37μg g^{-1} suggesting that some VA mycorrhizae may be more sensitive to Cu toxicity. VA mycorrhizal fungal colonisation and hyphal development was reduced with increasing Cu concentrations. Leaf P was less than 0.1% indicative of a deficiency, in colonised seedlings at Cu levels greater than and equal to 37μg g^{-1} and uncolonised plants had sufficient levels of leaf P at this concentration. At low levels of Cu, namely 2.7 and 18.8μg g^{-1}, plants with VA mycorrhizae had increased Cu in the roots but not the leaves. Copper concentration in the leaves increased gradually with increasing Cu concentrations in the soil, until the values of 160 and 248μg g^{-1} were attained and then leaf Cu increased rapidly. It was noted that Zn concentrations were increased in the plant with increasing Cu levels. Graham *et al.* (1986) obtained similar results from their field experiment. VA mycorrhizal colonisation was lower in soils with high levels of extractable Cu. They postulated that poor growth of VA mycorrhizal plants in soils with high Cu concentrations may be due to P deficiency induced by Cu on the mycorrhizae. Copper may reduce P uptake in the absence of P fertilizers either by inhibiting P uptake from the soil or by retarding extraradical hyphal development and colonisation.

The interaction of P and Cu was studied by Lambert *et al.* (1979) using soybean and corn colonised by *Gi. gigantea.* With increased levels of P, VA fungal colonisation and the concentration of Cu was decreased in mycorrhizal soybean and mycorrhizal and non-mycorrhizal corn but was not decreased in non-mycorrhizal soybean plants.

Whether it is Cu that induces P deficiency or P induces Cu deficiency is unresolved. Studies have shown that P fertilization reduces the Cu concentration in mycorrhizal plants (Gardiner and Christensen, 1991; Timmer and Leyden, 1980). A study was conducted by Timmer and Leyden (1980) on P induced Cu deficiency in citrus orchards. Sour orange (*Citrus aurantium* L.) colonised with *Gl. fasciculatum* did not exhibit Cu deficiency symptoms when P was added at 100, 200, 400mg l^{-1} and mild symptoms were seen when P was added at 800mg l^{-1} Addition of P or Cu did not increase the growth of non-mycorrhizal seedlings but when both elements were added the seedlings became Cu deficient. Increasing P concentrations resulted in decreasing foliar Cu when Cu was added at 4.3 and 8.6mg l^{-1}. Percentage VA mycorrhizal colonisation and spore production declined with Cu fertilization and increasing P fertilization. Mycorrhizal plants receiving 4.3mg l^{-1} Cu were taller than those receiving no or 8.6mg l^{-1} Cu. It appears that Cu at high rates may have an effect on the functioning of the mycorrhizae or the fungus, since reduced colonisation and spore production were evident with increasing Cu fertilization. This study proposed that at moderate levels of P, 200 and 400mg P l^{-1}, plant growth rate is increased such that Cu becomes a limiting factor. At higher P concentrations, e.g. 800mg P l^{-1}, mycorrhizal colonisation is reduced and Cu deficiency results from reduced uptake. Similar results were obtained by Gardiner and Christensen (1991) with mycorrhizal pear seedlings that had an increase in the uptake of Cu when the concentration of P added to the soil was between 0.25 and 0.40mg l^{-1}. Increased P reduces VA mycorrhizal colonisation and results in decreased uptake of micronutrients and possibly even to deficiency levels. Whether this increase is because of bigger roots or greater mycorrhizal effectiveness remains to be determined.

A field study of the effects of P and Cu nutrient uptake, photosynthesis and grain yields of wheat (*Triticum aestivum* L.) was conducted by Javadi *et al.* (1991). Super phosphate levels of 0, 560, 1680 and 3360kg ha^{-1} and Cu as copper sulphate at 0, 33.6 and 67.3kg ha^{-1} resulted in increased foliar P, reduced VA mycorrhizal spore populations, reduced photosynthetic rates but did not affect leaf Cu content or grain yields. Additional Cu to the same level of P resulted in reduced spore counts and had no other effect on the fungus or the plant. The presence of Cu has been shown to inhibit germination rates of VA mycorrhizal spores (Hepper, 1979).

Fungicidal Cu is commonly used to control some bacterial and fungal diseases and persistent usage may lead to the accumulation of this metal in the soil which may become toxic or harmful to VA mycorrhizal fungi. Menge (1982) however, reported that the non-systemic fungicide $CuSO_4$ has no harmful effects on VA mycorrhizal fungi and may even increase mycorrhizal development under certain conditions, such as when pathogenic fungi or bacteria are present in the soil. In a field experiment to test the effect of fungicides on the VA mycorrhizal fungus *Gl. etunicatum*, Nemec (1980) found that $CuSO_4$ (53% Cu) added to the soil at 56, 112 and 224kg ha^{-1} had no significant effect on fungal development. This was determined by counting vesicles, percentage colonisation and spore production.

A study conducted in *Citrus* orchards on four islands in the eastern Caribbean suggests that VA mycorrhizae may exhibit Cu tolerance when Cu is present at high concentrations (Michelini *et al.*, 1993). It appears that Cu delivered by hyphae to the roots remains there when P is low, but as P increases Cu is translocated to the shoots. This is indicative of a mechanism in VA mycorrhizae that regulates the release and transport of Cu from the root to the shoot depending on the shoot's demand (George *et al.*, 1992; Gildon and Tinker, 1983). The presence of the mycorrhizal fungus may also alleviate Cu deficiency at low to moderate P levels but not at high P since mycorrhizal colonisation is depressed (Li *et al.*. 1991).

5. Manganese

VA mycorrhizal plants tend to have the lowest shoot concentration of Mn when compared to the other types of mycorrhizae (Newman, 1991). Several authors (Gerdemann, 1968; Mosse, 1973; Raju *et al.*, 1987) have, however, reported increased Mn uptake by VA mycorrhizal plants. Recently, Heggo *et al.* (1990) and Arines and Vilarino (1991) reported that mycorrhizal plants had reduced Mn levels when toxic levels of this element were present in the soil. Plant colonisation by VA mycorrhizal fungi appears to decrease the concentration of Mn in both shoots and roots (Arines *et al.*, 1989). Kothari *et al.* (1990; 1991) reported that the decrease in mycorrhizal plant Mn is probably related to changes in the microorganism population in the rhizosphere caused by colonisation. They found that maize plants (cv. Tau) inoculated with only rhizosphere microorganisms had higher Mn concentrations in shoots and roots than mycorrhizal (*Gl. mosseae*) plants inoculated with the same microorganisms. Mycorrhizal plants had 20 to 30 times less microorganisms than non-mycorrhizal plants. It is thought that this observation may be related to the amount of root exudates in the rhizosphere and the redox potential of the microbial population in this region. Manganese availability in soils increases with increasing soil acidity (Arines *et al.*, 1989) and Mn readily complexes with organic matter in the soil. The availability of manganese in the rhizosphere depends on the Mn-oxidizing and Mn-reducing potential of the organisms in the rhizosphere (Marschner, 1992) and root colonisation by VA mycorrhizal fungi results in reduced root exudates several weeks after seed germination (Dixon *et al.*, 1989). Thus, VA mycorrhizal fungi may selectively affect certain rhizosphere microorganisms resulting in altered Mn acquisition by the plant. These fungi may affect rhizosphere organisms either by altering host physiology so that the plant root exudation is reduced or by altering the root exudate as it is released from the plant (Bagyaraj, 1984; Graham *et al.*, 1986).

It has been reported that manganese and iron may interact negatively or positively to affect the uptake of these nutrients by the plant (Chinnery and Harding, 1980). The nature of the interaction appears to depend on the plant species involved (Warden and Reisenauer, 1991). VA mycorrhizae may be protected from Mn toxicity as a result of colonisation by these fungi (Pacovsky, 1986; Ricken and Höfner, 1993). Griffioen and Ernst (1989) reported that the shoot/root ratio of Cu and Mn was altered as a result of VA mycorrhizal colonisation, however, the total uptake of these minerals remained the same. The underlying mechanisms regulating Mn uptake by VA mycorrhizal fungi deserve further investigation.

Nogueira and Harris (1993) concluded that the improvement in the growth of *Vigna radiata*, grown in a soil with toxic levels of Mn and colonised by *Gl. etunicatum*, was due to improved P nutrition. They grew plants in soil collected from a Mn ore bed in North Wales, U.K., and found that the mycorrhizal and non-mycorrhizal plants showed the same levels of Mn toxicity symptoms and had similar concentrations of Mn in shoot tissue (920 - 1000mg kg^{-1} dry weight).

The presence of Mn, like Cu and Zn, has been shown to have negative effects on VA mycorrhizal spore germination (Hepper, 1979).

6. Molybdenum

The function of molybdenum in plants is related to its ability to undergo valency changes as the metal component of enzymes. Information on the effects of Mo on VA mycorrhizae are scarce. Raju *et al.* (1987) found that mycorrhizal and non-mycorrhizal sorghum plants had similar shoot concentrations of Mo when the soil was supplemented with 0.0141mg kg^{-1}. Obviously this area requires further research.

7. Conclusion

It appears from the literature that the effects of micronutrients on VA mycorrhizae and VA mycorrhizal fungi are dependent on several factors. These include:- the soil type; the species of VA mycorrhizal fungus, whether native or introduced, used to colonise the plants (Trent *et al.*, 1993); the species of plant colonised; the quantity of the micronutrient in the soil; the interactions between the micronutrient and macronutrients, which may be modified by soil pH; and the interactions between VA mycorrhizal fungi and other rhizosphere organisms.

The principal advantage of VA mycorrhizae is the increased volume of soil explored for nutrients and, hence, enhanced efficiency of nutrient adsorption from the soil solution (Sieverding, 1991). The root morphology and the external hyphal network determine the nutrient uptake of the plant. The mycorrhizal hyphal network may have a lower K_m than the host roots which may account for their effectiveness (Cress *et al.*, 1979). VA mycorrhizal fungi may also selectively adsorb elements from the soil and may play a role in the amelioration of micronutrient toxicity.

This review highlights the need for more research in order to develop a better understanding of the interactions between micronutrients and VA mycorrhizae and VA mycorrhizal fungi. Research that is essential if we are going to fully exploit VA mycorrhizal fungi in the sustainable agriculture systems of the future.

REFERENCES

Arines, J. and Vilarino, A. 1991. Growth, micronutrient content and vesicular-arbuscular fungi infection of herbaceous plants on lignite mine spoils: A greenhouse pot experiment. *Plant and Soil.*, **135**:269-273.

Arines, J., Vilarino, A. and Sainz, M. 1989. Effect of different inocula of vesicular-arbuscular mycorrhizal fungi on manganese content and concentration in red clover (*Trifolium pratense* L.) plants. *New Phytol,.* **112**:215-219.

Bagyaraj, D.J. 1984. Biological interactions with VA mycorrhizal fungi. Pages 131-153: IN: Powell, C.Ll. and Bagyaraj, D.J., eds. VA Mycorrhizas. CRC Press, Boca Raton, Florida, U.S.A.

Bradley, R., Burt, A.J. and Read, D.J. 1982. The biology of mycorrhiza in the Ericaceae. VIII. Role of mycorrhizal infection in heavy metal resistance. *New Phytol.,* **91**:197-209.

Brundrett, M. 1991. Mycorrhizas in natural ecosystems. *Adv. Ecol. Res.*, **21**:171-313.

Caron, M. 1989. The potential of mycorrhizae in the control of soil-borne diseases. *Can. J. Pt. Pathol.*, **11**:177-179.

Chinnery, L.E. and Harding, C.P. 1980. The effect of ferrous iron on the uptake of manganese by *Juncus effusus* L. *Ann. Bot.,* **46**:409-412.

Christie, P. and Kilpatrick, D.J. 1992. Vesicular-arbuscular mycorrhiza infection in cut grassland following long-term slurry application. *Soil Biol. Biochem.*, **24**:325-330.

Cooper, K.M. and Tinker, P.B. 1978. Translocation and transfer of nutrients in vesicular-arbuscular mycorrhizas. II. Uptake and translocation of phosphorus, zinc and sulphur. *New Phytol.*, **73**:901-912.

Cress, W.A., Throneberry, G.O. and Lindsey, D.L. 1979. Kinetics of phosphorus absorption by mycorrhizal and non-mycorrhizal tomato roots. *Plant Physiol.,* **64**:484-487.

Dehn, B. and Schuepp, H. 1989. Influence of VA mycorrhizae on the uptak distribution of heavy metals in plants. *Agric. Ecosystems Environ.*, **29**:79-8..

Denny, H.J. and Wilkins, D.A. 1987. Zinc tolerance in *Betula* spp. IV. The mechanism of ectomycorrhizal amelioration of zinc toxicity. *New Phytol.*, **106**:545-553.

Dixon, R.K., Garrett, H.E. and Cox, G.S. 1989. Boron fertilization, vesicular-arbuscular mycorrhizal colonization and growth of *Citrus jambhiri* Lush. *J. Plant Nutr.*, **12**:687-700.

Ernst, W.H.O., Verkleij, J.A.C. and Schat, H. 1992. Metal tolerance in plants. *Acta Bot. Neerl.*, **41**:229-248.

Faber, B.A., Zasoski, R.J., Burau, R.G. and Uriu, K. 1990. Zinc uptake by corn as affected by vesicular-arbuscular mycorrhizae. *Plant and Soil.*, **129**:121-130.

Gardiner, D.T. and Christensen, N.W. 1991. Pear seedling responses to phosphorus, fumigation and mycorrhizal inoculation. *J. Hort. Sci.*, **66**:775-780.

George, E., Haussler, K., Kothari, S.K., Li, X.-L. and Marschner, H. 1992. Contribution of mycorrhizal hyphae in soil. Pages 42-47: IN: Read, D.J., Lewis, D.H., Fitter, A.H. and Alexander, I.J., eds. Mycorrhizas in Ecosystems. C.A.B. International, Wallingford, U.K.

Gerdemann, J.W. 1968. Vesicular-arbuscular mycorrhiza and plant growth. *Ann. Rev. Phytopath.* **6**:397-418.

Gerdemann, J.W. and Trappe, J.M. 1974. The Endogonaceae in the Pacific Northwest. Mycologia Memoir No. 5.

Gildon, A. and Tinker, P.B. 1983. Interactions of vesicular-arbuscular mycorrhizal infections and heavy metals in plants. II. The effects of infection on uptake of copper. *New. Phytol.*, **95**:263-268.

Gianinazzi, S. 1991. Vesicular-arbuscular (endo-) mycorrhizas: cellular, biochemical and genetic aspects. *Agric. Ecosystems Environ.*, **35**:105-119.

Gnekow, M.A. and Marschner, H. 1989. Role of VA-mycorrhiza in growth and mineral nutrition of apple (*Malus pumila* var. *domestica*) rootstocks cuttings. *Plant and Soil,*. **119**:285-293.

Graham, J., Leonard, R. and Menge, J.A. 1981. Membrane mediated decrease in root exudation responsible for phosphorus-inhibition of vesicular-arbuscular mycorrhiza formation. *Pl. Physiol.*, **68**:648-652.

Graham, J.H., Timmer, L.W. and Fardelmann, D. 1986. Toxicity of fungicidal copper in soil to citrus seedlings and vesicular-arbuscular mycorrhizal fungi. *Phytopath.*, **76**:66-70.

Griffioen, W.A.J. and Ernst, W.H.O. 1989. The role of VA mycorrhizas in the heavy metal tolerance of *Agrostis capillaris* L. *Agric. Ecosystems Environ.*, **29**:173-177.

Harley, J.L and Smith, S.E. 1983. Mycorrhizal Symbiosis. Academic Press, London, U.K..

Heggo, A., Angle, J.S. and Chaney, R.L. 1990. Effects of vesicular-arbuscular mycorrhizal fungi on heavy metal uptake by soybeans. *Soil Biol. Biochem.*, **22**:865-869.

Hepper, C.M. 1979. Germination and growth of *Glomus caledonius* spores: The effect of inhibitors and nutrients. *Soil Biol. Biochem.*, **11**:267-277.

Jackson, R.M. and Mason, P.A. 1984. Mycorrhiza. Edward Arnold, London, U.K.

Javadi, M., Beuerlein, J.E. and Arscott, T.G. 1991. Effects of phosphorus and copper on factors influencing nutrient uptake, photosynthesis and grain yield of wheat. *Ohio J. Sci.*, **91**:191-194.

Koide, R., Li, M., Lewis, J. and Irby, C. 1988. Role of mycorrhizal infection in the growth and reproduction of wild *vs* cultivated plants. I. Wild *vs* cultivated oats. *Oecologia*, **77**:537-543.

Kothari, S.K., Marschner, H. and Römheld, V. 1990. Direct and indirect effects of VA mycorrhizal fungi and rhizosphere microorganisms on acquisition of mineral nutrients by maize (*Zea mays* L.) in a calcareous soil. *New Phytol.*, **116** :637-645.

Kothari, S.K., Marschner, H. and Römheld, V. 1991. Effect of a vesicular-arbuscular mycorrhizal fungus and rhizosphere microorganisms on manganese reduction in the rhizosphere and manganese concentration in maize (*Zea mays* L.). *New Phytol.*, **117**:649-655.

Lambert, D.H., Baker, D.E. and Cole, H.Jr. 1979 The role of mycorrhizae in the interactions of phosphorus with zinc, copper and other elements. *Soil Sci. Soc. Am. J.*, **43**:976-980.

Leyval, C., Weissenhorn, I. and Berthelin, J. 1993. Heavy metal toxicity to AM fungi. Page 3: IN: Peterson, L. and Schelke, M., eds. Abstracts of the 9th North American Conference on Mycorrhizae. University of Guelph, Guelph, Ontario, Canada.

Li, X.-L., Marschner, H. and George, E. 1991. Acquisition of phosphorus and copper by VA mycorrhizal hyphae and root-to-shoot transport in white clover. *Plant and Soil.*, **136**:49-57.

Marschner, H. 1986. Mineral Nutrition of Higher Plants. Academic Press, San Diego, California, U.S.A.

Marschner, H. 1992. Nutrient dynamics at the soil-root interface (rhizosphere). Pages 3-12: IN: Read, D.J., Lewis, D.H., Fitter, A.H. and Alexander, I.J., eds. Mycorrhizas in Ecosystems. C.A.B. International, Wallingford, U.K.

McArthur, D.A.L. and Knowles, N.R. 1993. Influence of species of vesicular-arbuscular mycorrhizal fungi and phosphorus nutrition on growth, development, and mineral nutrition of potato (*Solanum tuberosum* L.). *Plant Physiol.*, **102**:771-782.

Menge, J.A. 1982. Effect of fumigants and fungicides on vesicular-arbuscular fungi. *Phytopath.* **72**:1125-1132.

Menge, J.A., Steirle, D., Bagyaraj, D.J., Johnson, E.L.V. and Leonard, R.T. 1978. Phosphorus concentrations in plants responsible for inhibition of mycorrhizal infection. *New Phytol.*, **80**:575-578.

Michelini, S., Nemec, S. and Chinnery, L.E. 1993. Relationships between environmental factors and levels of mycorrhizal infection of citrus on four islands in the Eastern Caribbean. *Trop. Agric.*, 70:135-140.

Michelsen, A. and Rosendahl, S. 1990. The effect of VA mycorrhizal fungi, phosphorus and drought stress on the growth of *Acacia nilotica* and *Leucaena leucocephala* seedlings. *Plant and Soil.*, **124**:7-13.

Mohandas, S. 1989. Response of papaya (*Carica papaya*) to VAM fungal inoculation. Pages 260-261: IN: Mahadevan, A., Raman, N. and Natarajan, K., eds. Mycorrhizae for Green Asia. Centre Adv. Studies Bot., Univ. Madras, India.

Morton, J.B. and Benny, G.L. 1990. Revised classification of arbuscular mycorrhizal fungi (Zygomycetes): A new order, Glomales, two new suborders, Glomineae and Gigasporineae and two new families, Acaulosporaceae and Gigasporaceae, with an emendation of Glomaceae. *Mycotaxon,* **37**:471-491.

Mosse, B. 1973. Advances in the study of vesicular-arbuscular mycorrhizas. *Ann. Rev. Phytopath,.* **11**:171-176.

Nemec, S. 1980. Effects of 11 fungicides on endomycorrhiz. . development in sour orange. *Can. J. Bot.*, **58**:522-526.

Newman, E.I. 1991. The role of fungi in plant communities. Pages 437-451: IN: Atkinson, D., ed. Plant Root Growth - An Ecological Perspective. Blackwell Scientific Publications, Oxford, U.K.

Nogueira, A.V. and Harris, P.J. 1993. Effect of an introduced VAM fungus on mungbean (*Vigna radiata* (L.) Wilczek) grown in a soil with a naturally high level of available manganese. Page 91: IN : Peterson, L. and Schelke, M., eds. Abstracts of the 9th North American Conference on Mycorrhizae. University of Guelph, Guelph, Ontario, Canada.

Pacovsky, R.S. 1986. Micronutrient uptake and distribution in mycorrhizal or phosphorus fertilized soybean. *Plant and Soil.*, **95**:379-388.

Potty, V.P. 1989. Response of cassava (*Manihot esculenta*) to VAM inoculation in acid laterite soil. Pages 246-249: IN: Mahadevan, A., Raman, N. and Natarajan, K., eds. Mycorrhizae for Green Asia. Centre Adv. Studies Bot., Univ. Madras, India.

Raju, P.S., Clark, R.B., Ellis, J.R. and Maranville, J.W. 1987. Vesicular-arbuscular mycorrhizal infection effects on sorghum growth, phosphorus efficiency and mineral element uptake. *J. Plant Nutr.*, **10**:1331-1339.

Rhodes, L.H. and Gerdemann, J.W. 1975. Phosphate uptake zones in mycorrhizal and non-mycorrhizal onions. *New Phytol.*, **75**:555-561.

Ricken, B. and Höfner, W. 1993. VA-mycorrhiza in heavy metal polluted soils. Page 24: IN : Peterson, L. and Schelke, M., eds. Abstracts of the 9th North American Conference on Mycorrhizae. University of Guelph, Guelph, Ontario, Canada.

Sharma, A.K. and Srivastava, P.C. 1991. Effect of vesicular-arbuscular mycorrhizae and zinc application on dry matter and zinc uptake of greengram (*Vigna radiata* L. Wilczek). *Biol. Fertil. Soils,.* **11**:52-56.

Shetty, K.G., Hetrick, B.A.D. and Figge, D.H. 1993. Relationships between mycorrhizal symbiosis and zinc tolerance in plants. Page 94: IN : Peterson, L. and Schelke, M., eds. Abstracts of the 9th North American Conference on Mycorrhizae. University of Guelph, Guelph, Ontario, Canada.

Sieverding, E. 1991. Vesicular-Arbuscular Mycorrhiza Management in Tropical Agrosystems. GTZ, Eschborn, Germany.

Smith, S.E. 1980. Mycorrhizas of autotrophic higher plants. *Biol. Rev.*, **55**:475-510.

Swaminathan, K. and Verma, B.C. 1979. Response of three crop species to vesicular-arbuscular mycorrhizal infection on zinc-deficient Indian soils. *New Phytol.*, **82**:481-487.

Taper, C.D. and Leach, W. 1957. Studies in plant mineral nutrition. III. The effects of calcium concentration in culture solutions upon the absorption of iron and manganese by dwarf kidney bean. *Can. J. Bot.,* **35**:773-777.

Timmer, L.W. and Leyden, R.F. 1980. The relationship of mycorrhizal infection to phosphorus-induced copper deficiency in sour orange seedlings. *New Phytol.,* **85**:15-23.

Trent, J.D., Svejcar, A.J. and Bethlenfalvay, G.T. 1993. Growth and nutrition combinations of native and introduced plants and mycorrhizal fungi in a semiarid range. *Agric. Ecosystems Environ.,* **45**:13-23.

Walker, C. and Sanders, F.E. 1986. Taxonomic concepts in the Endogonaceae: III. The separation of *Scutellospora* gen. nov. from *Gigaspora* Gerd. & Trappe. *Mycotaxon,* **27**:169-182.

Warden, B.T. and Reisenauer, H.M. 1991. Manganese-iron interactions in the plant-soil system. *J. Plant Nutr.,* **14**:7-30.

Weissenhorn, I., Leyval, C. and Berthelin, J. 1992. VA mycorrhizal colonization in an industrially polluted soil and heavy metal transfer to the plant. Page 409: IN: Read, D.J., Lewis, D.H., Fitter, A.H. and Alexander, I.J., eds. Mycorrhizas in Ecosystems. C.A.B. International, Wallingford, U.K.

Advancements In Micronutrient Research
Editor : A. Hemantaranjan
1996, Scientific Publishers, Jodhpur, pp. 383–399.

11

MICRONUTRIENT STRESS AND GENETIC VARIABILITY

INDRAVADAN CHHELSHANKER DAVE
Nuclear Agriculture Division,
Bhabha Atomic Research Centre,
Bombay, 400 085, INDIA

CONTENTS

1. INTRODUCTION

Adverse conditions for normal growth and development especially in crop plants affect the quality and yield of the economic product. The adversities are environmental, biotic and abiotic in nature. The first and foremost process that is affected is the gene expression in response to adverse conditions including water, environmental and mineral stress. Mineral stress includes deficient or toxic concentrations in the soil as also the process of their acquisition by plants through roots from rhizosphere.

It is known that plants respond to heat shocks or high temperature regimes by synthesizing different sets of proteins. It is also known that the same plant species synthesizes alternative sets of proteins under various sets of conditions of stress, viz., flooding, salinity and even exposure to UV radiations etc. However, meagre amount of information is available in relation to proteins synthesized or inhibition of synthesis under nutrient and especially micronutrient stress. This lack of information exists inspite of the fact that a large amount of data is available on the tolerant and susceptible as well as efficient and inefficient cultivars of most of the plants of economic importance, in relation to micronutrient stress.

Of these data, majority of the reports are confined to agronomic affiliations only. On the other hand, stress conditions like anaerobiosis and heat shock have been the subject of interest in the recent past.

This review deals with the area of micronutrient stress in relation to genetic variability in plants. It also deals with the response of plants to such stress in relation to altered gene expression in terms of synthesizing novel polypeptides and protein profiles.

2. Nutrient Stress in Crops and Other Plants

Symptoms of nutrient disorders arising out of nutritional stress in plants may occur in any part of the plant including leaves. Of the minor or micronutrients, boron, copper, iron, manganese, molybdenum, zinc and chloride belong to essential and beneficial class. The morphological and physiological characteristics of the plants as probable or sure indicators or deficiency symptoms are of great importance. These include relationship between shoot and root, uptake, transport and mobility of nutrients amongst various tissues and organs, damage to the growth and development of plant, susceptibility to various diseases and alterations in the pattern of the enzyme system.

Boron deficiency in agricultural soils are of two types; natural deficiency and induced one due to overliming. B deficiency symptoms in general include corky patches on leaf stalk, death of apical regions, general failure of bud to develop and sterility of flowers with reduction in fruit numbers. It also drastically alters metabolic rates, protein profiles, synthesis of nucleic acids and related enzymes as well as membrane permeability. Boron toxicity is of much concern in arid and semiarid zones, because of a fairly high boron content in the irrigation water. Symptomatic toxic effects are necrosis and chlorosis on margin of tip of leaves.

Typical copper deficiency symptoms include "white tip" or reclamation disease of cereals and "summer dieback" in trees as well as motted leaves, apical meristem necrosis and impairment of cell wall formation as a secondary symptom. Copper toxicity may cause iron deficiency induced chlorosis and disruption of thylakoid membranes.

Iron deficiency is marked through inhibition of chloroplast development in leaves and aberrations in the root growth like suppressed elongation, increased girth of apical root zones, abundant root hair formation as well that of rhizodermal cells. It also causes the enhanced mobilization of Mn leading to Mn toxicity. It also causes transformation of nicotianamine into avenic acid for release into rhizosphere to facilitate Fe(III) chelation and absorption by roots. It is considered as a global problem in loss of crop produce in calcareous zones causing chlorosis and is a major problem in terms of one of the most severe factors that limits the wetland rice.

When manganese concentration depletes below the critical level, which is surprisingly consistent irrespective of plant species or variety including environmental variations, chlorophyll content and net photosynthesis rate decrease without any perceptible change in respiration or transpiration in Mn deficient plants which also become susceptible to damage in temperate zones due to freezing temperature regimes. In contrast, Mn toxicity vary widely amongst plant species and even amongst cultivars. It causes symptomatic secondary damages due to induced deficiency of Fe, Mg and Ca.

Molybdenum occurring in the aqueous solution mainly as MoO^{2-}_4 is required in the minimum by plants than any other micronutrient. It is functionally related to changes in valance, as MoCO (Mo- cofactor) component of enzymes including nitrogenase and nitrate reductase in higher plants. Necrosis of "Yellow spot" type in citrus and "Whiptail" leaves in cruciferous species are typical visual symptoms of Mo deficiency. It also causes at toxic levels golden yellow discolouration of shoot tissues, leaf aberrations and formation of vacoular molybdocatechol complexes.

Zinc deficiency which is a problem of excessively withered acid and calcareous soils is quite often associated with iron deficiency as "lime chlorosis". Rosetting, i.e., shortening of internodes is accompanied by chlorotic leaves reduced in size. In some ecotypes of natural vegetation Zn toxicity is marked by the presence of Zn tolerant varieties. In Zn tolerant clones vacuoles contain sequestered Zn.

3. Stress and Genetic Variability

Micronutrient stress in the form of either deficient or toxic concentrations in the soil results due to a host of known and unknown factors present in the surrounding mileau as also due to plant species inefficient in nature. The interaction between the mileau in the soil as a passive phase and the plant roots as the active phase is of great importance, as this interaction alone can put into action a large number of physiological and biochemical reactions in a precise and sequential manner. These operations are strictly under the control of response mechanism which in turn is governed by the genetic make up of the plant. Under adverse circumstances the genetic adaptability mechanism will start operating to mitigate the situation arising. In the absence of such a self operating mechanism, the capability is acquired through many generations through chance mutations, which is a slow process. On the other hand the techniques in plant breeding, genetic engineering and their combinations when put to use create mutants having the tolerance or efficiency capabilities in relation to nutrient stress. This problem has been tackled in many ways; analysis of gene expression, genetic manipulation of various mechanisms against different types of stress to create tolerant phenotypes as well as mutation breeding of such cultivars viable agronomically in crop production. In genetic engineering endeavours, techniques called "antisense technology" have been found feasible and have been suggested as a complimentary technique to any overexpression study. The first such report concerned the genetic manipulation of superoxide dismutase enzyme (SOD) in tobacco and tomato plants which produced in excess a chloroplastic Cu/ZnSOD derived from *petunia* (Tepperman and Dunsmuir, 1990), which however did not prove more resistant to superoxide toxicity. But certainly it paved the way for opening new vistas in this direction. It surely is bound to make an impact creating newer species for the solution of stress problems.

4. Current Status of Micronutrient Stress and Genetic Variations in Plants

A survey of the literature reveals that of all the micronutrients, genetic variability response to Fe stress has been studied maximum. Based on 17 wheat cultivars grown over three years, Berg *et al.* (1993) concluded that the Fe-deficiency induced chlorosis in regrowing wheat cultivars after intensive grazing ranged from none to severe, establishing thus the response to Fe-stress correlated to the genetic make-up of the cultivars. They also suggested that moderate to extensive use of Fe-inefficient cultivars induces increased Fe-deficient conditions in the soil. Hence, it is essential to introduce Fe- efficient wheat cultivars only to maximize the produce as well to conserve soil Fe. Observations on the response to Fe-deficient conditions by five graminaceous monocots cultivars of corn, *Pennisetum*, sorghum, *Echinochloea*, wheat and *Lupinus* revealed that release of phytosiderophores induced by Fe-stress was found only in the case of wheat and sorghum, accompanied by a rise in pH of rhizosphere, which matched with the degree of severity of chlorosis and the rate of Fe-uptake (Onyezilli and Ross, 1993). The authors also concluded that the mechanisms for Fe-deficiency response suggested in dicots could only partially be seen in some monocots and not in all. Fe-stress induced chlorosis and its recovery have been extensively investigated by the colleagues in author's laboratory in relation to sorghum and other crop varieties. Two hybrids CSH-10 and 11 alongwith their parental lines 296-A and SB- 1055 and MR-715 were subjected to controlled Fe-stress and rate of Fe uptake and recovery were examined. Only the female parent 296-A showed very rapid recovery and to some extent the hybrids did so. Inspite of the total failure of the male parent to show any recovery, Fe uptake and transport was higher than in the female parent. The hybrids failed to express fully the trait of Fe-efficiency of the female parent. Such studies were also carried out in peanut, cotton, sesame, chickpea and lentil, wherein Fe-stress recovery was seen associated with a decrease in the pH of the medium, except in the case of lentil. Its two cultivars differed in the sense that in cv VL1 the recovery was remarkable without any decrease in the pH of the medium, while it was the other way round in cv. S-34 (Kannan and Ramani, 1987; Ramani, 1988). Soil temperature also seems to play a role in Fe- stress conditions. With an increase in the severity of the chlorosis due to elevated temperature, changes in the growth traits were ascribed to genotypic differences in sorghum irrespective of the response of the cultivar to Fe-stress resistance (Clark and Reinhard, 1991). Based on data on content of Fe, Zn, Cu and Mn in grains of 30 cultivars of rice (Velu *et al.,* 1989) and 73 cultivars of rice grown in acid sulfate soil, a classification of the rice cultivars has been suggested in relation to susceptibility and resistance to Fe stress (Rosmini and Mukhlis, 1990). Seedlings of 15 inbred genotypes of maize grown at different P levels exhibited genotypic variations in absorption, accumulation and utilization of Fe (Elliot and Lauchli, 1985). One hundred families from randomaly mated sorghum population grown in Fe-deficient soil were given foliar spray of Fe. Agronomical data showed significant recovery including improved grain yield (Williams *et al.,* 1986). Based on the studies on a large number of cultivars of legumes, sorghum, capsicum, fruit crops and forage plants, two cultivars of sorghum and *Phaseolus vulgaris* each, and five germplasm lines of sybean were found to be resistant to Fe-deficiency chlorosis (Rodriguez de Cianzo, 1991).

In recent past, considerable number of reports have appeared wherein, specific evidences have been forwarded from inheritance studies involving genotypes and strains that vary in their nutritional needs. In some cases a single gene and in most of the cases complex genetic apparatus are found to be involved in Fe stress resistance or efficiency (Gerloff and Gableman, 1983; Graham, 1984). Studies on nicotianamineless (NA) mutant of tomato showed large accumulation of citrates under Fe deficiency in contrast to a low accumulation in wild type plant. This was thought to be due to a possible activation of Fe uptake by NA, thus regulating Fe dependent processes at gene expression level (Pich *et al.,* 1991). A stable mutant E107 forming nodules exhibited pleiotropic properties through mutant alleles in terms of Fe accumulation in shoot. The recessive gene *brz* for bronze necrotic leaflet spots was non allelic with 2 other genes and was found on chromosome 4 by linkage (Kneen *et al.,* 1990). From the results seen in parents, F2S and F3 progenies, from genotypic crosses between susceptible x resistant cv of *Phaseolus vulgaris*, it was concluded that two genes act through physiological processes in relation to increased Fe solubility in the rhizosphere (Zaiter *et al.,* 1987). A single gene mutant E 107 of *P. vulgaris* accumulated toxic levels of Fe in older leaves with a 4-fold increase in Fe influx in comparison to + Fe Sparkle cv

(Grusak *et al.,* 1990). Culturing of 6 wheat varieties additional lines with chromosomes 1-4 and 6-7 from barley cv. *Betzes* showed mugineic acid and related compounds to be under the control of variable gene expression. Further, a cDNA clone specifically hybridizing to Fe-deficient barley roots was used to isolate a 503 nucleotide cDNA clone containing a putative open reading frame of 22 bp encoding a protein of 74 residues with MW 7500 D, containing regions with features of metallothionine domains (Mori and Nishizawa, 1989; Okumura *et al.,*1991). Elegant studies on progeny seggregated from two efficient and two inefficient lines for Fe deficiency stress, revealed that the capacity of the roots to lower the pH of the medium has higher heritability than that to reduce Fe(III). Low correlation values indicated that these two events mights be under the control of different genes or gene system without linkage. The authors believe that it should be possible to obtain homozygous inbred parental lines-Fe efficient by application of both the tests over several generations and then making the selections of parents from each progeny with the maximum capacity for Fe efficiency in an increasing order (Alcantara and Guardia, 1986).

The mechanism of ion uptake regulation from rhizosphere is very tricky, sophisticated and highly variable in nature. Interestingly, grasses have a different mode of Fe acquisition through a sharp increase in the release of nonproteinogenic amino acids known as phytosiderophores, which mobilize sparingly soluble Fe(III) through chelation. Roots of *Pyrus* sp. with a high tolerance to low Fe availability could reduce Fe (III) at a rate 12 times greater than *Cydonia oblongata* roots sensitive to low Fe. (Dolcet-Sanjuan *et al.,* 1992). Tissue culture experiments in soybean cells revealed that Fe availability is directly affecting the transcription of mRNA related to ferritin production (Lescure *et al.,* 1991). Genetic variability was the main cause in relation to release of phytosiderophores from roots of barley, wheat, maize and rye plants under Fe deficiency, indicating that PS release and return uptake of FePS complex were under different genetic control as also the activation events for the uptake of FePS (Romheld and Marschner, 1990). In fact the mechanisms for Fe absorption was under different genetic control for oat and corn. They further proposed that most of the Fe-efficient cultivars can initiate response to Fe stress mechanism with little or no Fe in the medium and that the factors like H^+ ion pump, release of reductants, Fe(III) reduction and immediate increase in Fe and chlorophyll content should act in concert as an adjunct for uptake and transport of Fe (Brown *et al.,* 1991 ; Jolly and Brown, 1987).

Stray reports connecting Fe nutrient status with pathogenicity have also appeared. Such relationship in tomatoes afflicted by *Verticillium* wilt disease has been shown from comparative studies of near isolines - Fe-deficient Pixie II(resistant) and Pixie (susceptible) varieties (Makur *et al.,*1991) and in crown gall tumor tissues having a higher Fe, Cu and P contents than in normal stem tissue in case of Fe efficient and inefficient cultivars of tomato (Jolley *et al.,*1993).

Population of insect *Sogatilla furcifera* on Fe stress resistant and susceptible varieties of rice cultivation was found to be higher in low or high concentrations in comparison to normal concentration of Fe (0.02, 40 and 2 ppm resp.) (Salim and Saxena, 1991).

In comparison to studies on Fe-stress and genetic variability, very meagre information is available on stress conditions due to B, Cu, Mn, Mo and Zn. Almost no information is available on Cl. In general, genotypic differences in chloride tolerance are closely associated to salt tolerance mechanism. The sensitivity of a genotype to salinity may change during ontogeny and such changes are governed by genetic constitution and environmental factors. There are instances wherein the response to salinity is paradoxical, e.g., tolerance of sugar beet during life cycle is high, but very low during germination. But it is the other way round in case of wheat, rice, tomato and barley, (Lynch *et al.*, 1982). Three species of *Sesbania* responded to salt tolerance in the order *S. rostrata* > *S. aculeata* > *S. speciosa* (Ramani *et al.*, 1989). Interspecific differences in Cl accumulation have been reported in soybean, sorghum, wheat, barley and other plants (Abel, 1969; Bernstein 1964 , Greenway and Munns, 1980).

Boron deficiency problem is ubiquitous in nature. Its availability is related to soil pH, water stress and its relative mobility. Plants grown in the same soil may differ in content widely (Gupta, 1979). Comparison of responses of wheat cv Chinese spring and Halbred and an amphiploid - Chinese spring x *Agropyron elongatum* to high B content in soil showed that the yield for the amphiploid was greater than in Chinese spring and this was presumed to be due to reduced B accumulation in the shoots (Paull *et al.*, 1991a). Toxicity effect was examined in 7 genotypes at organ and cellular levels in excised root culture. The resistant genotypes had more number of root axes and lateral roots and produced more callus in culture; resistance being linked to membrane permeability (Huang and Graham, 1990). Some hybrids of wheat exhibited ratios consistent with monogenic segregation for F2 and F3 generation in terms of B tolerance (Paull *et al.*, 1991b). Based on performance of 31 cultivars of sunflower, B - efficient cultivars exhibited a good correlation with leaf B content and leaf B\stem B ratio (Furlani *et al.*, 1990). Grain B content was not seen as a hindrance in testing the genotypes for their resistance to B toxicity (Nable and Paull, 1990). Based on the performance of 6 cultivars of sesame and mustard under 1.5 and 2.5 kg B\ha, sesame cv OMT- 11-6-3 and RT-4 and mustard cv. Pusa Bold were isolated as tolerant to B deficiency. They exhibited a positive correlation between yield and seed B uptake (Sakal *et al.*, 1991).

There is not much information on Cu stress and genotypic differences. Copper resistant lines could grow proficiently on medium with upto 3.0 mM Cu. (Tobita *et al.*, 1988). Pollen development in Cu deficient barley cultivars was highly irregular and endopolyploidy of tapetal nuclei and alteration in organalle composition were correlated with lower pollen fertility (Jewell *et al.*, 1988). Excised roots of rice cv. IR26 and tomato cv. Kewalo absorbed both Cu and Zn much more rapidly in comparison to that in rice cv. M101 and tomato cv. sel 625-2, without any difference to their affinity to the two elements (Bowen 1987). From studies on wheat-rye additional lines the gene(s) responsible for resistance to Cu deficiency was found to be located on long arm of rye chromosome 5 (5R1). These genes were suggested to be responsible for low Cu need, thus, raising the Cu efficiency (Podleska *et al.*, 1990). The gene or genes on this chromosome of rye was introduced in Australian wheat cv. Warigal, Oxpy, Timgalen and champion Pictic through crossing, backcrossing and finally selfing over two generations Lines with 5RL tested for grain yield on Cu-deficient soils in S. Australia exceeded their initial varieties by over 100% and also exhibited a 50-100% faster copper uptake without any perceptible change in Cu content in the plants (Graham *et al.*, 1987).

It is difficult to generalize the genetic variability effect in response to Mn deficiency or toxicity. Nine varieties of wheat and one variety of *Triticale* differed in their response to Mn stress. *Triticale* cv. TL419 and wheat cv. C306 were found to be most tolerant to Mn stress. A similar screening of wheat-barley additional lines revealed that Mn deficiency characteristics of the plants could be transferred to wheat from barley (Kaur and Takker, 1987; Marcar and Graham, 1987). Studies on comparative efficacy of 4 cultivars of *Spinacia oleracea* and *Phaseolus vulgaris* each showed that all the cultivars of bean significantly differed in their uptake of Mn as well as Fe and Zn and this was correlated to the number of root apices. Mn stress caused rapid increase in peroxidase in soybean with different rates of uptake in different cultivars. Cultivars *Bragg* and *Forrest* identified as Mn intolerant showed elevated peroxidase activity, while under Mn toxicity, Mn tolerant variety Lee and TO3 exhibited increased peroxidase activity under Mn deficiency (Helal *et al.*, 1990; Leidi *et al.*, 1987). Irrespective of cultivar differences amongst Fe and inefficient varieties of soybean, Mn toxicity was developed in presence of optimal Mn content (Jolley *et al.*, 1986).

Whatever work has been carried out in relation to Mo stress influencing the plant behaviour under genetic control is confined to activity of enzymes nitrogenase and nitrate reductase as their existence in higher plants has been well established. From agronomical point of view, it has been conclusively shown that the effect of Mn deficiency is widespread in genera of Crucifers, corn and legumes grown in acid mineral soils, due to an unusually high capacity of the iron oxide hydrates to absorb molybdate ions. Genotypical differences in relation to transport of Mo from roots to shoots is the main cause for its toxicity. In species of sunflower and bean, e.g., accumulation of Mo in xylem parenchyma tissues is due to steep gradiant taking place in Mo content from roots to leaves in contrast to tomato, wherein Mo gets readily transported to foliage region. This results into toxic conditions much earlier in tomato than in bean or sunflower (Hecht- Buchholz, 1973). *Phaseolus vulgaris* cv. Kabanima was superior to cv. Baseka in transport of Mo to seeds at maturity inspite of similar rate of uptake in both the varieties. The other notable differences were : a higher Mo content in the nodules of cv. Kabanima with 30-40% more nodular dry weight; a higher N_2- fixing ability and more efficient mobilization of Mo (Brodrick and Giller, 1991). Understanding of the functional aspects of Mo under variable genetic controls has considerably improved from the studies on the mutant behaviour of *Arabidopsis thaliana* in relation to their resistance to herbicide chlorates. They are defective in nitrate reductase activity(NR) due to mutation in Nrd genes or NR cofactor -molybdenum protein synthesis genes. Mutant Ch12 with low NR had low levels of Moco (mo -co-factor) which was far less than wild type factor and also lower than that of mutant Ch16 (B+3). Based on response to tungstate application on growth in mutants and wild type, the authors concluded that low Moco levels, chlorate resistance, low NR activity and tungstate sensitivity all cosegragate and hence points to a single mutation that maps to Ch12 locus on chromosome 2 (Labrie *et al.,* 1992; Wray and Kinghorn, 1989; Crawford and Campbell, 1990; Caboche and Rouze, 1990). One fourth of the F2 generation plants derived from crossing R9201, R11301 mutants from M2 population of barley with wild type Maris-Mink lacked NR activity, withered and died. All homozygous mutants lacked NR and xanthine dehydrogenase activity and these losses were considered to be due to effect on the synthesis of functional Moco factor (Steven *et al.,*1989). They futher showed that shoot cofactor from R9201 and R12202 failed to reconstitute NADPH-NR upto 40% in nit-1 mutant of *Neurospora crassa* as well as to dimerise the NR subunits in barley mutant. Shoot cofactor of R11301 could dimerise the subunits as also reconstitute NADPH-NR upto 40% of the wild type r levels. It was further shown that *nar2* gene locus encoding a step in Mo-protein biosynthesis as also mutant R11301 having a locus involved in functional cofactor synthesis. The loci governed MoCO synthesis in both root and shoot.

These findings as also the information on the ability of NIAI-1 point mutant to convert alanine residue into threonine of NIAI NR enzyme; and the importance of thiol group of cystine 191 acting as a ligand for Moco binding (Neame and Barber, 1989) are likely to explain a number of riddles associated with genetically directed functional aspects of Mo stress.

Observations demostrate that zinc deficiency symptoms are severe in maize, soybean and tomato grown in alkaline soil (pH 8.2) in constrat to wheat, barley, sugarbeet and potato grown in such soil (Marschner, 1986). A wide variance observed in Zn efficiency ability in bean and pigeonpea cultivars (Viets *et al.,*1954, Judy *et al.,*1965) clearly demostrates that sufficient genetic variability exists for making systematic selection for breeding programmes for improved adaptations of crop species to alkaline soils. In fact new breeding lines from International Rice Research Institute are at least moderately Zn-efficient (Ikehasi and Ponnamperuma, 1978). Based on evaluation of 16 varieties of rice as Zn-efficient or inefficient, it was shown that the tolerant varieties had higher Zn content than the susceptible ones. Use of Zn content in 3rd leaf for genetic variability studies and screening the lines has been suggested (Singh and Sakal, 1987). Similar findings have been reported for soybean also (Hartwig *et al.*, 1991). Five parental inbreds of maize with high, low or moderate accumulation of Zn were used to obtain F1 hybrids and tested for Zn accumulation. This was found to be genetically controlled. Four genes were found to be minimum seggregating factors in high x low Zn accumulating crosses (El-Bendary *et al.*, 1993). A significant decrease in the heterozygocity and degree of genetic polymorphism in seedling population of *Pinus sylvestris*, growing in Zn smelting area. It was found that due to Zn toxicity, enzymatic loci failed to maintain Hardy-Weinberg equilibrium, thus showing excess of homozygocity in relation to shikimate, malate and glutamate dehydrogenases (Prus-Glowaski and Godzik, 1991). A varied Zn tolerance pattern found in 8 populations of *Deschampsia cespitosa* (Gramineae) was ascribed to heterogenous soil-Zn content and various gene combination and recombination (Coulaud and McNeilly, 1992).

5. Conclusions

Essentially, existence of a large amount of inter and intrageneric and specific variations in relation to uptake and transport of micronutrient is a direct evidence for genetic control over the mineral nutritional aspects of the higher plants. Barring some cases, in most of the situations inheritance studies on cultivars and strains have shown the involvement of complex genetic systems rather than a single gene pair action for micronutrient needs. The definition that nutrient efficiency of a plant is reflected in the ability to produce a high yield in a soil that is limiting in one or more mineral nutrients for a standard genotype (Graham, 1984) can also be used for comparative studies on genotypes, cultivars or strains. Such a situation has successfully been exploited as a basis for genotypes with insufficient concentration of nutrients (Blair and Cordero, 1978).

Understanding of genotypical differences without any artifacts concerns acquisition, uptake, transport, utilization as also mobilization and remobilization for redistribution of micronutrients and is essential for a viable nutrition management efficiently. Acquisition of nutrients involves affinity for the uptake system(Km) and the threshold concentration(Cmin) will depend on genetic makeup even within a species. These genetic responses in terms of ability of a given genotype are intimately associated with many factors like pH of rhizosphere, cation exchange rate and kinetics, preferential selective uptake, mobility of the ion, active or passive transport pattern, ion diameter etc. to mention a few. The efficient acquisition, absorption and transport ultimately pave the way for metabolic activities which might be governed by interionic competition within the same tissue, nature of long or short distance transport as well as xylem vs. phloem transport etc. are some of the features the knowledge of which will aid in developing the required genotypes.

The survey of the information available in relation to genetic variation and micronutrient stress presented in the foregoing reflects total inadequacy in relation to experimental coverage of even few of the host of the features that are required to be investigated for a very efficient micronutrient management programme. It will be, therefore, not totally out of context to list some of the metabolic events wherein each micronutrient is essentially involved.

Rate of Fe uptake is predominantly in the apical region of the roots taking place in the rhizosphere root junction and its uptake is in the reduced form as a rule. The major metabolic function if Fe is the reversible redox reaction system - Fe(II) Fe(III) + e^-; the Fe containing constituents of which are hemoproteins and FeS proteins. The other important aspect is the localization of Fe in chloroplasts and the storage of Fe which is confined to stroma of the plastids. It is important to note that 80% of the iron content is localized in chloroplasts irrespective of the status of Fe nutrition. Under Fe deficiency, lamellar iron content goes up at the expense of stromal iron. Hence the ratio of lamellar Fe to stromal Fe, redox state of Fe in leaf extract as studied by Mossbauer spectrometry (Machold *et al.*, 1968), release of specific chelates from roots to rhizosphere and reduction in rhizosphere pH are some of the important markers helpful to look for, in breeding programmes to meet the challanges from Fe stress conditions.

Boron deficiency covers the largest area of the arable land globally and still attention paid to this problem is almost negligible. Withholding of boron from the nutrient medium results in drastic changes in RNase levels, nuclear DNA synthesis, tRNA concentration, membrane permeability and uptake and transport of Fe, Mn and Zn (Dave and Kannan, 1980, 1981; Dave, unpublished data; Tanada, 1978, Dave, 1993).

In most of the biological systems, oxidative stress resulting from the deleterious impact caused by reduced oxygen species, highly reactive in nature involves alterations in structure and function of superoxide dismutase (SOD) under any stress condition. Such conditions generate superoxide, hydroxyl radicals and their derivatives and through indiscriminate reactions they cause lipid peroxidation, protein denaturation and also mutate DNA apparatus. SOD enzymically protects the tissue metabolism against the aforesaid damaging radical species through their elimination. In fact, SOD is known to occur in three forms according to their metallic componentes, viz., Cu/Zn SOD, Mn SOD and Fe SOD (Bannister *et al.,* 1987). Higher eukaryotes within the plant kingdom contain a mitochondrial matrix localized Mn SOD and cytosol localized Fe SOD and/or Cu\Zn SOD. However, the number and relative abundance of their isozymes seem to be genus specific. This clearly implies the variable genetic control over them and hence the stress conditions resulting into deficiency of Fe, Cu, Zn or Mn will drastically affect the regime of these enzymes. An excellent review has been published on SOD and stress tolerance (Bowler *et al.,* 1992) in relation to photoinhibition, herbicidal effect, atmospheric pollutants, water-logging, drought and pathogenicity etc. However, it does not contain the information in relation to micronutrient stress directly, indicating thereby that research on this area is negligible or is being ignored. The very fact-the constituents of this group of enzymes are Fe, Mn, Cu and Zn, four out of the six essential micronutrients and their enzymic function deal with the destructive and metabolically disruptive oxygen species speaks volumes in itself, about their importance. And we have already seen the genetic apparatus controlling their synthesis and functions. It is a considered opinion of the reviewer that a true evaluation of SOD activity in plants can come only through genetic engineering manipulations.

From the foregoing discussion it is concluded that in proportion to the magnitude of micronutrient stress conditions, efforts to alleviate them through an interdisciplinary approach involving agronomical, physiological, biochemical, plant breeding- conventional as well as employing genetic engineering experimental designs is the need of the day and the horizon seems to be bright due to availability and rapid advancement in the sophisticated technology. The only caution needing to be exercised is that this concerted approach must be at any cost conducive to the concept of sustainable agriculture as also restricting the spread of what is known as invasive agriculture which has shaken the very foundations of ecological balance globally. The expert opinion is that inspite of unsurmountable difficulties in transferring the biotechnological aspects of genetic engineering into crop plants, near future does have a promise in this direction (Sandhu, 1992), e.g., protocols developed for *in vitro* regeneration of *Lathyrus sativus* and isolation of a plasmid with an Ox-Dapro degrading gene could allow the possibility of genetic engineering of *L. sativus* for toxin removal and thus rendering the plant non-toxic for fodder feeding (Mehta *et al.,* 1991).

REFERENCES

Abel, G.H. 1969. Inheritance of the capacity for chloride inclusion and chloride exclusion by soybeans. *Crop Sci.,* **9** : 697- 698.

Alcantara, E. 1986. Variability in the response of sunflower hybrids in parental lines to iron stress *J. Plant Nutr.,* **9** (3-7) : 443-451

Bannister, J.V., Bannister, W.H. and Rotilio, G. 1987. Aspects of the structure, function and application of superoxide dismutase. *CRC Crit. Rev., Biochem.,* **22** : 111-180

Berg, W.A., Hodges, M.E. and Krenzer, E.G. 1993. Iron deficiency in wheat grown on the southern plains. *J. Plant Nutr.,* **16** (7):1241-1248

Bernstein, L. 1964. Salt tolerance of plants. *Agri. info. Bull.,* (U.S. Dept. Agri.) No. 283.

Blair, G.J. and Cordero, S. 1978. The phosphorus efficiency of three annual legumes. *Plant and Soil,* **50** :387-398.

Bowen, J.E. 1987. Physiology of genotypic differences in zinc and copper uptake in rice and tomato. *Plant and Soil,* **99** (1):115-125.

Bowler, C., Montagu, M.V. and Inzc, D. 1992. Superoxide dismutase and stress tolerance. *Ann. Rev. Plant Physiol. Plant Mol. Biol.,* **43** : 83-116.

Brodrick, S.J. and Giller, K.E. 1991. Genotypic differences in molybdenum accumulation effects nitrogen fixation in tropical *Phaseolus vulgaris. J. Expt. Bot.,* **42** : 1339-1343.

Brown, J., Jolley, V.D. and Lytle, C.M. 1991. Comparative evaluation of iron-solubilizing substances (phytosiderophores) released by oats and corn: Iron-efficient and iron-inefficient plants. *Plant and Soil,* **130** (1-2):157-163

Caboche, M. and Rouge, P. 1990. Nitrate reductase - a target for molecular and cellular studies in higher plants. *Trends Genet.,* **6** : 187-192.

Clark, R.B. and Reinhard, N. 1991. Effect of soil temperature on root and shoot growth traits and iron deficiency chlorosis in sorghum genotypes grown on a low iron calcareous soil. *Plant Soil,* **130** (1-2): 97-103.

Couland, J. and McNeilly, T. 1992. Zinc tolerance in populations of *Deschampsia cespitosa* (Gramineae) beneath electricity pylons *Plant Systematics and Evolution,* **179** (1-2):175-185.

Crowford, N.M. and Campbell, W.H. 1990. Fertile fields. *Plant Cell,* **2**: 829-835.

Dave, I.C. 1993. Incorporation of ^{3}H- thymidine into nuclear membrane complex (M-band) and DNA polymerase activity in nuclei of boron deficient bean seedlings. *Proc. Intntl. Symp. on "Emerging Frontiers in Radiation Biology",* Dec.1-3., Bhabha Atomic Research Centre, Bombay, India. Abst. No. p-95.

Dave, I.C. and Kannan, S. 1980. Boron deficiency and its associated enhancement of RNase activity in bean plants. *Z. Pflanzenphysiol.,* **97** (3):261-263

Dave, I.C. and Kannan, S. 1981. Influence of boron deficiency in micronutrients absorption by *Phaseolus vulgaris* and protein contents in cotyledons. *Acta Physiol Planta.,* **3**(1): 27-32.

Dolect-Sanjaan, R., Mock, D.W.S. and Mock, M.C. 1992. Characterization and *in vitro* selection for iron efficiency in *Pyrus* and *Cyolonia.Acta Horti*, **300**:45-50

El-Bendary, A.A., El-Fouly, M.M., Rakha, F.A., Omar, A.A. and Abouyoussef, A.Y. 1993. Mode of inheritance of zinc accumulation in maize. *J.Plant Nutr.*, 16 :2043.

Elliot, G.C. and Lauchli, A. 1985. *Agron.J.* **77**(3): 399- 403.

Furlani, A.M. C., Ungaro, M.R.G. and Quaggio, J.A.1990. Differential performance of sunflower genotypes: efficiency of uptake and use of boron. *Revista Brasileira de Ciencia do Solo.*, **14** (2): 187-194.

Gablemann, W.H. and Gerloff, G.C. 1983. The search for and interpretation of genetic controls that enhance plant growth under deficient levels of a macronutrient. *Plant Soil*, **72** : 335-350.

Gerloff, G.C. and Gablemann, W.H. 1983. Genetic basis of inorganic plant nutrition. In "Encyclopedia of Plant Physiology, New Series", Vol. **15B** : 453-480.

Graham, R.D. 1984. Breeding for nutritional characteristics in cereals. In *"Advances in Plant Nutrition"* (eds. P.B. Tinker, A. Lauchli) Vol. **1**, pp. 57-102. Praeger. New York.

Graham, R.D., Ascher, J.S., Ellis, P.A.E. and Shepherd, K.W. 1987. Transfer to wheat of the copper efficiency factor carried on rye chromosome. 5RL *Plant Soil*, **99**(1) : 107-114.

Greenway, H. and Munns, R. 1980. Mechanism of salt tolerance in nonhalophytes. *Ann. Rev. Plant Physiol.*, **31** : 149-190.

Grusak, M.A., Kochian, L.V. and Welch, R.M. 1990. A transport mutant of pea (*Pisum sativas*) for the study of iron absorption in higher plant roots. *Proc. XIth Intntl. Plant Nutrition Colloquium*, Wageningen, Netherlands, July 30-Aug. 4, 1989. (Beuisichem, M.L. van, ed.) Kluwer Academic Publishers, pp. 219-222.

Gupta, U.C. 1979. Boron nutrition of crops. *Advances in Agronomy* , 31 : 273-307.

Hartwig, E.E., Jones, W.F. and Kilen, T.C. 1991. Identification and inheritance of inefficient zinc absorption in soybean. *Crop Sci.*, **31**(1) : 61-63.

Hecht- Buchhloz, C. 1973. Molybdanverteilung und -vertraglichkeit bei Tomate, Sonnenblume and Bohne, *Z. Pflanzenernaehr. Bodenkd.*, **136** : 110-119.

Helal, H.M., Arisha, H. and Rietz. E. 1990. The uptake of trace elements by spinach and bean varieties of different root parameters. *Plant Soil*, **123** (2) : 229-232.

Huang, C.Y. and Graham, R.D. 1990. Resistance of wheat genotypes to boron toxicity is expressed at cellular level. *Plant Soil*, **126**(2) : 295-300.

Ikehashi, H. and Ponnamperuma, F.N. 1978. Varietal tolerance of rice for adverse soils. *In Soil and Rice*. pp. 801-823. IRRI, Los Banos, Philippines.

Jewell, A.W., Murray, B.G. and Alloway, B.J. 1988. Light and electron microscopic studies on pollen development in barley (*Hordeum vulgare* L.) grown under copper sufficient and deficient conditions. *Plant, Cell and Environ*, **11**(4) : 273-281.

Jolley, V.D. and Brown, J.C. 1987. Soybean response to iron deficiency stress as related to iron supply in the growth medium. *J. Plant Nutri.*, **10**(6) : 637-651.

Jolley, V.D., Brown, J.C., Davis, T.D. and Walser, R.H. 1986. Increased Fe-efficiency in soybeans through plant breeding related to increased response to Fe-deficiency stress. II. Mineral nutrition. *J. Plant Nut.*, **9**(3-7) : 387-396.

Jolley, V.D., Terry, R.E. and Brown, J.C. 1993. Tumorous crown gall response to iron-deficiency stress in Fe-efficient T3238FER and Fe-inefficient T3238 fer tomatoes. *J. Plant Nut.*, **16**(6) : 1025-1037.

Judy, W., Melton, J., Lessmann, G., Ellis, B. and Davis, J. 1965. Zinc fertilization of pea, beans, corn and sugarbeet in 1964. *Mich. Agric. Exp. Stn., Farm Sci. Res. Rep.*, **33** : 1-8.

Kannan, S. and Ramani, S. 1987. Mechanisms of Fe-deficiency tolerance in crop cultivars : Effects of dibutyl phthalate and caffeic acid on Fe-chlorosis recovery. *J. Plant Nut.*, **10** (9- 16) : 1051-1056.

Kaur, N.P. and Takkar, P.N. 1987. Effect of Mn stress on the growth and yield parameters in wheat and triticale cultivars. *Wheat Info. Service.*, No. **64** : 24-33.

Kneen, B.E., La Rue, T.A., Welch, R.M. and Weedon, N.F. 1990. Pleiotropic effects of brz. A mutation in *Phaeolus sativus*. L. cv. "Sparkle" conditioning decreased nodulation and increased iron uptake and leaf neerosis. *Plant Physiol.*, **93**(2) : 717-722.

Labrie, S.T., Wilkinson, J.Q., Tsay, Y.F., Feldmann, K.A. and Crawford, N.M. 1992. Identification of two tungstate-sensitive molybdenum cofactor mutants, Chl 2 and Chl 7 of *Arabidopsis thallinna. Mol. Gen. Genet.*, **233** (1-2) : 169-176.

Leidi, E.O., Gomez, M. and Guardia, M.D. Dela. 1987. Soybean genetic differences in response to Fe and Mn : Activity of metalloenzymes. *Plant Soil*, **99**(1) : 139-146.

Le Scure, A.M., Proudhon, D., Pesey, H., Ragland, M., Theil, E.C. and Briat, J.F. 1991. Ferritin gene transcription is regulated by iron in soybean cell cultures. *Proc. Natt. Acad. Sc. of U.S.A.*, **88**(18) : 8222-8226.

Lynch, J., Epstein, E. and Lauchli, A. 1982. Na^+-K^+ relationship in salt stressed barley. *Proc. IXth Intntl. Plant Nutrition Colloquium,* Warwick, England (A. Scaife, ed.) pp. 347-352. Commonw. Agric. Bur. Farmham Royal, Bucks.

Machold, O. 1968. Einfluss der Ernahrungs bedingungen auf den Zustand des Eisens in der Blatter , den chlorophyllgehalt und die Katalase sowie Peroxyase activat. *Flora* (Jena), Abst. **A-159** : 1 -25.

Marcar, N.E. and Graham, R.D. 1987. Genotypic variations for manganese efficiency in wheat. *Austr. J. Agri. Res.*, **38**(3) : 501-511.

Marschner, H. 1986. Mineral nutrition of Higher Plants. Academic Press. London.

Makur, R.E., Mathre, D.E. and Olsen, R.A. 1991. Interaction between iron nutrition and *verticillium* wilt resistance in tomato. *Plant Soil,* **134**(2) : 281-286.

Mehta, S.L., Santha, I.M., Yadav, V.K., Roy, P.K. and Barat, G.K. 1991. Plant biotechnological approaches for elimination of oxalyldiaminopropionic acid neurotoxin from *Lathyrus sativus*. In "*Golden Jubilee Celebrations Symp. on Grain Legumes*", New Delhi, India. Indian Soc. Geneties and Plant Breeding. pp. 325-332.

Mori, S. and Nishizawa, N. 1989. Identification of barley chromosome No. 4, possible encoder of genes of mugineic acid synthesis from 2' deoxymugineic acid using wheat barley additional lines. *Plant Cell Physiol.*, **30**(7) : 1057-1061.

Nable, R.O. and Paull, J.G. 1990. Effect of excess grain boron concentrations on early seedling development and growth of several wheat (*Triticum aestivum*) genotypes with different susceptibilities to boron toxicity. *XIth Intntl. Plant Nutrition Colloquium*, Wageningen, Netherlands, July 30-Aug. 4, 1989. (Beuisichm, M.L. Van, ed.) Kluwer Academic Publishers. pp. 291-295.

Neame, P.J. and Barber, M.J. 1989. 2 cysteines in N=-terminal domain are probably involved in binding Mo cofactor. *J. Biol. Chem.*, **264** : 20894 -20901.

Okumura, N., Nishizawa, N.K., Umehara, Y. and Mori, S. 1991. An iron deficiency specific cDNA from barley roots having two homologus cysteine-rich MT domains. *Plant Mol. Biol.*, **17**(3) : 531-533.

Onyezilli, F.N. and Ross, J.D. 1993. Iron deficiency stress responses of five graminaceous monocots. *J. Plant Nut.*, 16(5) : 953-974.

Paull, J.G., Rathjen, A.J. and Cartwright, B. 1991a. Major gene control of tolerance of broad wheat (*Triticum aestivum* L.) to high concentration of soil boron. *Euphytica*, **55**(3): 217-228.

Paull, J.G. Rathjen, A.J. and Cartwright, B. 1991b. Tolerance to high concentration of boron for the amphiploid of *Triticum aestivum x Agropyron elongatum*. *Plant Soil*, **133**(2) 297-299.

Pich, A., Scholz, G. and Seifert, K. 1991. Effect of nicotianamine on iron uptake and citrate accumulation in two genotypes of tomato - *Lycopersicum esculentus* Mill. *J. Plant Physiol.*, **137**(3) : 323 - 326.

Podlesak, W., Werner, T., Grum, M. and Schlegel, R. 1990. Genetic differences in the copper efficiency of cereals. In "*Proc. XIth Intntl. Plant Nutrition Colloquium*, Wageningen, Netherlands, July 30, -Aug. 4, 1989. (Beusicherm, M.L. van, ed.) Kluwer Academic Publishers, pp. 297-301.

Prus-Glowacki, W. and Godzik, S. 1991. Changes induced by zinc smelter pollution in genetical structure of Pine (*Pinus radiata* L.) seedling populations. *Silvae Genetica*, **40**(5-6) : 184-188.

Ramani, S. 1988. An examination of Fe-deficiency stress response in relation to Fe-uptake in two sorghum hybrids and their parent cultivars. *J. Plant Nutri.*, **11**(1) : 107-115.

Ramani, S., Joshna, D.C. and Shaikh, M.S. 1989. Response of three *Sesbania* species to salinity when grown hydroponically. *J. Plant Nut.*, **12** : 1447-1455.

Rodriguez de Cianzo, S.R. 1991. Recent advances in breeding for improving iron utilization by plants. *Plant Soil*, **130**(1- 2) : 63 - 68.

Romheld, V. and Marschner, H. 1990. Genotypical differences among graminaceous species in release of phytosiderophores and uptake of iron-phytosiderophores. *Plant Soil*, **123** (2) : 147-153.

Rosmini, H. and Mukhlis. 1990. Iron toxicity tolerance of rice cultivars in acid sulfate soils of Indonesia. *Intntl. Rice Res. Newslet.*, **15**(6) : 10.

Sakal, R., Chaudhari, L.B., Sinha, M.J. and Singh, B.P. 1985. Differential response of some cheena (*Panicum miliaceum* L.) germplasms to zinc in calcareous soil. *J. Ind. Soc. Soil. Sci.*, **33**(3) : 591-595.

Sakal, R., Singh, A.P., Sinha, R.B. and Bhogal, N.S. 1991. Relative susceptibility of some important varieties of seasamum and mustard to boron deficiency in calcareous soil. *Fertilizer News.*, **36**(3) : 43-49.

Salim, M. and Saxena, R.C. 1991. Iron, silica and aluminium stresses and varietal resistance in rice : Effects on white backed plant hopper. *Crop Sci.*, **32**(1) : 212-219.

Sandhu, T.S. 1992. Potential of genetic resources and biotechnology in pulse breeding. *Plant Breeding Abst.*, **62**(2) : 131-139.

Singh, B.N. and Sakal, R. 1987. Screening for zinc deficiency tolerance in rice. *Intntl. Rice Res. Newslet.*, **12**(6) : 15- 16.

Steven, B.J., Kirk, D.W., Bright, S. W.J. and Wray, J.L. 1989. Biochemical genetics of further chlorate resistant, molybdenum cofactor defective conditional lethal mutants of barley. *Mol. Gen. Genet.*, **219**(3) : 421-428.

Tanada, T. 1978. A rapid gravitational effect on the translocation of bluorescein in mung bean hypocotyl sections. *Planta,* **142** : 221-223.

Tepperman, J.M. and Dunsmuir, P. 1990. Transformed plants with elevated levels of chloroplastic superoxide dismutase are more resistant to superoxide toxicity. *Plant Mol. Biol.,* **14** : 501-511.

Tabita, S., Takahashi, H., Miyake, H. and Totsuka, T. 1988. Selection and partial characterization of copper resistant line of rice *(Oryza sativa)* Callus culture. *J. Plant Physiol.,* **133**(5) : 545-549.

Velu, G., Savithri, P. and Manie Ram, T.S. 1989. Variability in micronutrient contents of seed in rice varieties. *Madras Uni. Agri. J.*, **76**(11) : 642-644.

Viets, F.G. Jr., Boawn, L.C. and Crawford, C.L. 1954. Zinc contents and deficiency symptoms of 26 crops grown on a zinc deficient soil. *Soil Sci.,* **78** : 305-316.

Williams, E.P., Ross, W.M., Clark, R.B., Herron, G.M. and Witt, M.D. 1986. Iron deficiency chlorosis; Its heritability and effects on agronomic traits in a sorghum population. *J. Plant Nut.,* **9** (3-7) : 423-433.

Wray, J.L. and Kinghorn, J.R. (eds.) 1989. Molecular and Genetic Aspects of Nitrate Assimilation. Oxford Sci. Publi. Oxford Uni. Press. 410 pp.

Zaiter, H.Z., Coyne, D.P. and Clark, R.B. 1987. Genetic variability and inheritance of resistance of leaf iron-deficiency chlorosis in dry beans. *J. Amer. Soc. Hort. Sc.,* **112**(6) : 1019-1022.

Advancements In Micronutrient Research
Editor: A. Hemantaranjan
1996, Scientific Publishers, Jodhpur, pp. 401–452.

12

TECHNIQUES IN MICRONUTRIENT RESEARCH

S.C. AGARWALA AND CHITRALEKHA CHATTERJEE
Department of Botany, Lucknow University,
Lucknow 226 007, INDIA.

CONTENTS

1. INTRODUCTION

Over the years, important techniques have been developed, largely, by chemists, physicists, engineers and biochemists which have been utilised for furthering research on micronutrient elements. Some of these methods have facilitated the discovery of new micronutrient elements and have helped in the elucidation of their role in different facets of plant metabolism. With further improvement in the technology many new elements which are required at a lower concentration than known essential elements may be added to the list of micronutrients. The essentiality of the latter has been established for a limited number of plant species. This information should be made available for all plant species, specially for varieties of crop plants, which are growing wild in nature. This will help in categorising them into efficient and inefficient varieties in respect to a micronutrient. This information could be utilised in growing efficient varieties in areas marginally deficient in that micronutrient. An effort should be made to mark out the genes responsible for efficient uptake and/or translocation or utilisation of the micronutrient with a view to transfer them to inefficient varieties. There are many elements whose essentiality has been shown for one or a limited number of plant species. By the use of improved technology it is necessary to critically examine whether the element is essential for a larger group of plants. So far major contributions in micronutrient research have come from Australia, U.K., U.S.A., Europe and Japan. In developing countries the research on micronutrient elements is steadily picking up. But this research is comparatively expensive and requires the use of sophisticated technology which is not readily available to scientists and it is becoming increasingly difficult to acquire because of financial problems. One of the important ways to determine the quantitative requirement of micronutrient elements is to find the critical values of deficiency of each micronutrient at least for each plant species, if not varieties. This can be accomplished by growing plants in solution, sand or flowing culture at graded levels of a micronutrients, or in field grown plants showing acute deficiency of a micronutrient and supplying increasing levels of that element. In these plants dry weight yield should be determined at various stages of growth and tissue content of the micronutrient determined by atomic absorption spectrophotometry or inductive plasma analyser in plant part which shows the maximum difference between deficient and adequate plants. This will help in diagnostic work. The technique for growing plants in sand or solution culture has been described briefly. For further details, some standard book like 'Sand and water culture methods used in the study of plant nutrition' by E.J. Hewitt (1966) should be consulted. To get an insight in the mechanism of the uptake of the micronutrient element, the radioisotopes of the elements should be used in whole plants and isolated organelles of plants. There is considerable information on the *in vitro* enzyme activities of plants deficient in micronutrients. But there is hardly any information on *in situ* enzyme activities in a plant deficient in a micronutrient.

The techniques for separation of different fractions of nitrogen, nucleic acids, carbohydrates, fats, organic acids, etc., which are or are likely to be influenced by micronutrient supply are briefly described. These could be estimated by the well known methods.

2. Culture Methods

It is possible to produce acute deficiency of micronutrients in controlled culture. But it is essential that the components used in culture work, namely, water, nutrient salt solutions, containers, sand in case of sand culture and other materials being used are freed from micronutrient elements. Broadly speaking, culture methods may be classified into three groups, namely (i) water culture (ii) sand culture and (iii) tissue culture.

Water culture which many scientists like to refer to as nutrient culture seems to be easier to handle than sand culture as in the latter purification of sand is involved, which is an extremely tedious process. But in water culture (nutrient culture) there are some problems one has to face. These are (i) of aeration of the rooting medium - (nutrient solution) which has only a limited supply of oxygen (ii) of iron chlorosis if non-chelated form of iron is being used (iii) of infection of roots.

Water culture seems to be more convenient than sand culture as in the former, change in the nutrient composition and pH can be made easily. But inspite of these drawbacks, many scientists think that sand culture simulates the conditions of soil culture to a great extent in that the plants are anchored in sand particles and excess solutions is allowed to drain off and roots are aerated in a natural way. In tissue culture, in addition to macro and micronutrient salts, carbohydrate source, necessary vitamins and hormones have to be provided in sterile conditions. The techniques used in tissue culture will be explained after those involved in sand or water culture (nutrient culture).

Some of the problems that one may have to encounter in sand and water culture (nutrient culture) are common. These are purification of water and macronutrients so that they are free from micronutrient elements. The second problem is the choice of containers and other materials used in culture work such that after vigorous cleaning, they do not contribute any micronutrient to the nutrient solution. In sand culture there is additional problem of purifying the sand.

Containers

Containers used in culture are made of different types of materials, e.g., clay, metals, porcelain, glazed porcelain polyethylene, soft galss, borosilicate glass like hysil, corning or pyrex, teflon etc. For obtaining deficiency effects of micronutrients except boron, it is preferable to use containers made of borosilicate glass like pyrex, hysil or corning make. For work on boron and silicon deficiencies containers made of polyethylene or of clay painted thrice with bitumen solution could be used. The latter could also be used for obtaining acute deficiency of iron and mild effects of manganese deficiency. But instead of painting bitumen solution, the interior surface of the pots could be lined with alkathene. To minimise chances of contamination from containers, from accidental tear in the alkathene lining, two or three sheets of alkathene should be placed one upon another in the interior surface of the container made of clay or polyethylene. These pots made of polyethylene used with alkathene could be used for obtaining acute effects of deficiency of iron and manganese, zinc and copper. For obtaining deficiency of molybdenum, containers of pyrex or corning glass should be used. For sand culture work it is desirable to have a hole in the central region of the bottom of the pot through which solution could flow. The hole should be covered with an inverted watch glass. The space between the rim of the watch glass and the rim of the hole should be covered with glass wool. The use of containers made of teflon (polytetrafluroethylene) might be superior to pyrex glass.

Covers

It is not necessary to use covers for experimental containers. But where they are used, they serve several purposes, e.g., reduction of water loss, exclusion of the dust in the atmosphere and the most practical means of support to plants in water cultures. The dust particles in the atmosphere are a major source of contamination. When a glass slide of 7 cms X 2.5 cms is exposed in a green house meant for culture work, the elements which accumulated on the surface of the glass during one month period were as follows :

(Element accumulated as mg/m^2/month).

Fe 2, Mn 0.4, Cu 0.3, Zn 0.3, B 0.6, Mo 0.04.

When plastic or glass containers are being used for controlled culture it is necessary to exclude light, this can be done by the use of thick brown paper which is to be fixed on the outer walls of the containers.

It is more difficult to provide covers for sand than for water culture (nutrient culture). The composition of covers is important for micronutrient work as moisture might condense on the covers dissolving traces of micronutrient present in the cover, which may drop into the culture solution. A suitable cover can be made by moulding it from plaster of Paris leaving holes for insertion of the plant. The hole could be impregnated with paraffin wax. Hardwood covers impregnated with wax can also be used. Even plastic covers can be used.

Cleaning of Glassware and Containers

All glasswares including reagent bottle and containers of glass and polyethylene used in micronutrient work should be cleaned by a detergent solution and water, followed by 1 : 1 HCl or 1 : 1 HNO_3 and then plenty of water of the highest purity. The deposits in reagent bottles of glass or polyethylene can be removed by shaking paper pulp along with 1 : 1 HCl. For removing the sulphides, aqua regia could be used. For work on molybdenum deficiency more vigorous cleaning of pyrex or corning glass is necessary. These should be additionally treated with vapour of boiling HCl. After

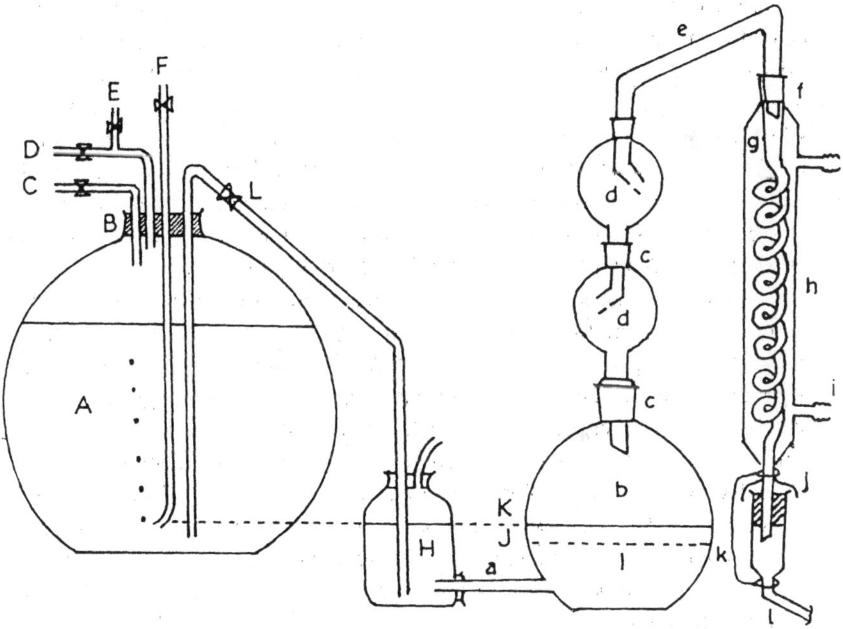

Fig. 1. Diagrammatic sketch of pyrex glass distillation assembly.

A Reservoir
B Air tight rubber cock
C Refill inlet
D Attachment to suction pump
E Pressure release
F Air inlet
G Siphon with stop cock
H Stabilizing vessel with air outlet
I Basal park of still
J Level when boiling
K Level when cold
L Stopcock closed when refilling A

a Inlet from constant level stabilizing vessel (H)
b 3 litre flask for heating
c B 34 joints
d Spray trap
e Wide bore vapour bends
f B 24 joints
g Inlet tube
h Condenser
i Water inlet
j Dust cover
k Supporting string
l Delivery tube

steaming with HCl vapour the containers should be treated with steam and finally washed several times, with glass distilled water (G.D.W.).

Purification of water

The water which is obtained for distillation from stills made of metals is not suitable for micronutrient work. This water has to be distilled in pyrex or 'corning' glass stills fitted with quick-fit joints as shown in the fig. 1. To prevent creeping of water from the boiling flask two splash traps, which are heated with electrical thermal tape, should be introduced. For finer work on micronutrients the water obtained by distillation in all pyrex assembly should be distilled again. This water is not suitable for work on boron, sodium and silicon. The water obtained from distillation in all pyrex vessels can be tested for the presence of traces of heavy metals by putting a drop of water in purified 0.1% dithiozone (in chloroform) solution. If its colour changes from green to orange, that means heavy metals are present. Water should be stored in clean containers made up of pyrex/'corning' glass or polyethylene. For large scale culture work, a large amount of water has to be stored. Instead of feeding water distilled from metal stills to the pyrex glass stills, the water obtained by passing rain water through resin columns can be fed. The rain water has to be passed through several columns of cation exchange resins alternating with anion exchange resins arranged in series,the last column should be a cation exchange one. The resins should be packed in a pyrex glass container or containers made of polyethylene. These columns should be connected with clean tubing of polyethylene. The water which has passed through resin columns is satisfactory for work with boron and iron deficiency. For the last many years, mixed bed resins of high purity, are also available. The demineralised water passed through these resins is of as good a quality as glass distilled water. But the removal of cations or anions by resins would depend on their contents in water which is fed to the resin columns. If it contains high amounts of salts which is present usually in municipal waters the resin would get exhausted very soon and will have to be regenerated very frequently. For finer work instead of regenerating resins which is a cumbersome process, the exhausted resins may be replaced by new resins.

For obtaining acute deficiency of each micronutrient, glass distilled water used should not contain more than the concentration of each nutrient given in the table 1.

TABLE 1 : Micronutrient composition (μg/L) of demineralized water distilled in all glass pyrex stills.

Iron	Manganese	Copper	Zinc	Molybdenum
1.0	1.0	0.1	0.05	0.001

For obtaining acute effects of deficiency of boron, sodium and silicon, water should be distilled in distilling units made of stainless steel or silver. But satisfactory results can also be obtained if only condensing unit is made of stainless steel or silver.

Purification of sand

Sand from river beds is generally not suitable for culture as it is often mixed with large amounts of silt and mica. Silica sand or powdered quartz is a suitable material for the culture work. In India, sand beds, near Shankargarh, Allahabad, can be used. The other source from where the sand can be obtained in India is from Ganganagar, Rajasthan. First sand is sieved and particle size 0.25 to 0.84 mm is taken for culture work. In the first step, the sand is freed from silt, clay etc. by releasing a stream of running water under pressure at the bottom of the sand. Further purfification of sand is carried out in (i) high density polyethylene bins or (ii) 'Keebush' sand digester as described by Hewitt (1966).

(i) *Purification of sand using polyethylene buckets*

The water washed sand is further purified in polyethylene bins by (a) *Cold hydrochloric acid treatment* : The air dried water washed sand is saturated with 4% w/v C.P. HCl and is kept for 3 days. (b) *Hot hydrochloric acid treatment* : 1 L of 17% w/v HCl is added per 10 kg sand. The sand acid mixture is agitated by blowing steam at a pressure of 20 lbs/sq inch in the acid sand mixture for 6-8 hrs and then washed with deionised water.

(c) *Hot hydrochloric acid oxalic acid treatment* : In addition to 1L of 17% HCl (w/v), 10 g oxalic acid/L of HCl was added per 10 kg of sand and steam blown as in treatment (b) and then washed with deionised water (d) *Hot alkali treatment* : To the sand obtained after treatment (c) 1 L of 2% KOH was added per 10 kg of acid washed sand after thorough leaching with water. Alkali-sand mixture was agitated as in the treatments (b) by blowing steam for 4-6 hrs. The sand treated after treatment (c) is thoroughly washed with deionised water and treatment (c) is repeated again.

(ii) *Purification of sand in a 'Keebush' sand digester for micronutrient work*

Vigorous purification of the sand is carried out in special equipment designed at Long Ashton (England) in collaboration with Kestner Evaporator and Engineering Company. The equipment is known as Kestner Long Ashton "Keebush" sand digester (Fig. 2). It is made of acid-alkali resisting resin bonded asbestos material. The capacity of this digester varies from 100-350 kg depending upon the model.

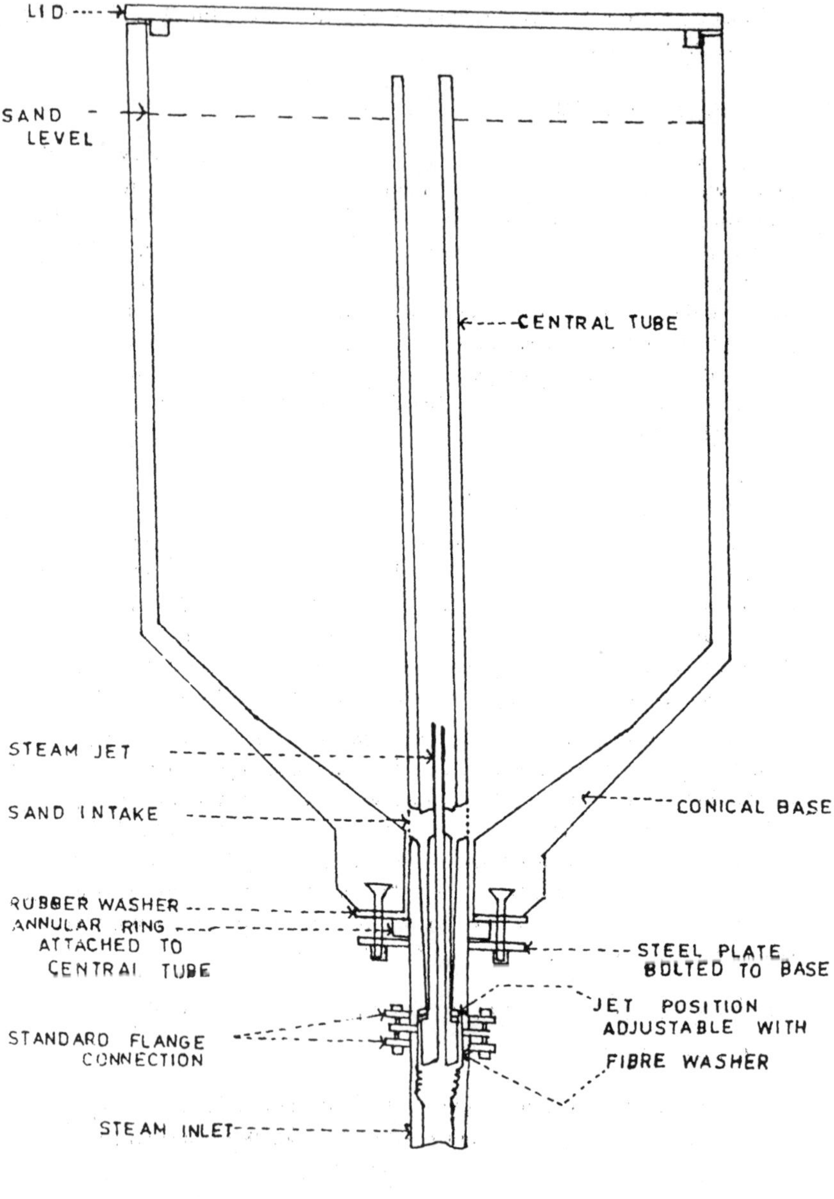

Fig. 2. Diagrammatic representation of 'Keebush' Unit.

About 120 kg sand which has already been thoroughly washed with water and is loaded in "Keebush" sand digester for further purification. One litre of acid mixture containing 17% HCl and 1% oxalic acid is added per 20 kg of sand. The sand acid mixture is churned in sand digester by passing steam generated by an electric boiler at a pressure of 20 lbs/sq inch for 4-6 hrs. During operation the sand acid mixture is drawn in central tube through holes present at the base of the central tube by suction pump action of the steam jet also called as "Venturi effect" and then transported up the tube by force of steam which is issued from the jet. This sand is deposited over the untreated sand, which by force of gravity is displaced downwards. The circulation of sand in the sand digester is therefore continuous and automatic. After the desirable time (4-6 hrs) the steam boiler is switched off and acid is drained out. Sand is then unloaded and washed thoroughly in appropriately purified water. For finer work for producing acute deficiency of micronutrients this operation may be repeated twice or thrice. Each such operation is altered with alkali treatment in which the sand is treated with 5% Na_2CO_3 or 2% KOH and ultimately with acid mixture. The acid is removed by thorough washing of the sand with deionised or distilled water. After each of the above treatments the sand should be thoroughly washed with deionised or distilled water. Finally the thoroughly washed sand should be filled in containers which are to be used for culture work and then it is again washed with distilled water. Finally the sand is leached with 0.06% $Ca(NO_3)_2$, till the pH of the sand ranges between 6.5 to 7. The calcium nitrate solution should be appropriately purified before use. The sand is usually considered pure only when the micronutrient in question is excluded to that extent so that the deficiency of the nutrient could be produced.

Purification of macronutrients and iron salts

Even analytical reagent grade salts are contaminated with many micronutrient elements. The use of macronutrients and iron salts without freeing them from micronutrient element would give erroneous results in culture meant for micronutrient work. The micronutrient salts are also contaminated with other trace elements. But these impurities in micronutrient salts are not likely to interfere in any way as the quantity of these salts used in culture work is very low. But to make sure that these impurities are removed even from micronutrient salts, these salts should be recrystallised several times. Alternatively, 'specpure' elements obtained from Mesors Johnson Matthey Ltd. Hatton Garden, London, could be used to make sure that there will be no contamination from micronutrient salts.

The methods used for freeing macronutrient and iron salts are described in the following pages. Molar or two molar stock solutions of macronutrient salts could be made before starting the purification process.

Preparing stock solutions of macronutrients free from Fe, Mn, Cu and Zn

The volumes of different reagents added to 1 L of stock solution of each macronutrient are as follows : 10 ml 2M $Ca(NO_3)_2$ (32%) except to Ca-salts; 10 ml 1M K_2HPO_4 or Na salt if K is omitted; 10 ml 1M $NaHCO_3$ (8.4%); 20 g A.R. grade $CaCO_3$. Solutions should be shaken vigorously when phosphate is added. Again shaken repeatedly after addition of sodium bicarbonate. The flasks (conical) containing the macronutrients solution are plugged with cotton wool or capped by a small beaker when autoclaved at a pressure of 20 lbs/inch for 60 minutes. As soon as the solution becomes a little cool, it is filtered through sintered glass filter funnel of No. 4 porosity. When layers of calcium phosphate and carbonate are settled on the funnel, the filtrate emerging from the filter funnel should be very clear. This solution can be used for obtaining Fe, Mn, Cu and Zn deficiencies.

To ensure that Cu and Zn have been removed completely by the above method the dithizone extraction is necessary. The whole procedure may be repeated again to be sure than most of the iron, Mn, Cu and Zn have been removed from the stock solution. Further purification of the autoclaved solution which has been filtered to make it free from Cu and Zn is affected by extraction with purified 0.01% dithizone solution in chloroform according to the method of Parks, Hedde, Hurwitz and Ellic (1943). At pH 5, Zn and Cu can be removed by extraction with dithizone. After repeated extraction with dithizone, the solution is shaken with very dilute (0.005%) solution of dithizone, the colour of dithizone would remain pale green for considerable time if all copper and zinc had been removed. Traces of chloroform are removed by shaking with redistilled CCl_4 and then CCl_4 is removed by volatilization.

Preparation of stock solution of macronutrients free from boron, chlorine, and sodium

Boron can be satisfactorily removed from macronutrient stock solutions by recrystallisation. Chloride and sodium can also be removed by this process, in which recrystallisation has to be repeated several times. The salt which has to be purified is dissolved (1 or 2 kg) in nearly boiling water, so as to produce a saturated solution. Some solid salt is added in portions with continued stirring till nearly all salt dissolves. The solution is heated and filtered through a whatman filter paper no, 51 and transferred to a large beaker. Crystallisation would commence but in this process temperature drops to about 40°C, 96% redistilled ethanol is added in a volume equal to that of the solution and the mass is stirred till very cool. The crystals are collected by filtration. The mother liquor is put in a frigidaire and cooled. The crystals formed in this process is also separated by filtration, recrystallisation process is repeated. The crystals which are formed in recrystallization should be washed with ethanol and dried at low temperature (40°C). After addition of alcohol sodium phosphate and magnesium sulphate may form two immiscible phases. This can be avoided by continuous stirring as calcium chloride does not crystallise early. Addition of sodium chloride assists in the formation of crystals. Calcium nitrate which is required for studies of boron or chlorine deficiencies is prepared from purest calcium carbonate available by dissolving in analytical grade nitric acid diluted 1 : 1 with water. Calcium chloride required for boron deficiency is prepared by action of redistilled A.R. grade hydrochloric acid with pure $CaCO_3$.

Ammonium salt can be prepared by passing ammonia into water through a water bubbler. This solution is mixed with redistilled HCl, HNO_3 or sulphuric acid until the solution attains a pH of 7.

Molybdenum free stock solutions of macronutrients

Two methods can be used for this purpose. First method is based on the use of 8-hydroxyquinoline and iron. This co-precipitates molybdenum. But the second method which is based on co-precipitation of copper sulphide and molybdenum. This method is more efficient than the first method in removing molybdenum from the stock solution.

In the second method to 1 L of stock solution (2 M) prepared from Analytical grade salts, 10 ml of 19.5% w/v $CuSO_4.5H_2O$ is added. In this solution H_2S is passed for thirty minutes after filtration through water and glass wool. The solution and the precipitate are filtered through a clean pyrex glass sintered funnel of porosity size no. 4. The precipitation is repeated a second time and left over night and filtered through a clean pyrex glass sintered funnel as before. Excess hydrogen sulphide is removed by boiling the stock solution and boiling continued till no smell of H_2S can be detected. Disodium monohydrogen phosphate and sodium carbonate form a precipitate with copper sulphate. But H_2S decomposes these precipitates and molybdenum is removed from the solution efficiently. Stock solution of Na_2CO_3 may be slightly alkaline after passing H_2S. Under these conditions polysulphides are formed. These are decomposed to S and H_2S by prolonged boiling, and can be removed by filtration. The stock solutions purified in the aforesaid way against molybdenum also removes nearly all copper from the solution and can be used for obtaining Cu deficiency effects.

Removal of molybdenum and other elements by co-precipitation with 8-hydroxyquinoline

Scott and Mitchell (1943) first used this for method for concentration of metals in ash of plants prior to their spectographic analysis. By addition of redistilled hydrochloric acid to 1800 ml of stock solutions of macronutrient (2 M), the pH of solution is reduced to 1.8. To this solution is added 56 g of specpure iron dissolved in HCl or ether extracted ferric chloride solution and 455 mg of 8-hydroxyquinoline is added. A 2N solution of analytical reagent grade KOH is added slowly till the pH of the solution rises to 5 or 6. The precipitate is left over night and filtered through a clean pyrex sintered funnel of the porosity size 4. The filtrate is extracted with 1% solution of 8-hydroxyquinoline in 4 L pyrex funnel. The solution is finally extracted twice with redistilled ether which removes the last traces of chloroform. The ether is removed by aeration at room temperature.

This method has been improved by Mitchell and Scott (1947) by inclusion of thionalid and tannic in a solution of 8-hydroxyquinoline. This method can also be utilised for removal of vanadium.

Removal of cobalt from stock solutions of macronutrients

Prepare 0.5 M to 1 M stock solution of each macronutrient salt. Add to this solution 16 ml w/v sodium acetate solution and adjust the pH to about 6.5. Add 16 ml 0.1% w/v solution of nitroso-R-salt of the best grade available and 8 g washed Darco-G-60 charcoal. The solution should be stirred. Filter as soon as possible through a clean pyrex sintered funnel of porosity size 4. About 99.9% cobalt would be removed by this process.

Removal of vanadium from stock solution of macronutrients

Either of two methods can be used for the removal of vanadium from stock solutions of macronutrients.

First method

In the first method, to 2 L of molar or 2 molar stock solution is added 4 ml redistilled 6N HCl, 10 ml 10% w/v ferrous sulphate ($Fe\ SO_4.7H_2O$) and 40 ml saturated solution of cupferron in water. The precipitate is removed by filtration in a pyrex glass sintered funnel porosity size 4. Remove the precipitate, extract the filtrate 4-5 times in 100 ml redistilled chlorofrom, add 2 g acid extracted darco-G-60 charcoal, stir and filter. Vanadium is removed to the extent of 99.9% for amount upto 400 μg per litre. Repeat the whole process.

The second method is that of extraction of vanadium by presence of auxin in presence of thionalid and tannic acid. To each litre of molar stock solution of macronutrient, add 4 ml 10N redistilled HCl, 160 g of ferric iron, 40 mg of aluminium (Al^{+++}) as specpurechloride solution and 40 ml of 5% w/v 8-hydroxyquinoline solution in 2N acetic acid. Adjust the pH to 1.8 with ammonium solution. Add 100 ml 2N ammonium acetate with continuing stirring followed in the following order by 8 ml 10% w/v tannic acid and 4N acetic acid, and 8 ml 1% w/v thionalid in glacial acetic acid. Thoroughly mix, leave it to stand at over night at room temperature. Filter through a pyrex sintered funnel of porosity size 4.

Purification of compounds used to supply iron

Iron salts contain significant amount of several micronutrients. If these iron salts are included in the culture solution without purification these micronutrients are likely to interfere with the development of deficiency of micronutrient which were present in the iron salts. Therefore, before these salts are used in culture work, these should be purified and made free of particular micronutrient whose deficiency, it is desired to produce.

Removal of molybdenum from iron salts

It is possible to eliminate molybdenum from a solution of ferric chloride (which has been freed of Cu, Zn, Mn and cobalt) by forming complex with dithiol (solution made in amylacetate) at pH.[2]

Molybdenum could be removed from potassium EDTA and ferrous sulphate by co-precipitation with CuS. Then the solution of the two compounds should be mixed in equivalent amounts and aerated for 24 hrs. Thus, one can get Fe-EDTA free from molybdenum.

In our experience the tedious process of removing micronutrients from iron salts which are to be used for culture work could be avoided if one could use specpure iron which used to be supplied by Mesors Jhonson Matthey Ltd., Hatton Garden, London.

Spec-pure iron could be dissolved in very pure redistilled HCl or HNO_3 and thus one could obtain iron salts free of micronutrients. Possibly this will help also in the search of new micronutrients. But, spec-pure elements are expensive. For routine culture work on micronutrients it would be more practical if iron salts could be purified.

Removal of copper, zinc, manganese and cabalt from solutions of iron salts

In ether, in presence of 6N HCl, the solubilities of $FeCl_3$ differs from that of zinc, copper, manganese and cobalt salts. Filter through whatman No. 44 filter paper a solution of ferric ammonium sulphate 500 g in 350 ml of hot glass distilled water. The filtrate should be transferred to a clean pyrex or corning brand separating funnel. Add 1250 ml analytical grade HCl saturated with ether. Shake thoroughly. Run off the aqueous phase in another separating funnel and retain the ether phase. Wash the ether phase with glass distilled water. The aqueous phase should be extracted with ether twice and all ether phases should be combined. Ether phase should be washed with HCl saturated with ether. Ether phase is retained. Shake the ether phase with glass distilled water and the aqueous phase is retained in a reagent bottle. This aqueous phase contains ferric chloride free from Cu, Zn, Mn and Co. This method was originally recommended by Piper (1941).

Preparation and purification of iron citrate and iron EDTA

To ferric chloride solution in which iron has been freed from Mn, Cu, Zn and Co add an equivalent amount of recrystallised citric acid. Citric acid may be added in slight excess. Heat the mixture of $FeCl_3$ and citric acid in a clean silica evaporating dish till all hydrochloric acid is eliminated. Redissolve the residue and make it upto volume. Citric acid is usually pure enough for this purpose. But if further purification of citric acid is desired so as to make it completely free of Cu and Zn, it may be converted into ammonium citrate and then purified by extracting Cu and Zn with dithizone at pH 5.

Preparation of Fe-EDTA

Mix a solution of disodium or potassium EDTA with equivalent amounts of ferric citrate prepared from $FeCl_3$ which has been freed from Fe, Mn, Cu, Zn or Mo. Autoclave this mixture for 1 hour. Iron will chelate with EDTA. Potassium EDTA or sodium EDTA of good quality does not contain any appreciable amount of Mn, Cu and Zn which could interfere in culture experiments in which it is desired to obtain acute effects of Mn, Cu or Zn deficiencies.

But it is possible to free EDTA of Cu and Mo by co-precipitation with CuS and filtering the solution through a pyrex sintered funnel of porosity 4.

2.1. Water and Sand-culture of Plants for Micronutrient Work

Macronutrients and iron salts which have been freed from micronutrient elements by the techniques described in the preceding pages could be used in culture work on micronutrients. The latter could be provided by Analytical reagent grade salts manufactured by standard companies without purification as they are needed in small amounts and impurities present in these salts would not vitiate the culture. But, still, if it is desired to purify A.R. salts used for supplying micronutrients, they could be re-crystallised several times. For very fine work on micronutrients, iron and other micronutrient elements could be obtained in 'spec-pure' form, from Mesors Johnson Matthey Ltd., Hatton Garden, London. These elements could be dissolved in A.R. HCL or HNO_3 after re-distillation of acids in all pyrex or corning assembly. In describing further, the water or sand culture techniques for micronutrients, it will be assumed that macronutrients and iron salts have been freed from micronutrient elements. Plants can be grown at the desired level(s) of a micronutrient by omitting that element from the culture solution and then separately adding the micronutrient at known level(s). Luxuriant growth of plants would take place in a culture solution which has all the known macro and micronutrient elements in a balanced form. The composition of a solution meant for water culture (nutrient culture) is not different from that used for sand culture. The terminology used in describing the composition of the culture solution deserves mention. Various terminologies have been used in describing the composition of nutrient media such as g/L, parts pre million, mg equivalent per litre, milligram atom or microgram per litre. Molar (M) or mM, or µM. The parts per million terminology coveys to the mind concentrations of separate elements in million parts of a solution. This terminology is still used by some physiologists and biochemists, specially in work related to micronutrients. Use of some of these terms has been criticised by scientists as unscientific and confusing. These days most scientists like to use the term M, mM or µM in describing the composition of macro and micronutrients in solution used in nutrient culture or sand culture. In earlier days prominent workers in micronutrient research used the terminology g or mg/L.

The range of macronutrient concentrations in solutions used for the study of micronutrient problems in water culture (nutrient culture) or in sand culture is as follows in mM; K 2-8, Ca 3 to 5, Mg, 0.5 to 2, NO_3 5 to 15, NH_4 0 to 3, PO_4 0.66 to 1.33, SO_4 0.5 to 2. This range may vary according to the age of plants or plant species. The solutions used in water culture work should have an osmotic pressure of 0.32 to 0.78 atmosphere and for sand culture it could be 0.98 to 1.45 atmosphere. Usually plants are grown well at an osmotic pressure of 0.5 atmosphere at a pH of 5 or 6. The known micronutrient elements Fe, Mn, Cu, B, Mo and Cl_2 should be included in the culture solution. The use of A-Z micronutrient supplement of Hoagland and Synder (1933) in culture solution with and without omitting one or more micronutrient elements has helped in the discovery of new micronutrients. The A-Z micronutrient supplement has the following composition :

1 ml/L of solutions of the following compostion as g/18 L.

A $Al_2(SO_4)_3$, 1.0; KI, 0.5; K Br 0.5; TiO_2 1.0; $SnCl_2.2H_2O$ 0.5; LiCl 0.5; $MnCl_2.4H_2O$ 7.0; H_3BO_3 11.0; $ZnSO_4.7H_2O$; 1.0; $CuSO_4.5H_2O$, 1.0; $NiSO_4.6H_2O$ 1.0; $Co(NO_3)_2.6H_2O$ 1.0.

B 1 ml/L of solution containing as g/18L
As_2O_3, 0.1; $BaCl_2$, 0.5; $CdCl_2$ 0.1, $Bi(NO_3)_2$, 0.1; Rb_2SO_4 0.1; K_2CrO_4, 0.5, KF, 0.1, $PbCl_2$, 0.1, $HgCl_2$, 0.1; MoO_3 0.425; H_2SeO_4, 0.1; $SrSO_4$, 0.5; VCl_3 0.1.

But these days physiologists like to include only well established micronutrients in the culture solution as could be seen by the following table. The micronutrients used in culture solutions by some scientists are given in the following table.

Micronutrient supplement except iron used by some scientists

	Minimum Micronutrient supply in ppm						
	Mn	Cu	Zn	B	Mo	Cl_2	CO
Arnon (1940)	0.5	0.02	0.05	0.5	0.05		
Lewis and Power (1941)	0.5	0.02	0.05	0.5	0.05		
Piper (1941)	0.5	0.1	0.2	H_3BO_3 0.5 mg/L	0.2		
Gregory (1943)	0.5	0.02	0.05	0.5	0.02		
Hayward *et al.* (1946)	0.4	0.01	0.1	0.5	0.1		
Johnson *et al.* (1957)	0.274	0.032	0.13	0.27	0.0096	1.77	
Ahmad and Evans (1960, 1961)	0.15	0.01	0.05	0.1	0.02	0.2	0.001
Hewitt (1966)	0.55	0.064	0.065	0.55	0.048	3.5	
Agarwala and Sharma (1976)	0.55	0.064	0.065	0.33	0.05	3.5	

In water culture (nutrient culture) seeds should be germinated separately on a solid medium made free of micronutrients and the seedlings should be transferred to water culture (nutrient culture) through holes in the covers. The space between the holes and seedlings should be filled with glass wool or polyethylene of nylon gauze so as to hold the plant firmly but allowing room for the expansion of the stem. If the volume of the container used in water culture is small - 1 to 2 L, the nutrient solution should be changed at intervals of two or three days. But if the volume of the solution is 40-50 L, a change after several weeks would be appropriate. Too frequent changes would perhaps keep the composition of solution without much change but would introduce impurities, besides being expensive. The solution must be aerated which will also keep the culture solution stirred. The aeration could be accomplished by the use of a Mariotte bottle or a cylinder filled with air at a pressure. The air should pass through water which would dissolve contaminants from the cylinder and then passed on to culture solution. A force pump could also be used. Various other devices which have been described by Hewitt (1966) could be used for keeping the culture solution aerated. Asher and Edwards (1983) have described the flowing solution culture technique in which the solution is continuously circulated. By the use of this technique the optimum concentration of micronutrients for growth of plants was found to be much less than that determined by ordinary water culture (nutrient culture) or sand culture. But the use of this technique requires a lot of paraphernalia which is difficult to maintain and the chances of some micronutrient elements being introduced in culture solution increases. Asher and Edwards (1983) have stated the micronutrient deficiency symptoms could be developed in ordinary sand or water culture rather than in flowing solution culture perhaps because it is difficult to avoid completely the contamination. Not many scientists have adopted this technique.

In sand culture the seeds of the experimental plant could be sown in sand directly. The culture solution could be applied at the surface of the sand and the excess solution should be allowed to drain through hole at the bottom of the pot.

Periodicity in change of Culture Solutions

Ideal means of maintaining the nutrient composition almost constant would be to ensure that the nutrient solution is changed before any appreciable change takes place in the composition of nutrient solution. It would be best to continuously renew the culture solution. But this procedure is very expensive and involves considerable increase in equipment and maintenance. It may be feasible for water culture (nutrient culture) but in sand culture it may cause water logging which may lead to the development of iron deficiency symptoms (Hewitt, 1966). In sand culture experiments it may be useful for short experiments where small containers are used. In continuous renewal of culture solutions small contaminations in culture solutions may accumulate and delay or even prevent the appearance of deficiency symptoms. In sand culture the normal procedure adopted for change in culture solution is that the solution is applied to the culture pot every day except on sunday when the pot is flushed with distilled water to remove deleterious substances and accumulated salts from the rooting medium. Numerous devices have been described by Hewitt (1966) for periodical or continuous change in culture solution. Hewitt (1966) referred to various other forms of cultures, atomised culture, the wick cultures, etc., which are not popular with scientists. He has also referred to sterile culture in sand or water (nutrient) culture. But it is very difficult to manage them.

Sand or water culture (nutrient culture) of plants can be used for finding out new micronutrients, for obtaining micronutrient deficiency symptoms in plants in which these symptoms have not been observed so far, specially in plants growing wild. Crop plants could be grown to maturity in sand or water culture at graded levels of micronutrients and the dry weight yield of these plants could be measured at various stages of growth and maturity. Tissue concentration of the element could be determined in different parts of the plants and these results could be plotted. This will give the basic idea of threshold of deficiency, adequate concentration for optimum yields and threshold of toxicity. This will enable us to identify the nutrient deficiencies in fields.

2.2. Tissue Culture of Plants

The tissue culture of plants differs from sand or solution culture in that sucrose, vitamins, myoinositol, plant growth regulators and some amino acids have to be provided in addition to the usual macro and micronutrient salts for the growth of explants under sterile condition. This technique was almost simultaneously initiated by Gautheret (1939), Nobecourt (1939) and White (1939). Since then remarkable progress has been made in the use of this technique and achievements made are astounding. For tissue culture work a separate air conditioned room which can be maintained in sterile condition is essential. A laminar flow cabinet should also be provided which will provide a gentle flow of ultrafiltered air to pass across the working area. The composition of a typical solution for growing explants is given separately (Table 2).

TABLE 2 : Composition of MS (Murashige and Skoog, 1962) Medium

Constituents		mg/L
Macronutrients :	NH_4NO_3	1,650
	KNO_3	1,900
	$CaCl_2 . 2H_2O$	440
	$MgSO_4.7H_2O$	370
	KH_2PO_4	170
Iron : Na_2 EDTA		37.3
$FeSO_4.7H_2O$		27.8
Micronutrients :	$MnSO_4.4H_2O$	22.3
	$ZnSO_4.7H_2O$	8.6
	H_3BO_3	6.2
	KI	0.83
	Na_2MoO_4	0.25
	$CuSO_4 .5H_2O$	0.025
	$CoCl_2 .6H_2O$	0.025
Glycine		2
Vitamins :		
Nicotinic acid		0.5
Pyridoxin HCl		0.5
Thiamin HCl		0.1
Cytokinin : Kinetin		0.1
	Myo-inositol	100
	IAA	10
	Sucrose Agar 1 % (w/v)	30,000
	pH 5.7 - 5.8	

The explants may be surface sterilised by dipping them in 90% v/v ethyl alcohol for 5 to 60 sec. washing with sterile glass distilled water then immersing them in 0.525% w/v NaOCl for 5-30 min and thoroughly washing them in sterile glass distilled water. These sterilised explants should be transferred to sterilised liquid or semisolid agar medium under aseptic conditions in a sterile room. The solution of thermostable substances, namely, salts of macro and micronutrient elements, glycine, kinetin, GA_3, agar, water, glass containers, glass distilled water could be sterilised by autoclaving them at 15 lbs/sq inch for 30 to 60 min. The thermolabile substances, sucrose, vitamins, auxins (103.4 KPa) - should be filter sterilised. After sterilisation thermolabile substances should be mixed in appropriate amounts with thermostable substances, poured in sterile containers and plugged with cotton wool under aseptic conditions. Aluminium foil should be covered on the cotton plug. These containers should then be transferred to a suitable room or an incubator maintained at 27°C for a suitable time. From the composition of the culture solution for tissue work it could be seen that many micronutrients are being supplied at a much higher concentration than in sand or solution culture and some elements are being routinely supplied although they are not considered essential for all plant species. Therefore it is necessary that indepth study of micronutrient requirements in tissue culture be made.

3.1. Atomic Absorption Spectrophotometry

Even traces of metallic elements can be determined by Atomic Absorption Spectrophotometry. The instrument is based on the principle that atoms of an element in ground state are able to absorb electromagnetic radiation over a series of very narrow sharply defined wavelengths. The solution containing the sample is aspirated into a flame and thus it is converted into an atomic vapour. Most of the atoms remain in the ground state. Therefore, these atoms can absorb radiation of a specific wavelength. A hollow cathode lamp has a cathode containing the element to be determined along with a tungsten anode. It is a sharp line source of the radiation supply (Fig. 3).

When enough voltage is applied the filler gas present in cathode lamp is ionised and the ions are accelerated towards the cathode. When these ions bombard the cathode, the cathode material forms an atomic vapour in which atoms exist in an excited electronic state. When these atoms return to the ground state, the lines characteristic of the element are emitted and pass through the flame where they are absorbed by the atomic vapour of the element present in the test solution. This method is very specific in addition to being sensitive, because the test element can absorb the radiation emitted by the cathode lamp.

For each element there is a separate cathode lamp. Some firms have started manufacturing multi-element cathode lamps which can be used for

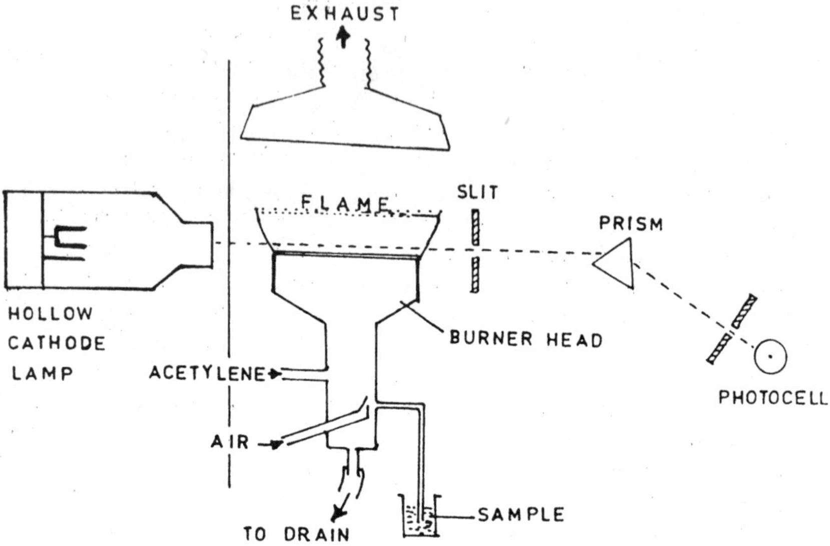

Fig. 3. Digrammatic representation of different components of Atomic Absorption Spectrophotometry

measuring two or more elements without the cumbersome process of the change of cahode lamps. An idea of the sensitivity of this instrument can be made by the following table (Table 3) which gives a picture of the sensitivity obtained in chemical estimation of the element and its estimation by Atomic Absorption

TABLE 3 : Minimum detectable limits of micronutrients by different methods

Element	ICP-AES (µg/L)	AAS (flame source (mg/L)	Chemical methods (mg/L)
Iron	2	0.04	5
Manganese	1	0.02	0.05
Copper	1	0.03	1
Zinc	2	0.008	1
Molybdenum	4	0.2	0.2 0.000001*
Boron	3	9.0	0.5

* by bioassay method (Nicholas and Fielding, 1951).

Spectrophotometry : A double beam atomic absorption spectrophotometer is preferable to single beam spectrophotometer as this is more stable.

Instead of the flame techniques use can be made of electrically heated graphite furnace. The sensitivity obtained by the use of this techniques is much higher than that obtained by the flame technique. The sample is placed in the graphite furnace which lies in light path. The temperature of the graphite furnace is increased in stages. The atomisation of the sample takes place at 3000°C. The rate of heating must be fast so that the atoms are present in dense cloud in the light path. Direct analysis of a solid sample is possible. A small sample is required for analysis.

3.2. Inductively coupled plasma analyser

In this technique the sample is converted into its atoms at a temperature of about 10,000°C. These atoms are excited to a higher electronic state. These atoms emit radiation and fall to their ground state. The radiation emitted by these atoms is of a specific wavelength. Only radiation of specific wavelength is allowed to enter the detector which emits current proportional to the intensity of the specific radiation. This gives a quantitative estimation of the element. This method is more sensitive than atomic absorption spectrophotometry. About 80% of the elements in the periodic table can be detected and analysed by this method. Many of these elements can be analysed simultaneously.

4. Latent Deficiency Of Micronutrients

The deficiency of an essential element in a plant could be identified by observing the visual pathological effects. In case of doubt regarding the diagnosis, recourse may be taken to plant analysis. The deficient element would be found to be very much depressed as compared to its concentration in comparable plants supplied adequate amount of that element. Even in the absence of the deficiency symptoms, the plant analysis could help in identifying the latent deficiency of an element provided the tissue concentration could be compared with the values of critical deficiency obtained from a separate pot culture experiment as explained later. Goodal and Gregory (1947) have traced the history of plant analysis which could determine the nutrient requirement of a soil to maintain its fertility. A good deal of work has been carried out in the last thirty or forty years which has established that the chemical composition of a plant can reflect the nutrient requirements of plants themselves. It has become apparent that increased growth would depend on the increased supply of the element provided all other factors for plant growth were optimal. Under these circumstances, the concentration of the element would also increase in the dry matter of plants at higher rates of supply. Thus a relationship between nutrient supply and dry matter yield and also between dry matter yield and nutrient concentration in plants could be established (Bould, 1968; Reuter and Robinson, 1986; Bates, 1971). One could establish these relationships by growing plants in pot, soil, sand or solution culture at increasing levels of a micronutrient from acute deficiency of excess or in field experiments by suitable micronutrient amendments. The results obtained by pot culture experiments could be plotted to show the relation between nutrient supply and dry weight yield. A separate graph could be plotted to show the relation between tissue concentration of the concerned element and dry weight obtained at varying levels of nutrient supply. From the latter it would be possible to find out the critical value for deficiency, threshold of deficiency and threshold of toxicity. Critical value of deficiency is the nutrient concentrations at which yield is 90% of the maximum yield. Deficiency value is the nutrient concentration at which dry matter yield is 66% of the optimum yield. Threshold of toxicity is the supra-optimal concentration at which dry matter yield is 90% of the optimum yield. In this connection Bould (1963) has stated that "leaf analysis offers the only means of applying directly to field work the results of sand and solution culture experiments". He further states that the interpretation of the nutritional status on the basis of leaf composition is not concerned with the means by which that composition was obtained but only with the effect on growth or changes in leaf composition.

For locating deficiency of an element on the basis of plant analysis some soil scientists like to rely on the critical deficiency values obtained in soil pot culture experiments only, rather than on solution or sand culture experiments. These soil scientists conveniently forget that strictly speaking the critical deficiency values obtained by soil pot culture experiments would be applicable to that soil or comparable soils which had similar chemical and physical properties but not to soils different in chemical and physical properties.

Instead of analysing all parts of a plant it is desirable to analyse the tissue concentrations of that part of the plant which shows greatest difference between adequate and deficient supply of an element. Typical graphs obtained by these studies are given below (Fig. 3a).

These values could be used to predict deficiencies or toxicity of an element in field conditions. But the values obtained should not be applied too rigorously. In case of multiple deficiency or in cases where some interaction seems to have taken place, the critical values of deficiency or toxicity may not apply. The critical values of several micronutrients obtained by different workers associated with senior author at the Botany Department, Lucknow University, India is given below (Table 4).

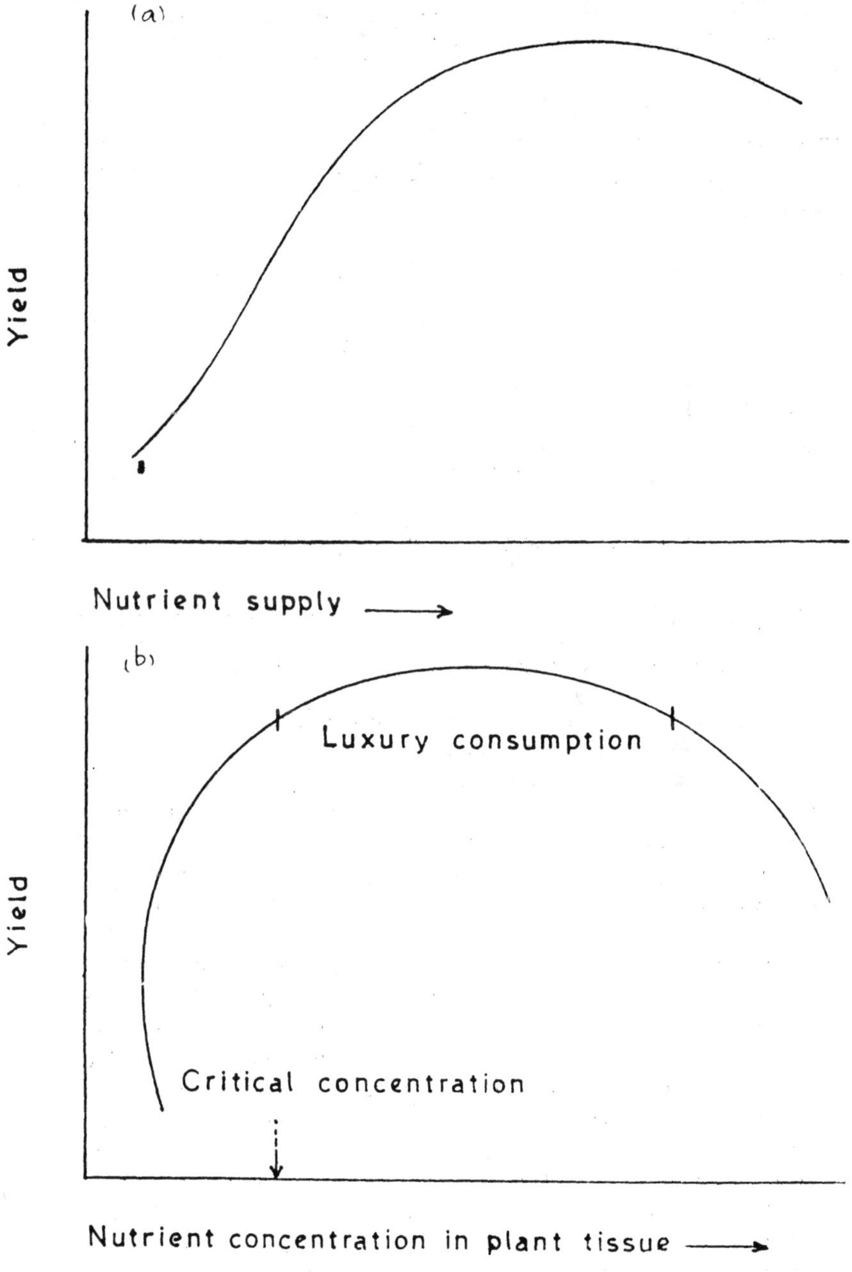

Fig. 3a. Curves representing a relationship between yield and nutrient supply (a) or nutrient concentration (b)

Table 4 : Critical deficiency value of micronutrients in various crop plants at which 50% yield response would be obtained.

Crop plants	Mn	Cu	Zn	Mo	B
Green gram	14	6	–	1.7	9
Black gram	15	5.9	15.5	1.7	15
Gram	10	5	22-32	–	17
Pigeon pea	5	–	18	0.32	–
Wheat	25	4	15	0.81	4
Barley	25	3	20	0.4	8
Rice	25	2.5	15	–	12
Maize	10	2.5	15	–	7
Mustard	10	6	22-25	0.9	20
Sesamum	30	6.5	30	–	2.5
Sunflower	15	4.2	–	–	30

Source : Data obtained from Unit of Plant Nutrition Botany Department, Lucknow University, Lucknow.

This may help in diagnosing latent deficiency of micronutrient elements.

5. Physiological and Biochemical Methods Used for Identifying Micronutrient Deficiency in Fields

The deficiency of a micronutrient in a plant growing in the fields could also be assessed by the use of physiological or biochemical methods. In these methods, the micronutrient whose deficiency is suspected is introduced in plants or part of a plant and the physiological or biochemical changes that are induced, are measured (Bouma, 1983). Bouma and Dowling (1962; 1966; 1967) have introduced a method which is based on the differential leaf area response obtained by transfer of a plant to the complete nutrient solution and to solutions each without a different plant nutrients. The responses were greater in leaf area than in the dry weight or in the nutrient concentrations in the plant tissues. Greewood (1976) have suggested the concept of nutrient stress. Although this method has been applied to determine the status of plants in nitrogen stress but could be adapted to nutrient other than nitrogen. Bouma (1967) have measured photosynthesis in detached leaves from a suspected P deficient field and compared it with that of a comparable detached leaf supplied adequate P. The principle embodied by Bouma for detcting P deficiency could be extended to include micronutrient elements as well. Randall (1969) has suggested that molybdenum deficiency could be identified by measuring the nitrate reductase activity of detached leaves of a plant suspected to be molybdenum deficient after suspending it in a solution of adequate KNO_3 with or without molybdenum in light. In case of deficiency of molybdenum there will be larger difference in the nitrate reductase activity of plant parts supplied molybdenum and those without molybdenum. Similarly, iron deficiency which is difficult to identify on the basis of status of iron in leaves (Agarwala *et al.*, 1965). Bar-Akiva *et al.* (1967) have suggested that the measurement of peroxidase per mg protein nitrogen could provide the answer but in the opinion of the authors, measurement of catalase activity per mg of protein could be a better indicator of Fe deficiency. The estimation of the activity of peroxidase can perhaps distinguish iron deficiency from manganese deficiency. In iron deficiency the peroxidase activity decreases but in Mn deficiency its activity increases. Brown (1953) found that the activity of ascorbic acid oxidase was a good indicator of copper status of a plant. Agarwala *et al.* (1994) also found that in sugarcane, ascorbic acid oxidase activity was a good measure of Cu status of sugarcane grown in refined sand. Similarly the activity of carbonic anhydrase could be used to indicate the Zn status of plants (Bar-Akiva and Lavon, 1969; Dwivedi and Randhwa, 1974).

6. SPECTROPHOTOMETRY

When a beam of light in the visible or ultraviolet region is passed through a solution, the solute in the solution absorbs light of specific wavelength. But if light of wavelengths other than absorbed by solute molecules are allowed to pass through the solution these may be absorbed by substances other than the solute and may be passed on to the detector. Therefore it is necessary that the light which passes through the solution should be of narrow wavelength. To some extent this can be achieved by using filters which absorb most of the light but allow light of particular wavelength to pass through and this process may be better accomplished by the use of monochromators. The monochromator disperses the radiation into component wavelength and the radiation light of required wavelength is allowed to pass through the monochromator through a narrow slit. The size of slit is very important forgetting wavelength of narrow range. Narrow beam of light which has passed through the solution is transmitted to the detector, it relays a current proportional to the intensity of wavelength. This current can be measured by the usual procedure. The detectors are of many kinds, viz., photovoltaic cell, phototubes, photomultiplier tubes etc. In a good spectrophotometer the latter is used as it multiplies the small current many-fold. Advantage is taken of laws propounded by Lambert and Beer to convert intensity of current into qunatitative estimate of the substances which has absorbed that specific radiation. The law propounded by Lambert states "when a beam of light is allowed to pass through a transparent medium, the rate of decrease of intensity with thickness of medium is directly proportional to the intensity of light."

It follows from Lambert's law

that $-\frac{dI}{dt} = KI$

Where I denotes the intensity of incident radiation light of specific wavelength, t denotes the thickness of the medium and k denotes the proportionality factor.

Beer's law states "that the intensity of beam of monochromatic light decreases exponentially with increase in concentration of the absorbing substance.

Integrating this equation and when

$i = Io$ and t is zero, $In \frac{Io}{It} = kt$

$It = Ioe^{-kt}$

Where Io denotes the intensity of the incident light. It denotes intensity of the transmitted light and k is a constant, which depends upon the wavelength and the absorbing medium used. On changing equation from natural to common logarithem, we get,

$It = Io,\ 10^{-0.433\,kt} = Io,\ 10^{-kt}$.........E.1

where $K = k/2.3026$

K is the absorbance co-efficient which is defined as :

"It is the reciprocal of the thickness which is required to reduce the light to 1/10th of its intensity"

The above definition follows from E.1

$It/Io = 0.1 = 10^{-kt}$ or $Kt = 1$

or $K \frac{1}{t}$

The ratio Io/It is called as transmittance, I, and the ratio log It/Io is termed as absorbance A, of the sample. Previously absorbance was known as optical density, D or extinction co-efficient, E. The ratio Io/It is termed as opacity.

$$A = \log \frac{Io}{It}$$

Cells which hold the sample usually a solution, should be transparent to the incident wavelength for visible region, can be made from colour corrected fused glass. Thickness of the cell is generally 1 cm (internal distance).

For ultraviolet measurements, the cells which are to contain sample solution, should be made of materials of uniform thickness and the surface facing incident light must be optically flat. The cells should be inert to solvent and they must transmit the light of the wavelength used. Then most commonly used cells are made up of quartz or fused silica. They are readily available in matched pairs.

These days both ultraviolet and visible range spectrophotometry measurements can be made in the same instrument by changing the radiation source by a suitable switch.

7. MICROSCOPY

Human eye is not able to distinguish or resolve two points closer to 0.1 mm. Structural details of the cell can only be seen with the help of magnifying glasses or a simple light microscope. The limit of resolution (R) of a microscope can be defined as

$$R = \frac{0.61\ \lambda}{NA}$$

Where λ is the wavelength of light and NA is the numerical aparture of the lens system, and

$$NA = \underline{n} \sin \underline{u}$$

Where $\underline{n}$ is the refractive index of the medium between specimen and the lens and $\underline{u}$ is the angle of light accepted by the objective.

Light microscopy

The resolving power of a light microscope is limited. Using a good objective lens with NA of 1.40 and violet filter (λ= 400 nm) a resolving power of 0.175 µm or 175 nm can be obtained in simple light microscopes. Therefore, the cell organelles smaller than this value, such as ribosomes, membranes and microbodies are difficult to observe by a light microscope. Their resolution power can be slightly increased either by the use of Abbe's substage condenser, which projects a wide angle cone of light or immersion oil with a refractive index of 1.5. Light microscopy has following applications :

(i) Bright field microscopy

Separate light intensity lamps are used as source of light with lamp codenser and iris diaphragm. Many microscopes have built in Kohler illumination.

(ii) Dark field microscopy

Structural details of aquatic organisms can be studied by placing a central circular stop below the condenser in simple light microscopes. No light rays are gathered by the objective lens and light rays diffracted from the specimen cause a bright image in black background.

Phase contrast microscopy

It is also known as interference microscopy which enhances the contrast between the cells and their environment between internal organelles and their surroundings by the use of a specially designed optical system, which is able to cause a phase shift in light waves.

Two light waves when pass side by side through and encounter the media having same refractive index will emerge in phase and cause a greater amplitude and much brighter light. But, when one of them is encountered by a cell organelle and not the other, the two will emerge slightly out of phase, the resultant amplitude would be zero and nothing could be seen. Normally a phase shift of 0.25 λ results in partial destruction of light. The phase contrast microscope is equipped with a phase plate, which selectively retards the out of phase light wave to another 0.25 λ and a condition of completely destructive interference is obtained. This results in complete darkness which is achieved by a combination of two waves which cancel eath other. Thus the contrast at the boundaries of cells or cell organelles is increased and organelles such as mitochondria, mitotic chromosomes, nucleoli, etc., can be easily seen.

Fluorescent microscopy

Many substances when exposed to blue violet or ultraviolet light fluoresce in the visible region. The fluorescing objects appear intensely bright in uniformely dark background. Some celluaIr components such as chlorophyll and collagen possess natural fluorescence and appear in different colours. Other non-fluorescing objects can be made to fluoresce when stained with fluorescent dyes or fluorochromes.

Electron Microscopy

Light microscopes with all its variations and modifications remained a powerful research tool till 1930, but they have a limited resolution power. Greater resolutions, i.e., magnifications upto several hundred thousand times can be possible by an electron microscope. A beam of electrons with a wavelength of 0.005 nm or 0.05 A° is used in place of light to obtain a resolution nearly 5A°.

Transmission electron microscopy (TEM)

In transmission electron microscope the source of electrons is a heated tungsten filament which functions as a cathode. The electrons are then accelerated towards positively charged anode. The application of a potential difference of 100 kV produces a beam of electrons with extremely short wavelength. The beam projects downwards through a hole in the centre of the anode and passes through a column from which interfering air has been removed. Large circular electromagnets of the electron microscope column focus the beam on the specimen placed in the path. The electrons transmitted through the specimen are viewed on a screen coated with a fluorescent material such as zinc sulphide. The electrons excite this material, which produces visible light, which can be viewed directly on a screen. Electrons which encounter the electron dense regions cause an image on the viewing plate. The image can be captured directly on a film. The speed of electrons can be controlled by regulation of voltage, increasing the voltage between 50 kV to 100 kV results in resolving power of 1A° to 10A°. Magnifications upto 3,00,000 times are possible using an electron microscope which can be further enlarged photographically upto 1×10^6 times. Following techniques are involved in electron microscopy :

Fixation and staining

Osmium tetraoxide (OsO_4) has a very high electron density so it is used as a fixative in electron microscopy. Other fixatives used are $KMnO_4$ and glutaraldehyde. Ultra thin sections are required for electron microscopy, therefore a harder embedding and infiltrating material, e.g., methacrylate, epon or vestopal is used. These are obtained in a liquid form but upon heating the liquid undergoes polymerization and forms a hard plastic. More elaborate microtomes are required to obtain ultra thin (100 to 500 A° thick) section. These microtomes are equipped with diamond knives or knives prepared by fracturing glass plate in place of polished steel knives used in light microscopy.

Mounting

The sections are not generally stained as no colours are seen but are stained with electron dense materials such as uranyl acetate, uranyl nitrate and lead citrate after mounting to obtain good contrasts. The sections are mounted on small perforated discs or copper grids, which are coated with a very thin film of carbon. The beam of electrons before striking the fluorescent screen is passed through the spaces between the grid and the supporting film and the section.

High voltage electron microscopy

In TEM a potential of 100 kV acts as an accelerating voltage. Electron microscopes using a potential of approx. 3×10^3 kV were developed in 1960's in hv EM electrons produced have sufficient kinetic energy to penetrate thick specimen or whole cells. There is no need of ultra thin sections. This instrument is used for the study of the cytoskeleton and cell division under detail.

Scanning electron microscopy (SEM)

This technique has become increasingly popular after its introduction in 1960's. With this technique surface topography of a specimen may be examined in considerable detail. The resolution power of a SEM is 30 A°. The organization of SEM is very similar to TEM, but in SEM relatively low-energy beam of electrons is allowed to strike on the surface of the specimen instead of being passed through a specimen as in TEM. The electrons are scattered by the specimen and they cause secondary electrons to be emitted from the specimen. These secondary electrons are gathered by a Faraday cage. This instrument contains a scintillator which produces photons as a result of collision with the electrons. A system of coils within the column moves the primary beam across the specimen producing a series of images created by the secondary electrons. The scanning of primary beam is synchronised with the scanning of the cathode ray viewing screen. Specimens to be examined by are first coated with a metal generally a gold-palladium alloy and then mounted on a disk or sample holder which is placed in the path of electron beam. The specimen can be rotated, so it is possible to obtain views from different angles.

Scanning electron microscopy has following applications :

a. *Shadow casting* : Vapours of chromium or platinum produced by high temperature treatments are deposited on viruses or macromolecules mounted on coated grids at a precise angle. The metal is deposited cnly in front of the sample particles and not behind them. Areas which are formed by deposition of electron-dense metal appear light and areas behind the particles which are electron-transparent appear as dark shadow.

b. *Negative staining* : In this procedure a heavy metal salt such as phosphotungstic acid is used as a stain instead of metal. The electron-dense material which is used as a stain, accumulates around the object and adheres to it on drying. The electron beam will be scattered by the stain and will pass unaffected through the specimen. This produces a light object in dark background.

c. *Freeze fracture* : Recently this technique is introduced for studying the internal structure of the cell especially the membranes. In this technique cells are kept in glycerol and frozen quickly in liquid nitrogen at -196°C and exposed to a microtome knife in an evacuated chamber. They tend to fracture along the lines of weakness. The fractured plane often follows the internal membrane system. When the plane of fracture intersects the membrane of a vesicular structure or a cell organelle, the membrane splits along its centre producing two half membranes, the external, the E or exterier half faces cells external phase and the internal P or half (protoplasmic) faces the internal phase or the protoplasm.

A replica of the surface is then prepared after shadowing with a platinum and carbon mixture. The application of carbon directly from above strengthens the structure and allows it to be removed for observation under the electron microscope. The resulting images give a three dimensional impact. This technique is used in the study of the cell membranes.

Centrifugation

Particles of varying size and density which are suspended in solution can be separated, on application of centrifugal force. Under the influence of gravity the suspended particles would settle at the bottom of the container and an increasing the gravitational force, the rate at which the particle could settle would be greatly increased.

A centrifuge consists of a rotor in which samples are placed and a motor to spin the rotor. A sample in a spinning rotor is subjected to a force F, in an outward direction, the magnitude of which depends on w (omega), the angular velocity in radians of the spinning rotor per second and r, the radius of rotation, F can be expressed as :

$$F = w^2 r$$

When the value of the force applied on a sample is compared to the earth's gravitational force it is called as relative centrifugal force (RCF) expressed as number times g. The value can be derived as :

$$RCF = \frac{W^2 r}{980}$$

Since w can be expressed as revolutions per minute (rpm), RCF can be calculated if the radius or distance between the sample from the axis of rotation is known by applying

$$w = \frac{\pi (rpm)}{30}$$

There are two different types of centrifugation which are usually used,

(i) High speed centrifugation

(ii) Ultra centrifugation

H[illegible]g[illegible] speed centrifugation

[illegible]h[illegible] typ[illegible] of centrifugation is used to prepare material for additional study [illegible] ec[illegible] is called preparative centrifugation. The desk top clinical c[illegible] tr[illegible] e i[illegible] nplest example of this, with a maximum speed of 3000 rpm.

More sophisticated version of the clinical centrifuge is used for cell fractionation and sedimentation in which the temperature can be controlled and the duration of the run is automatically timed and terminated. These centrifuges have maximum speed of 20,000 rpm which can generate a force upto 50,000 g at which most micro-organisms and cell organelles can be sedimented.

Preparative centrifuges may have fixed angler rotors, in which the samples are maintained, at a constant angle (usually 15° to 45° from verticality) throughout centrifugation or swinging bucket rotor in which the line of force is parallel to the side of the tube containing the sample. The swinging bucket rotor has a central harness or the rotor, the centrifuge tubes containing the sample are inserted into the buckets which swing upward from vertical position to horizontal position.

Ultracentrifugation

These centrifuges are operated at a very high speed (upto 75,000 rpm) which could generate a force upto 5,00,000 g. These were designed and built by T. Svedberg in 1920. High centrifugal forces are applied to sediment cellular macromolecules such as proteins and nucleic acids. A unit called as Svedberg unit denoted as S is used to describe the sedimentation co-efficient of 1×10^{-13} sec. Most cellular proteins have sedimentation co-efficients between 1 and 200S. Both fixed angle and swinging bucket rotors are used in ultracentrifuges.

Differential centrifugation

This is the most common technique used in the ultracentrifuges. This technique was introduced in early 1940s by the Noble-Laureate Albert Cladue for isolating subcellular particles by stepwise removal of particles with increasing RCF. Particles suspended in the medium are sedimented at rates related to their size, shape and density.

The material to be fractionated is at first subjected to low speed centrifugation to sediment the largest particles present. The unsedimented material (or the supernatant) is then transferred to another tube and subjected to a higher speed to sediment particles of somewhat smaller size. The sequence is repeated several times unless all particles in the medium have sedimented. Each sediment is then used for further analyses.

Different cell organelles can be fractionated by differential centrifugation. Fresh tissue which is homogenized in cold buffer can be centrifuged for 10 min at 700 g to sediment the nuclear fraction. Some unbroken cells and large cell fragments may be present in this fraction. The supernatant obtained after pelleting the nuclear fraction is transferred to another tube and centrifuged at a higher speed at 2,000 g for 10 min to sediment a fraction rich in chloroplast. The chloroplast supernatant (or the supernatant obtained after sedimentation of chloroplasts) is subjected to a speed of 20,000 g for 15 min to obtain mitochondria rich fraction as a sediment. Again the mitochondrial supernatant in centrifuged at a still higher speed of centrifugation at 105,000 g for 60 min to sediment the microsomes which include ribosomes and fragments of intracellular membranes. The microsomal supernatant refers to the soluble phase of the cell or cytosol and includes soluble proteins, nucleic acids soluble polysaccharides lipid droplets and other smaller particles.

This technique has certain disadvantages as the particles sedimented first may have relatively smaller particles as contaminants; only the smallest particles which sediment most slowly are the ones to be obtained in pure form. Particles having similar sedimentation co-efficients are difficult to separate by this technique, e.g., mitochondrial fraction also contains lysosomes.

Density gradient centrifugation :

This technique was developed in late 1940s. In this centrifugation special type of rotors are required which could operate at a very high centrifugal force. Centrifuge tubes are filled in such a way that the density of the fluid decreases from base to the top of the tube in steps. It is referred to as step or dis-continuous or linear gradient. Continuous gradients are prepared by mixing solutions of high density and low density in varying proportions and delivering the mixture to the tube in an uninterrupted stream. Samples containing the mixture to be resolved is layered to the top of the gradient and centrifuged.

Various solutes used to provide solutions of density gradients, e.g., sucrose, Ficoll (a copolymer of sucrose and epichlorohydrin), Percoll (a colloidal form of silica) are most commonly used which provide a density upto 1.3 g/cm^3. Cesium chloride and other denser salts are used to obtain a considerable high density (upto 1.8 g/cm^3). Density gradient centrifugation is of two types :

(i) *Rate sedimentation*

If the density of some particles is more than the density of the solution used for preparing the density gradient, all particles reach the bottom of the tube if centrifuged for a long time. But if the centrifugation is carried out for a limited time, the particles will be distributed in the density gradient according to their sedimentation co-efficients. This is known as rate sedimentation as the particles sediment according to their size and density.

(ii) *Isopycnic sedimentation*

If the density of the densest particle is less than the density of most dense portion of the density gradient no particle could reach the bottom of the tube even on prolonged centrifugation. The particles will sediment in the gradient at a point where the density of gradient corresponds to that of the particle, i.e., at isodense or isopycnic position. Thus particle mixture can be separated according to their density.

This way cell organelles and cellular macromolecules can be separated to a large extent free from other entities.

8. Isolation And Characterisation Of Cellular Macro-molecules

Over the years a number of techniques have been developed which can separate various macromolecules present in cells and tissues. Broadly speaking some of these techniques are electrophoresis, chromatography, gel filtration, ultra centrifugation etc. For separation it is necessary that the molecules be present is a soluble form. Some difficulty is encountered in separating macro-molecules present in nuclei, cellular membranes, mitochondria, etc. where they constitute an integral part of the cell organelle. But some techniques have been developed which can separate the macromolecules from cell organelles, e.g., homogenisation, sonification, chemical procedures. Chloroform methanol mixture - or acetone have been used to isolate lipids from cells or from organelles which have been previously isolated. Urea and mercaptoethanol, chelating agents (e.g., ethylene diamine tetracetic acid, EDTA), Triton X, Sodium deoxycholate, Sodium dodecyl sulphate can be used for isolating proteins from membranes. Many proteins which are insoluble in water become soluble on addition of a small amount of salt. The interaction between ions and proteins which have charged groups in side chains makes the protein soluble in salt solution which is known as salting in procedures. But if in a solution the concentration of salts is gradually increased a point may reach at which the proteins begin to precipitate. This phenomenon is known as salting out. As different proteins precipitate at different salt concentrations, this procedure could be adopted for separation of proteins. Ammonium sulphate $(NH_4)_2SO_4$, which is highly soluble and has high ionic strength and does not denature proteins even on precipitation is the method of choice for separating the proteins. To separate proteins by this method, the salt concentration is gradually increased, the protein that would precipitate at a given salt concentration is removed by filtration or centrifugation and then salt concentration is further increased and the whole procedure repeated till all the proteins in the solution are separated.

Isoelectric precipitation

Polar groups in proteins and nucleic acids are influenced by pH. There is a definite pH at which there are equal amounts of positive and negetive charges. This pH is known as the isoelectric point (or isoelectric pH). The solubility of proteins is least at this pH the different proteins which have different isoelectric points can be easily precipitated and removed from the solution by ultra centrifugation or filtration.

Dialysis or Ultrafiltration

Grahan in 1861 described the technique called dialysis. Semipermeable membranes are used to separate solutes of different molecular size. Solute mixture is placed in a semipermeable bag which is immersed in distilled water. Molecules which are larger than the pore size of the membrane are retained in the bag and the rest diffuse out in distilled water. The size of the pores of the semi-permeable membrane can be altered by chemical or physical means. This property can be used to separate molecules of different sizes by the method of dialysis.

But generally dialysis is used to separate low molecular weight substances with less than 5000 molecular weight from larger molecules. In ***Ultrafiltration*** smaller molecules along with the solvent are forced out through the semipermeable membrane. This process has an advantage over dialysis in that substances which cannot pass through the semi-permeable memmbranes are also concentrated.

Electrophoresis

Under the influence of an electric potential the particles which are charged migrate through a liquid or semisolid medium and this process is known as electrophoresis. Now it is referred to any technique which can separate particles of different net charges, size and shape under the influence of electrical potential gradients. Proteins which have charges on the side chains, and polynucleotides can be separated by this process. If two electrodes are immersed in an electrolytic solution containing macro-molecules of different net charges. The charged macro-molecules will proceed towards the opposite pole with a force proportional to applied electrical potential. If applied electrical potential is constant, the velocity of movements of different charged macro-molecules will be different. In addition, the movement of macro-molecules will be influenced by their moleculr size and shape and nature of the medium through which macro-molecules are migrating.

Electrophoratic mobility (usually expressed as cm^2 Sec^{-1} $volt^{-1}$) is a term used to express the rate at which proteins migrate towards positive or negative electrode. The alteration in pH will alter the charge of the protein particles. The concentration of H^+ is increased when the pH is lowered and that of OH is increased when the pH is increased. On lowering the pH, concentration of H^+ increases and negative charged groups become protonated consequently negative charges are neutralised. At low pH, protein will carry a net positive charge and migrate towards the cathode whereas at high pH the proteins will migrate to the anode.

Moving Boundary Electrophoresis

It is known as moving boundary electrophoresis. A mixture of proteins of different types can be separated by dissolving this mixture in an appropriate buffer solution which maintains the pH of the electrolyte. This mixture of proteins in a buffer is placed in a glass tube which is connected to electrodes. The proteins migrate to the cathode or anode according to the charge they hold. Different proteins would migrate at different rates. This is known as moving boundary technique and has been used for analysis of a complex mixture of proteins.

Zone Electrophoresis

This technique is both analytical and preparative. The resolution of proteins is better than that in moving boundary electrophoresis. In this technique the proteins are speparated in a semisolid or porous supporting medium,

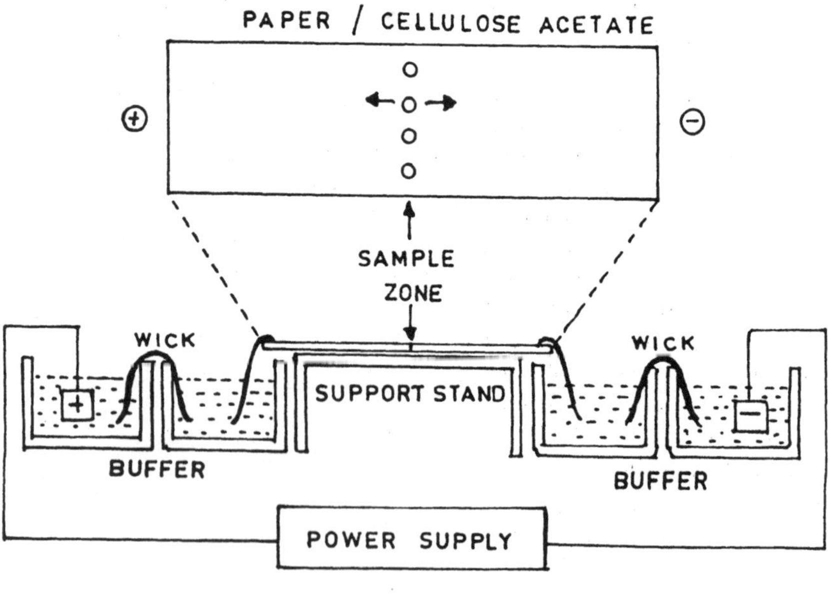

Fig. 4. Different components of zone electrophoresis assembly (diagrammatic)

e.g., filter paper, gels of various sorts, polyacrylamide, cellulose acetate etc. when filter paper was used as supporting medium, the filter paper strip is saturated with electrolyte solution, usually a buffer, and suspended between two baths which contain buffer solution Fig. 4. These baths are connected with two outer baths which contain the electrodes. The electrodes are inserted into outer baths which contain buffer solution and thus the pH in the filter paper and inner bath remain more or less constant. On applying an electrical potential the proteins which are put in the centre of the filter paper migrate through the filter paper towards the appropriate electrode and a number of discrete zones are formed. These zones can be identified by suitable staining. The separated proteins can also be eluted.

These zones can be further separated by what is known as two dimensional electrophoresis. After the separation made in zone electrophoresis, the supporting medium is rotated 90°, the pH is changed and electrophoresis carried on again. Thus, the components are resolved much better than in the simple zone electrophoresis.

Disc electrophoresis

This form of electrophoresis was developed by Ornstein and Davis (1960). It is carried out in polyacrylamide gel. The concentration of polyacrylamide gel would determine the pore size. This gel is spread in sheets or is contained in columns. The gel also acts like a molecular sieve. The electrophoretic separations are made on the basis of the size of pores and charge of the constituents which are to be separated. In disc electrophoresis the polyacrylamide gels are in the form of cylindrical blocks. A mixture of proteins which has to be separated is loaded into the region which has a low concentration of polyacrylamide so that the pore size is large and does not hinder the movement of any protein. This region of polyacrylamide gel may be called 'sample gel'. Next follows the 'stacking gel' which is formed by the same concentration of polyacrylamide as in the sample gel but to start with is free of proteins. This is followed by what is known as separating gel has a higher pH 9.5 than in sample gel and stacking gel which are maintained at pH of 8.3. This region of the gel has also a higher concentration of polyacrylamide. The two ends of cylinders containing the polyacrylamide gel are immersed in reservoirs as seen in the figure 5. When an electrical potential is applied ultimately the mixture of proteins is separated into individual proteins each of which forms a sharply defined band. The separated protein bands can be identified by spraying suitable dyes or by chemical reactions or physical methods. Then the column polyacrylamide in which proteins have separated is treated with substrates of an enzyme reaction. The enzyme proteins react with the substrates and give coloured reaction. In this way many isoenzymes of enzyme protein can be identified. These protiens could also be eluted for further study.

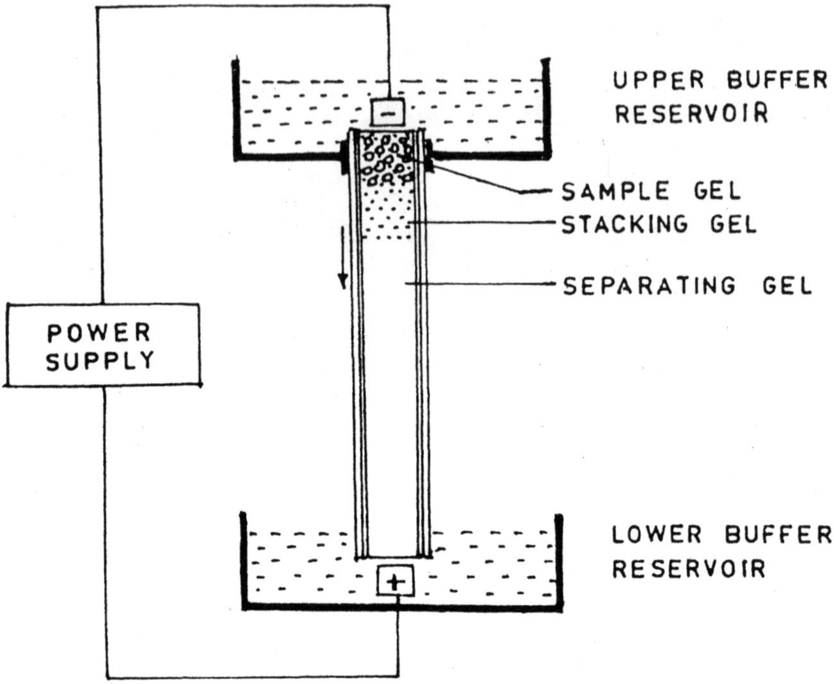

Fig. 5. Various components of discontinuous gel electrophoresis (digrammatic)

SDS gel electrophoresis

Sodium dodecyl sulphate (SDS) is an ionic detergent which can solublise membrane proteins. These protein can be separated by electrophoresis. Membrane proteins are not usually soluble in aqueous solution. SDS is incorporated in electrophoresis gel. Then electrophoresis is carried out which separates individual proteins.

Iso electric focussing

This is a special form of electrophoresis. A mixture of proteins can be separated into individual proteins on the basis of the differences in the isoelectric points. It is extremely sensitive method. A difference of 0.0025 units in the isoelectric pH, is sufficient to separate proteins by this process.

Paper partition chromatography

It is a technique of separating different solutes in a mixture. The differ solutes discrete zones on the basis of differences in partition of solutes between aqueous phase, which is stationary and is bound to the filter paper and mobile organic liquid phase. The liquids in the mobile phase passes through the filter paper by capillary action. In actual practice a large rectangular sheet of filter paper is saturated with aqueous phase and allowed to dry in the air. The sample is applied as a narrow zone, on a line drawn near one of the edges of the paper. This paper is then hung in a chromatographic chamber which has been saturated with vapours of mobile organic phase. The edge of the paper containing the sample is immersed in a bath containing the solvent mixture. The solvent mixture slowly percolates through the paper. The movement of the solvent may be downward (descending chromatography) or upward (ascending chromatography). As the solvent mixture advance and nearly reaches the opposite edge of the paper (the side which does not contain the solute mixture) further flow of the solvent is stopped. The filter paper is taken out from the chamber and dried in air. The separated solutes can be located on the filter paper by physical methods or spraying solution which reacts with the solutes and forms a coloured compound. If the solutes are left far behind the solvent front it may be desirable to allow the solvent to run off the edge of the paper. This will allow solutes to move to a longer distance and the solutes will be separated into component parts. The ratio between the distance travelled by each constituent of the solute mixture and the distance travelled by solvent front is known as Rf. This applies to the condition in which the solvent is not allowed to run off the edge of the chromatographic paper.

If the chromatographic sheet is dried in air and rotated 90° and then chromatographed a second time, using a different solvent system and chromatogram is developed. Solutes which could not be separated using the first sovlent system are likely to be separated in a second sovlent system is used. This process in known as two dimensional chromatography.

Thin layer chromatorgraphy

It is a modification of paper partition chromatography. Glass plates are used instead of filter papers. Glass plate is covered with thin a uniform layer of the adsorbent in water. This plate is dried. The sample is applied in a thin line near one end of the long axis of the glass plate. When the sample has dried, the plate is supported vertically so that the end near the sample line is immersed in a tray containing a shallow layer of the solvent as is done in the ascending chromatography. The solvent rises in the plate by capillary action and the sloutes become distributed on the plate according to their partition in the stationary and mobile phases.

When the solvent front had reached atleast half the length of the plate it is taken out from the chromatographic chamber and dried in air. If the constituent parts of the solute can fluroesce in UV they can be recognised on exposing the plate to ultraviolet light. If the separated zones, do not fluoresce in ultraviolet light, the constituent parts of the solute, can be made visible by spraying the plate with a suitable reagent and drying. The separated solute particles can be eluted for further examination.

Ion exchange column chromatography

It is carried out in a tall column made of glass which is packed with a synthetic resin-cation exchanger or anion exchanger. Resin which have negative charges are known as cation exchangers and those which have positive charges are known as anion exchangers. When a solution of proteins or nucleic acids passes through a column of resins, the constitutent proteins or nucleotides compete with each other for charged sites on the resin. The movement of constituent proteins or nucleotides in the mixture would be determined by their affinity to charged sites of resins. This is the basis of the separations of proteins and nucleic acids. Diethylaminoethyl cellulose (DEAE-cellulose) an anion exchanger and carboxymethyl cellulose (DEAE-cellulose) a cation exchanger are usually used in the ion exchange chromatography. At an alkaline pH, aspartic acid, glutamic acid etc., would form a number of bonds with DEAE cellulose. A change in pH would alter the affinity of a protein for the resin. When the pH is lowered, hydrogen will bind to the side chains of proteins which bore negative charges and the resin particles would be displaced. Addition of salt solutions will produce similar effect. The procedure used in the separation of proteins on a resin column is briefly described. A slurry of resins is made in a buffer, or in a salt solution. This is poured in a glass column. At the base of the glass column a glass filter is fixed. The mixtures of proteins is suspended in the same solution in which a slurry of resin was made. This is applied in narrow zone at the top of the resin column. Some proteins present in the mixture would form electrostatic bond with the resin particles. When a solution of different pH is poured at the top of the resin column, the proteins which had formed electrostatic bonds with resins will be released from resins, while other proteins will form electrostatic bond with resins. The pH of the solution and the strength of salt solution will determine which proteins will form electrostatic bonds with the resins. When a fresh solution of different pH or salt strength is poured on the resin column, the electrostatic bond between proteins and resin will be broken and fresh bonds would be formed with different proteins. This will happen throughout the length of the resin column where some proteins will form transient bonds with resin. Thus some proteins will reach the bottom of the resin column and will be eluted in a more or less pure form. This process of ion exchange chromatography could be repeated with isolated proteins or recourse taken to separation by electrophoresis. In a manner similar to separation of proteins, the nucleic acids could also be separated by this technique. *Affinity chromatography* is a form of column chromatography. In this process proteins or nucleic acids are retarded in their passage through the column because of specific biological reaction having taken place between some proteins or nucleic acids and column matrix. This specifically may be due to antigen antibody reaction or may be due to interaction of an enzyme with its substrate, allosteric effector or coenzyme or due to some similar reaction. The column is packed with gel particles. These particles are covalently coupled to specific biological material. When a solution in which proteins or nucleic acids are dissolved or suspended is poured at the top of the column some proteins or nucleic acids which have a high affinity with column matrix are bound, and the rest of the constituents are allowed to pass as such. The bound molecules are released when a solution of different pH or different salt strength is applied and get eluted in a more or less pure form. This technique can be used to separate specific enzymes of specific immunoglobins or messenger RNA.

Gel filtration (Molecular seiving)

This technique was introduced by Porat and Flodin in 1959. The molecules of different size and molecular weights are separated by this technique. Gels produced by reaction of dextrans with epichlorohydrin or acrylamide are used. These gels can be produced in a variety of pore size. The dry gel particles are first put in a buffer or water. These particles will get swollen and then they are packed into a glass column. The solvent in the column is present in two phases. One phase is the solvent surrounding the gel particles and the other is the solvent present in the gel particles. The passage of solute particles through the column is determined by their size and molecular weight. Molecules whose size is greater than the size of the pore in the gel cannot pass into the gel phase and can be said to be above the exclusion limit of the gel. Smaller molecules which are within exclusion limit can enter the gel phase. When a mixture of molecules of different sizes and molecular weight is applied at the top of the column, molecules which are above the exclusion limit of the gel pass through the column rapidly. But molecules below the exclusion limit pass through pores in the gel. The moelcules of lower molecular weight and size will be retained by the gel particles for a long time and emerge after the particles of higher molecular weight and size have emerged from the bottom of the column. Cross linked dextrans can be obtained with a variety of exclusion limits and can be used for separation of proteins, nucleic acid and polysaccharides. This technique can be used to determine also the molecular weight of substances.

Gas chromatography

This may be used to separate amino acids, lipids, oligosaccharides etc. after these compounds have been converted into gaseous form. This is a form of column chromatography in which a gas is used in a mobile phase. The stationary phase could be a liquid (gas-liquid chromatography) referred to as GLC or a solid (gas-solid chromatography) referred to as GSC. The separation can be accomplished by the differential partition or adsorption of different molecules between stationary phase (liquid or solid) and the gas phase. The basic components of the gas chromatograph are (i) source of a gas (ii) a chamber through which sample could be introduced (iii) chromatographic chamber and (iv) the recorder as seen in the diagram (Fig. 6).

The gas usually N_2 or CO_2 is contained in a metal cylinder which is connected to the column through a metal tubing. Various devices are used to accurately regulate the flow of gas. The sample is introduced through a

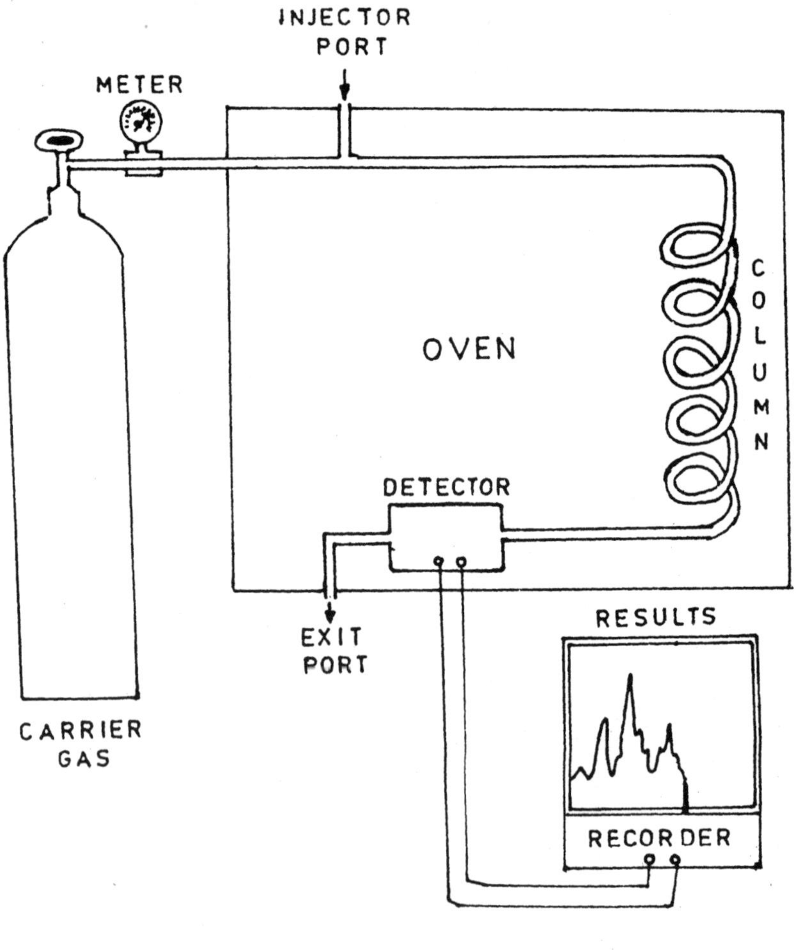

Fig. 6. Digrammatic representation of gas chromatography apparatus

microsyringe in the sample chamber which has a diaphragm which gets sealed automatically after the insertion of the needle of the microsyringe. The sample chamber is enclosed in heating block so that the sample is converted or maintained in a gaseous form. The sample in the gaseous form passes through the chromatographic columns which are made of metal or preferably stainless steel. As the length of these columns is several feet these columns are enclosed in the chamber in the form of a spiral. For GLC the packing material is an inert material, e.g., kieselgur which is impregnated with a suitable liquid which has a low volatility so that the liquid will not be emitted from the column along with gas. For GSC the packing material is an adsorbent. The molecules of different substances contained in the sample will pass through the column at different rates according to their partition of adsorption characteristics. The separated components pass through a detector. This detector relays electrical signals proportional to amount of the separated molecules. Flame ionisation detector is most common form of detectors.

9. Radioactive Isotopes

Those chemical elements which have the same atomic number but different atomic masses are called isotopes. Some of the known isotopes are unstable and undergo spontaneous nuclear changes (transmutation) accompanied by the emission of particulate and sometime with electromongenetic radiations. These isotopes are radioactive and are known as radiosisotopes or radionuclides. The use of radioisotopes have helped in solving many problems as these elements can be conveniently employed as tracers. In case the radioisotope is a part of a molecule, then the fate of the entire or part of that molecule can be followed. Results obtained from experiments involving the use of radioisotopes could provide quantitative information. Each radioactive isotope decays at a characteristic rate and therefore has a specific half life, which should be taken into account in the use and measurement of the radioisotopes.

Detection and Measurements of Radiation

On the basis of the type and energy of the emitted radiation by the isotope, various instruments can be used to detect and measure these radiations. The most commonly used detectors are :

(i) *Geiger, Muller counters* (GM counters) : these are employed primarily in isotopes emitting beta particles of intermediate or high energy and can also be used at low efficiency for the measurement of gamma radiation.

(ii) *Solid Scintillation counters* : these are usually employed with gamma ray emitting isotopes.

(iii) *Liquid Scintillation counters* : these are employed with isotopes emitting low energy beta particles.

The most common type of radiations are alpha particles, beta particles and gamma rays. Only those isotopes which emit netative beta particles or gamma rays are usually used by biologists. The energy acquired by a beta particle during nuclear transmutation may vary over a specific range, whereas gamma rays have discrete energy values. Beta particles and gamma rays interact with matter by ionizing or exciting atoms in their path and it is this property that is used in instruments designed to detect and measure the radioisotope present in a sample.

Geiger Muller Counters

The detector is also called as GM tube, consisting of a long cyliner containing two electrodes and filled with a readily ionisable gas, e.g., helium or argon. The metallic internal surface of the cylinder serves as a cathode. A narrow wire passing down the centre end of the G M tube serves as an anode. A thin material such as myler plastic or mica, forms the end window at one end of the G M tube. The anode and cathode terminals are connected to high voltage source and a scaler, which counts electrical pulses. When the radiation from a radioactive isotope enters a GM tube some of the gas molecules are ionised and form a number of ion pairs. On application of sufficient high electric potential to the electrodes, the ion-pairs will migrate towards the appropriate electrode. During this migration, the ions collide with and ionize additional gas molecules, so that a large number of ions are collected at the electrode. These events produce an electrical pulse that is recorded by the scaler as a count. Each ionizing ray entering the GM tube is registered as count and the amount of radioactivity is expressed as counts per minute (cpm).

Several isotopes, e.g., ^{3}H, ^{14}C, ^{35}S, ^{45}Ca etc. emit energy of particles and can be detected by G M counter.

A small count is recorded even when no radioactive isotope is placed below the window of the G M tube. This is known as background count and results from the cosmic radioaction (primarily gamma rays). This background count is always subtracted from the counts obtained for a radioactive sample. The basic units of measurement is the Curie (Ci) and is defined as the quantity of radioisotopes undergoing 2.22×10^{12} dpm (disintergration per minute). In most tracer experiments the quantity of radioisotope is generally at the level of millicurie (mCi) or micro-curie (μCi).

Solid Scintillation Counter

This is used to detect and measure radioisotopes emitting gamma rays. A large crystal of thallium activated sodium iodide and a photo-multiplier tube encased in an aluminium housing serves as a detector. The detector is interfaced with a pre-amolifier tube, a source of high voltage and a scaler. The sample is placed into a well shaped opening in the crystal. The crystal absorb energy of the gamma rays, and raises electrons of atoms composing the crystal, to higher energy orbitals. These electrons return again to their lower energy orbital and flashes of light or scintillation are emitted, proportional to the number and energy of the gamma rays existing the crystal. The light photons are converted by the photomultiplier tube into electrical pulses of corresponding magnitude and frequency and these are relayed to the scaler. In contrast to G M counters few problems involving self absorption are incurred when solid scintillation methods are used.

Liquid Scintillation Counter

The energies of the beta particles emitted by the hydrogen (^3H) and carbon (^{14}C) containing compounds, could be detected by liquid Scintillation counter and cannot be counted accurately in G M counters. In this case, the radioactive sample is mixed in a glass or plastic vial with a "Scintillation fluid." Certain molecules dissolved in the fluid indirectly absorb the energy dissipated by the beta particles and respond by scintillating. The resulting photons of light are detected by photo-multiplier tubes that relay proportional electrical pulses to the scaling units. The immediate advantage of the procedure is that if the sample is thoroughly mixed with the Scintillation fluid no self absorption occurs. The scintillation fluid contains three major components (i) an organic solvent constituting the bulk of the mass; (ii) a primary solute or fluer (iii) a secondary solute or fluer. The energy of the beta particles is transferred to the solvent, causing ionization or excitation of solvent molecules. The electrical pulse produced by the photomultiplier tube depends upon the energy dissipated by the beta particles in the scintillation fluid and is proportional to it.

10. Enzyme Studies

Enzyme activities cannot be measured accurately in intact plants. It is, therefore, necessary to disrupt the tissue or cell. Various techniques are being used to disrupt tissues such as grinding, mechanical shaking french press, osmotic shock etc. For cell disruption in unicellular systems, cells suspended in aqueous environment are subjected to high frequency sonic oscillation which cause the cell breakage. The homogenization is generally carried out by grinding in a glass pestle and mortar for comparative studies but a homogeniser could be used.

Enzymes are relatively fragile. They are inactivated under unfavourable conditions. The conditions of their stability vary in different enzymes. To avoid thermal inactivation during grinding the homogenization should be carried out at low temperature (0-4°C). The media of homogenization should be buffered according to the pH optimum of the enzyme. Centrifugation at low temperature should be preferred to filtration of the crude enzyme extract. Assay of enzymes should follow their extraction or the enzyme solutions could be stored in a deep freeze till their activity can be estimated. Some enzymes lose their activity even in a deep freeze.

In micronutrient studies an almost quantitative assessment of enzyme activity can be made in their crude solution form and it is not always necessary to purify the enzymes, as during the process some part of the enzyme may be lost and the extent of this loss may vary in different samples.

Methods of enzyme assays

Two types of methods are generally followed for measuring the enzyme activities (i) sampling and (ii) continuous methods. The sampling method is based on the chemical estimation either of the substrate or the product of the reaction at a fixed time. If the enzyme reaction is simple either of the two methods can measure the enzyme activity. But if the reaction is a two stage reaction disappearance of the substrate gives a correct measure of the enzyme activity. In the continuous method a number of readings are taken at fixed intervals. For measurement of the rate of reaction it is necessary that the reaction should be carried out in a water bath controlled thermostatically and all reactants used in the reaction should be brought to the required temperature before starting the assay. Adequate buffering should be provided and the pH should be at nearly optimum level. Some of the methods commonly used for enzyme assays are given below :

1. *Spectrophotometric methods* : Many substrates and products of enzymes reactions absorb light either in the visible or ultraviolet region of the spectrum. The substrate and the products of reactions do not ordinarily have identical absorption spectrum.

The progress of reaction can be measured in a Beckman Spectrophotometer or a similar instrument (which is the most conveniently and widely applicable to all the methods of following enzyme reaction). The spectrophotometric methods of estimating the activity of enzymes are preferred to other methods on account of their simplicity and sensitivity. Besides, no reagent is required except the actual reactants.

According to Dixon and Webb (1979) "An unit of enzyme is defined as that amount which will catalyse the transformation of one micromole of substrate per minute or where more than one bond of each substrate molecule is attacked, one micro equivalent of the group concerned per minute under defined conditions".

In some oxidising enzymes the second reactant (hydrogen acceptor) undergoes change in absorption when reduced by the substrate, e.g., the reduced forms of NAD and NADP can be measured at 340 nm. Flavoproteins can be measured in oxidised form at 450 nm, reduced froms of cytochromes at specific wave length in the visible region. Therefore, the assay of oxidizing enzymes can be accurately followed spectrophotmetrically.

2. *Fluorescence methods* : The measurement of fluorescence is more sensitive than absorption measurement. Flavin compounds fluoresce strongly in oxidized form and lose fluorescence after reduction. NAD and NADP on reduction show blue fluorescence and no fluorescence in oxidised form.

3. *Manometric methods* : This method can be followed if one of the components of the enzyme reaction is a gas. This is used for the measurement of O_2 uptake in oxidases and CO_2 output in decarboxylases and for some other similar reactions.

4. *Chemical methods* : Many enzyme reactions are followed by withdrawing a sample at fixed intervals and estimating the amount of substrate or product in the sample by chemical methods. For example, the colorimetric method of Fiske and Subba Row (1925) is generally followed for the estimation of inorganic phosphate while measuring the activity of phosphatases, phosphorylase, nuclectidases, kinases, etc. Acyl-phosphates can be estimated by the method of Ochoa *et al.* (1951). In some enzyme reactions the reactions or products can be estimated chromatographically.

Enzyme assay : For the assay of enzyme activities fresh leaves are ground in a glass pestle and mortar using purifed sand as an abrasive in an appropriate medium at a temperature not exceeding 4°C.

Catalase : The enzyme activity is measured by a modification of the method of Euler and Josephson (1927). The assay mixture contains 10 ml substrate mixture (0.5 m moles H_2O_2, 100 μ moles phosphate buffer pH 7.0 previously standardised against 0.1 N $KMnO_4$), is stabilised at 25°C. The reaction is initiated by the addition of 1 ml suitably diluted enzyme and the whole mixture is shaken vigorously. The reaction is allowed to proceed for 5 min and then stopped by 2 ml 5N H_2SO_4. Corresponding blanks are run simultaneously with sulphuric acid added before the addition of enzyme extract. The amount of H_2O_2 reduced by the enzyme is determined by titrating the reaction mixture against 0.1 N $KMnO_4$. The enzyme activity is expressed as m moles of H_2O_2 decomposed per mg protein.

Peroxidase : The activity of enzyme is measured as an increase in OD at 475 nm as a result of the oxidation of p-phenyldiamine in a reaction mixture (Luck, 1963) containing 500 μ mol phosphate buffer pH 6, 0.001% (w/v) H_2O_2 and 0.05% (w/v) p-phenyldiamine in a final volume of 10 ml, after 5 min of enzyme addition and expressed as the change in OD mg^{-1} protein.

Polyphenol oxidase : The enzyme extract is prepared in 0.1M phosphate buffer pH 6.0 containing 10 mg/ml PVP (polyvinyl pyrealidone) and is centrifuged at 2°C for 15 min at 800xg. The enzyme activity is assayed in the supernatant (Henry and Richards, 1979) using DL-DOPA as substrate and the activity is expressed as the change in OD per min at 470 nm.

Cytochrome c oxidase : The enzyme is assayed in mitochondrial fraction (Pierpoint, 1959) by the method of Cooperstein and Lazarow (1951). The reaction is initiated by adding 0.045 μ mol of reduced cytochrome c to the reaction mixture containing 0.1 m mol phosphate buffer pH 7.5, 0.03% (w/v) digitonin and 0.1 ml enzyme extract in a final volume of 3 ml. The change in OD per min at 550 nm was measured at 15 sec intervals. The initial rate of reaction was calculated according to the method of Yonetani (1962) and expressed as μ mol cytochrome c oxidised per min per mg protein.

Nitrate reductase : Enzyme is assyed *in vitro* by a modified method of Sanderson and Cocking (1964) in an enzyme extract of fresh leaves prepared in 0.1M Tris-HCl buffer, pH 7.6 containing 1mM EDTA and 1mM cysteine after centrifugation at 2°C. The suitable diluted enzyme extract was added to an assay system containing 0.1 m mol phosphate buffer pH 7.6, 0.01 m mol KNO_3 and 0.3 μ mol NADH and after 20 min incubation at 25°C the reaction is stopped with 0.2 ml 1M zinc acetate and 4 ml ethanol. Appropriate blanks without NADH are run simultaneously. Nitrite was estimated in an aliquot drawn from the supernatant of the reaction mixture after centrifugation at 800 g according to the method of Snell and Snell (1949). The activity of the enzyme is expressed as n mol nitrite formed h^{-1} mg^{-1} protein.

Acidphosphatase : The assay mixture (Schmidt, 1955) contains 50 μ mol acetate buffer pH 5, 0 μ mol sodium B glycerophosphate in 0.6 ml. The reaction is stated by adding 0.4 ml of suitably diluted enzyme extract and is stopped by the addition of 1 ml 10% TCA after 20 min. Inorganic phosphorus thus formed was estimated colorimetrically (Fiske and Subba Row, 1925). For each sample a blank at zero time is run simultaneously and the temperature is maintained at 30°C.

Ribonuclease : The reaction mixture (Tuve and Anfinsen, 1960) contains 0.1 m mol citrate buffer pH 6.0 and 0.125% (w/v) RNA in 2.0 ml. The reaction is started by adding 0.5 ml homogenate (10%) and is stopped after 30 min by adding 0.5 ml of 0.75% uranyl acetate in 25% (w/v) perchloric acid. The OD is measured at 260 nm after suitable dilutions of the supernatant obtained by centrifugation at 800 g at 2°C. Simultaneous blanks are also run, in which uranyl acetate is added prior to the addition of enzyme extract and the substrate.

Invertase : The enzyme activity is measured by a modification of the method of Hatch and Glasziou (1963). The assay mixture final volume 2.5 ml contains 1 ml homogenate, 100 μ mol citrate buffer pH 5.4 and 10 μ mol sucrose solution. The reaction is allowed to proceed for 30 min at 30°C and was stopped by 1.5 ml saturated solution of sodium sulphate and 1.0 ml hot ethanol. Inverted sugars are measured in a suitable aliquot colorimetrically by the method of Nelson (1944). The activity is expressed as μ mol inverted sugars formed per mg protein.

Adenosine triphosphatase : The homogenate of fresh leaf tissue is prepared in 0.1M Tris-HCl buffer pH 7.5 containing 0.5M sucrose and 1mM EDTA and is centrifuged at 20,000 g at 2°C for 30 min and suitable aliquots from the supernatant are drawn for the assay of enzyme, the reaction mixture contains 0.1 m mol Tris-HCl buffer pH 7.5, 50 μ mol KCl/Mg/$CaCl_2$ (depending on the ion involved), 10 μ mol ATP in 3.5 ml final volume. The reaction is run at 38°C for 30 min and is stopped by the addition of 1.0 ml 20% (w/v) TCA. Phosphorus is estimated in a suitable aliquot from the supernatant after centrifugation (Fiske and Subba Row, 1925). The activity is expressed as μg Pi released per hour.

Carbonic anhydrase : For the assay of the activity of carbonic anhydrase enzyme homogenate is prepared in 0.02M veronal buffer pH 8.15 containing 0.1M 2-mercaptoethanol and 1mM EDTA.

The activity of the enzyme is measured on the basis of time taken for the reduction in pH of veronal buffer containing bromothymol blue by cold CO_2 saturated water at 0°C in the presence and absence of the enzyme (Rickli *et al.*, 1964). Unit of carbonic anhydrase activity is expressed as $10 \times \frac{(to - te)}{te}$, where to = time taken with the enzyme extract and te = time taken without the enzyme extract (blank).

Protein estimation in enzyme extracts : Protein of enzyme extracts is precipitated with an equal volume of 20% TCA. After centrifugation, supernatant is discarded and protein is estimated in the residue by the method of Lowry *et al.* (1951). After removing the pigments from the residue, it is solubilised in 0.1N NaOH for 10 min at 80°C in a waterbath. Protein is estimated in a suitable aliquot by treating it with Folin Ciocalteau reagent and the colour intensity is measured at 660 nm. Results are expressed as per cent in fresh weight.

ACKNOWLEDGEMENTS

The authors would like to thank Professor C.P. Sharma, Head, Department of Botany, Lucknow University, Lucknow for allowing us to use the departmental facilities. The authors would also like to thank Dr. (Mrs.) Nirmala Nautiyal, Dr. (Mrs.) Neena Khurana, Dr. (Miss) Pratima Sinha and Dr. (Miss) Radha Jain for their help in the preparation of the manuscript.

REFERENCES

Agarwala, S.C. and Sharma, C.P. 1976. Pot and sand culture technique for the study of the mineral nutrient deficiencies under Indian condition. *Geophytol.*, **6** : 356-367.

Ahmed, S. and Evans, H.J. 1960. Cobalt : a micronutrient element for the growth of soybean plants under symbiotic conditions. *Soil Sci.*, **90** : 205-210.

Ahmed, S. and Evans, H.J. 1961. The essentiality of cobalt for soybean plants grown under symbiotic conditions. *Proc. Natl. Acad. Sci. U.S.*, **47** : 24-36.

Arnon, D.I. 1940. Vitamin B_1 in relation to the growth of green plants. *Science*, **92** : 264.

Asher, C.J. and Edwards, D.G. 1983. Modern Solution Culture Techniques. In : *Encyclopedia of Plant Physiology. New Series Vol. 15A.* (A. Lauchli and R.L. Bieleski eds.) Springer-Verlag, Berlin, New York.

Bates, T.E. 1971. Factors affecting critical nutrient concentration in plants and their evaluation : review. *Soil Sci.*, **112** : 116-130.

Bould, C. 1968. Leaf analysis as a diagnostic method and advisory aid in crop nutrition. *Exp. Agric.*, **4** : 17-27.

Bouma, D. 1983. Diagnosis of Mineral Deficiencies Using Plant Tests. In : *Encyclopedia of Plant Physiology, New Series Vol. 15A.* (A. Lauchli and R.L. Bieleski, eds.) pp. 120-141. Springer-Verlag, Berlin, New York.

Bouma, D. and Dowling, E.J. 1962. The phyisological asssessment of the nutrient status of plants. I. Preliminary experiments with phosphorus. *Aust. J. Agric. Res.*, **13** : 791-800.

Bouma, D. and Dowling, E.J. 1966. The physiological assessment of the nutrient status of plants. II. The effect of the nutrient status of the plant with respect to phosphorus, sulphur, potassium, calcium or boron on the pattern of leaf area response following the transfer to different nutrient solutions. *Aust. J. Agric. Res.*, **17** : 633- 646.

Bouma, D. and Dowling, E.J. 1967. The physiological assessment of the nutrient status of plants. III. Experiments with plants raised at different nitrogen levels. *Aust. J. Agric. Res.*, **17** : 647-655.

Cooperstein, S.J. and Lazarow, A. 1951. A micro-spectrophotometric method for determination of cytochrome oxidase. *J. Biol. Chem.*, **189** : 665-670.

Dixon, M. and Webb, E.C. (Ed.) 1979. Enzymes, Longman Group Limited, London.

Euler, H.V. and Josephsen, K. 1927. Uber Katalase I. Justus Liebigs Annden der chemie, **452** : 158-87.

Fiske, C.H. and Subba Row, Y. 1925. The colorimetric determination of phosphorus. *J. Biol. Chem.*, **66** : 275-400.

Gautheret, R.J. 1939. Surla possibilite de realiser la culture indefinie des tissu de tubercules de carotte. *C.R. Acad. Sci.* (Paris), **208** : 118-21.

Goodall, D.W. and Gregory, F.G. 1947. Chemical composition of plants as an index of their nutritional status. Tech. Commun. 17, Imp. Bur. Hortic. Plant Crops, E. Malling.

Gregory, F.G. 1943. Report on experiments on minor element deficiencies in flax in water culture carried out in 1943. Rept. Mineral Deficiencies Conf. Agric. Res. Coun. 7221.

Hatch, M.D. and Glaszion, K.T. 1963. Sugar accumulation cycle in sugarcane. II. Relationship of invertase activity to sugar content and growth rate in storage tissue of plants grown in controlled environments. *Plant Physiol.,* **38** : 344-348.

Hayward, H.E., Long E.M. and Uhvitis, R. 1946. Effect of chloride and sulphate salts on the growth and development of the Elberta Peach on Shalil and Lovell rootstroks. Tech. Bull. U.S. Dep. Agric. 922.

Henry, F.W. and Richard, L.W. 1979. A study of the effect of applied enthophom on enzyme (Polyphenol oxidase, peroxidase, catalase & ATPase) activity in tabacoo (*Nicotiana tabacum*) apical tissue chloroplasts. *Zeitschrift fur Pflanzenphysiologie,* **92** : 11-22.

Hewitt, E.J. 1966. Sand and Water Culture Methods Used in the Study of Plant Nutrition. Tech. Commun. No. 22. Commonwealth Bureau of Horticulture and Plantation Crops, East Malling, Maidstone, Kent.

Johnson, C.M., Stout, P.R., Broyer, T.C. and Carlton, A.B. 1957. Comparative chlorine requirements of different plant species. *Plant Soil,* **8** : 337.

Katsumi, M. and Fukuhara, M. 1969. The activity of α-amylase in the shoot and its relation to gibberellin-induced elongation. *Physiol. Plant.,* **22** : 68-75.

Lewis, J.C. and Powers, W.L. 1941. Anatagonistic action of chlorides on the toxicity of iodides to corn. *Plant Physiol.,* **16** : 393.

Lowry, O.H., Rosebrough, N.J., Farr, A.L. and Randall, R.J. 1951. Protein measurement with Folin phenol reagent. *Journal of Biological Chemistry,* **193** : 265-275.

Luck, H. 1963. Peroxidase. In Methods of enzymatic analysis. Ed. H.U. Bergmeyer. Academic Press, New York, pp. 95-7.

Mitchell, R.L. and Scott, R.O. 1947. Concentration methods in spectrographic analysis. II. Recovery of trace constituents in plant materials and soil extracts by mixed organic reagents. *J.Soc.Chem.Ind.Lond.,* **66** : 330.

Nobecourt, P. 1939. Sur la perennita et l' angmentation de volume des cultures de tissue vegetaux. *C.R. Soc. Biol.* (Paris), **130** : 1270-1271.

Ochoa, S. Stern, J.R. and Schneider, M.C. 1951. *J.Biol.Chem.,* **193** : 691.

Pierpoint, W.S. 1959. Mitochondrial preparations from the leaves of tabacco (*Nicotiana tabacum*). *Biochem. J.,* **71** : 518-528.

Piper, C.S. 1941. Marsh spot of peas. A manganese deficiency disease. *J.Agric. Sci.,* **31** : 44.

Randall, P.J. 1969. Changes in the nitrate and nitrate reductase levels on restoration of molybdenum to molybdenum-deficient plants. *Aust.J.Agric.Res.,* **20** : 635-642.

Reuter, D.J. and Robinson, J.B. 1986. Plant Analysis. An Interpretation Manual. Inkata Press, Melbourne, Sydney.

Rickli, E.E., Ghazanfar, B.H., Gibbons, B.H. and Edsall, J.T. 1964. Carbonic anhydrase from human erythrocytes. Preparation and properties of two enzymes.

J.Biol.Chem., **239** : 1065-1078.

Sanderson, G.W. and Cocking, E.C. 1964. Enzymatic assimilation of nitrate in tomato plants. 1. Reduction of nitrate to nitrite. *Plant Physiol.*, **39** : 416-422.

Schmidt, G. 1955. Acid Prostatic Phosphatase. In Methods in Enzymology Vol. 11 (Ed. by S.P. Colowick and N.O. Kaplan). pp. 523-530. Academic Press, New York.

Scott, R.O. and Mitchell, R.L. 1943. Concentration methods in spectrographic analysis. I. Recovery of cobalt, nickel, molybdenum, copper and zinc from plant materials and soil extracts by 8- hydroxyquinoline. *J.Soc.Chem. Ind. London*, **62** : 5.

Snell, F.D. and Snell, C.T. 1949. Colorimetric Methods of Analysis Vol. 11, pp. 802-807. Van Nastrand and Co., New York.

Tuve, T.V. and Anfinsen, C.B. 1960. Preparation and properties of spinach ribonuclease. *J.Biol.Chem.*, **237** : 550-559.

White, P.R. 1939. Potentially unlimited growth of excised plant callus in an artificial medium. *Am.J.Bot.*, **26** : 59-64.

Yonetani, T. 1962. Studies on cytochrome oxides. IV. The cytochrome oxidase activity. *J.Biol.Chem.*, **237** : 550-559.

GENERAL INDEX

A

D

E

F

P

W

X

Z